COHERENT
SPREAD SPECTRUM
SYSTEMS

COHERENT SPREAD SPECTRUM SYSTEMS

Jack K. Holmes
Holmes Associates

A Wiley-Interscience Publication

John Wiley & Sons   New York · Chichester · Brisbane · Toronto · Singapore

Library of Congress Cataloging in Publication Data:

Holmes, Jack K. (Jack Kenneth), 1936–
  Coherent spread spectrum systems.

  "A Wiley-Interscience publication."
  Includes index.
  1. Spread spectrum communications. I. Title.

TK5102.5.H5885   621.38′043   81-4296
ISBN 0-471-03301-4   AACR2

Printed in the United States of America

10 9 8 7 6 5 4 3 2 1

To My Family
Fontayne, Julie, and Rachel
And to My Parents

PREFACE

Nowadays almost every NASA or military satellite communication system uses one or more types of spread spectrum modulation methods. Yet in light of this fact, and although many communication system conferences include two or more sessions on spread spectrum systems, only one textbook now exists in this country on the subject. It was for this reason that I embarked on my 5 (plus) year adventure in book writing.

This book focuses primarily on the synchronization of direct sequence spread spectrum systems using coherent carrier demodulation. Clearly this is only one aspect of spread spectrum communication systems. For example, frequency hopping, a common method of spread spectrum modulation is not discussed here nor are optimum jamming strategies for frequency hopped or direct sequence systems.

The portion of the theory developed in this book that deals with direct sequence spread spectrum systems has two aspects: one deals primarily with direct sequence acquisition, tracking, and lock detection along with associated losses in despreading; the other with the generation and properties of pseudonoise and Gold codes.

In addition to the spread spectrum topics, carrier loops and bit synchronizers are considered in the book along with bit demodulation and carrier modulation alternatives.

About half of the material in this book was presented in graduate communication system courses at the University of California Los Angeles (Extension) and California State University, Northridge. While most chapters are relatively self-contained, the reader is assumed to have had an introduction to the theory of random variables and random processes, as well as a first course in communication systems.

The primary motivation and basis for this work has been provided by my association (both as an employee and later as a consultant) with TRW Space and Defense division of TRW in Redondo Beach, California, my association at the Jet Propulsion Laboratory in the communication systems division in Pasadena, California, and with my consulting work at the Axiomatix Corporation in Los Angeles.

Problems included in the book vary in difficulty from trivial to challenging. In part, they extend the theory presented in the text and should be looked over even if the solutions are not attempted.

Although it is probably impossible to do so fully, every attempt was made to give credit where credit was due. To those authors who feel slighted, I apologize.

I would like to mention Dr. W. C. Lindsey, Dr. M. K. Simon, Dr. K. T. Woo, Mr. M. Huang, and Dr. H. Osborne who encouraged me as the book progressed as well as Dr. Osborne who reviewed a good portion of the manuscript.

The typing of the manuscript was done by numerous typists including Mrs. Jan Northlund, Mrs. Carole Ziff, Miss Lesley Paul, and Mrs. Sandy Parker.

Any errors that are found herein may be brought to the author's attention by writing him at 1338 Comstock Ave., Los Angeles, California 90024.

JACK K. HOLMES

Westwood (L.A.), California
September 1981

CONTENTS

COHERENT
SPREAD SPECTRUM
SYSTEMS

SYNCHRONOUS PSEUDONOISE CODED SPREAD SPECTRUM SYSTEMS

A communication system is said to be synchronous if a (near) exact copy of the frequency and phase of the received signal (immersed in noise) is required to demodulate the data properly, which is modulated onto the carrier. The purpose of this chapter is to outline a typical spread spectrum pseudonoise coded system and to discuss the types of problems arising in the design of such a system. In addition, a brief summary of each chapter is given to aid the reader in locating the topics of interest. The chapter is concluded with a short history of spread spectrum communications.

1.1 A TYPICAL COHERENT PN SPREAD SPECTRUM SYSTEM

Consider Figure 1-1, which illustrates a basic pseudonoise (PN) encoded transmitter and receiver. The data from the source is coded by an appropriate coding scheme and modulated onto the carrier, which is in turn modulated by the spread spectrum PN code. This modulated signal, which has a bandwidth many times larger than its data bandwidth, is then amplified and sent from the transmitting antenna to the receiver antenna. The signal intercepted by the receiver antenna is amplified by the preamplifier and fed to the PN acquisition circuitry whose function is to test sequentially each possible PN code phase until the received and locally generated codes are synchronized within some small nominal timing error. When the two codes are synchronized, the code tracking loop maintains the locally generated PN code in synchronism with the received PN code. This locally generated code is then used to "wipe-off" the PN code of the received signal, thus reducing the spectrum to the data bandwidth. Typically a lock detector is utilized to indicate whether or not the locally generated PN code is synchronized with the received PN code.

The despread signal is fed to the carrier loop, which has provisions for

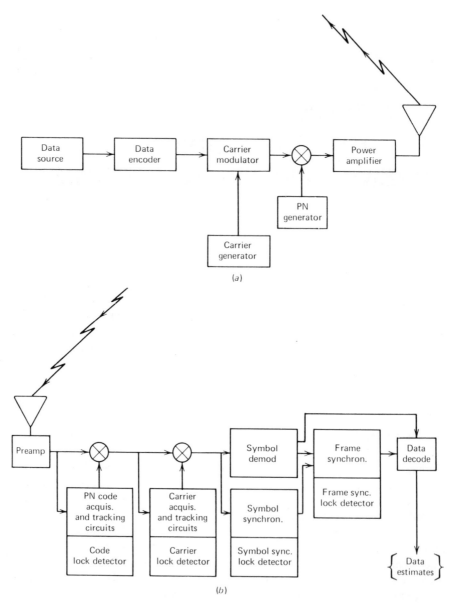

Figure 1.1 PN encoded spread spectrum transmitter (*a*) and coherent receiver (*b*).

both acquisition and tracking of the suppressed carrier signal. In addition, a lock detector is normally employed to monitor the carrier loop status.

After carrier demodulation, the baseband signal is fed to a symbol synchronizer that provides the clock used to control the symbol demodulator normally implemented as a matched filter. The symbol demodulator extracts the symbol estimates, and the frame synchronizer uses these estimates to obtain frame synchronization. Finally, data bits are obtained from the decoder, which translates the encoded symbol stream to the decoded bit stream.

With the exceptions of the preamplifier and the frame synchronizer, all the blocks in Figure 1.1, will be investigated in detail in this book.

1.2 ADVANTAGES OF PN SPREAD SPECTRUM SYSTEMS

The question arises, why utilize a spread spectrum in communication systems? The answer is multifaceted. Perhaps the primary motivation for utilizing spread spectrum systems is the capability of the system to reject intentional or unintentional jamming. Another important reason for its use is its capability of low probability of intercept (LPI). This is accomplished by generating a very broad spectral bandwidth signal that is therefore hard to detect in noise. Code division multiplexing is a method in which a group of carriers operate at the same nominal center frequency but are separable from the others by the low cross-correlation of the codes used. Therefore, spread spectrum code division multiplexing provides multiple access communication channels. Furthermore, having a spread spectrum modulation imposed on the data bearing carrier provides message privacy. The casual listener will be unable to decode the data. In addition, if there is a group of users with a unique code for each user, one is capable of identifying the user by the particular code. Another important use of spread spectrum codes, one utilizing its wide bandwidth, occurs in navigation and ranging, where the range errors can be made to be a small fraction of a chip and are, therefore, very accurate. And finally, spread spectrum signals can be used for providing multipath rejection.

1.3 OUTLINE OF THE TEXT

We now outline the major topics included in each chapter. Chapter 2 discusses pulse code modulation (PCM) formatting and develops some results useful for computing the power spectral density of baseband digital data signals as well as modulated signals.

In Chapter 3 multichannel coherent modulation schemes are considered. Topics addressed include determining the spectral character of the modulated signal as well as determining the power levels of the data and residual carrier components. Also two-channel interplex is compared with the conventional two-channel subcarrier scheme.

The topic of Chapter 4 is the performance of first-, second-, and third-order phase locked loops. Noise-free properties are considered initially; then their performance in noise is derived. Stability, noise bandwidth, and transient response to various inputs are considered. In addition, cycle slipping and acquisition times are addressed. Also developed is the effect of a bandpass limiter preceding the phase locked loop, carrier loop lock detectors, and the effect of phase noise on tracking.

Chapter 5 deals with suppressed carrier tracking loops. Topics include tracking performance of Costas loops, squaring loops, decision directed feedback loops with and without delay, and the modified four-phase Costas loop. Both the Costas loop and the squaring loop are dealt with in some detail. Included are optimum filter bandwidths, false lock levels, and tracking performance in white thermal noise.

The effects of a residual carrier component are also developed in this chapter for the case of a Costas loop. Sweep rate estimates for acquisition are provided along with the performance of lock detectors for phase shift keyed (PSK) signals. Finally the chapter is concluded with a section on the mean slip time formula for multiphase loops.

In Chapter 6 coherent detection of signals is discussed. Block coded signal structures are presented along with their correlation properties. The bit error rate is determined for the general case of block coded signals in terms of the signal correlation properties. The word and bit error rates of orthogonal and bi-orthogonal signals are developed. A section on coherent frequency shift keying, coherent phase frequency shift keying, and differential multiple phase shift keying is also presented. Convolutional codes with Viterbi decoding are also discussed.

The effect of an imperfect phase reference on bit error rate is considered for both residual carrier and suppressed carrier tracking loops and for both uncoded and Viterbi decoded systems. Another topic included in Chapter 6 is the effect of an imperfect carrier reference on uncoded unbalanced QPSK systems. Finally a short discussion of intersymbol interference is presented.

Chapter 7 is an introduction to linear PN sequences. Single and multiple return shift registers are discussed along with the state diagram. Both the matrix and the generating function characterization of the PN sequences are developed. Numerous properties of these PN sequences are derived based on both the matrix and the generating function approach. The randomness properties are discussed and shown to be satisfied by maximal length codes. A list of PN code (maximal length) generators is given at the end of the chapter.

The subject of PN coded spread spectrum systems is discussed in Chapter 8. First a basic PN coded spread spectrum modulator and demodulator are discussed. Then biphase and quadriphase modulations are treated in terms of a noncoherent tone jammer and processing gain. Other topics include the correlation despreading loss due to bandpass filtering of the received PN code and phase distortion of the PN filter due to both amplitude and phase distortion of the PN bandpass filter transfer function. Next, reference code filtering is considered and a discussion of the effects of AGC follows. Noise spreading by correlating noise is developed. Quadriphase and offset quadriphase phase shift keying signals are compared after they are passed through a bandpass filter. Another topic of importance, post- and predetection signal combining, is developed for quadriphase signals.

The effects of spurious signals on PN coded systems are considered, and a development of the spectral density of a PN waveform times a shifted version of itself is given. Finally, surface acoustic wave (SAW) devices are briefly discussed along with charge coupled devices.

In Chapter 9 the topic of PN acquisition and lock detection is addressed. Signal flow graph theory is introduced and then utilized in discrete, time invariant Markov processes. Generating functions are developed from the flow graph of the system under study. The theory developed is used to derive the acquisition time for single dwell time acquisition systems. These results are developed for the effects of doppler and to extend the theory to the double dwell time system. Next some approximations are developed for the statistics of the output of the postdetection filter for the case of an integrate-and-dump filter and an RC low pass filter.

Lock detector theory for PN code spread spectrum systems is developed based on the use of absorbing Markov chains. This topic is followed by an introduction to sequential detection as applied to PN acquisition. A discussion of passive matched filters in the next section is followed by a theory of an optimum search strategy for active correlation acquisition. Finally the mechanism of false lock on PN code acquisition circuits is developed.

Chapter 10 presents the theory of PN code tracking loops, including full-time, time-shared, and τ-dither loops, as well as baseband loops. As a standard of performance the baseband loop is first analyzed. Next, the full-time code tracking loop is analyzed, followed by a quadriphase tracking loop, a time-shared code tracking loop, a τ-dither loop, and then the time gated code tracking loop. Another topic of interest included is self-noise in code tracking loops, which is followed by the effect of tone jamming on code tracking.

Noise-free performance, including acquisition via the phase plane for various conditions, is considered in some detail. Another topic that is treated and that depends on noise is the mean time to lose lock. Finally the theory of image noise on code tracking is developed and how it degrades tracking performance is shown.

Gold codes are the subject of Chapter 11. First, cyclic correlation properties and partial period correlation properties of codes are presented. Then a bound on the cyclic cross correlation in terms of the cyclic autocorrelation of each code is developed. Next, preferred pairs of codes are introduced, which leads to Gold codes. Various useful results for Gold codes are presented. Then some properties of Gold codes are developed with the help of the characteristic function. Properties of balanced Gold codes are then presented. Finally false code lock for both maximal length and Gold codes are derived.

The final chapter, Chapter 12, is concerned with bit synchronizers for PSK modulation. First, optimum type synchronizers are discussed and then near-optimum DTTLs and early-late gate synchronizers are analyzed. Next suboptimum bit synchronizers based on harmonic generation are discussed, and a promising one, the filter and square synchronizer, is analyzed in detail.

Another important loop, the incremental phase modulator (IPM) is analyzed in depth for both tracking performance and mean time to cycle slip. Next, bit error rate performance for bit synchronizers is developed for both NRZ data and Manchester data. Finally a bit synchronizer lock detector analysis is presented followed by an SNR estimator.

As can be seen by the topics included, primary emphasis is in the area of synchronization of PN coded spread spectrum systems. Frequency hopping or hybrid frequency hopping–direct sequence modulation have not been addressed. Also, jamming strategies for jamming frequency hopped schemes have not been dealt with. It was felt that the inclusion of these topics would have made the book unmanageably large.

1.4 BRIEF HISTORY OF SPREAD SPECTRUM SYSTEMS

A communication system incorporating noiselike signals and correlation detection apparently was first built and conceived (in this country at least) around 1949 by deRosa and Rogoff [1] at the Federal Telecommunications Laboratories (subsidiary of ITT). This system operated successfully in a communication link between New Jersey and California. The system used a 10-kHz direct sequence code for spreading.

A theoretical study along with experimental verification for noiselike modulation was pursued in 1950 by Basore [2], who coined the acronym NOMACS (Noise Modulation And Correlation detection System).

The Army Signal Corps, in the spring of 1951, requested that a NOMAC system be developed by the Research Laboratory of Electronics of MIT for application in a long-range, high frequency radio teletype communication link exposed to enemy jamming. The project was transferred to Lincoln Laboratory of MIT later in 1951. The first of these experimental NOMAC systems were of the transmitted reference type [3, 4]. The Lincoln P9D system was built and field-tested in 1952. This type of NOMAC system was abandoned in favor of the stored reference type because of the greater immunity of the latter to jamming [5]. The Lincoln F9C system was built and field-tested in 1953–1955 [6]. An operational prototype, designated F9C-A, was then built for the Signal Corps by Sylvania's Electronic Defense Laboratory.

Soon the Navy and the Air Force began development of their spread spectrum systems using names such as "Phantom" and "Hush-Up" (Air Force) and "Blades" (Navy). Frequency hopping was utilized on both Phantom versions. These original systems used vacuum tubes and consequently required rooms full of equipment. Spread spectrum systems really did not become practical until the advent of the transistor. With the development of integrated circuits, systems became even more manageable in package size and ease of circuit design.

The original interest in spread spectrum included jamming resistance, low detectability, and low interference to other systems operating in the same band. Later, ranging became a new application of direct sequence systems, which is a very important use now.

Some well-known recent systems include the global positioning system (GPS or NAVSTAR), which is a satellite navigational system utilizing direct sequence spreading in which up to 24 satellites will be used to provide signals that can be processed by users to derive position, accurate time, and velocity information for navigational purposes.

Another recent spread spectrum system is the tracking and data relay satellite system (TDRSS), which is a NASA program. The basic idea is to provide support to a number of earth orbiting vehicles by giving them a relay station (in the sky!) to communicate to CONUS (continental United States). The spread spectrum modulation is code division multiple access (CDMA) on some channels.

REFERENCES

1 deRosa, L. A., and Rogoff, M., "Application of Statistical Methods to Secrecy Communication Systems," Section I: *Communications*, Federal Telecommunication Labs., Inc., Nutley, N.J., Proposal 946, August 28, 1950.

2 Basore, B. L., "Noiselike Signals and Their Detection by Correlation," Sc.D. dissertation, MIT, Cambridge, Mass., May 26, 1952.

3 Quarterly Progress Reports, Division 3 Communications and Components, Lincoln Lab., MIT, Cambridge, Mass., DDC Docs. AD 4907, AD 12 270, AD 17 898, AD 22 246, January 15, 1953 to October 15, 1953.

4 Pankowski, B. J., "Multiplexing a Radio Teletype System Using a Random Carrier and Correlation Detection," Research Lab. of Electronics and Lincoln Lab., MIT, Cambridge. Mass., Tech. Report No. 5, DDC Doc. AD 168 857, May 16, 1952.

5 Green, P. E. Jr., "Correlation Detection Using Stored Signals," Lincoln Lab., MIT, Cambridge, Mass., Tech. Report No. 33, DDC Doc. AD 20 524, August 4, 1953.

6 Green, P. E. Jr. et al., "Performance of the Lincoln F9C Radioteletype System," Lincoln Lab., MIT, Lexington, Mass., Tech. Report No. 88, DDC Doc. AD 80 345, October 28, 1955.

7 Dixon, R. C., *Spread Spectrum Systems*, Wiley-Interscience, New York, 1976.

8 Alem, W. K., Huth, G. K., Holmes, J. K., and Udalov, S., "Spread Spectrum Acquisition and Tracking for Shuttle Communication Links," *IEEE Communications*, Part I, November 1978.

9 Cahn, C. R., "Spread Spectrum Applications and State-of-the-Art Equipments," Paper No. 5, AGARD-NATO Lecture Series No. 58, May 28 to June 6, 1973.

10 Smith, M., private communication on the history of spread spectrum systems.

11 Dixon, R. C., "Spread Spectrum Techniques," *IEEE Press Book of Selected Reprints*, New York, 1976.

2

PCM CODE FORMATTING
AND SPECTRA

In this chapter we discuss various modes of pulse code modulation (PCM) code formatting along with some of their advantages. Additionally, we consider the power spectral density associated with the PCM data stream. We assume the data stream, PCM coded, can be modeled as a Markov sequence. The results are first obtained for baseband (without a carrier) schemes and then modified for the case when a carrier is employed.

2.1 PCM DATA FORMATS

Before we discuss the various PCM formats, we mention that there are a number of criteria that a system designer considers when choosing a PCM format:

1 Spectral characteristics—wideband or narrowband, does it possess a dc component?
2 Synchronization capabilities—high density of zero crossings?
3 Noise immunity—signals that are opposite in the signal space sense are best.*
4 Complexity and cost—depend on the above characteristics.

With these thoughts in mind, consider a number of PCM formats that are commonly used in communication systems.

The Inter-Range Instrumentation Group (IRIG) of the Range Commanders' Council [1, 8], recognizes seven permissible digital formats for PCM. Figure 2.1 illustrates these data formats plus four additional data formats of interest [2].

Perhaps the simplest format is *NRZ-L*, in which a binary "one" is represented by one signal level and a "zero" by a second level. When the

*See Chapter 6 for more details.

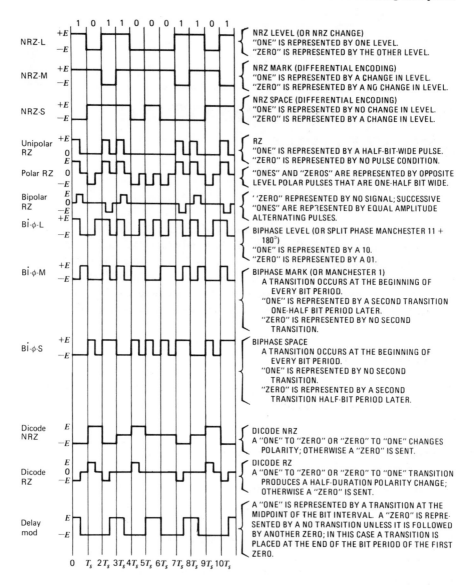

Figure 2.1 PCM code formatting.

symbol "one" is represented by a signal level and "zero" by a zero level, the waveform is denoted as *unipolar NRZ-L*. In addition, *polar NRZ-L* denotes a signal in which equal positive and negative signal amplitudes correspond to the two binary symbols "one" and "zero."

A modification of the basic NRZ-L waveform, the *NRZ mark*, is obtained by encoding the information in the signal transitions. This encoding procedure is more commonly known as differential encoding. The structure of NRZ mark is such that a "one," or mark, is represented by a change in amplitude, and a "zero," or space, by no change in amplitude. The term *NRZ space* refers to the opposite; that is, a "zero" is represented by a change in amplitude and a "one" by no change.

The structure of *polar RZ* is shown in Figure 2.1, where binary "ones" and "zeros" are represented by opposite level polar pulses that are one-half a bit period wide. *Unipolar RZ* has a binary "one" represented by a pulse one-half period wide and a "zero" by the absence of a signal. The *bipolar RZ* is a three-level signaling scheme whereby a "zero" is represented by the absence of a signal, and successive "ones" are represented by equal magnitude opposite polarity pulses that are one-half a bit period wide. This scheme is utilized by the Bell System in the T1 carrier system [3]. The various polar schemes can be NRZ waveforms also.

In the *biphase level* (Bi-ϕ-L) signaling scheme both "ones" and "zeros" are represented as bilevel signals. The "ones" are represented by signals that have the higher of the two levels during the first half of a bit period and the lower level during the last half of the bit period. The "zeros" are represented by signals that are the inverses of the "one" signal. (This signal is very desirable for its synchronization capabilities, see Figure 2.1.) In the *biphase mark* (Bi-ϕ-M) signal a transition occurs at the beginning of every bit period. A "one" is represented by a second transition one-half a bit period later, and a "zero" is represented by no transition. As in Bi-ϕ-L, the synchronization capabilities are very desirable, since there is at least one "zero" crossing per bit. The spectra of Bi-ϕ-M is the same as Bi-ϕ-L. In a *biphase space* (Bi-ϕ-S) a transition occurs at the beginning of every bit period. A "zero" is represented by a second transition one-half a bit period later, and a "one" is represented by no second transition.

In the *dicode RZ* (Meacham's twinned binary) a "one to zero" or a "zero to one" transition produces a change of polarity of the half-bit duration symbols, as shown in Figure 2.1. When there is no change in data bits, then no signal is sent. In *dicode NRZ* the pulses also indicate transitions in the digital information, but the pulse lasts for a full bit time.

The *duobinary* coding scheme [4, 5] is a method by which the binary data are transformed into a three-level signal. In duobinary, each of the three resulting levels is associated with the existing binary digit and preceding bits. In a duobinary signal there are three levels, designated 0, 1, and 2. The signal is coded so that if either the level 0 or 2 results in the duobinary signal, a binary "zero" is being transmitted. On the other hand, if the signal is at the level 1, a binary "one" is being transmitted. A "zero" that follows an even number of consecutive "ones" is assigned the same level as the last "zero." A "zero" that follows an odd number of consecutive "ones" is assigned the alternate level. An example of the duobinary signal and its generation is shown in Figure 2.2. A unique characteristic of the duobinary signal is that for any two consecutive bit

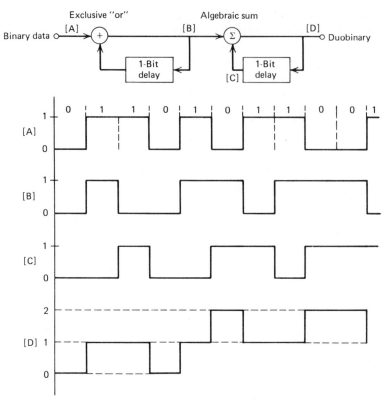

Figure 2.2 A representative circuit that produces a duobinary signal [D] from the sum [B] + [C].

periods the signal can differ in value by only one value. Therefore the code can detect errors.

Delay modulation (Miller code), [6, 7, 16] is an encoding procedure that maps a "one" into a signal transition at the midpoint of the bit interval. A "zero" maps into a no transition unless it is followed by another "zero." In this case a transition is placed at the end of the bit period of the first "zero." The power spectral density is very narrow, which makes it desirable for tape recording [6].

2.2 POWER SPECTRA FOR RANDOM SEQUENCE PCM

The following analysis is based in part on references 9, 10, and 11 for the general formula of the power spectra of PCM waveforms modulated by Markov random sequences. Weber [10] noted that Titsworthe and Welch overlooked a term in their derivation. This term vanishes in many, but not all, important cases. The simpler case of binary signaling is discussed in

Bennett and Davey [12] under the assumption that the signals are statistically independent from bit to bit.

Consider a sequence of random digits, m_i, $i = \ldots, -2, -1, 0, 1, 2, \ldots$,
with each digit drawn from the set $\{1, 2, \ldots, M\}$, the source alphabet
shown in Figure 2.3. The transmitter transmits one of M waveforms
$\{s_1(t), s_2(t), \ldots, s_M(t)\}$ translated to the appropriate time interval. We
assume that:

1 $\quad s_i(t) = 0 \qquad \forall t \notin [0, T_s)$
2 $\quad$ If $m_i = j$, then $s_j(t)$ is sent during $t \in [iT_s, (i + 1)T_s)$

Hence the signal can be represented for $i = -K + 1, -K + 2, \ldots, K$ by

$$U_K(t) = \sum_{i=-K+1}^{K} s_{m_i}(t - iT_s) \qquad (2.2\text{-}1)$$

where $s_{m_i}(t)$ denotes one of M signals realized at time interval i, that is,
$t \in [iT_s, (i + 1)T_s)$.

We now assume some source statistics in order to proceed to obtain the
spectral properties of the signals:

1 $\quad$ The source probabilities of m_i are assumed to be

$$P(m_i = j) = \pi_j \qquad (2.2\text{-}2)$$

2 $\quad$ Define the average signal as

$$E[s(t)] = \sum_{j=1}^{M} \pi_j s_j(t) \triangleq p(t) \qquad t \in [0, T_s) \qquad (2.2\text{-}3)$$

where $p(t)$ is the periodic part of the signal sequence.

3 $\quad$ Define

$$E[U_K(t)] = \sum_{i=-K+1}^{K} p(t - iT_s) \triangleq p_K(t) \qquad (2.2\text{-}4)$$

as a known periodic function of period $2KT_s$ in the interval $[(-K + 1)T_s, (K + 1)T_s]$, which contains $2K$ terms.

4 $\quad$ Define the nonperiodic component of the signal and its $2K$ extension by

$$\begin{aligned} s_j'(t) &= s_j(t) - p(t) \qquad t \in (0, T_s) \qquad j = 1, \ldots, M \\ U_K'(t) &= U_K(t) - p_K(t) \qquad t \in (0, T_s) \end{aligned} \qquad (2.2\text{-}5)$$

It follows that

$$E[s'(t)] = 0 \qquad \text{and} \qquad E[U_K'(t)] = 0 \qquad (2.2\text{-}6)$$

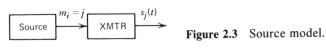

Figure 2.3   Source model.

Therefore it follows that the Fourier transform satisfies

$$E\{\mathscr{F}[s_j'(t)]\} = 0 \quad \text{and} \quad E\{\mathscr{F}[U_K'(t)]\} = 0 \tag{2.2-7}$$

where $\mathscr{F}[a]$ denotes the Fourier transform of $a(t)$.

An equivalent system to generate the transmitted waveform is given in Figure 2.4. In the figure

$$\delta_j(t) = \delta\text{-function for the }j\text{th signal} = \sum_i \delta(t - m_i T_s)$$
$$\in (\text{set of times where }j\text{th signal appears})$$

Note that $\Sigma_{j=1}\,\delta_j(t)$ is a sequence of $2K$ δ-functions, one at each signal starting point, that is, at $(-K+1)T_s, (-K+2)T_s, \ldots, KT_s$ sec. A possible realization for $K = 5$ and $M = 2$ is shown in Figure 2.5.

Now define

$$S_j(f) = \mathscr{F}[s_j(t)] \tag{2.2-8}$$

$$S_j'(f) = \mathscr{F}[s_j'(t)] \tag{2.2-9}$$

$$\Delta_j(f) = \mathscr{F}[\delta_j(t)] \tag{2.2-10}$$

The power spectral density of the signal $U_K(t)$ is defined by (averaging over all $2K$ signals):

$$\mathscr{S}(f) = \lim_{K \to \infty} \frac{E|\mathscr{F}[U_K(t)]|^2}{2KT_s} \tag{2.2-11}$$

Upon breaking the signal into nonperiodic and periodic parts we have

$$\mathscr{S}(f) = \lim_{K \to \infty} \frac{E\{|\mathscr{F}[U_K(t) - p_K(t)] + \mathscr{F}[p_K(t)]|^2\}}{2KT_s} \tag{2.2-12}$$

PROBLEM 1

Show that the spectra can be written as

$$\mathscr{S}(f) = \lim_{K \to \infty} \frac{E[|\mathscr{F}[U_K(t) - p_K(t)]|^2]}{2KT_s} + \lim_{K \to \infty} \frac{|\mathscr{F}[p_K(t)]|^2}{2KT_s}$$

or

$$\mathscr{S}(f) = \mathscr{S}_c(f) + \mathscr{S}_d(f)$$

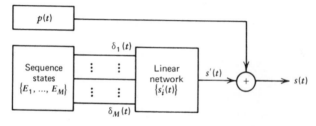

Figure 2.4 Equivalent signal generator.

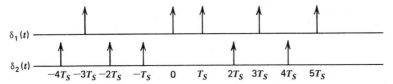

Figure 2.5   Example of the $\delta_i(t)$ functions for $k = 5$ and $M = 2$.

The second term is the power spectral density of the periodic component $p_K(t)$, and the first the continuous component.

From Problem 1 we see that the problem can be broken down into a continuous spectral component $\mathcal{S}_c(f)$ and a discrete component $\mathcal{S}_d(f)$. The discrete component can be found in three steps:

1   Find the Fourier series of $p_K(t)$.
2   Find the autocorrelation function of $p_K(t)$.
3   Transform the autocorrelation function of the periodic function to obtain the power spectra.

The complex Fourier series for $p_\infty(t) = \lim_{K\to\infty} p_K(t)$ is

$$p_\infty(t) = \sum_{n=-\infty}^{\infty} p_n \exp\left(-\frac{j2\pi nt}{T_s}\right) \tag{2.2-13}$$

where

$$p_n = \frac{1}{T_s} \int_0^{T_s} p_\infty(t) \exp\left(-\frac{j2\pi nt}{T_s}\right) dt \tag{2.2-14}$$

or

$$p_n = \frac{1}{T_s} \sum_{j=1}^{M} \pi_j S_j\left(\frac{n}{T_s}\right) \tag{2.2-15}$$

where $S_j(n/T_s)$ is the Fourier transform of $s_j(t)$ at $f = n/T_s$.

PROBLEM 2

The time autocorrelation function of $p(t)$ is

$$R_p(\tau) = \frac{1}{T_s} \int_0^{T_s} p(t)p(t+\tau) \, dt$$

Show that

$$R_p(\tau) = \sum_{n=-\infty}^{\infty} |p_n|^2 \exp\left(-\frac{j2\pi n\tau}{T_s}\right)$$

From Problem 2 we have that the discrete-spectra component is given by

$$\mathcal{S}_d(f) = \sum_{n=-\infty}^{\infty} |p_n|^2 \delta\left(f - \frac{n}{T_s}\right) \tag{2.2-16}$$

using (2.2-15) for p_n, we obtain

$$\mathcal{S}_d(f) = \frac{1}{T_s^2} \sum_{n=-\infty}^{\infty} \left| \sum_{j=1}^{M} \pi_j S_j \left(\frac{n}{T_s}\right) \right|^2 \delta\left(f - \frac{n}{T_s}\right) \tag{2.2-17}$$

which is the power spectral density of the periodic part.

Now consider the continuous part of the spectra, which is given by

$$\mathcal{S}_c(f) = \lim_{K \to \infty} \frac{E|\mathcal{F}[U'_K(t)]|^2}{2KT_s} \tag{2.2-18}$$

or since

$$U'_K(t) = \sum_{j=1}^{M} \delta_j(t) * s'_j(t) \tag{2.2-19}$$

we have (the averaging being over the Δ_j's)

$$\mathcal{S}_c(f) = \lim_{K \to \infty} \frac{1}{2KT_s} \sum_{j=1}^{M} \sum_{k=1}^{M} E[\Delta_j^*(f)\Delta_k(f)]S_j'^*(f)S_k'(f) \tag{2.2-20}$$

since the Fourier transform of the convolution of $s'_j(t)$ and $\delta_j(t)$ is the product of $S'_j(f)$ and $\Delta_j(f)$. Define

$$G_{jk}(f) = \lim_{K \to \infty} \frac{E[\Delta_j^*(f)\Delta_k(f)]}{2KT_s} \tag{2.2-21}$$

Note that $\Delta_j(f)$ and $\Delta_k(f)$ are defined on the $2K$ signal duration, and hence, depend on K. Now define

$$R_{jk}(\tau) = [\mathcal{F}^{-1}\{G_{jk}(f)\}] \tag{2.2-22}$$

so that

$$R_{jk}(\tau) = E\left[\lim_{K \to \infty} \frac{1}{2KT_s} \int_{(-K+1)T_s}^{KT_s} \delta_j(t)\delta_k(t + \tau)\, dt \right] \tag{2.2-23}$$

since the inverse Fourier transform of a product is the convolution of the respective time functions.

Note that the power spectral densities of two sources modulating a set of signals that exist in the interval $[0, T_s)$ are equal if the correlation matrix $R_{jk}(\tau)$ is the same for both sources.

From equations 2.2-20 and 2.2-21 the power spectral density becomes

$$\mathcal{S}_c(f) = \sum_{j=1}^{M} \sum_{k=1}^{M} G_{jk}(f)S_j'^*(f)S_k'(f) \tag{2.2-24}$$

What remains is the computation of $G_{jk}(f)$ or, equivalently, $R_{jk}(\tau)$ for the $2K$ symbol input sequence. Define the quantity

$N_{jk}^{(n)}$ = number of times state E_k occurs at the nth transmission
after state E_j in the interval $2KT_s$. (This includes all
starting times in the $2KT_s$ interval.)   (2.2-25)

The total number of times state E_j occurs in the time interval $2KT_s (K \gg n)$

is

$$N_j^{(n)} \cong \sum_{k=1}^{M} N_{jk}^{(n)} \qquad (2.2\text{-}26)$$

Define the following quantities when they exist:

1 Transition probabilities in n steps:

$$p_{jk}^{(n)} = \lim_{K \to \infty} \left(\frac{N_{jk}^{(n)}}{N_j^{(n)}} \right) \qquad \left\{ \begin{array}{c} n \text{ step transition} \\ \text{probabilities} \end{array} \right\} \qquad (2.2\text{-}27)$$

which is analogous to

$$p^{(n)}(k \mid j) = \frac{p^{(n)}(j, k)}{p(j)}$$

2 Stationary probabilities:

$$\pi_j = \lim_{K \to \infty} \left(\frac{N_j^{(n)}}{2K} \right) \qquad \left\{ \begin{array}{c} \text{stationary probabilities} \\ \text{independent of } n \end{array} \right\} \qquad (2.2\text{-}28)$$

under the assumption they exist.

The sequence correlations are

$$R_{jk}(\tau) = \lim_{K \to \infty} \frac{E}{2KT_s} \int_{(-K+1)T_s}^{KT_s} \delta_j(t)\delta_k(t + \tau) \, dt \qquad (2.2\text{-}29)$$

These correlation functions are impulses at those values of τ such that both τ and t are integral multiples of T_s, E_j having occurred at t and E_K at $t + \tau$. If $\tau = nT_s$, the number of times this criterion is met by our given sequence in a given interval $2KT_s$ is $N_{jk}^{(n)}$ for τ positive, $N_{kj}^{(n)}$ for τ negative, and $N_{jj}^{(0)} = N_j$ for $\tau = 0$. Therefore, the sequence functions become formally

$$R_{jk}(\tau) = \lim_{k \to \infty} E \left\{ \overbrace{\frac{N_j}{2KT_s} \delta_{jk}\delta(\tau)}^{j = k} + \overbrace{\sum_{n=1}^{2K} \frac{N_{jk}^{(n)}}{2KT_s} \delta(\tau - nT_s)}^{j \neq k,\, \tau > 0} \right.$$

$$\left. + \overbrace{\sum_{n=1}^{2K} \frac{N_{kj}^{(n)}}{2KT_s} \delta(\tau + nT_s)}^{j \neq k,\, \tau < 0} \right\} \qquad (2.2\text{-}30)$$

where δ_{jk} is the Kronecker delta function

$$\delta_{jk} = 1 \quad \text{for } j = k$$
$$= 0 \quad \text{otherwise}$$

Taking limits, we have (using $N_{jk}^{(n)}/2K \to p_{jk}^{(n)}\pi_j$)

$$R_{jk}(\tau) = \frac{1}{T_s} \left[\pi_j\delta_{jk}\delta(\tau) + \sum_{n=1}^{\infty} \pi_j p_{jk}^{(n)}\delta(\tau - nT_s) \right.$$

$$\left. + \sum_{n=1}^{\infty} \pi_k p_{kj}^{(n)}\delta(\tau + nT_s) \right] \qquad (2.2\text{-}31)$$

Taking Fourier transforms, we have that $G_{jk}(f)$ is given by

$$G_{jk}(f) = \frac{1}{T_s}\left[\pi_j\delta_{jk} + \sum_{n=1}^{\infty}\pi_j p_{jk}^{(n)}\exp(-i2\pi fnT_s) + \sum_{n=1}^{\infty}\pi_k p_{kj}^{(n)}\exp(i2\pi fnT_s)\right]$$

(2.2-32)

Note that

$$G_{jk}(f) = G_{kj}^*(f)$$

(2.2-33)

Define $Q_{jk}(f)$ by

$$Q_{jk}(f) = \sum_{n=1}^{\infty}p_{jk}^{(n)}\exp(-i2\pi nfT_s)$$

(2.2-34)

so that, using expression (2.2-24) for $\mathscr{S}_c(f)$, we have

$$\mathscr{S}_c(f) = \frac{1}{T_s}\sum_{k=1}^{M}\pi_k|S_k'(f)|^2 + \frac{1}{T_s}\sum_{j=1}^{M}\sum_{k=1}^{M}[\pi_jQ_{jk}(f) + \pi_kQ_{kj}^*(f)]S_j^{*'}(f)S_k'(f)$$

(2.2-35)

Since the last two terms are complex conjucates, we have

$$\mathscr{S}_c(f) = \frac{1}{T_s}\sum_{k=1}^{M}\pi_k|S_k'(f)|^2$$

$$+ \frac{2}{T_s}\text{Re}\left[\sum_{j=1}^{M}\sum_{k=1}^{M}\pi_jS_j^{*'}(f)S_k'(f)Q_{jk}(f)\right]$$

(2.2-36)

Letting $z = \exp(-i2\pi fT_s)$ we see* that we can write $Q_{jk}(f)$ in matrix form:

$$Q(z) = \sum_{n=1}^{\infty}[P]^nz^n$$

(2.2-37)

with $[P]$ being the transition matrix $\{p_{jk}\}$.

Finally summing the discrete and continuous components produces the final general result for the two-sided power spectral density:

$$\mathscr{S}(f) = \frac{1}{T_s^2}\sum_{n=-\infty}^{\infty}\left|\sum_{j=1}^{M}\pi_jS_j\left(\frac{n}{T_s}\right)\right|^2\delta\left(f - \frac{n}{T_s}\right)$$

$$+ \frac{1}{T_s}\sum_{j=1}^{M}\pi_j|S_j'(f)|^2$$

$$+ \frac{2}{T_s}\text{Re}\left[\sum_{j=1}^{M}\sum_{k=1}^{M}\pi_jS_j'^*(f)S_k'(f)Q_{jk}(f)\right]$$

(2.2-38)

This general result gives the complete power spectral density of a Markov process. If there is a jump at some f, then $Q_{jk}(f)$ is required (see reference 9) to be defined by

$$\bar{Q}_{jk}(f) = \lim_{\epsilon\to 0}\tfrac{1}{2}[Q_{jk}(f + \epsilon) + Q_{jk}(f - \epsilon)]$$

---

*Hence $Q_{jk}(z)$ is the generating function of the sequence [9].

We conclude that our general result holds for homogenous Markov sources where $p_{jk}^{(n)}$ is the transition probability that signal $s_k(t)$ is transmitted n steps after signal $s_j(t)$. The probabilities (π_j) are the stationary probabilities.

PROBLEM 3

Show that the line spectrum disappears when

$$\sum_{j=1}^{M} \pi_j S_j(t) = 0$$

which implies

$$\sum_{j=1}^{M} \pi_j S_j \left(\frac{n}{T_s}\right) = 0 \qquad \forall n$$

2.2.1 Purely Random Source

A *purely random source* [16] is one in which a signal, in a given interval of T_s sec, is independent of the previous signaling intervals. A purely random (statistically independent) source has the property that

$$[P]^n = [P] \qquad \forall n \geq 1 \tag{2.2-39}$$

or

$$p_{jk}^{(n)} = p_{jk} \qquad \forall n \geq 1 \tag{2.2-40}$$

so that

$$Q_{jk}(z) = \sum_{n=1}^{\infty} p_{jk} z^n = p_{jk} \frac{z}{1-z} \tag{2.2-41}$$

Also for a purely random source

$$p_{jk} = \pi_k \qquad \text{(stationary probabilities)} \tag{2.2-42}$$

so the transition matrix is

$$[P] = \{p_{jk}\} = \begin{bmatrix} \pi_1, \pi_2, \pi_3, \ldots, \pi_M \\ \pi_1, \ldots, \qquad \pi_M \\ \pi_1, \ldots, \qquad \pi_M \end{bmatrix} \tag{2.2-43}$$

Hence we have

$$Q_{jk}[\exp(-i2\pi f T_s)] = \pi_k \left[\frac{\exp(-i2\pi f T_s)}{1 - \exp(-i2\pi f T_s)}\right] \tag{2.2-44}$$

Simplifying, we have

$$Q_{jk}[\exp(-i2\pi f T_s)] = -\frac{\pi k}{2}[1 + i \cot(\pi f T_s)] \tag{2.2-45}$$

Hence

$$\mathrm{Re}[Q_{jk}\exp(-i2\pi fT_s)] = -\frac{\pi k}{2} \qquad (2.2\text{-}46)$$

$$\mathrm{Im}[Q_{jk}\exp(-i2\pi fT_s)] = -\frac{\pi k}{2}\cot(\pi fT_s) \qquad (2.2\text{-}47)$$

Now consider the third term T_3 in our general spectral result. First consider the third term when $j = k$:

$$T_3^{jj} = \frac{2}{T_s}\mathrm{Re}\left\{\sum_{j=1}^{M}\sum_{k=1}^{M}\pi_j S_j'^*(f)S_k'(f)Q_{jk}[\exp(-i2\pi fT_s)]\right\}\Bigg|_{j=k} \qquad (2.2\text{-}48)$$

or

$$T_3^{jj} = \frac{2}{T_s}\sum_{j=1}^{M}\pi_j|S_j'(f)|^2\left(-\frac{\pi_j}{2}\right) \qquad (2.2\text{-}49)$$

since the imaginary part is removed by the "real part of" operator. We have finally

$$T_3^{jj} = -\frac{1}{T_s}\sum_{j=1}^{M}\pi_j^2|S_j'(f)|^2 \qquad (2.2\text{-}50)$$

Now consider the third term when $j \neq k$:

$$T_3^{jk} = \frac{2}{T_s}\mathrm{Re}\left\{\sum_{\substack{j=1\\j\neq k}}^{M}\sum_{k=1}^{M}\pi_j S_j'^*(f)S_k'(f)Q_{jk}[\exp(-i2\pi fT_s)]\right\} \qquad (2.2\text{-}51)$$

or

$$T_3^{jk} = -\frac{1}{T_s}\mathrm{Re}\left\{\sum_{\substack{j=1\\j<K}}^{M}\sum_{k=1}^{M}\pi_j S_j'^*(f)S_k'(f)\pi_k[1 + i\cot(\pi fT_s)]\right\}$$

$$-\frac{1}{T_s}\mathrm{Re}\left\{\sum_{\substack{j=1\\j>k}}^{M}\sum_{k=1}^{M}\pi_j S_j'^*(f)S_k'(f)\pi_k[1 + i\cot(\pi fT_s)]\right\} \qquad (2.2\text{-}52)$$

After interchanging j and k in the second term, we have

$$T_3^{jk} = -\frac{1}{T_s}\mathrm{Re}\sum_{\substack{j=1\\j<k}}^{M}\sum_{k=1}^{M}\Big(\pi_j S_j'^*(f)S_k'(f)\pi_k[1 + i\cot(\pi fT_s)]$$

$$+ \pi_k S_k'^*(f)S_j'(f)\pi_j[1 + i\cot(\pi fT_s)]\Big) \qquad (2.2\text{-}53)$$

This can be written, using $z + z^* = 2\,\mathrm{Re}[z]$, as

$$T_3^{jk} = -\frac{2}{T_s}\sum_{\substack{j=1\\j<k}}^{M}\sum_{k=1}^{M}\pi_j\pi_k\,\mathrm{Re}[S_j'(f)S_k'^*(f)] - \frac{1}{T_s}\sum_{\substack{j=1\\j<k}}^{M}\sum_{k=1}^{M}\pi_j\pi_k\,\mathrm{Re}[W_{jk}i\cot(\pi fT_s)$$

$$+ W_{jk}^*i\cot(\pi fT_s)] \qquad (2.2\text{-}54)$$

where $W_{jk} = S_j'(f)S_k^{*'}(f)$.

PROBLEM 4

Show that the second term in (2.2-54) is equal to zero.

We then have, using the result of Problem 4, that

$$
\begin{aligned}
T\,_3^{jk} = &-\frac{1}{T_s} \sum_{j=1}^{M} \pi_j^2 |S_j'(f)|^2 \\
&-\frac{2}{T_s} \sum_{j=1}^{M} \sum_{k=1}^{M} \pi_j \pi_k \, \mathrm{Re}[S_j'(f)S_k^*(f)]
\end{aligned}
\tag{2.2-55}
$$

Consequently, we have for the purely random source case, the two-sided spectral density:

$$
\begin{aligned}
\mathscr{S}(f) = &\frac{1}{T_s^2} \sum_{n=-\infty}^{\infty} \left| \sum_{j=1}^{M} \pi_j S_j \left(\frac{n}{T_s}\right) \right|^2 \delta\left(f - \frac{n}{T_s}\right) \\
&+\frac{1}{T_s} \sum_{j=1}^{M} \pi_j(1 - \pi_j)|S_j'(f)|^2 \\
&-\frac{2}{Ts} \sum_{\substack{j=1 \\ j<k}}^{M} \sum_{k=1}^{M} \pi_j \pi_k \, \mathrm{Re}[S_j'(f)S_k'^*(f)] \qquad \forall f
\end{aligned}
\tag{2.2-56}
$$

PROBLEM 5

Show that in the binary case $(M = 2)$ that the spectral density reduces to

$$
\begin{aligned}
\mathscr{S}(f) = &\frac{1}{T_s^2} \sum_{n=-\infty}^{\infty} \left| pS_1\left(\frac{n}{T_s}\right) + (1-p)S_2\left(\frac{n}{T_s}\right) \right|^2 \delta\left(f - \frac{n}{T_s}\right) \\
&+\frac{1}{T_s} p(1-p)|S_1'(f) - S_2'(f)|^2 \qquad \forall f
\end{aligned}
$$

PROBLEM 6

If $p = \frac{1}{2}$ and $s_1(t) = -s_2(t)$, then show:

1 The discrete component is zero and

$$
\mathscr{S}(f) = \frac{1}{T_s} |S_1(f)|^2 \qquad \forall f
$$

2 If the baseband signal is NRZ with amplitude A, then $(p \neq \frac{1}{2})$:

$$
S_1(f) = AT_s \exp(-i\pi f T_s) \frac{\sin \pi f T_s}{\pi f T_s}
$$

and

$$
\mathscr{S}(f) = \frac{E_s}{T_s} (1 - 2p)^2 \delta(f) + 4E_s p(1-p) \left(\frac{\sin \pi f T_s}{\pi f T_s}\right)^2
$$

where $E_s = A^2 T_s$. Note the periodic signal portion is a constant.

2.3 POWER SPECTRA FOR VARIOUS DATA FORMATS

Now we consider various data format spectra.

2.3.1 Unipolar RZ Format

The two signals are shown below:

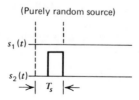

Hence we have

$$S_1(f) = 0$$

$$S_2(f) = \int_{-\infty}^{\infty} e^{-i\omega t} s_2(t)\, dt$$
(2.3-1)

or

$$S_2(f) = \frac{AT_s}{2} \exp\left(-\frac{3i\pi f T_s}{2}\right) \frac{\sin(\pi f T_s/2)}{\pi f T_s/2}$$
(2.3-2)

Using the results for the binary, purely random case, we have

$$\mathscr{S}(f) = \frac{E_s}{4T_s}(1-p)^2\delta(f) + \frac{E_s}{4T_s}(1-p)^2 \sum_{\substack{n=-\infty \\ n\neq 0 \\ n \text{ odd}}}^{\infty}\left(\frac{2}{n\pi}\right)^2 \delta\left(f - \frac{n}{T_s}\right)$$

$$+ \frac{E_s}{4} p(1-p)\frac{\sin^2(\pi f T_s/2)}{(\pi f T_s/2)^2}$$

2.3.2 Manchester Format (Biphase Format)

In this format

$$s_1(t) = A \qquad 0 \le t \le T_s/2$$
$$s_1(t) = -A \qquad T_s/2 \le t \le T_s$$
$$s_2(t) = -s_1(t)$$

The results are given in Problem 7.

PROBLEM 7

Show that the resulting power spectral density is given by

$$\mathscr{S}(f) = \frac{E_s}{T_s}(1-2p)^2 \sum_{\substack{n=-\infty \\ n\neq 0 \\ n \text{ odd}}}^{\infty}\left(\frac{2}{n\pi}\right)^2 \delta\left(f - \frac{n}{T_s}\right) + 4p(1-p)\frac{\sin^4(\pi f T_s/2)}{(\pi f T_s/2)^2}$$

See Figure 2.6 for the case $p = \frac{1}{2}$.

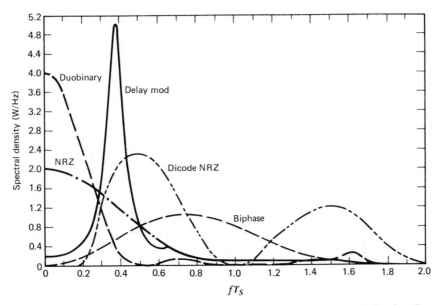

Figure 2.6 Spectral density of a few PCM schemes (arbitrary relative level).

2.3.3 Dicode RZ

In dicode RZ signals [2] there is no dc signal component and the spectral density becomes (amplitude A)

$$\mathcal{S}(f) = 2p(1-p)A^2(1 - \cos \pi f T_s)\left[\frac{\sin(\pi f T_s/2)}{\pi f T_s/2}\right]^2 \tag{2.3-4}$$

2.3.4 Dicode NRZ

The power spectra of the NRZ dicode signal [2] is the same as the RZ dicode case except that the pulsewidth is T_s sec, so that

$$\mathcal{S}(f) = 2p(1-p)A^2(1 - \cos \pi f T_s)\frac{\sin^2 \pi f T_s}{(\pi f T_s)^2} \tag{2.3-5}$$

This spectrum is plotted in Figure 2.6 for the case $p = \frac{1}{2}$.

2.3.5 Duobinary Signaling

According to Lender [5] the power spectral density of the duobinary signal is

$$\mathcal{S}(f) = \frac{A^2}{8T_s}(1 + \cos 2\pi f T_s)\frac{\sin^2 \pi f T_s}{(\pi f T_s)^2} \tag{2.3-6}$$

and is plotted in Figure 2.6.

2.3.6 Delay Modulation (Miller Coding)

The power spectra of delay modulation, obtained by Hecht and Guida [6], is given by

$$
\mathscr{S}(f) = \frac{2A^2}{(2\pi f)^2 T_s [17 + 8\cos(8\pi f T_s)]} [23 - 2\cos(\pi f T_s)
$$
$$
- 22\cos(2\pi f T_s) - 12\cos(3\pi f T_s) + 5\cos(4\pi f T_s)
$$
$$
+ 12\cos(5\pi f T_s) + 2\cos(6\pi f T_s) - 8\cos(7\pi f T_s) + 2\cos(8\pi f T_s)]
$$

$$(2.3\text{-}7)$$

where T_s is the signal duration and A is the signal amplitude. The normalized $(T = A = 1)$ spectra is plotted in Figure 2.6. Note that delay modulation is very peaked near $f = 0.4/T_s$ and is very suitable for magnetic recording since it has a very small dc component (which is desirable because magnetic recording has no dc response) and a smaller bandwidth than NRZ, for example.

2.4 POWER SPECTRA FOR CORRELATED SEQUENCE PCM

The purpose of this section is to provide a derivation of a power spectral density formula for the important case of correlated PCM symbols such as the duobinary encoding scheme. The derivation is different from, and simpler than, the one given in Section 2.2. On the other hand, the derivation is more restrictive in that only one basic transmitted signal waveform is allowed in the analysis. The resulting expression for the power spectra is, however, a useful complement to the ones derived in Section 2.2.

Following Bennett [15], we derive the continuous and discrete power spectra component for a correlated M-valued sequence. We utilize a simpler notation since only one signal is involved.

Let the encoded signal be of the form

$$
x(t) = \sum_{n=-\infty}^{\infty} a_n g(t - nT_s) \tag{2.4-1}
$$

with a_n a random sequence taking on any one of M distinct values. The time T_s is the inverse of f_s, the signaling rate, and $g(t)$ is a pulse waveform of arbitrary duration. It will be assumed that the message ensemble $\{a_n\}$ is ergodic so that an average over time in any member of the ensemble is equal to an average over the ensemble of fixed times except for a set of probability zero.

Define the mean (ensemble average) of any element of the sequence by m_1:

$$
m_1 = E[a_n] \tag{2.4-2}
$$

The autocovariance $R(n)$ is defined by the ensemble average

$$R(n) = E[a_k a_{k+n}] \qquad \text{independent of } k \qquad (2.4\text{-}3)$$

If the message source is purely random, then $R(0) = m_2$ and $R(n) = m_1^2$ $(n \neq 0)$. Now consider the average waveform as was done in Section 2.2. We have

$$E[x(t)] = m_1 \sum_{n=-\infty}^{\infty} g(t - nT_s) \qquad (2.4\text{-}4)$$

Since this is a periodic function, it can be expanded in a Fourier series:

$$E[x(t)] = \sum_{m=-\infty}^{\infty} C_m \exp(im\omega f_s t) \qquad \omega = 2\pi f \qquad (2.4\text{-}5)$$

where

$$C_m = \frac{1}{T_s} \int_0^{T_s} E(x(t)) \exp(-jm\omega f_s t)\, dt \qquad (2.4\text{-}6)$$

or

$$C_m = m_1 f_s \sum_{n=-\infty}^{\infty} \int_0^{T_s} g(t - nT_s) \exp(-jm\omega f_s t)\, dt \qquad (2.4\text{-}7)$$

The combination of summation and integration produce

$$C_m = m_1 f_s \int_{-\infty}^{\infty} g(u) \exp(-jm\omega f_s u)\, du \qquad (2.4\text{-}8)$$

But the Fourier transform of a pulse is given by

$$G(f) = \int_{-\infty}^{\infty} g(t) \exp(-2\pi f j t)\, dt \qquad (2.4\text{-}9)$$

so that

$$C_m = m_1 f_s \{G(mf_s)\} \qquad (2.4\text{-}10)$$

Therefore

$$E[x(t)] = m_1 f_s \sum_{m=-\infty}^{\infty} G(mf_s) \exp(jm2\pi f_s t) \qquad (2.4\text{-}11)$$

or

$$E[x(t)] = m_1 f_s \sum_{m=-\infty}^{\infty} G(mf_s) \cos(m2\pi f_s t) \qquad (2.4\text{-}12)$$

since $\sin(x)$ is an odd function. The discrete power spectra then is given by

$$\mathcal{S}_d(f) = m_1^2 f_s^2 \sum_{n=-\infty}^{\infty} |G(nf_s)|^2 \delta(f - nf_s) \qquad \forall f \qquad (2.4\text{-}13)$$

or for positive frequencies only:

$$\mathcal{S}_d(f) = m_1^2 f_s^2 |G(0)|^2 \delta(f) + 2m_1^2 f_s^2 \sum_{n=1}^{\infty} |G(nf_s)|^2 \delta(f - nf_s) \qquad f \geq 0 \qquad (2.4\text{-}14)$$

Now consider the continuous part of the spectra. As discussed in Section 2.2, the mean or periodic signal is subtracted from the total signal to produce a continuous spectrum. Considering an ensemble $y_N(t)$ that includes only the pulses from $-N$ to N, we write for the zero mean signal

$$y_N(t) = \sum_{n=-N}^{N} (a_n - m_1)g(t - nT_s) \qquad (2.4\text{-}15)$$

Assuming the Fourier transform exists, we have

$$S_N(f) = \sum_{n=-N}^{N} (a_n - m_1)G(f)\exp(-jn\omega T_s) \qquad (2.4\text{-}16)$$

The continuous part of the spectral density is then defined, as before, by

$$\mathscr{S}_c(f) = \lim_{N\to\infty} \frac{E\{|S_N(f)|^2\}}{(2N+1)T_s} \qquad (2.4\text{-}17)$$

Consider the term $E\{|S_N(f)|^2\}$:

$$E\{|S_N(f)|^2\} = \sum_{m=-N}^{N}\sum_{n=-N}^{N} E[(a_n - m_1)(a_m - m_1)]G(f)G^*(f)\exp[j\omega T_s(m-n)] \qquad (2.4\text{-}18)$$

Let $k = m - n$. Then we have

$$E\{|S_N(f)|^2\} = \sum_{m=-N}^{N}\sum_{k=m-N}^{m+N} [R(k) - m_1^2]|G(f)|^2\exp(jk\omega T_s) \qquad (2.4\text{-}19)$$

From Problem 8 we have

$$\sum_{m=-N}^{N}\sum_{k=m-N}^{m+N} = \sum_{k=-2N}^{0}\sum_{m=-N}^{k+N} + \sum_{k=1}^{2N}\sum_{m=k-N}^{N} \qquad (2.4\text{-}20)$$

PROBLEM 8

Show that the following double sum can be written as

$$\sum_{m=-N}^{N}\sum_{k=m-N}^{m+N} = \sum_{k=-2N}^{0}\sum_{m=-N}^{k+N} + \sum_{k=1}^{2N}\sum_{m=k-N}^{N} \qquad (2.4\text{-}21)$$

Hence, using the results of Problem 8, we obtain

$$E\{|S_N(f)|^2\} = \sum_{k=-2N}^{0} (2N+1+k)[R(k) - m_1^2]|G(f)|^2\exp(jk\omega T_s)$$

$$= \sum_{k=1}^{2N} (2N+1-k)[R(k) - m_1^2]|G(f)|^2\exp(jk\omega T_s) \qquad (2.4\text{-}22)$$

Finally, combining, we have

$$E\{|S_N(f)|^2\} = (2N+1)|G(f)|^2\left\{ R(0) - m_1^2 \right.$$

$$\left. + 2\sum_{k=1}^{2N}\left(1 - \frac{k}{2N+1}\right)[R(k) - m_1^2]\cos(k\omega T_s)\right\} \qquad (2.4\text{-}23)$$

Using the definition of the spectra, we have for the continuous spectrum

$$\mathscr{S}_c(f) \cong \frac{1}{T_s} |G(f)|^2 \left\{ R(0) - m_1^2 + 2 \sum_{k=1}^{\infty} [R(k) - m_1^2] \cos(k\omega T_s) \right\}$$

(2.4-24)

Combining the discrete and continuous spectral densities, we obtain our desired result

$$\mathscr{S}(f) \cong m_1^2 f_s^2 \sum_{n=-\infty}^{\infty} |G(nf_s)|^2 \delta(f - nf_s)$$

$$+ f_s |G(f)|^2 \left\{ R(0) - m_1^2 + 2 \sum_{k=1}^{\infty} [R(k) - m_1^2] \cos(k\omega T_s) \right\} \qquad \forall f$$

(2.4-25)

The approximation in the above equation comes from the fact that the continuous spectral term contains a sum of the form

$$\sum_{k=1}^{2N} \left(1 - \frac{1}{2N+1}\right) \xi(k)$$

(2.4-26)

with

$$\xi(k) = [R(k) - m_1^2] \cos(k\omega T_s)$$

(2.4-27)

It is not necessarily true that

$$\lim_{N \to \infty} \sum_{k=1}^{2N} \left(1 - \frac{k}{2N+1}\right) \xi(k) = \sum_{k=1}^{\infty} \xi(k)$$

(2.4-28)

unless $R(k)$ is sufficiently well behaved. For example, if $\xi(k) = 0$ for $k \geq k_0$, then the above equality is true.

In the special case of independent messages with $a_n = 1$ with probability p, and $a_n = 0$ with probability $1 - p$ we have

$$m_1 = p$$
$$m_2 = p$$
$$R(k) = m_1^2 = p^2 \qquad k \neq 0$$

(2.4-29)

so that

$$\mathscr{S}(f) = p^2 f_s^2 \sum_{n=-\infty}^{\infty} |G(nf_s)|^2 \delta(f - nf_s) + f_s |G(f)|^2 p(1-p) \qquad \forall f$$

(2.4-30)

which is seen to be equivalent to the results in Section 2.2 (Problem 5).

PROBLEM 9

Consider a sequence $\{a_n\}$ that is statistically independent from sample to sample and satisfies

$$P(a_n = 1) = p$$
$$P(a_n = -1) = (1-p)$$

Find the signal spectral density.

Consider now an application for which the derived formula applies directly; for bipolar signaling. Recall from Section 2.1 that bipolar RZ, for example, is characterized by a three-level signaling scheme in which a "zero" is represented by no signal and successive "ones" are represented by equal magnitude opposite polarity pulses that are one-half bit period wide. An example of the waveform is shown in Figure 2.7. We model

$$P(a_n = 1) = \tfrac{1}{4}$$
$$P(a_n = -1) = \tfrac{1}{4} \tag{2.4-31}$$
$$P(a_n = 0) = \tfrac{1}{2}$$

Therefore

$$m_1 = 0(\tfrac{1}{2}) + 1(\tfrac{1}{4}) - 1(\tfrac{1}{4}) = 0$$
$$R(0) = m_2 = E[a_n^2] = 1^2(\tfrac{1}{4}) + 1^2(\tfrac{1}{4}) + 0 = \tfrac{1}{2} \tag{2.4-32}$$

Since successive ones are transmitted with alternating a_n's, the interdigit correlation must be accounted for. Starting at the ith digit and going up to the $(i + k)$th digit there are $k + 1$ total digits. The total number of possible binary patterns is 2^{k+1}. Any pattern with a zero at the beginning or the end of the sequence would have a product $a_i a_{i+k} = 0$. Hence, we need consider only those patterns that start and end in ± 1, of which there are 2^{k-1}. Let $k = 1$; then

$$E[a_n a_{n+1}] = 0 + 0 - \tfrac{1}{4} = -\tfrac{1}{4} \tag{2.4-33}$$

since there are four possible sequences, all with probability $\tfrac{1}{4}$ and with only the sequence $1, -1$ having a nonzero value. Let $k = 2$; then

$$E[a_n a_{n+2}] = \tfrac{1}{8} - \tfrac{1}{8} = 0 \tag{2.4-34}$$

since in this case there are a total of eight sequences, each with probability $\tfrac{1}{8}$, but with only two having "ones" at the start and finish, that is, $1, -1, 1$, and $1, 0, -1$. Using the results of Problem 10, it is possible to show that $R(k) = 0 \; |k| \ge 2$. We conclude that

$$\mathcal{S}(f) = \tfrac{1}{2} f_s |G(f)|^2 (1 - \cos \omega T_s) \qquad \forall f \tag{2.4-35}$$

PROBLEM 10

Using induction show that the correlation between a_n and a_{n+k} for the bipolar sequence $(p = \tfrac{1}{2})$ is zero when k is larger in magnitude than one.

Conclude that

$$\mathcal{S}(f) = \tfrac{1}{2} f_s |G(f)|^2 (1 - \cos \omega T_s) \qquad \forall f$$

1 0 1 1 0 0 0 1 1 0 1 0 0 1

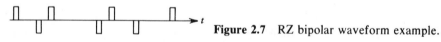

Figure 2.7   RZ bipolar waveform example.

PROBLEM 11

Using the same approach as in the bipolar message case obtain the spectra for the duobinary signaling case by showing

1 That $R(k) = 0, k \geq 2$ $(p = \frac{1}{2})$.
2 That $R(1) = +\frac{1}{4}$.
3 And finally that $\mathcal{S}(f) = \frac{1}{2} f_s |G(f)|^2 (1 + \cos \omega T_s)$.

PROBLEM 12

Show that the power spectral density for the dicode signal is given by

$$\mathcal{S}(f) = \frac{1}{2} f_s |G(f)|^2 (1 - \cos \omega T_s)$$

which is the same as the bipolar signal case.

PROBLEM 13

Analyze the delay modulation scheme by showing that:

(a) $R(\tau + 4T_s) = -\frac{1}{4} R(\tau)$ $\tau > 0$
(b) $R(0) = 1, R(\frac{1}{2}T_s) = \frac{1}{4}, R(T_s) = -\frac{1}{2}$
   $R(\frac{3}{2}T_s) = -\frac{1}{2}, R(2T_s) = 0, R(\frac{5}{2}T_s) = \frac{3}{8}$
   $R(3T_s) = \frac{1}{4}, R(\frac{7}{2}T_s) = -\frac{1}{8}$
(c) $\mathcal{S}(f) = \int_{-\infty}^{\infty} R(\tau) e^{-j\omega\tau} \, d\tau$

$$= 2\mathrm{Re}\left\{ \frac{1}{1 + [\exp(-j\omega 4 T_s)/4]} \int_0^{4T_s} R(\tau) \exp(-j\omega\tau) \, d\tau \right\}$$

Finally, obtain the result given in Section 2.3.6 and plotted in Figure 2.6 (see [6]).

2.5 CARRIER MODULATED SIGNALS

Many cases of interest are based on heterodyning the baseband pulse code modulation up to a carrier frequency. For example, heterodyning the baseband signal $s(t)$ up to frequency ω_0 yields

$$C(t) = \sqrt{2} A s(t) \cos(\omega_0 t + \theta) \tag{2.5-1}$$

where θ is the carrier phase. Any particular component of $s(t)$, say, $\cos \omega t$, produces

$$\tilde{C}(t) = \sqrt{2} A \cos \omega t \cos(\omega_0 t + \theta)$$

$$= \frac{A}{\sqrt{2}} \cos[(\omega + \omega_0)t + \theta] + \frac{A}{\sqrt{2}} \cos[(\omega - \omega_0)t - \theta] \tag{2.5-2}$$

Hence, an element of power at frequency ω in the baseband produces a component of power at ω_0 and another at $-\omega_0$. If the carrier frequency is not harmonically related to the signaling frequency, the direct and folded sideband contributions will have incoherent phases and consequently power addition will be valid. The resulting power spectral density will be

$$\mathscr{S}_{RF}(f) = \tfrac{1}{2}\mathscr{S}(f + f_0) + \tfrac{1}{2}\mathscr{S}(f - f_0) \tag{2.5-3}$$

where $\mathscr{S}(f)$ is the spectral density of the baseband signal $s(t)$.

In many communication systems of interest the foldover sideband is negligible and then the phase relation is of no consequence. When the baseband and carrier frequencies are harmonically related the above equation does not apply, and the exact solution is best obtained by considering the composite rf signal process itself rather than decomposing into baseband and carrier waveforms.

We conclude that in most applications the rf spectral density can be found by obtaining the baseband spectral density and shifting it up to ω_0 and down to $-\omega_0$ with the appropriate factor to make the resulting power correct.

It should be reemphasized that the basic formulas derived earlier in this chapter can be used directly to obtain heterodyned signal spectral densities.

PROBLEM 14

Consider a quadriphase signal transmitted on two independent signals in phase quadrature:

$$s(t) = As_1(t) \cos(\omega_0 t + \theta) + As_2(t) \sin(\omega_0 t + \theta) \tag{2.5-4}$$

Compute the spectral density assuming two independent, purely random sources. The two baseband signals $s_1(t)$ and $s_2(t)$ are clocked at the same time and have the same bit rate.

For additional viewpoints on the computation of the power spectral density see references 14 and 17–20.

REFERENCES

1 "Telemetry Standards," IRIG Document No. 106-66, March 1966.

2 Deffebach, H. L., and Frost, W. O., "A Survey of Digital Baseband Signaling Techniques," NASA Technical Memorandum X-64615, June 30, 1971; Marshall Space Flight Center, Ala.

3 Fultz, K. E., and Penick, D. B., "The T_1 Carrier System," *Bell System Tech. J.* Vol. 44, September 1965, pp. 366–372.

4 Lender, A., "Correlative Level Coding for Binary-Data Transmission," *IEEE Spectrum*, February 1966, pp. 104–115.

5 Lender, A., "The Duobinary Technique for High Speed Data Transmission," *IEEE Trans. Communication Systems*, Vol. 82, May 1963, pp. 214–218.

6 Hecht, M., and Guida, A., "Delay Modulation," *Proc. IEEE*, July 1969, pp. 1314–1316.

7   Booye, M. A., "An Engineering Evaluation of the Miller Code in Direct PCM Recording and Reproducing," Prepared by Custom Products Engineering Department, Ampex Corp.

8   Batson, B. H., "An Analysis of the Relative Merits of Various PCM Code Formats," NASA Internal Note No. MSC-EB-R-68-5, November 1, 1968.

9   Titsworthe, R. C., and Welch, L. R., "Power Spectra of Signals Modulated by Random and Pseudorandom Sequences," JPL Technical Report No. 32-140, October 10, 1961.

10   Weber, C., unpublished class notes entitled, "Signaling with Message Sequences," USC, September 1973.

11   Lindsey, W. C., and Simon, M. K., based on unpublished notes, March 1974.

12   Bennett, W. R., and Davey, J. R., *Data Transmission*, McGraw-Hill, New York, 1965, Chap. 8.

13   Golomb, S. W., *Shift Register Sequences*, Holden Day, San Francisco, Cal., 1967.

14   Tausworthe, R. C., "Correlation Properties of Cyclic Sequences," Jet Propulsion Laboratory, Tech. Report No. 32-388, July 1, 1963.

15   Bennett, W. R., "Statistics of Regenerative Digital Transmission," *Bell System Tech. J.*, November 1958, pp. 1501–1542.

16   Lindsey, W. C., and Simon, M. K., *Telecommunication Systems Engineering*, Prentice-Hall, Englewood Cliffs, N.J., 1973.

17   Betts, J. A., *Signal Processing, Modulation and Noise*, American Elsevier, New York, 1971.

18   Bosik, B. S., "The Spectral Density of a Coded Digital Signal," *Bell System Tech. J.*, April 1972, pp. 921–932.

19   Scholtz, R. A., "How Do You Define Bandwidth?" Proceedings of the International Conference, Los Angeles, Cal., October 1972, pp. 281–288.

20   Simon, M. K., The Computation of Power Spectral Density for Synchronous Data Pulse Streams, JPL memo 3391-79-015, March 21, 1979.

3

MODULATION OF MULTICHANNEL COHERENT SYSTEMS

This chapter is concerned with conventional type single-channel, two-channel, and multiple-channel phase shift keyed communication systems utilizing subcarriers.

Topics include the modulation system, the demodulation system, and the determination of the relative power components of the modulated signal. We start the discussion with single-channel systems.

3.1 SINGLE-CHANNEL SYSTEMS

We consider both the sinewave and the square subcarrier cases of single-channel systems.

3.1.1 Binary Phase Modulation

Consider the basic single-channel system shown in Figure 3.1. Binary data (which may be encoded) is used to modulate the carrier using binary valued signals. The modulated carrier is then amplified and transmitted to the receiver, which "locks on" to the carrier by means of a PLL if there is a residual carrier, or by means of a suppressed carrier loop if there is not. The coherent reference is then used to beat the data coherently down to the baseband, where it can be synchronized and demodulated. If the data were encoded, then the data would have to be decoded.

The signal can be modeled as

$$s(t) = \sqrt{2}A \sin[\omega_0 t + \theta\, d(t)] \tag{3.1-1}$$

where A is the rms voltage ($A^2 = P$ the transmitted power), ω_0 is the carrier center frequency in radians per second, θ is the modulation index in radians, and $d(t)$ is the binary valued data sequence. Expanding (3.1-1), we

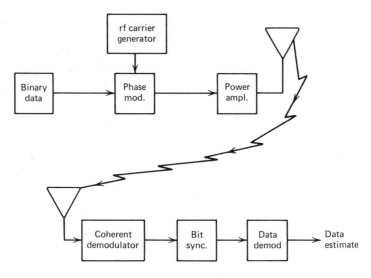

Figure 3.1 Single-channel modulator and demodulator.

have

$$s(t) = \overbrace{\sqrt{2A}\cos\theta\sin\omega_0 t}^{\text{carrier}} + \overbrace{\sqrt{2A}d(t)\sin\theta\cos\omega_0 t}^{\text{data}} \qquad (3.1\text{-}2)$$

where we have used

$$\cos[\theta d(t)] = \cos\theta$$
$$\sin[\theta d(t)] = d(t)\sin\theta \qquad (3.1\text{-}3)$$

It is seen that the carrier power, denoted by P_C, and the data power, denoted by P_D, are given by

$$\frac{P_C}{P} = \cos^2\theta \qquad (3.1\text{-}4)$$

$$\frac{P_D}{P} = \sin^2\theta \qquad (3.1\text{-}5)$$

Hence if $\theta = \pi/2 \pm n\pi$, with n an integer, then the carrier is suppressed. Notice that in this case the total power is given by the data power, and consequently suppressed carrier modulation places all power into the data. Later we will see that the total power is not the sum of data power and carrier power for multichannel systems.

Generally the carrier power must be large enough to support the operation of a phase locked loop*; the remaining power is left for data demodulation, synchronization, and detection.

---

*Unless suppressed carrier operation is employed.

3.1.2 Single-Channel Systems with a Squarewave Subcarrier

When it is desired to utilize residual carrier telemetry systems it is some-times advantageous to employ a subcarrier to place the data away from the carrier in order to eliminate interference effects. A block diagram of the system is shown in Figure 3.2. In this system the data are modulated onto the subcarrier, which is in turn modulated onto the carrier. The carrier demodulation produces the data modulated by the subcarrier leaving the baseband data signal.

The signal can be modulated in the form

$$s(t) = \sqrt{2}A \sin[\omega_0 t + d(t) \, \text{sq} \, (\omega_{SC} t) + \theta_0] \tag{3.1-6}$$

where θ is the phase modulation index, $d(t)$ is the data sequence, and sq$(\omega_{SC}t)$ represents the squarewave subcarrier at frequency ω_{SC}. Expand-ing, we see that the same power relationship exists for this case as the previous one, that is,

$$\frac{P_C}{P} = \cos^2 \theta$$
$$\tag{3.1-7}$$
$$\frac{P_D}{P} = \sin^2 \theta$$

which is due to the fact that sq$(\omega_{SC}t)$ is a binary valued (± 1) function.

3.1.3 Single-Channel Systems with a Sinewave Subcarrier

Many times it is convenient to utilize a sinusoidal subcarrier, for ranging, for example, or for band limited systems that would not pass a squarewave

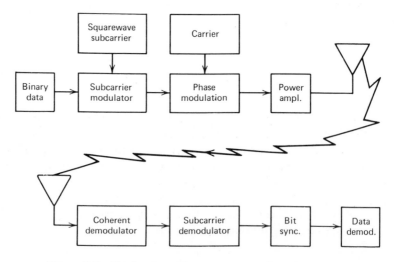

Figure 3.2 Single-channel squarewave subcarrier system.

subcarrier without losing the harmonics anyway. Figure 3.2 is representative of the system except that the subcarrier is sinusoidal instead of
squarewave. We model the signal as

$$s(t) = \sqrt{2}A \sin\{\omega_0 t + \theta \sin[\omega_{SC}t + \theta(t)] + \theta_0\} \qquad (3.1\text{-}8)$$

where A^2 is the transmitted power, ω_0 is the carrier frequency, θ is the
subcarrier modulation index, and θ_0 is the carrier phase.

Expanding, we have

$$s(t) = \sqrt{2}A \sin(\omega_0 t + \theta_0) \cos\{\theta \sin[\omega_{SC}t + \theta(t)]\}$$
$$+ \sqrt{2}A \cos(\omega_0 t + \theta_0) \sin\{\theta \sin[\omega_{SC}t + \theta(t)]\} \qquad (3.1\text{-}9)$$

Using the identities

$$\cos(\theta \sin x) = J_0(\theta) + \sum_{\substack{n=2 \\ n\ \text{even}}}^{\infty} 2J_n(\theta) \cos(nx) \qquad (3.1\text{-}10)$$

$$\sin(\theta \sin x) = \sum_{\substack{n=1 \\ n\ \text{odd}}}^{\infty} 2J_n(\theta) \sin(nx) \qquad (3.1\text{-}11)$$

we obtain

$$s(t) = \sqrt{2}A \sin(\omega_0 t + \theta_0)\left(J_0(\theta) + \sum_{\substack{n=2 \\ n\ \text{even}}}^{\infty} 2J_n(\theta) \cos\{n[\omega_{SC}t + \theta(t)]\}\right)$$
$$+ \sqrt{2}A \cos(\omega_0 t + \theta_0)\left(\sum_{\substack{n=1 \\ n\ \text{odd}}}^{\infty} 2J_n(\theta) \sin\{n[\omega_{SC}t + \theta(t)]\}\right) \qquad (3.1\text{-}12)$$

Hence we see that the carrier and information bearing components have
the following relative power:

$$\frac{P_C}{P} = J_0^2(\theta)$$
$$P = A^2 \qquad (3.1\text{-}13)$$
$$\frac{P_D}{P} = 2J_1^2(\theta)$$

with the remaining power contained in intermodulation components. Notice
when the modulation angle is 2.405 we have $J_0(2.405) = 0$, so that the carrier
is completely suppressed. This also occurs when the modulation index is
5.520, 8.654, 11.794, and so on.

3.2 TWO-CHANNEL MODULATION

Many times two independent data signals must be communicated to a
remote location. One method is to utilize two carriers in phase quadrature
(quadriphase), and another is to utilize two subcarriers. We consider both

schemes in some detail. Sections 3.2.2–3.2.4 follow reference 2 closely which should be consulted for more details along with references 8–13.

3.2.1   Quadriphase Signaling

A quadriphase signal is a four-phase phase shift keyed carrier containing two information bits per encoded symbol, assuming equally likely phases. The signal is described by

$$s(t) = \sqrt{2} A_1 d_1(t) \sin(\omega_0 t + \theta_0) + \sqrt{2} A_2 d_2(t) \cos(\omega_0 t + \theta_0) \qquad (3.2\text{-}1)$$

where A_1^2 and A_2^2 are the respective powers in each carrier phase, ω_0 is the carrier frequency, θ_0 is the rf phase, and $d_1(t)$ and $d_2(t)$ are the two binary valued data signals that are NRZ, RZ, Manchester, etc. The total power in the signals is given by

$$P = A_1^2 + A_2^2 \qquad (3.2\text{-}2).$$

To demodulate quadriphase signals a coherent reference must be established. Basically two references are needed: $\sin[\omega_0 t + \hat{\theta}_0(t)]$ and $\cos[\omega_0 t + \hat{\theta}_0(t)]$, where $\hat{\theta}_0(t)$ is an estimate of θ_0. A demodulator is shown in Figure 3.3. Chapter 5 has a discussion of the type of loop needed to obtain the sine and cosine references. Clearly the power in channel 1 is A_1^2, and the power in channel 2 is A_2^2.

3.2.2   Power Allocation in a Convention Two-Channel Squarewave Subcarrier System

The transmitted signal in a conventional two-channel squarewave subcarrier system, illustrated in Figure 3.4, is of the form

$$\overset{\text{channel 1}}{\qquad} \quad \overset{\text{channel 2}}{\qquad}$$
$$s(t) = \sqrt{2P}\ \sin[\omega_0 t + \overbrace{\theta_1 d_1(t)\ \mathrm{sq}(\omega_1 t)} + \overbrace{\theta_2 d_2(t)\ \mathrm{sq}(\omega_2 t)}] \qquad (3.2\text{-}3)$$

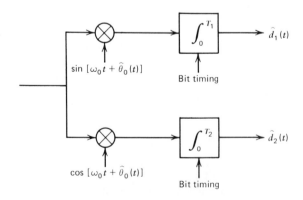

Figure 3.3   Quadriphase demodulator.

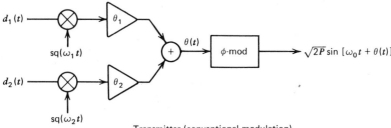

Transmitter (conventional modulation)

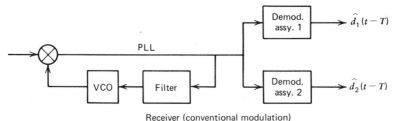

Receiver (conventional modulation)

Figure 3-4   Conventional two-channel squarewave subcarrier system.

where

P is total transmitted power.

θ_1, θ_2 are modulation angles of channel 1 and 2.

ω_0 is the carrier frequency (rad/sec).

ω_1, ω_2 are subcarrier frequencies (squarewave subcarriers).

$d_1(t), d_2(t)$ are the two data streams $[d_i(t) = \pm 1]$.

$\mathrm{sq}(\omega_i t) = \pm 1$ at the frequency ω_i.

Expanding (3.2-3) twice, we have

$$s(t) = \sqrt{2P} \, \sin \omega_0 t \{\cos[\theta_1 d_1 \, \mathrm{sq}(\omega_1 t)] \cos[\theta_2 d_2 \, \mathrm{sq}(\omega_2 t)]$$
$$- \sin[\theta_1 d_1 \, \mathrm{sq}(\omega_1 t)] \sin[\theta_2 d_2 \, \mathrm{sq}(\omega_2 t)]\}$$
$$+ \sqrt{2P} \, \cos \omega_0 t \{\sin[\theta_1 d_1 \, \mathrm{sq}(\omega_1 t)] \cos[\theta_2 d_2 \, \mathrm{sq}(\omega_2 t)]$$
$$+ \cos[\theta_1 d_1 \, \mathrm{sq}(\omega_1 t)] \sin[\theta_2 d_2 \, \mathrm{sq}(\omega_2 t)]\} \qquad (3.2\text{-}4)$$

Simplifying, we arrive at

$$s(t) = \sqrt{2P} \, \sin \omega_0 t [\cos \theta_1 \cos \theta_2 - d_1 d_2 \, \mathrm{sq}(\omega_1 t) \, \mathrm{sq}(\omega_2 t) \sin \theta_1 \sin \theta_2]$$
$$+ \sqrt{2P} \, \cos \omega_0 t [d_1 \, \mathrm{sq}(\omega_1 t) \sin \theta_1 \cos \theta_2 + d_2 \, \mathrm{sq}(\omega_2 t) \cos \theta_1 \sin \theta_2]$$
$$(3.2\text{-}5)$$

From (3.2-5) we see that the first term is the carrier signal, having power

$$P_C = P(\cos \theta_1 \cos \theta_2)^2 \qquad (3.2\text{-}6)$$

The second term is the intermodulation (loss) term, having power

$$P_L = P(\sin\theta_1 \sin\theta_2)^2 \tag{3.2-7}$$

The third term is the first data channel term, with power

$$P_1 = P(\sin\theta_1 \cos\theta_2)^2 \tag{3.2-8}$$

The fourth term is the second data channel term, with power

$$P_2 = P(\cos\theta_1 \sin\theta_2)^2 \tag{3.2-9}$$

PROBLEM 1

Show that the total transmitted power P is given by

$$P = P_C + P_L + P_1 + P_2$$

PROBLEM 2

Normally (1) $P_1 > P_2$ and (2) $P_1 > P_C$ in two-channel systems. Show that

(a) Condition (2) implies $P_L > P_2$, that is, losses are greater than P_2.
(b) Condition (1) implies $\theta_1 > \theta_2$ for $\theta_1\theta_2 \in (0, 90°)$.

In the Mariner Mars 1969 Spacecraft the modulation index was set at $\theta_1 = 65°$ and $P_L/P_2 = +6.63\,\text{dB}$. The modulator and demodulator are shown in Figure 3.4. The efficiency of this communication transmission system can be defined as

$$\eta = \frac{P_1 + P_2}{P} \tag{3.2-10}$$

Now

$$P = P_1 + P_2 + P_C + P_L \tag{3.2-11}$$

Now we show that

$$P_L = \frac{P_1 P_2}{P_C} = \frac{P^2 \sin\theta_1 \cos^2\theta_2 \cos^2\theta_1 \sin^2\theta_2}{P \cos^2\theta_1 \cos^2\theta_2} \tag{3.2-12}$$

or

$$P_L = P(\sin\theta_1 \sin\theta_2)^2 = P_L \tag{3.2-13}$$

from (3.2-7).
   From (3.2-11) and (3.2-12) we have

$$P = P_1 + P_2 + P_C + \frac{P_1 P_2}{P_C}. \tag{3.2-14}$$

Usually $P_C < P_1$, so $P_L > P_2$. In fact the loss (cross modulation terms) increases as P_C decreases. Let

$$P = P_1 + P_2 + P_C + \frac{P_1 P_2}{P_C} \tag{3.2-15}$$

We now find the value of P_C such that the total power is minimized. Set

$$\frac{dP}{dP_C} = 1 - \frac{P_1 P_2}{P_C^2} = 0 \qquad (3.2\text{-}16)$$

or

$$1 = \left(\frac{P_1 P_2}{P_C^0}\right) \frac{1}{P_C^0} \Leftrightarrow P_C^0 = P_L$$

also

$$P_C^0 = \sqrt{P_1 P_2} \quad \text{(optimum value)} \qquad (3.2\text{-}17)$$

Now $P_L = P_C$ implies

$$\sin^2 \theta_1 \sin^2 \theta_2 = \cos^2 \theta_1 \cos^2 \theta_2 \qquad (3.2\text{-}18)$$

or

$$\cos(\theta_1 + \theta_2) = 0 \Rightarrow \theta_1 + \theta_2 = 90° \qquad (3.2\text{-}19)$$

for minimum total power and θ_1, $\theta_2 \in (0, 90°)$. The total minimum total power is given by

$$P_0 = P_1 + P_2 + \sqrt{P_1 P_2} + \sqrt{P_1 P_2} = (\sqrt{P_1} + \sqrt{P_2})^2 \qquad (3.2\text{-}20)$$

so max η, denoted by η_M, is given by

$$\eta_M = \frac{P_1 + P_2}{(\sqrt{P_1} + \sqrt{P_2})^2} = \frac{P_0 - 2\sqrt{P_1 P_2}}{P_0} \qquad (3.2\text{-}21)$$

or

$$\eta_M = 1 - \frac{2\sqrt{P_1 P_2}}{(\sqrt{P_1} + \sqrt{P_2})^2} \qquad (3.2\text{-}22)$$

Let $\alpha = P_2/P_1$, then

$$\eta_M = 1 - \frac{2\sqrt{\alpha}}{(1 + \sqrt{\alpha})^2} \qquad (3.2\text{-}23)$$

So conventional PSK/PM cannot be 100% efficient unless $\alpha = 0$, which implies the single-channel case! For two equal channels ($P_1 = P_2$), $\eta_M = 50\%$.

The conclusions for two-channel squarewave subcarrier transmissions are:

1   P_C usually more than required since usually $P_C^0 = \sqrt{P_1 P_2} \geq P_C^{\min}$
2   $P_L \geq P_2$ if $P_1 > P_2$ and $P_1 > P_C$ (from Problem 2)
3   $P_L = 0$ only in the single-channel case
4   $P_C = 0$ suppressed carrier (impossible)

For example, consider the downlink telemetry two-channel system, such as that used in the Mariner Mars 1969 application where science information is transmitted on channel 1 at 16.2 KBPS and the engineering information is transmitted over channel 2 at $33\frac{1}{3}$ BPS. The phase modulation index was set at $\theta_1 = 65°$. So

$$\frac{P_L}{P_2} = \frac{P \sin^2 \theta_1 \sin^2 \theta_2}{P \cos^2 \theta_1 \sin^2 \theta_2} = \tan^2 \theta_1 = 4.6 = 6.63 \text{ dB} \qquad (3.2\text{-}24)$$

so that the cross modulation power is 6.63 dB greater than the low data rate channel!

We will see in the next section that it is possible to increase the efficiency of the two-channel system by interchanging the roles of the second channel and the cross modulation term.

3.2.3   Power Allocation in an Interplex Two-Channel Squarewave Subcarrier System

In interplex [10] (Figure 3.5) the first data channel is modulated in the usual way and the second channel is transmitted as a cross modulation term, that is, $d_1(t)d_2(t) \text{ sq}(\omega_1 t) \text{ sq}(\omega_2 t)$, so that the transmitted wave is of the form

$$s(t) = \sqrt{2P} \, \sin[\omega_0 t + \theta_1 d_1(t) \text{ sq}(\omega_1 t) + \theta_2 d_1(t)d_2(t) \text{ sq}(\omega_1 t) \text{ sq}(\omega_2 t)] \qquad (3.2\text{-}25)$$

expanding and letting $s_i = d_i(t) \text{ sq}(\omega_i t)$ we obtain

$$s(t) = \sqrt{2P} \, \sin \omega_0 t [\cos(\theta_1 s_1) \cos(\theta_2 s_1 s_2) - \sin(\theta_1 s_1) \sin(\theta_2 s_1 s_2)]$$
$$+ \sqrt{2P} \, \cos \omega_0 t [\sin(\theta_1 s_1) \cos(\theta_2 s_1 s_2) + \cos(\theta_1 s_1) \sin(\theta_2 s_1 s_2)] \qquad (3.2\text{-}26)$$

Simplifying, we have

$$s(t) = \sqrt{2P} \, \sin \omega_0 t [\cos \theta_1 \cos \theta_2 - d_2(t) \text{ sq}(\omega_2 t) \sin \theta_1 \sin \theta_2]$$
$$+ \sqrt{2P} \, \cos \omega_0 t [d_1(t) \text{ sq}(\omega_1 t) \sin \theta_1 \cos \theta_2$$
$$+ d_1 d_2 \text{ sq}(\omega_1 t) \text{ sq}(\omega_2 t) \cos \theta_1 \sin \theta_2] \qquad (3.2\text{-}27)$$

The various components of power are shown in Table 3.1.

Table 3.1 Comparison of interplex and conventional two-subcarrier modulation

Interplex	Conventional
$P_C = P \cos^2 \theta_1 \cos^2 \theta_2$	$= P_C$
$P_2 = P \sin^2 \theta_1 \sin^2 \theta_2$	$= P_L$
$P_1 = P \sin^2 \theta_1 \cos^2 \theta_2$	$= P_1$
$P_L = P \cos^2 \theta_1 \sin^2 \theta_2$	$= P_2$

The transmitter and receiver for interplex are shown in Figure 3.5. The only change from the conventional transmitter is a multiplier (exclusive-or) to combine (or cross-multiply) both channels. The receiver is modified to demodulate the second data channel by generating a coherent reference signal shifted 90° from the carrier.

Now note that

$$P_L = P_C P_2 / P_1$$

as in the conventional modulation scheme. Also

$$P_L = \alpha P_C \qquad \alpha = \frac{P_2}{P_1} \qquad (3.2\text{-}28)$$

The power in the cross modulation signal is proportional to the power in the carrier component. Hence there is no advantage in transmitting more carrier power than required for tracking. Further, if $P_C \leq P_1$, we see that $P_L \leq P_2$ for interplex; but we found $P_L \geq P_2$ for conventional two-channel systems. Now

$$P = P_1 + P_2 + P_C + P_L \qquad (3.2\text{-}29)$$

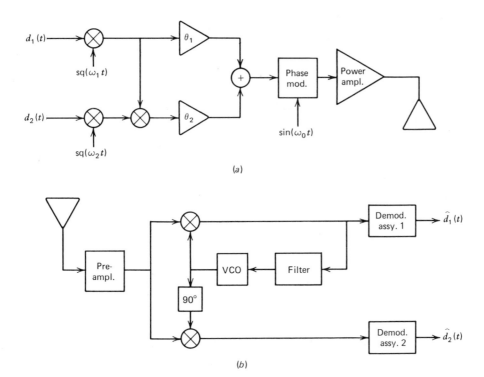

(a)

(b)

Figure 3.5 Interplex transmitter (a) and receiver (b).

or

$$P = P_1 + P_2 + P_C + \frac{P_C P_2}{P_1} \tag{3.2-30}$$

so that

$$P = (1 + \alpha)(P_1 + P_C) \qquad \alpha = \frac{P_2}{P_1} = \tan^2 \theta_1 \tag{3.2-31}$$

Now the efficiency is shown in Problem 3 to be

$$\eta = \frac{P_1 + P_2}{P} = \sin^2 \theta_1$$

PROBLEM 3

Show that for interplex

(a) $\quad \eta = 1 - \dfrac{P_C}{P_1 + P_C} = \sin^2 \theta_1$

(b) $\quad \eta = 1 - \dfrac{P_C(1 + \alpha)}{P}$

Hence we see that if $\theta_1 = 90°$, both the carrier power and the cross modulation are zero. Furthermore, η can approach 100% as long as P_C is small compared to P_1 [$\eta = 1 - P_C/(P_C + P_1)$]. In Figure 3.6 the data efficien-

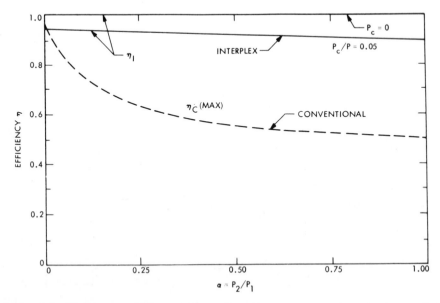

Figure 3.6 Data power efficiency of two-channel squarewave subcarrier ($P_2 = \alpha P_1$) systems (from Butman and Timor [2] with permission).

cies of the two-channel squarewave subcarrier systems are shown. Note that the carrier power allocated by optimizing the conventional system is more than the amount required by the interplex scheme except for very small α. The improvement gained by the interplex scheme approaches 3 dB when $P_1 = P_2$.

PROBLEM 4

Show that the interplex scheme is capable of increasing the channel efficiency by 3 dB over the conventional scheme.

Although our discussion was based on squarewave subcarriers, the same methods can be applied to sinewave subcarriers. The performance with sinusoidal subcarriers is inferior to that with squarewave subcarriers, however.

3.2.4 Power Allocation in Two-Channel Sinusoidal Subcarrier Systems

Sinewave subcarriers are sometimes used when high frequencies are employed when the system cannot pass the squarewave subcarrier. In this case our modulated signal is of the form

$$s(t) = \sqrt{2P} \, \sin(\omega_0 t + \theta_1 d_1 \sin \omega_1 t + \theta_2 d_2 \sin \omega_2 t) \tag{3.2-32}$$

where θ_1 and θ_2 are the two modulation angles, ω_1 and ω_2 are the two subcarrier frequencies, $d_1(t)$ and $d_2(t)$ are the two binary valued data signals, and P is the total power. Expanding yields

$$s(t) = \sqrt{2P} \, \sin(\omega_0 t) \cos(\theta_1 d_1 \sin \omega_1 t + \theta_2 d_2 \sin \omega_2 t)$$

$$+ \sqrt{2P} \, \cos(\omega_0 t) \sin(\theta_1 d_1 \sin \omega_1 t + \theta_2 d_2 \sin \omega_2 t) \tag{3.2-33}$$

Expanding, we have

$$s(t) = \sqrt{2P} \, \sin \omega_0 t [\cos(\theta_1 \sin \omega_1 t) \cos(\theta_2 \sin \omega_2 t)$$

$$- d_1 d_2 \sin(\theta_1 \sin \omega_1 t) \sin(\theta_2 \sin \omega_2 t)]$$

$$+ \sqrt{2P} \, \cos \omega_0 t [d_1 \sin(\theta_1 \sin \omega_1 t) \cos(\theta_2 \sin \omega_2 t)$$

$$+ d_2 \cos(\theta_1 \sin \omega_1 t) \sin(\theta_2 \sin \omega_2 t)]$$

$$\tag{3.2-34}$$

where we have used $\sin(d\theta) = d \sin \theta$ and $\cos(\theta d) = \cos \theta$.
Now we use the identities

$$\cos(\theta \sin \omega t) = J_0(\theta) + \sum_{\substack{n=2 \\ (n \text{ even})}}^{\infty} 2J_n(\theta) \cos n\omega t \tag{3.2-35}$$

$$\sin(\theta \sin \omega t) = \sum_{\substack{n=1 \\ (n \text{ odd})}}^{\infty} 2J_n(\theta) \sin n\omega t \tag{3.2-36}$$

So that we obtain

$$\text{carrier}$$

$$s(t) = \sqrt{2P} \sin \omega_0 t \left[\overbrace{J_0(\theta_1)J_0(\theta_2)} + O_1(n\omega_1, m\omega_2) \right.$$

$$\left. - d_1 d_2 \sum_{\substack{n \\ \text{odd}}}^{\infty} 2J_n(\theta_1) \sin n\omega_1 t \sum_{\substack{m \\ \text{odd}}}^{\infty} 2J_m(\theta_2) \sin n\omega_2 t \right]$$

$$+ \sqrt{2P} \cos \omega_0 t [2d_1(t)J_1(\theta_1)J_0(\theta_2) \sin \omega_1 t$$

$$+ O_2(n\omega_1, m\omega_2) + 2d_2(t)J_0(\theta_1)J_1(\theta_2) \sin \omega_2 t$$

$$+ O_3(n\omega_1, m\omega_2)] \tag{3.2-37}$$

where O_1, O_2, and O_3 are terms that produce interference and O_1 is defined by

$$O_1 = J_0(\theta_1) \sum_{\substack{n=2 \\ n \text{ even}}}^{\infty} 2J_n(\theta_2) \cos n\omega_1 t + J_0(\theta_2) \sum_{\substack{m=2 \\ m \text{ even}}}^{\infty} 2J_m(\theta_1) \cos m\omega_2 t$$

$$+ 4 \sum_{\substack{n=2 \\ n \text{ even}}}^{\infty} \sum_{\substack{m=2 \\ m \text{ even}}}^{\infty} J_n(\theta_1)J_m(\theta_2) \cos n\omega_1 t \cos m\omega_2 t \tag{3.2-38}$$

O_2 and O_3 are given by similar type expressions.

We see that

$$P_C = PJ_0^2(\theta_1)J_0^2(\theta_2)$$
$$P_1 = 2PJ_1^2(\theta_1)J_0^2(\theta_2)$$
$$P_2 = 2PJ_0^2(\theta_1)J_1^2(\theta_2) \tag{3.2-39}$$
$$P_L = P - P_C - P_1 - P_2$$

which generalizes Section 3.1.3 to the case of two sinewave subcarriers. It is clear now how to write down the general n subcarrier case; it will be treated in a later section.

It has been mentioned that sinewave subcarriers are less efficient than squarewave subcarriers. For one subcarrier, $J_1(\theta)$ is maximum at $\theta = 1.84$ rad, which leads to

$$\frac{P_1}{P} = 0.68, \qquad \frac{P_C}{P} = 0.1, \qquad \frac{P_L}{P} = 0.22$$

For squarewave subcarriers we have

$$\frac{P_1}{P} = \sin^2 \theta_1$$

$$\tag{3.2-40}$$

$$\frac{P_C}{P} = \cos^2 \theta_1$$

If we let $P_C/P = 0.1$, then $P_1/P = 0.9$ and $P_L/P = 0$ ($\theta_1 = 1.25$ rad). Then we see that the squarewave subcarrier case yields a 1.22-dB increase in data power!

3.2.5 Power Allocation in Two-Channel Sinewave-Squarewave Systems

We now consider the case that a single sinewave and single squarewave subcarrier are used. The signal is modeled as

$$s(t) = \sqrt{2P} \, \sin\{\omega_0 t + m_1 d(t) \, \mathrm{sq}(\omega_1 t) + m_2 \sin[\omega_{SC} t + \theta(t)]\} \quad (3.2\text{-}41)$$

where

ω_0 is the carrier frequency.

$d(t)$ is a binary valued data signal.

$\mathrm{sq}(\omega_1)$ is the binary valued squarewave subcarrier at frequency ω_1.

m_1 is the squarewave modulation index.

ω_{SC} is the sinewave subcarrier frequency.

$\theta(t)$ is the information bearing signal phase modulated onto the subcarrier.

From Problem 5 we have (assuming that the two subcarriers are not coherent) that

$$\frac{P_C}{P} = \cos^2 m_1 J_0^2(m_2)$$

$$\frac{P_D}{P} = \sin^2 m_1 J_0^2(m_2) \quad (3.2\text{-}42)$$

$$\frac{P_{SC}}{P} = 2 \cos^2(m_1) J_1^2(m_2)$$

Notice that if $\theta_1 = \pi/2$ in order to suppress the carrier, the sinewave subcarrier is also suppressed.

PROBLEM 5

(a) Derive the relative power components in the above two-subcarrier system resulting in (3.2-42).

(b) Let $m_1 = 1.1$ and $m_2 = 1.0$ (as in the Space Shuttle system and show that $P_C/P = -9.2$ dB, $P_D/P = -3.3$ dB, and $P_{SW}/P = -11$ dB.

(c) Show that the modulation scheme is 67% efficient, with the carrier and the two subcarriers as desired components and the remaining terms creating intermodulation loss.

PROBLEM 6

Prove that

$$\frac{P_i}{P} = 2 \frac{P_C}{P} \left[\frac{J_1(\theta_i)}{J_0(\theta_i)} \right]^2$$

for the power distribution of an N sinewave subcarrier channel system, where P_i denotes the ith subcarrier power level.

3.3 GENERAL CASE

3.3.1 Power Allocation with N Squarewave and M Sinewave Subcarriers

The most general case, that of N squarewave subcarriers and M sinewave subcarriers is considered [3]. Consider a signal of the form

$$s(t) = \sqrt{2P} \, \sin\left(\omega_0 t + \sum_{i=1}^{N} \theta_i d_i(t) \, \mathrm{sq}(\omega_i t) + \sum_{j=1}^{M} \phi_j \cos[\omega'_j t + \gamma_j(t)]\right)$$

(3.3-1)

where

θ_i is the modulation index of the ith squarewave subcarrier.

$d_i(t)$ is the ith binary value (± 1) data sequence.

$\mathrm{sq}(\omega_i t)$ is the squarewave subcarrier at frequency ω_i.

ϕ_j is the modulation index of the jth sinewave subcarrier.

$\gamma_j(t)$ is the jth sinewave subcarrier modulation at frequency ω'_j.

In what follows it is assumed that all the frequencies $\{\omega_i, \omega'_j\}$ are sufficiently well separated to prevent interference. Denoting $\theta(t)$ as the weighted sum of sinewave subcarriers and $\phi(t)$ as the weighted sum of squarewave subcarriers, we have

$$s(t) = \sqrt{2P} \, \mathrm{Im}\{e^{j\omega_0 t} \, e^{j\theta(t)} \, e^{j\phi(t)}\}$$

(3.3-2)

where

$$\theta(t) = \sum_{i=1}^{N} \theta_i d_i(t) \, \mathrm{sq}(\omega_i t)$$

(3.3-3)

$$\phi(t) = \sum_{k=1}^{M} \phi_k \cos[\Omega_k t + \gamma_k(t)]$$

utilizing

$$e^{j\theta(t)} = \prod_{i=1}^{N} [\cos \theta_i + j \sin \theta_i \, \mathrm{sq}(\omega_i t)]$$

(3.3-4)

and the Fourier-Bessel series ($j = \sqrt{-1}$)

$$e^{jx \cos z} = \sum_{n=-\infty}^{\infty} j^n J_n(x) \, e^{jnz}$$

(3.3-5)

we have

$$s(t) = \sqrt{2P} \; \text{Im}\Big(e^{j\omega_0 t} \prod_{i=1}^{N} [\cos \theta_i + j \sin \theta_i \; \text{sq}(\omega_i t)]$$

$$\times \prod_{k=1}^{M} \sum_{n_k=-\infty}^{\infty} (j)^{n_k} J_{n_k}(\phi_k) e^{j n_k [\Omega_k t + \gamma_k(t)]} \Big) \qquad (3.3\text{-}6)$$

where n_k is the kth n index (n_1 through n_M). Therefore the carrier component is obtained from $s(t)$ by taking all the terms in (3.3-6) that yield a constant times $\text{Im}(e^{j\omega_0 t})$. Thus we obtain

$$\frac{P_C}{P} = \prod_{i=1}^{N} \cos^2 \theta_i \prod_{k=1}^{M} J_0^2(\phi_k) \qquad (3.3\text{-}7)$$

Clearly if any $\phi_k = 2.4048$ or any $\theta_i = (\pi/2) \pmod{2\pi}$, then $P_C = 0$. In the same manner the power in the pth squarewave subcarrier is given by the appropriate term that produces a signal of the form

$$\text{sq}(\omega_p t) \cos \omega_0 t \qquad (3.3\text{-}8)$$

This term is given by

$$s_p(t) = \sqrt{2P} \; \text{sq}(\omega_p t) \cos \omega_0 t \prod_{i=1}^{N} \cos \theta_i \tan \theta_p \prod_{k=1}^{M} J_0(\phi_k) \qquad (3.3\text{-}9)$$

Therefore the relative power is given by

$$\frac{P_p}{P} = \prod_{i=1}^{N} \cos^2 \theta_i \tan^2 \theta_p \prod_{k=1}^{M} J_0^2(\phi_k) \qquad (3.3\text{-}10)$$

The power in the qth sinewave subcarrier, obtained in the same manner as is the squarewave subcarrier case, produces the desired signal:

$$s_{SNq}(t) = \sqrt{2P} \prod_{i=1}^{N} \cos \theta_i \prod_{k=1}^{M} J_0(\phi_k) 2 \cos \omega_0 t \sin[\omega_q t + \gamma_q(t)] \qquad (3.3\text{-}11)$$

Hence the relative power is given by

$$\frac{P_q}{P} = 2 \prod_{i=1}^{N} \cos^2 \theta_i \prod_{k=1}^{M} J_0^2(\phi_k) \left[\frac{J_1^2(\phi_q)}{J_0^2(\phi_q)} \right] \qquad (3.3\text{-}12)$$

It is seen that this general result encompasses all previous results.

3.3.2 N Channel Squarewave Interplex Systems

Butman and Timor [2] have shown that the N squarewave subcarrier interplex signal is of the form

$$s(t) = \sqrt{2P} \; \sin\Big\{ \omega_0 t + \Big[\theta_1 + \sum_{n=2}^{N} \theta_n d_n(t) \; \text{sq}(\omega_n t) \Big] d_1(t) \; \text{sq}(\omega_n t) \Big\} \qquad (3.3\text{-}13)$$

Whether interplex or the conventional modulation system is more efficient depends on the power allocations required as well as the number of channels employed. In general Butman and Timor found that interplex is more efficient when N is not too large ($N \le 4$ if the channels are equal in power, but larger values of N can be used when there is one high-power channel with many low-power channels). Two cases are shown in Figure 3.7. Notice that for large N the conventional system is superior, whereas for small $N(\le 4)$ interplex is superior.

3.3.3 Multichannel Sinewave Interplex

La Frieda [7] has considered two-channel sinewave subcarriers, and Timor [8] has shown that for a large number of subcarriers both squarewave and sinewave subcarrier systems have about the same efficiency. In addition Timor has shown that in conventional systems both sinewave and square-wave subcarrier systems converge to the same efficiency ($1/e$).

PROBLEM 7

Using the results for N-channel squarewave systems show that the efficiency η defined by

$$\eta = \frac{\displaystyle\sum_{i=1}^{N} P_i}{P}$$

where P_i is the power in the ith squarewave subcarrier, satisfies

$$\lim_{N \to \infty} \eta = e^{-1} = 36.79\%$$

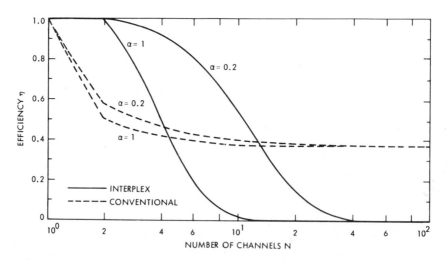

Figure 3.7 Efficiency for N channels with $P_k/P_1 = \alpha$, $k = 2, \ldots, N$ (from Butman and Timor [2] with permission).

PROBLEM 8

Show the same limiting result for efficiency (as in Problem 7) applies to the sinusoidal subcarrier signals case.

3.4 DEMODULATION SYSTEMS

In the next sections we briefiy describe the demodulators used for demodulation.

3.4.1 Demodulation of Conventional Two-Channel Systems

Consider the demodulator used for two-channel squarewave subcarrier systems shown in Figure 3.8.

The modulated signal is of the form

$$s(t) = \sqrt{2P} \, \sin[\omega_0 t + \theta_1 d_1(t) \, \mathrm{sq}(\omega_1 t) + \theta_2 d_2(t) \, \mathrm{sq}(\omega_2 t) + \theta] \qquad (3.4\text{-}1)$$

After expanding we have

$$s(t) = \sqrt{2P} \, \sin(\omega_0 t + \theta)[\cos \theta_1 \cos \theta_2$$
$$- d_1(t) d_2(t) \, \mathrm{sq}(\omega_1 t) \, \mathrm{sq}(\omega_2 t) \sin \theta_1 \sin \theta_2]$$
$$+ \sqrt{2P} \, \cos(\omega_0 t + \theta)[d_1(t) \, \mathrm{sq}(\omega_1 t) \sin \theta_1 \cos \theta_2$$
$$+ d_2(t) \, \mathrm{sq}(\omega_2 t) \cos \theta_1 \sin \theta_2] \qquad (3.4\text{-}2)$$

Therefore both subcarriers are obtained at the phase detector output. The demodulated output is given by

$$D(t) = [\sqrt{P_1} d_1(t) \, \mathrm{sq}(\omega_1 t) + \sqrt{P_2} d_2(t) \, \mathrm{sq}(\omega_2 t)] \cos \phi$$
$$+ [\sqrt{P_C} + \sqrt{P_L} d_1(t) d_2(t) \, \mathrm{sq}(\omega_1 t) \, \mathrm{sq}(\omega_2 t)] \sin \phi + n'(t)$$
$$(3.4\text{-}3)$$

with P_1 and P_2 the power in channels 1 and 2, respectively. $\phi = \theta - \hat{\theta}$ is the

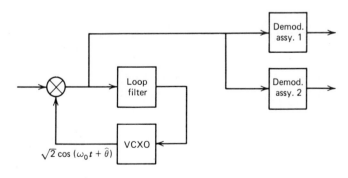

Figure 3.8 Squarewave demodulator.

tracking phase error of the loop; $d_i(t)$ is the ith data sequence; sq($\omega_i t$) the ith squarewave subcarrier; and finally $n'(t)$ is the equivalent white Gaussian noise (WGN) present at the phase detector output. If the phase error is small, the intermodulation power is small and distortion is not a problem for relatively (power) balanced channels.

For sinewave subcarriers the demodulator is of the same form as shown in Figure 3.8.

3.4.2   Demodulation of Two-Channel Interplex With Squarewave Subcarriers

In order to demodulate the interplex system with squarewave subcarriers it is necessary to add an extra correlator at 90° to the phase detector, as shown in Figure 3.9.

From (3.2-27) we see that the received signal, with two-channel interplex squarewave subcarriers, is given by

$$s(t) = \sqrt{2P}\,\sin(\omega_0 t + \theta_0)[\cos\theta_1\cos\theta_2 - d_2(t)\,\text{sq}(\omega_2 t)\sin\theta_1\sin\theta_2]$$
$$+ \sqrt{2P}\,\cos(\omega_0 t + \theta_0)[d_1(t)\,\text{sq}(\omega_1 t)\sin\theta_1\cos\theta_2$$
$$+ d_1(t)d_2(t)\,\text{sq}(\omega_1 t)\,\text{sq}(\omega_2 t)\cos\theta_1\sin\theta_2]$$

$$(3.4\text{-}4)$$

Since a phase locked loop will provide a phase detector input signal of the form

$$\sqrt{2}\cos(\omega_0 t + \hat{\theta}_0) \qquad\qquad (3.4\text{-}5)$$

it is clear that the first channel will appear at the output of the phase detector, and the second channel can be obtained with a quadrature carrier of the form

$$\sqrt{2}\sin(\omega_0 t + \hat{\theta}) \qquad\qquad (3.4\text{-}6)$$

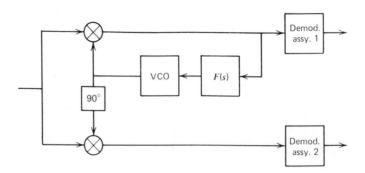

Figure 3.9   Demodulator used for two-channel interplex.

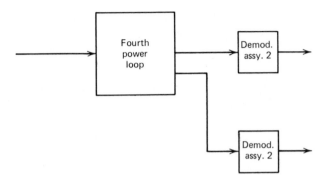

Figure 3.10   Demodulator of a quadriphase signal using a fourth power loop.

3.4.3   Demodulation of Quadriphase Signals

Quadriphase is a phase quadrature scheme that can be viewed as a four-phase signal taking on the values $\pi/2$, π, $3\pi/2$, and 2π. It is necessary to use a fourth power loop or a four-phase In-phase quadrature (I–Q) loop, as discussed in Chapter 5, to obtain the phase reference. One possible demodulation scheme is shown in Figure 3.10; more are discussed in Chapter 5.

REFERENCES

1 Edelson, R. E., Ed., "Telecommunication Systems Design Technique Handbook," *JPL Tech. Memo.* 33–571, July 15, 1972.

2 Butman, S., and Timor, U., "Interplex Modulation," *JPL Quarterly Tech. Rev.*, Vol. 1, No. 1, April 1971.

3 Lindsey, W. C., and Simon, M., *Telecommunication Systems Engineering*, Prentice-Hall, Englewood Cliffs, N.J., 1973.

4 Martin, B. D., "The Pioneer IV Lunar Probe: A Minimum-Power FM/PM System Design," *JPL Tech.* Report No. 32–215, March 1962.

5 Giacolleto, L. J., "Generalized Theory of Multitone Amplitude and Frequency Modulation," *Proceedings of the IRE*, July 1947.

6 Timor, U., "Optimum Configurations for PSK/PM Systems," *Space Program Summary*, Vol. III, pp. 33–36, *JPL*, Pasadena, Cal., December 1970.

7 La Frieda, J., "Optimum Performance of Two-Channel High-Rate Interplex Systems," *JPL SPS* 37–65, Vol. III.

8 Timor, U., "Efficiency of Bi-Phase Modulated Subcarriers N-Channel Telemetry Systems," *JPL SPS* 37–65, Vol. III, October 1970.

9 La Frieda, J., "Optimum Modulation Indexes and Maximum Data Rates for the Interplex Modern," *JPL SPS* 37–64, Vol. III, 1970.

10 Butman, S., and Timor, U., "An Efficient Two-Channel Telemetry System for Space Exploration," *JPL SPS* 37–62, Vol. III, April 30, 1970.

11   Butman, S., and Timor, U., "Efficient Multichannel Space Telemetry," *JPL SPS*, Vol. III, June 30, 1970.

12   Butman, S., and Timor, U., "Suppressed-Carrier Tracking for Two-Channel Phase Modulated Telemetry," *Proceedings of the National Electronics Conference*, Vol. 26, December 1970.

13   Tausworthe, R. C., "A Boolean-Function Multiplexed Telemetry System," *IEEE Trans. Space Electronics Telemetry*, Vol. Set-9, No. 2, June 1963, pp. 42–45.

4

PHASE-LOCKED LOOPS

The phase-locked loop (PLL) is a closed loop tracking device that is capable of tracking the phase of a received signal that has a residual carrier component. The PLL is not adequate for tracking suppressed carrier signals, as we will see in the next chapter. Its great success has been due, in large part, to its ability to track, with great accuracy, a very weak signal, immersed in noise having many times the power of the carrier signal being tracked. Basically, a PLL, used for carrier tracking, is a very narrow band tracking filter that, by virtue of its narrow bandwidth, can produce loop signal-to-noise ratios (SNRs) that are considerably greater than 1. It appears the earliest description of the phase-locked loop was published by DeBellescize in 1932 [1] for synchronous reception of radio signals.

Widespread use of phase-locked loops came with the synchronization of the horizontal and the vertical scan in television receivers [2]. A major impetus to scientific use of PPLs coming with the space program, included orbiting satellites and deep space probes.

Other applications include FSK demodulation, frequency synthesis, timing recovery, television color carrier regeneration, and FM stereo decoding. Most of the above loops can be realized with integrated circuit technology [3]. Frequency multiplication and division can be achieved using PLLs also. Bit synchronization and PN code tracking utilize modified forms of PLLs.

4.1 STOCHASTIC DIFFERENTIAL EQUATION OF OPERATION

Before we develop the stochastic differential equation of operation we shall define a long and a short loop. The term *short loop* indicates that there is only VCO feedback after the IF filters. A *long loop* is a loop that is not a short loop. Since all long loops can be modeled as short loops, we will consider them in what follows. Figure 4.1 illustrates a double heterodyne long loop. There is a free-running oscillator plus a loop controlled oscillator along with three phase detectors or multipliers and a programmed ephemeris to track out doppler effects. A short loop is shown in Figure 4.2. The short loop consists of a phase detector (multiplier), a loop

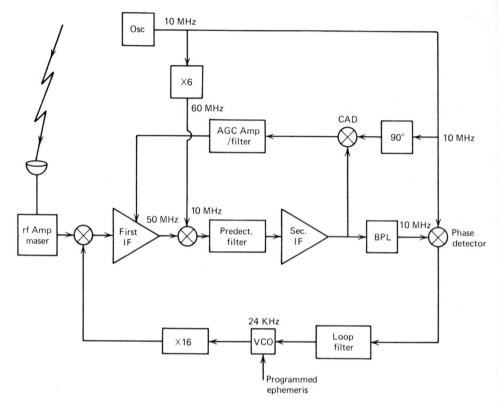

Figure 4.1   A double-heterodyne long loop.

filter, and a voltage controlled oscillator [or voltage controlled crystal oscillator (VCXO)] arranged in a feedback loop.

We model the input process $y(t)$ as the sum of white Gaussian noise (WGN) plus a carrier with phase modulation process $\theta(t)$ so that

$$y(t) = \sqrt{2}\, A \, \sin[\omega_0 t + \theta(t)] + n(t) \qquad (4.1\text{-}1)$$

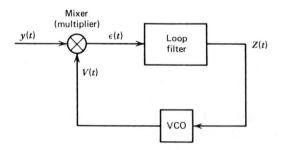

Figure 4.2   A short-loop phase locked loop.

where $n(t)$ has two-sided power spectral density $N_0/2$, ω_0 is the carrier frequency (rad/sec), $\theta(t)$ is the input phase process, and A is the rms voltage. The voltage controlled oscillator (VCO) output is represented by

$$V(t) = \sqrt{2} \, K_1 \cos[\omega_0 t + \hat{\theta}(t)] \qquad (4.1\text{-}2)$$

where $\hat{\theta}(t)$ is the loop estimate of $\theta(t)$. We write the WGN in terms of the in-phase and quadrature components;

$$n(t) = \sqrt{2} n_c(t) \cos \omega_0 t + \sqrt{2} n_s(t) \sin \omega_0 t \qquad (4.1\text{-}3)$$

We have that the input to the loop filter is

$$\begin{aligned}
\epsilon(t) = K_m V(t) y(t) = &AK_1 K_m \{\sin(\theta - \hat{\theta}) + \sin[2\omega_0 t + \theta(t) + \hat{\theta}(t)]\} \\
&+ K_1' n_c(t) \cos\hat{\theta}(t) - K_1' n_s(t) \sin \hat{\theta}(t) \\
&+ K_1' n_s(t) \sin[2\omega_0 t + \hat{\theta}(t)] \\
&+ K_1' n_c(t) \cos[2\omega_0 t + \hat{\theta}(t)] \qquad (4.1\text{-}4]
\end{aligned}$$

where $K_1' = K_1 K_m$ and K_m is the mixer gain. The product $AK_1 K_m$ is the *phase detector gain* with dimensions of volts per radian. Since normally, the mixer (and loop filter) cannot respond to the $2\omega_0$ terms, neglecting $2\omega_0$ terms, we have

$$\epsilon(t) = AK_1 K_m \sin(\theta - \hat{\theta}) + K_1 K_m n'(t) \qquad (4.1\text{-}5)$$

with

$$n'(t) = n_s(t) \cos \hat{\theta} - n_c(t) \sin \hat{\theta} \qquad (4.1\text{-}6)$$

Both Viterbi and Sakrison [4, 5] have shown that $n'(t)$ is essentially WGN with spectral density $N_0/2$ if the loop bandwidth is small compared to the input noise bandwidth. We define the phase error as $\phi = \theta - \hat{\theta}$, so that

$$\epsilon(t) = AK_1 K_m \sin \phi(t) + K_1 K_m n'(t) \qquad (4.1\text{-}7)$$

The VCO produces a signal of the form $\omega_0 + K_{VCO} Z(t)$, with K_{VCO}, the VCO gain constant, having dimensions of radians per second per volt. So the VCO output phase is

$$\hat{\theta}(t) = \int_0^t K_{VCO} Z(u) \, du \qquad (4.1\text{-}8)$$

Using Heaviside operator notation, with s the Laplace transform variable, we have

$$\hat{\theta}(t) = \frac{K_{VCO}}{s} Z(t) \qquad (4.1\text{-}9)$$

But

$$Z(t) = F(s)\epsilon(t) = F(s)K_m[AK_1 \sin \phi(t) + K_1 n'(t)] \qquad (4.1\text{-}10)$$

and

$$\hat{\theta}(t) = \frac{K_{VCO}}{s} F(s)[AK_1 K_m \sin \phi(t) + K_1 K_m n'(t)] \qquad (4.1\text{-}11)$$

and $\hat{\theta}(t) = \theta - \phi$, so that we have

$$\theta(t) = \phi(t) + \frac{F(s)}{s} [AK \sin \phi(t) + Kn'(t)] \qquad (4.1\text{-}12)$$

which is the basic equation defining loop operation, and

$$K = K_1 K_m K_{VCO} \qquad (4.1\text{-}13)$$

and where AK is called the *loop gain* (or open loop gain), which has dimensions of frequency (Hz). The term $AKF(0)$ is called the *dc loop gain*. We note the Heaviside operator symbolism

$$s \leftrightarrow \frac{d}{dt} \quad \frac{1}{s} \leftrightarrow \int_0^t (\,\cdot\,) \, dt', \quad F(s) \leftrightarrow F\left(\frac{d}{dt}\right) \qquad (4.1\text{-}14)$$

Multiplying our basic equation by s, we have an equivalent form*

$$\dot{\theta}(t) = \dot{\phi}(t) + F(s)[AK \sin \phi + Kn'(t)] \qquad (4.1\text{-}15)$$

The first-order phase locked loop equation $[F(s) = 1]$ is then

$$\dot{\theta}(t) = \dot{\phi}(t) + AK \sin \phi + Kn'(t) \qquad (4.1\text{-}16)$$

since $F(s) = 1$, and since the resulting stochastic differential equation is first order. In general, when $F(s)$ has n poles, the loop is said to be an $(n + 1)th$ order phase locked loop. We will show later that this is due to the fact that the closed-loop response has $n + 1$ poles. Equation 4.1-12 leads directly to our nonlinear baseband model shown in Figure 4.3. Notice that this model is independent of the carrier frequency and depends only on the phase process of the input signal and baseband parameters of the loop.

The baseband model defines a Markov process for the phase error, $\phi(t)$, and has been analyzed using Fokker-Planck techniques (see the Appendix and references 4, 6, and 7).

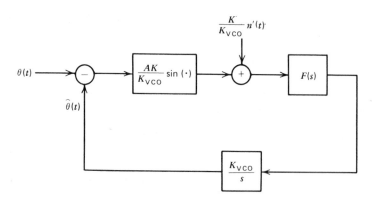

Figure 4.3   The baseband nonlinear model of a PLL.

*We note that $\dot{\theta}(t) = \dfrac{d}{dt}\theta(t)$ etc.

4.2 LOOP FILTERS AND THE PHASE PLANE DESCRIPTION

In order to understand the behavior of the loop during lock-in, it is useful to study the phase plane description of the loop. The phase plane defines the admissible trajectories when $\dot{\phi}(t)$ is plotted versus $\phi(t)$, with $n'(t)$ set equal to zero.

4.2.1 First-Order Phase Locked Loop

Consider an input phase process of the form

$$\theta(t) = \theta_0 + \Omega_0 t \qquad (4.2\text{-}1)$$

where Ω_0 and θ_0 are constants. Then when $n'(t) = 0$, we have that ϕ and $\dot{\phi}$ satisfy the differential equation from (4.1-15):

$$\Omega_0 = \dot{\phi}(t) + AK \sin \phi \qquad (4.2\text{-}2)$$

Since $\Omega(t) = \dot{\phi}(t)$ is the frequency error as a function of time, we have

$$\Omega_0 = \Omega(t) + AK \sin \phi(t) \qquad (4.2\text{-}3)$$

The phase plane plot is shown in Figure 4.4.

Whenever $\dot{\phi}(t)$ is positive ϕ, tends to increase; and whenever $\Omega(t)$ is negative, ϕ tends to decrease. If $|\Omega_0| < AK$, there are points on the trajectory where $\Omega = 0$; that is, the loop is locked. These points are denoted as *stable lock points*; that is, if the frequency error was slightly positive, this would drive $\phi(t)$ positive, making the frequency error go to zero and conversely. At *unstable lock points* any slight frequency error causes the

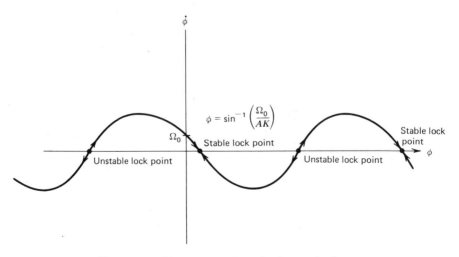

Figure 4.4 Phase plane plot of a first-order loop.

phase to diverge from the lock point, making the frequency error even larger, etc.

At the stable lock points $\dot{\phi} = 0$, so that it follows that the steady-state phase error for a first-order loop is given by

$$\phi_{ss} = \sin^{-1}\left(\frac{\Omega_0}{AK}\right) + 2n\pi \tag{4.2-4}$$

where n is an integer which is normally unknown. If $|\Omega_0| > AK$, we see that the trajectory never crosses the ϕ axis and hence the loop never locks up. Therefore, the <u>maximum pull-in range</u> is then

$$|\Omega_{pi}| \equiv AK \tag{4.2-5}$$

We conclude that the first-order loop ultimately tracks the incoming signal when $\theta(t) = \Omega_0 t + \theta_0$, when $|\Omega_0| < AK$ with a static phase error ϕ_{ss} given by

$$\phi_{ss} = \sin^{-1}\left(\frac{\Omega_0}{AK}\right) \tag{4.2-6}$$

where we have dropped the $2n\pi$ term since in almost all applications only the phase offset from zero is of no consequence.

4.2.2   Passive Loop Filter Second-Order Phase Locked Loop

The most commonly used loop filter produces a second-order PLL. There are two types that are used: the passive filter and the active filter. The passive loop filter that produces a second-order PLL has the form

$$F(s) = \frac{1 + \tau_2 s}{1 + \tau_1 s} \tag{4.2-7}$$

This filter can be implemented as shown in Figure 4.5 (neglecting loading effects).

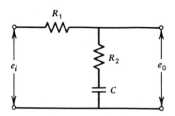

Figure 4.5   Passive second-order loop filter implementation.

PROBLEM 1

(a)   Show that the transfer function of the above circuit (neglecting loading), is given by

$$F(s) = \frac{1 + \tau_2 s}{1 + \tau_1 s} \qquad s = i\omega$$

where $\tau_2 = R_2 C$ and $\tau_1 = (R_1 + R_2)C$

(b)   Obtain the transfer function of the following filter (assume $A \to \infty$).

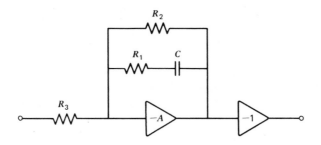

It will be shown later that <u>the maximum-degree polynomial in t that a passive second-order PLL can track is</u> $\theta(t) = \theta_0 + \Omega_0 t$. The (zero noise) differential equation for the above passive filter is, from (4.1-15)

$$s(1 + \tau_1 s)\theta = (1 + \tau_1 s)s\phi + AK(1 + \tau_2 s) \sin \phi \qquad (4.2\text{-}8)$$

or

$$\Omega_0 = \tau_1 \ddot{\phi} + (1 + AK\tau_2 \cos \phi)\dot{\phi} + AK \sin \phi \qquad (4.2\text{-}9)$$

when $\theta(t) = \theta_0 + \Omega_0 t$. In the steady state (<u>stable lock point</u>) we have

$$\phi_{ss} = \sin^{-1}\left(\frac{\Omega_0}{AK}\right) \qquad (4.2\text{-}10)$$

so that the loop never locks when $|\Omega_0| > AK$. However, even for values of Ω_0 less than AK, <u>pull-in</u> may not occur. In fact, Viterbi [4] has shown that lock will occur when $|\Omega_0/AK| < 0.693$. One reason for this is that the phase plane is not a simple sinusoid as in the first-order loop case, as is clear from the second-order differential equation that describes the solution ϕ. In order to plot the solution on a phase plane $\ddot{\phi}$ must be eliminated from the equation. To this end note that

$$\ddot{\phi} = \frac{d\Omega}{dt} = \frac{d\Omega}{d\phi}\frac{d\phi}{dt} = \Omega\frac{d\Omega}{d\phi}, \qquad \Omega = \dot{\phi} \qquad (4.2\text{-}11)$$

Hence, the phase plane equation becomes

$$\Omega_0 = \left(\tau_1 \frac{d\Omega}{d\phi} + 1 + AK\tau_2 \cos \phi \right)\Omega + AK \sin \phi \qquad (4.2\text{-}12)$$

If we normalize Ω in (4.2-12) by $2\zeta\omega_n$ we obtain

$$\frac{\Omega_0}{AK} = \left(r\frac{d\Omega'}{d\phi} + \frac{\tau_2}{\tau_1} + r \cos \phi \right)\Omega' + \sin \phi \qquad (4.2\text{-}13)$$

where

$$\Omega' = \frac{\Omega}{2\zeta\omega_n} \qquad r = \frac{AK\tau_2^2}{\tau_1}$$

with ζ the damping factor ($\zeta = \sqrt{r}/2$) and ω_n the loop natural frequency ($\omega_n = \sqrt{r}/\tau_2$). These parameters will be discussed later.

It being a second-order differential equation, two initial conditions are needed, namely, $\Omega'(0)$ and $\phi(0)$, to describe the solution. Two sets of solutions are shown in Figures 4.6 and 4.7. The trajectories traverse from left to right in the upper half plane and vice versa in the lower one, since when $\dot\phi$ is positive, ϕ increases; and when $\dot\phi$ is negative, ϕ decreases. Starting a trajectory in the upper half plane, one follows it to the right until $\phi = \pi$, then skips back to $\phi = -\pi$ at the same value of $\dot\phi$ found at $\phi = \pi$,

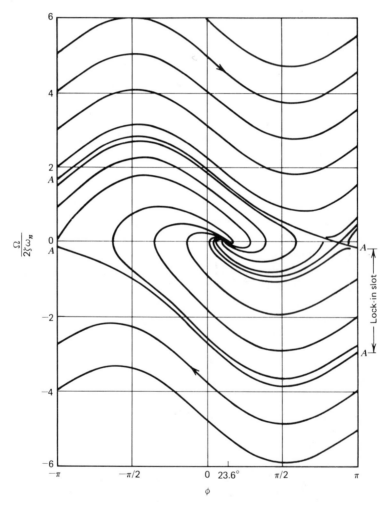

Figure 4.6 Lock-in behavior of a second-order loop with imperfect integrator: $F(s) = (1 + \tau_2 s)/(1 + \tau_1 s)$, for $\Omega_0/AK = 0.4$, $AK\tau_2^2/\tau = 2$, $\zeta = 0.707$, and $\tau_2/\tau_1 \simeq 0$ (from Tausworthe [8] with permission).

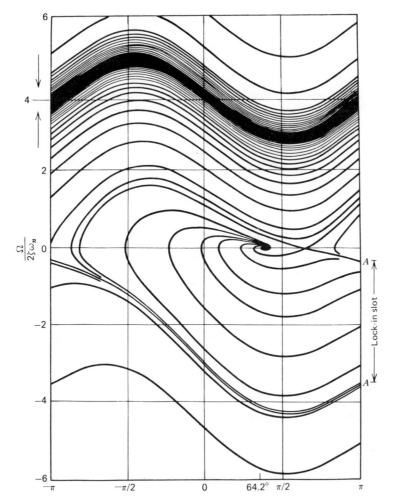

Figure 4.7 Lock-in behavior of a second-order loop with imperfect integrator: $F(s) = (1 + \tau_2 s)/(1 + \tau_1 s)$, for $\Omega_0/AK = 0.9$, $AK\tau_2^2/\tau_1 = 2$, $\zeta = 0.707$, and $\tau_2/\tau_1 \approx 0$. (from Tausworthe [8] with permission).

continuing until lock is obtained. Between the upper and lower lines A–A of Figure 4.6, the loop will stop skipping cycles, and the phase and frequency error will decay towards $\dot{\phi} = 0$ and $\phi = 0 \bmod 2\pi$. Otherwise, at least another cycle will be slipped prior to lock-in. The same behavior ensues when the frequency error is negative, except the trajectory moves from right to left. When the system trajectory is above the lower line the system stops skipping cycles and pulls into lock.

In Figure 4.7 note that lock-in occurs only when the trajectory happens to pass through the A–A region; otherwise, the loop has a trajectory that enters the periodic frequency lag region indicated by the two arrows on the left of the diagram.

4.2.3   Active Loop Filter Second-Order Phase Locked Loop

The active loop filter that produces a "perfect integrator" second-order PLL has a loop filter of the form shown in Figure 4.8. It is shown in Problem 2 that the transfer function is given by

$$F(s) = \frac{A(sCR_2 + 1)}{sCR_2 + 1 + (1 + A)sCR_1}$$

$$\simeq \frac{1 + \tau_2 s}{\tau_1 s} \qquad \text{for large } A \qquad (4.2\text{-}14)$$

where

$$\tau_2 = R_2 C$$
$$\tau_1 = R_1 C$$

PROBLEM 2

(a)   Show that the transfer function of the active filter of Figure 4.8 produces an open circuit transfer function given by

$$F(s) = \frac{1 + \tau_2 s}{\tau_1 s}$$

when the gain A is large.

(b)   Obtain the finite gain static phase error ϕ_{ss} from the differential equation. Assume the noise is zero.

The noise-free differential equation is given, using (4.1-14) with $n'(t) = 0$, by

$$\tau_1 \ddot{\theta} = \tau_1 \ddot{\phi} + AK\tau_2 \cos(\phi)\dot{\phi} + AK \sin(\phi) \qquad (4.2\text{-}15)$$

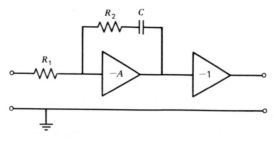

Figure 4.8   Perfect integrator second-order loop filter.

Consequently, in steady state with a frequency offset $\theta = \Omega_0 t + \theta_0$, the lock point is $\phi_{ss} = 0$ mod 2π, as can be seen from the above equation (4.2-15). However, if the input has a doppler rate, that is, when

$$\theta(t) = \frac{\Lambda_0}{2} t^2 + \Omega_0 t + \theta_0 \qquad (4.2\text{-}16)$$

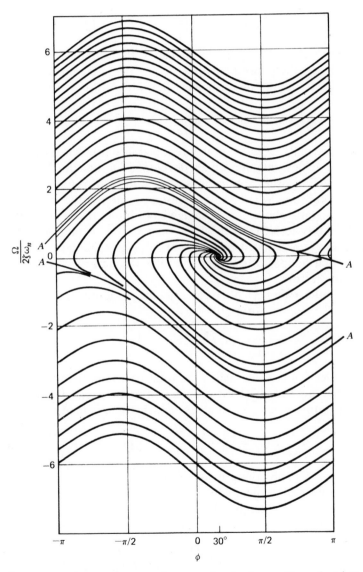

Figure 4.9 Phase plane trajectory of a second-order loop with perfect integrator to a doppler rate input Λ_0 for $AK\tau_2^2/\tau_1 = 2$, $\zeta = 0.707$, and $(\tau_1\Lambda_0)/AK = \frac{1}{2}$ (from Tausworthe [8] with permission).

then, as is shown in Problem 3, the steady state value is given by

$$\phi_{ss} = \sin^{-1}\left(\frac{\tau_1\Lambda_0}{AK}\right) = \sin^{-1}\left(\frac{\Lambda_0}{\omega_n^2}\right) \tag{4.2-17}$$

with ω_n the loop natural frequency, which will be discussed later.

PROBLEM 3

Establish the equality for ϕ_{ss} in (4.2-17).

When $\theta(t) = \Omega_0 t + \theta_0$, the phase plane for the active second-order loop is very much like the phase plane for the passive loop, except that $\phi_{ss} = 0$. Figure 4.9 illustrates the phase plane when $\theta(t) = \frac{1}{2}\Lambda_0 t^2 + \Omega_0 t + \theta_0$.

4.2.4 Third-Order Phase Locked Loop

While third-order PLLs have an inherent stability problem associated with low loop gain values, they have been developed for difficult [10] deep-space missions as well as other applications. The authors of reference 10 suggest that third-order loops, when properly designed, offer better tracking performance (lower static phase error), have wider pull-in, faster acquisition, and less susceptibility to VCO drift when compared to second-order loops. In addition, there is a reduced requirement for high loop gain and long time constants when compared to the passive filter second-order PLL to maintain small tracking errors.

The optimum (Wiener) filter for tracking an input phase acceleration $\theta(t) = (\Lambda_0 t^2)/2$ is of the form [10]

$$F(s) = \frac{1 + \tau_2 s}{\tau_1 s} + \frac{1}{2\tau_1\tau_2 s^2} \tag{4.2-18}$$

when the objective is to minimize the total transient distortion plus noise variance. (See Section 4.3.3 for a definition of transient distortion.) Since perfect integrators cannot be synthesized, this is modified to

$$F(s) = \frac{1 + \tau_2 s}{1 + \tau_1 s} + \frac{1}{(1 + \tau_1 s)(\delta + \tau_3 s)} \tag{4.2-19}$$

This filter approximates the previous one when τ_1 and τ_3 are large, except in the region near the origin. This loop transfer function approximation to the optimum Wiener filter has been used by Tausworthe and Crow [10].

The loop filter $F(s)$ can be expressed

$$F(s) = \frac{K_1 K_2 (1 + T_2 s)(1 + T_4 s)}{(1 + T_1 s)(1 + T_3 s)} \tag{4.2-20}$$

One implementation of this filter is shown in Figure 4.10. The relationship between the two sets of parameters and the resistors and capacitors are easily

$$F(s) = \frac{R_2 R_5}{R_1 R_4} \frac{(1 + R_3 C_1 s)(1 + R_6 C_2 s)}{[1 + R_2 + R_3)C_1 s][1 + (R_5 + R_6)C_2 s]}$$

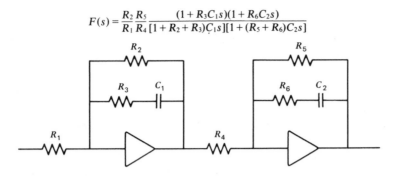

Figure 4.10 Proposed implementation for third-order loop filter $F(s)$.

obtained by equating (4.2-19) to (4.2-20).

$$T_1 = \tau_1 \qquad T_1 = (R_2 + R_3)C_1$$

$$T_3 = \tau_3/\delta \qquad T_2 = R_3 C_1$$

$$T_2 T_4 = \frac{\tau_2 \tau_3}{1 + \delta} \approx \tau_2 \tau_3 \qquad T_3 = (R_5 + R_6)C_2$$

$$T_2 + T_4 = \frac{\tau_3 + \tau\delta_2}{1 + \delta} \approx \tau_3 \qquad T_4 = R_6 C_2 \qquad (4.2\text{-}21)$$

$$\frac{T_2 T_4}{(T_2 + T_4)^2} = \frac{\tau_2}{\tau_3} = k \qquad K_1 = \frac{R_2}{R_1}$$

$$K_1 K_2 = \frac{1 + \delta}{\delta} \approx \frac{1}{\delta} \qquad K_2 = \frac{R_5}{R_4}$$

Other approximations to the "optimum" filter have been made by Gupta and also Mallinckrodt [11]; although we will not discuss them here, the responses are quite similar.

Again using (4.1-15), with $n'(t)$ set equal to zero, (4.2-19) and the input of $\theta(t) = \theta_0 + \Omega_0 t$, we obtain

$$\delta\Omega_0 = (1 + \delta)AK \sin \phi \qquad (4.2\text{-}22)$$

and consequently the steady state phase error is given by

$$\phi_{ss} = \sin^{-1}\left[\frac{\delta}{1 + \delta}\left(\frac{\Omega_0}{AK}\right)\right] \qquad \text{mod } 2\pi \qquad (4.2\text{-}23)$$

Conventional phase plane analysis for third-order PLLs is not applicable since three variables are now functionally related: ϕ, $\dot{\phi}$, and $\ddot{\phi}$. Viterbi [4] has plotted projections of phase plane trajectories for third-order loops. This will not be attempted here.

4.3 LINEAR MODEL RESULTS

Considerable insight into the operation of the PLL and a number of important results can be obtained by linearizing the basic stochastic differential equation (4.1-12). We first consider steady state behavior by employing the final value theorem.

4.3.1 Steady-State Tracking When the Loop Error Is Small

Under tracking conditions a design goal is to maintain a small phase error due to noise, doppler components, and modulation. Under the assumption ϕ is small (say, less in magnitude than $\pi/6$) we can make the linearizing approximation

$$\sin \phi(t) \cong \phi(t) \tag{4.3-1}$$

From the basic loop equation, (4.1-12), when the noise is zero we have (under this linearization) that

$$\theta(t) = \frac{s + AKF(s)}{s} \phi(t)$$

alternatively we have

$$\phi(t) = \frac{s}{s + AKF(s)} \theta(t) \tag{4.3-2}$$

or using $\phi = \theta - \hat{\theta}$, we have

$$\hat{\theta}(t) = \frac{AKF(s)/s}{1 + AKF(s)/s} \theta(t) \tag{4.3-3}$$

The linear PLL model is shown in Figure 4.11. There are normally four major components that contribute to the loop error. One is due to the signal modulation, call it $\theta_m(t)$; another is due to relative dynamics (doppler effect) between transmitter and PLL receiver, which we will call $\theta_d(t)$; the third is

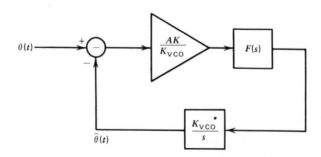

Figure 4.11 A linear baseband model of a PLL.

due to receiver noise; the fourth one is carrier and VCXO phase noise. Normally for carrier tracking loops the modulation is designed not to interfere (overlap) with the discrete carrier component, so we consider only the doppler term $\theta_d(t)$. Our primary goal in this section is to determine the long-term tracking error, so we consider a doppler input of the form

$$\theta_d(t) = \theta_0 + \Omega_0 t + \tfrac{1}{2}\Lambda_0 t^2 \qquad (4.3\text{-}4)$$

where upon differentiating we see that the instantaneous frequency (doppler frequency) is given by

$$\dot\theta_d(t) = \Omega_0 + \Lambda_0 t \qquad (4.3\text{-}5)$$

and the rate of change of doppler is given by

$$\ddot\theta_d(t) = \Lambda_0 \qquad (4.3\text{-}6)$$

The steady state error due to doppler effects can be determined by the final value theorem of the Laplace transform theory; that is, if $\phi(t)$ is the loop phase error due to doppler effects, then the steady state phase error ϕ_{ss} is given by

$$\phi_{ss} = \lim_{s\to 0}\,[s\Phi(s)] \qquad (4.3\text{-}7)$$

where $\Phi(s)$ is the Laplace transform of $\phi(t)$. If we express both $\phi(t)$ and $\theta_d(t)$ in terms of their Laplace transforms, we have, from (4.3-2) and (4.3-7) that

$$\phi_{ss} = \lim\left[\left(\frac{s^2}{s + AKF(s)}\right)\Theta_d(s)\right] \qquad (4.3\text{-}8)$$

where $\Theta_d(s)$ is the Laplace transform of $\theta_d(t)$. Using our three-term polynomial for $\theta_d(t)$, we find that

$$\Theta_d(s) = \frac{\theta_0}{s} + \frac{\Omega_0}{s^2} + \frac{\Lambda_0}{s^3} \qquad (4.3\text{-}9)$$

Consider now the first-order PLL, where $F(s) = 1$. We find that

$$\phi_{ss} = \frac{\Omega_0}{AK} \qquad \text{when } \Lambda_0 = 0$$

$$\text{does not exist} \qquad \text{when } \Lambda_0 \neq 0 \qquad (4.3\text{-}10)$$

which agrees with the nonlinear result obtained earlier for small phase errors. The above result indicates that when $\Lambda_0 \neq 0$, the loop will eventually lose lock.

For the passive second-order loop we have

$$\phi_{ss} = \lim_{s\to 0}\left[\frac{s^2}{s^2 + AK\left(\dfrac{1+\tau_2 s}{1+\tau_1 s}\right)}\left(\frac{\theta_0}{s} + \frac{\Omega_0}{s^2} + \frac{\Lambda_0}{s^3}\right)\right] \qquad (4.3\text{-}11)$$

so that, as with the first-order loop, we have

$$\phi_{ss} = \frac{\Omega_0}{AK} \qquad \Lambda_0 = 0$$

$$\phi_{ss} = \text{does not exist} \qquad \Lambda_0 \neq 0.$$

However, by using the fact that $\lim_{t\to\infty} \dot{f}(t) = \lim_{t\to 0}[s^2 F(s)]$, it is not difficult to show that ϕ_{ss} is asymptotic to

$$\phi_{ss} = \frac{\Omega_0}{AK} + \frac{\Lambda_0}{\omega_n^2} + \frac{\Lambda_0}{AK} t \qquad (4.3\text{-}12)$$

Now consider the <u>active second-order loop</u>. Using our final value result, we obtain

$$\phi_{ss} = \lim_{s\to 0} \left[\frac{s^2}{s^2 + AK\left(\dfrac{1 + \tau_2 s}{\tau_1 s}\right)} \left(\frac{\theta_0}{s} + \frac{\Omega_0}{s^2} + \frac{\Lambda_0}{s^3}\right) \right] \qquad (4.3\text{-}13)$$

or

$$\phi_{ss} = \frac{\tau_1 \Lambda_0}{AK} = \frac{\Lambda_0}{\omega_n^2} \qquad (4.3\text{-}14)$$

It will be shown later that this result can also be written as

$$\phi_{ss} = \frac{\Lambda_0 (r + 1)^2}{4rW_L^2}$$

where r is related to the loop damping factor and W_L is the two-sided loop noise bandwidth.

PROBLEM 4

(a) Show that if the static phase error magnitude is less than $\pi/6$, then to a good approximation

$$\phi_{ss} = \frac{\delta}{1 + \delta} \frac{\Omega_0}{AK} \qquad \Lambda_0 = 0$$

for the single operational amplifier third-order loop filter when $\Lambda_0 = 0$ [see (4.2-19)]; otherwise, ϕ_{ss} is unbounded.

(b) Show that the steady state asymptotic error ϕ_{ss} is given by

$$\phi_{ss} = \frac{(\Omega_0 + \Lambda_0 t)}{AK} (\delta) + \frac{\Lambda_0 \tau_1}{AK} \left(\frac{\epsilon}{k} + \delta\right)$$

where $\Omega_0 = $ the initial radian frequency offset and Λ_0 is frequency rate in rad/sec².

It should be emphasized that the steady state value is only meaningful when obtained from a linearized loop model when sin $\phi \simeq \phi$ or when $\phi_{ss} \leq \pi/6$, say. For larger values, the results become inaccurate.

4.3.2 Loop Stability via Root Locus Plots

A basic necessity of an operable system is that it must be stable under operating conditions so that unwanted oscillations will not occur. A useful means to ascertain whether a system is stable is the root locus method [12]. The *root locus* is a plot of the poles of the closed-loop transfer function as a function of the loop gain. The root locus method allows a quick determination of the poles of the closed-loop response from the locations of the known open-loop poles and zeros by using some basic rules of locus construction. Briefly, the root locus starts on the open-loop poles at zero gain and terminates on the open-loop zeros with unbounded gain. The open-loop transfer function for any PLL is

$$G(s) = \frac{AKF(s)}{s} \qquad (4.3\text{-}15)$$

The *closed-loop transfer function*, from (4.3-3), for any PLL is given by the ratio

$$\frac{\hat{\theta}(s)}{\theta(s)} = \frac{AKF(s)/s}{1 + AKF(s)/s} = H(s) \qquad (4.3\text{-}16)$$

Note, in terms of the usual control theory notation, forward loop gain is $G(s) = AKF(s)/s$ and the feedback gain is $H_F(s) = 1$. The closer the root locus is to the $\sigma \geq 0$ half plane, the more the system approaches instability. When the root locus just passes into the right half plane the system is unstable at that respective value of gain.

Now consider the first-order loop with $F(s) = 1$.

Since the open-loop transfer function pole is at $s = 0$, and the zero is at infinity, the root locus has the form ($s = \sigma + j\omega$) shown in Figure 4.12, where the small × denotes an open-loop pole and a ○ denotes an open-loop zero. The simplest second-order PLL filter has a single lag so that the loop filter is of the form

$$F(s) = \frac{1}{1 + \tau_1 s}$$

The root locus is shown in Figure 4.13.

Next consider the passive integrator second-order PLL, where

$$F(s) = \frac{1 + \tau_2 s}{1 + \tau_1 s}$$

The root locus is shown in Figure 4.14.

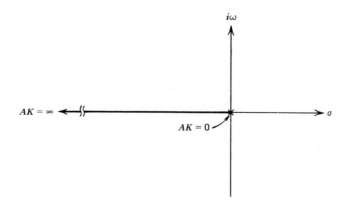

Figure 4.12   Root locus plot for the first-order PLL, with $F(s) = 1$.

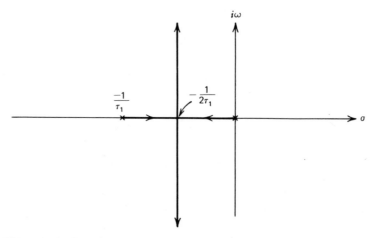

Figure 4.13   Root locus for a second-order PLL, with $F(s) = 1/(1 + \tau_1 s)$.

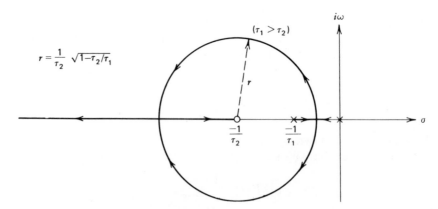

Figure 4.14   Root locus for a passive filter second-order PLL.

The root locus for the active integrator second-order PLL, where

$$F(s) = \frac{1 + \tau_2 s}{\tau_1 s}$$

is essentially similar to the passive case, except for the double pole at the origin, as seen in Figure 4.15.

The Tausworthe-Crow third-order PLL has two poles and two zeros in its loop filter. Letting

$$k = \frac{\tau_2}{\tau_3}, \qquad \epsilon = \frac{\tau_2}{\tau_1}, \qquad r = AK\frac{\tau_2^2}{\tau_1}, \qquad k_0 \simeq \frac{1}{2(2 + \delta)},$$

and

$$k_{\max} = \frac{1}{(1 + \delta)r} \left(\frac{r + \epsilon + \delta k_{\max}}{3} \right)^3$$

the root locus is sketched below for five cases ([10] and [13]), with $1/\tau_1$ and δ/τ_3 assumed very small using the filter function of (4.2-19).

The main point of the sketches is that unless $\tau_3 = 0$ (second-order loop), the third-order loop is unstable for small values of loop gain. In the figures $r = (AK\tau_2^2)/\tau_1$ is used as a "loop gain" variable and the small boxes denote the closed loop poles. When $k > \frac{1}{3}$ (Figure 4-16a), there are two underdamped (complex) roots and one overdamped (real) root for all $r > k$. When $k = \frac{1}{3}$, as in Figure 4.16b, there are two underdamped roots and one overdamped root for all $r > k$, except at $r = 3.0$, at which point all three roots become equal. In the case $\frac{1}{4} < k < \frac{1}{3}$, as shown in Figure 4.16c, there is a region where two roots pass from underdamped to critically damped to overdamped, to critically damped, and finally to underdamped. As shown in Figure 4.16d, when $k = \frac{1}{4}$, the system roots are always critically damped or overdamped for $r > 3.375$. Finally, in Figure 4.16e, we show the case when $k < \frac{1}{4}$. This case is similar to the $k = \frac{1}{4}$ case

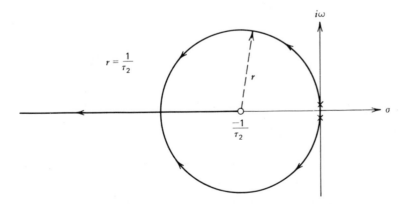

Figure 4.15 Root locus for an active filter second-order PLL.

except that there is a root nearer the origin, which indicates a more sluggish response. Tausworthe and Crow have noted that it is best that the loop be overdamped for practical design. They recommend $k = \frac{1}{4}$ and r set equal to 3.375 (see Figure 4.16d) at a minimum signal strength expected.

Hybrid designs, in which acquisition is achieved in second-order form and then switched to third-order for tracking, are strong system design candidates.

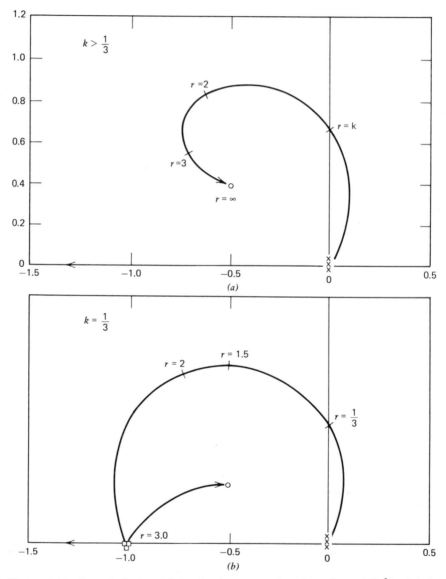

Figure 4.16 Root loci of a third-order loop as a function of $r = AK\tau_2^2/\tau_1$ (a) for $k > k_{max}$, (b) for $k = k_{max}$ (from Tausworthe and Crow [10], with permission).

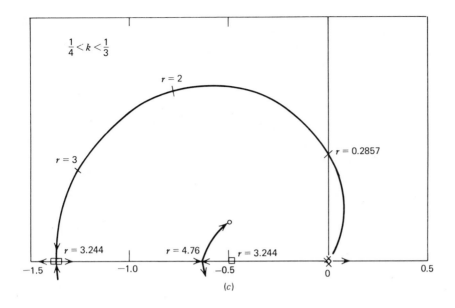

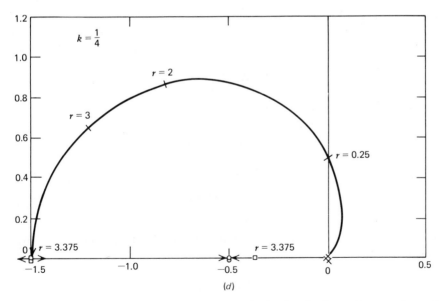

Figure 4.16 Continued. (*c*) for $k_0 < k < k_{max}$, (*d*) for $k = k_0$ (from Tausworthe and Crow [10], with permission).

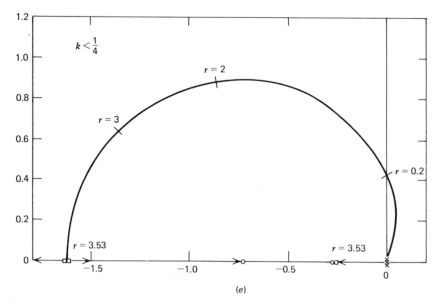

Figure 4.16 Continued. (*e*) for $k < k_0$ (from Tausworthe and Crow [10], with permission).

4.3.3 Tracking Error Due to Thermal Noise and Modulation

From the linearized version of (4.1-15) we have that

$$\hat{\theta}(t) = \frac{AKF(s)}{s} \left[\phi(t) + \frac{n'(t)}{A} \right]. \tag{4.3-17}$$

Solving for $\hat{\theta}$, we find that

$$\hat{\theta}(t) = \frac{AKF(s)}{s + AKF(s)} \left[\theta(t) + \frac{n'(t)}{A} \right] \tag{4.3-18}$$

where

$$H(s) = \frac{AKF(s)}{s + AKF(s)} \tag{4.3-19}$$

was defined as the closed-loop transfer function. Rearranging the equation above produces

$$\phi(t) = [1 - H(s)]\theta(t) - H(s)\frac{n'(t)}{A} \tag{4.3-20a}$$

neglecting oscillator noise. If we include oscillator noise, we obtain

$$\phi(t) = [1 - H(s)]\theta(t) - H(s)\frac{n'(t)}{A} - [1 - H(s)]\theta_{\text{osc}}(t) \tag{4.3-20b}$$

in terms of the input phase process $\theta(t)$, the input noise process $n'(t)$, and the oscillator phase noise process $\theta_{\text{osc}}(t)$. The loop response $H(s)$ and the error response $1 - H(s)$ are sketched in Figure 4.17 for a typical second-order PLL system in terms of the loop natural frequency, which we will define shortly.

The basic point is that the phase modulation or carrier phase noise appears as a highpass function and the loop response to noise appears as a lowpass function. The *total mean squared tracking error* can be written as

$$\sigma_T^2 = \overbrace{\mu^2(t)}^{} + \overbrace{\sigma_m^2 + \sigma^2}^{\text{modulation} \quad \text{noise}} \quad \text{rad}^2 \; , \qquad (4.3\text{-}21)$$

where the first term is due to transient distortion, the second term is due to modulation, and the third term is due to thermal noise. The first term $\mu(t)$ is a function of time. However, the *total transient distortion* is not and is defined by

$$\epsilon_T^2 = \int_0^\infty \mu^2(t) dt \; , \qquad (4.3\text{-}22)$$

Following Tausworthe [8], this doppler induced term ϵ_T^2 can be modified to include an initial uncertainty in θ_0 [modeled as a uniform random variable (rv) in $(-\pi, \pi)$], so that the average total transient distortion is given, with the aid of Parseval's theorem and the fact that $\phi(t) = [1 - H(s)]\theta_d(s)$, by

$$\overline{\epsilon_T^2} = \frac{1}{2\pi} \int_{-\infty}^{\infty} |1 - H(i\omega)|^2 E|\Theta_d(i\omega)|^2 \, d\omega \; . \qquad (4.3\text{-}23)$$

Recall $\Theta_d(s)$ is the Laplace transform of the doppler term $\theta_d(t) = \theta_0 + \Omega t + \cdots$ to the appropriate degree. The second term of (4.3-21), using $\phi(t) = [1 - H(s)]\theta_m(s)$, is given by

$$\sigma_m^2 = \frac{1}{2\pi} \int_{-\infty}^{\infty} |1 - H(i\omega)|^2 \mathscr{S}_m(\omega) d\omega \; , \qquad (4.3\text{-}24)$$

where $\mathscr{S}_m(f)$ is the spectral density of the modulation. Notice $\overline{\epsilon_T^2}$ and σ_m^2 are of similar form. The final component, due to thermal noise of the receiver, using (4.3-20), is given by

$$\sigma^2 = \frac{1}{2\pi} \int_{-\infty}^{\infty} |H(i\omega)|^2 \frac{S_n(\omega)}{A^2} \, d\omega \qquad (4.3\text{-}25)$$

Notice that $H(i\omega)$ is a lowpass function. Under the assumption of WGN with two-sided spectral density $N_0/2$, we obtain

$$\boxed{\sigma^2 = \frac{N_0 B_L}{A^2} \qquad \text{rad}^2} \qquad (4.3\text{-}26)$$

where $2B_L$ is the two-sided loop noise bandwidth, with

$$2B_L = \frac{1}{2\pi} \int_{-\infty}^{\infty} |H(i\omega)|^2 \, d\omega$$

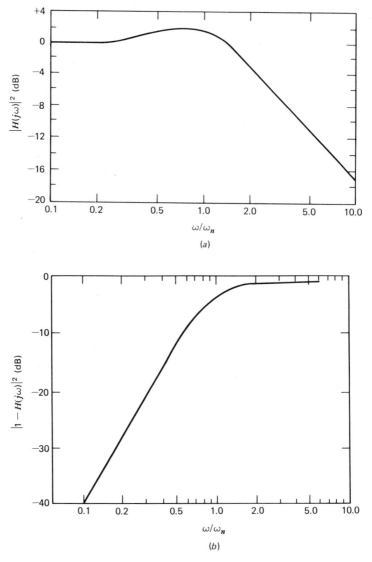

Figure 4.17 (*a*) Typical response of $|H(j\omega)|^2$ for a second-order loop. (*b*) Response of $|1 - H(j\omega)|^2$ for a second-order loop.

Equation 4.3-26 states that under the linear model assumption the phase error variance due to thermal noise is the product of the one-sided noise spectral density and the one-sided (unnormalized) loop bandwidth,* divided by the signal power A^2. Note that this is the linearized loop result. Without

---

*Note that therefore $\sigma^2 = 1/\rho$ and $\rho = 1/\sigma^2$, where ρ is the SNR in the baseband bandwidth.

linearization it is not hard to show that the actual variance is unbounded. If one views the signal at rf, the signal-to-noise (SNR) is given by

$$\mathrm{SNR}_{rf} = \frac{A^2}{2N_0 B_L}$$

so that

$$\sigma^2 = \frac{1}{2} \cdot \frac{1}{\mathrm{SNR}_{rf}} \qquad \mathrm{rad}^2 \qquad (4.3\text{-}27)$$

Tausworthe [8] calls the unnormalized noise bandwidth the *fiducial bandwidth* since the normalization $|H_{max}|^2$ has been left out of the definition of the two-sided loop noise bandwidth. We will henceforth call B_L the one-sided loop noise bandwidth for consistency with the literature.

Lindsey [6] has tabulated some well-known integrals up to order 3; we include one more based on [14]. Let $s = i\omega$, so that the closed-loop transfer function is $H(i\omega) = H(s)|_{s=i\omega}$. Then if we denote $H_n(s)$ by

$$H_n(s) = \frac{C_0 + C_1 s + C_2 s^2 + \cdots + C_{n-1} s^{n-1}}{d_0 + d_1 s + d_2 s^2 + \cdots + d_n s^n} \qquad (4.3\text{-}28)$$

we obtain

$$2B_{L_n} = \frac{1}{2\pi i} \int_{-j\infty}^{j\infty} |H_n(s)|^2 ds \qquad (4.3\text{-}29)$$

where

$$|H_n(s)|^2 = H_n(s)H_n(-s) \qquad (4.3\text{-}30)$$

The loop noise bandwidth $2B_{L_n}$ is tabulated in Table 4.1.
The two-sided loop noise bandwidth is usually denoted W_L, that is, $W_L = 2B_L$.

PROBLEM 5
Show that $2B_{L_4} \to 2B_{L_3}$ if $C_3 = 0$ and $d_4 = 0$.

Table 4.1 Loop noise bandwidth for first- to fourth-order loops

$$2B_{L1} = \frac{C_0^2}{2d_0 d_1} \qquad 2B_{L2} = \frac{C_0^2 d_2 + C_1^2 d_0}{2d_0 d_1 d_2}$$

$$2B_{L3} = \frac{C_2^2 d_0 d_1 + (C_1^2 - 2C_0 C_1)d_0 d_3 + C_0^2 d_2 d_3}{2d_0 d_3 (d_1 d_2 - d_0 d_3)}$$

$$2B_{L4} = \frac{C_3^2(d_0 d_1 d_2 - d_0^2 d_3) + (C_2^2 - 2C_1 C_3)d_0 d_1 d_4 + (C_1^2 - 2C_0 C_2)d_0 d_3 d_4 + C_0^2(d_2 d_3 d_4 - d_1 d_4^2)}{2d_0 d_4(d_1 d_2 d_3 - d_1^2 d_4 - d_0 d_3^2)}$$

For the <u>first-order loop</u> we have $F(s) = 1$, hence

$$H(s) = \frac{AK}{s + AK} \tag{4.3-31}$$

We conclude, using Table 4.1, that

$$2B_{L_1} = \frac{AK}{2} \tag{4.3-32}$$

From Problem 6 we have that the <u>second-order passive filter PLL</u> $[F(s) = (1 + \tau_2 s)/(1 + \tau_1 s)]$ has the transfer function

$$H(s) = \frac{1 + \tau_2 s}{1 + (\tau_2 + 1/AK)s + (\tau_1/AK)s^2} \tag{4.3-33}$$

and

$$W_L = \frac{r + 1}{2\tau_2 \left(1 + \frac{\tau_2}{r\tau_1}\right)} \cong \frac{r + 1}{2\tau_2} \tag{4.3-34}$$

(under "normal" design conditions) and

$$r = \frac{AK\tau_2^2}{\tau_1} \tag{4.3-35}$$

Notice that the loop noise bandwidth depends upon the loop gain and not the dc loop gain! If we set

$$\omega_n^2 = \frac{AK}{\tau_1} \qquad \left(\tau_2 + \frac{1}{AK}\right) = \frac{2\zeta}{\omega_n}$$

with ω_n the natural frequency (rad/sec) and ζ the loop damping factor, then the transfer function can be written as (using standard control theory notation)

$$H(s) = \frac{s(2\zeta\omega_n - \omega_n^2/AK) + \omega_n^2}{s^2 + 2\zeta\omega_n s + \omega_n^2} \tag{4.3-36}$$

where the loop natural frequency ω_n is given by

$$\omega_n = \sqrt{\frac{AK}{\tau_1}} \tag{4.3-37}$$

and where the loop damping ζ is given by

$$\zeta = \frac{\sqrt{AK}}{2\sqrt{\tau_1}} \left(\tau_2 + \frac{1}{AK}\right) \tag{4.3-38}$$

PROBLEM 6

For the passive second-order PLL show that

$$W_L = \frac{r + 1}{2\tau_2 \left(1 + \frac{\tau_2}{r\tau_1}\right)} \cong \frac{r + 1}{2\tau_2} \qquad \text{for} \quad \tau_1 \gg \tau_2$$

where

$$r = \frac{AK\tau_2^2}{\tau_1}$$

PROBLEM 7

An imperfect integrating filter is used with $\tau_1 = 4500$ sec, $\tau_2 = 115$ sec, and $AK = 2$ sec^{-1}.

(a) Determine the loop bandwidth.
(b) Let $C = 250 \, \mu F$ and find R_1 and R_2.

PROBLEM 8

Compute the loop bandwidth of the second-order PLL having a loop filter of the form

$$F(s) = \frac{1}{1 + \tau s}$$

Comment on the dependence on τ.

We conclude the section on the passive filter second-order PLL by noting that the transfer function may also be written as

$$H(s) = \frac{1 + \left(\frac{r+1}{2W_L}\right)s}{1 + \left(\frac{r+1}{2W_L}\right)s + \left(\frac{r+1}{2W_L\sqrt{r}}\right)^2 s^2} \qquad (4.3\text{-}39)$$

Based on the standard second-order control system nomenclature, the natural frequency ω_n and the damping factor are given by

$$\omega_n = \sqrt{\frac{AK}{\tau_1}} = \frac{\sqrt{r}}{\tau_2} = \frac{2\sqrt{r}W_L}{r+1} \qquad (4.3\text{-}40)$$

$$\zeta = \frac{1}{2}\left(1 + \frac{\tau_2}{r\tau_1}\right)r^{1/2} \cong \frac{\sqrt{r}}{2} \qquad (4.3\text{-}41)$$

If an FM signal, with a sinusoidal input, is the input to a second-order PLL, the phase error is maximum at $\omega = \omega_n$. Furthermore, the steady state phase error increases as ζ decreases.

The closed-loop transfer function of the second-order active filter PLL becomes

$$H(s) = \frac{1 + \tau_2 s}{1 + \tau_2 s + (\tau_1/AK)s^2} \qquad (4.3\text{-}42)$$

and the loop noise bandwidth using Table 4.1 becomes

$$W_L = \frac{r+1}{2\tau_2} \qquad (4.3\text{-}43)$$

By comparison with the passive loop filter PLL, we see that essentially the same performance is achieved when $r\tau_1 \gg \tau_2$.

This loop can also be written in the standard control system notation. We have

$$H(s) = \frac{2\zeta\omega_n s + \omega_n^2}{s^2 + 2\zeta\omega_n s + \omega_n^2} \tag{4.3-44}$$

where

$$\omega_n = \sqrt{\frac{AK}{\tau_1}} \tag{4.3-45}$$

$$\zeta = \frac{\tau_2}{2}\sqrt{\frac{AK}{\tau_1}} \tag{4.3-46}$$

PROBLEM 9

Show that for the ideal (operational amplifier version) loop of (4.3-44) that the following relationship holds:

$$2B_L = W_L = \omega_n\left(\zeta + \frac{1}{4\zeta}\right)$$

It is left as an exercise to the reader to show that the closed-loop transfer function for the Tausworthe-Crow third-order loop is given by

$$H(s) = \frac{rk + r\tau_2 s + r(\tau_2 s)^2}{rk + r\tau_2 s + r(\tau_2 s)^2 + (\tau_2 s)^3} \qquad \frac{\tau_2}{\tau_1} \ll 1 \qquad \delta \ll 1 \tag{4.3-47}$$

where, as before, $r = AK\tau_2^2/\tau_1$. Tausworthe and Crow [10] have shown that

$$W_L \cong \frac{r}{2\tau_2}\left(\frac{r - k + 1}{r - k}\right) \qquad \frac{\tau_2}{\tau_1} \ll 1 \qquad \delta \ll 1 \tag{4.3-48}$$

is the two-sided loop noise bandwidth, with $k = \tau_2/\tau_3$. The optimum Weiner closed-loop filter (to be discussed shortly) is of the form

$$H_0(s) = \frac{2B_0(s^2 + B_0 s + B_0^2/2)}{s^3 + 2B_0 s^2 + 2B_0^2 s + B_0^3} \tag{4.3-49}$$

so that the corresponding loop filter is of the form

$$F_0(s) = \frac{2B_0(s^2 + B_0 s + B_0^2/2)}{AKs^2} \tag{4.3-50}$$

which contains a single and a double integrator. For the above it can be shown [11] that two-sided loop noise bandwidth satisfies

$$2B_L = \tfrac{5}{3} B_0 \tag{4.3-51}$$

The approximate Weiner [11] closed-loop filter response is of the form

$$H_a(s) = \frac{2B_1(s^2 + \sqrt{2}B_1 s + B_1^2/2)}{s^3 + 2B_1 s^2 + 2\sqrt{2}B_1^2 s + B_1^3} \tag{4.3-52}$$

with loop bandwidth satisfying

$$2B_L = 1.859B_1 \qquad (4.3\text{-}53)$$

and finally the Mallinckrodt [11] closed-loop response has a transfer function given by

$$H_m(s) = \frac{\frac{9}{4}B_m(s^2 + \frac{2}{3}B_m s + B_m^2/9)}{s^3 + \frac{9}{4}B_m s^2 + \frac{3}{2}B_m^2 s + \frac{1}{4}B_m^3} \qquad (4.3\text{-}54)$$

where

$$2B_L = 1.485B_m \qquad (4.3\text{-}55)$$

Now we turn to the transient response of the linearized second- and third-order loops.

4.3.4 Transient Response of Second- and Third-Order Loops

Often in practice it is useful to estimate the (linearized) transient response of a PLL. We first consider the response of a second-order PLL to the following signals: (1) a step phase input θ_0, rad; (2) a step in frequency Ω_0, rad/sec; (3) a step in acceleration Λ_0, rad/sec^2.

PROBLEM 10

Show, for an ideal second-order loop, that when $\zeta = 0.707$ that

(a)
$$|H(f)|^2 = \frac{1 + 2(f/f_n)^2}{1 + (f/f_n)^4}$$

(b)
$$|1 - H(f)|^2 = \frac{1}{1 + (f/f_n)^2}$$

The responses to these outputs are obtained from the closed-loop error frequency response $1 - H(s)$ which, for the ideal second-order loop, is given by

$$1 - H(s) = \frac{s^2}{s^2 + 2\zeta\omega_n s + \omega_n^2} \qquad (4.3\text{-}56)$$

By multiplying $[1 - H(s)]$ by the Laplace transform of the input phase function and then obtaining the inverse Laplace transform, the transient phase error may be obtained. The results are shown in Table 4.2 (reference 15).

Note that in the frequency ramp plus doppler offset case if the loop is not perfect (finite dc gain), then there will be a steady state (long-term) response of

$$\phi_{ss}(t) \rightarrow \frac{\Lambda_0}{\omega_n^2} + \frac{\Lambda_0}{AKF(0)}t + \frac{\Omega_0}{AKF(0)} \qquad (4.3\text{-}57)$$

which applies to the passive loop filter case also.

Also, if the loop has finite gain when subjected to only a frequency step

Table 4.2 Transient phase error of a linearized perfect second-order loop, $\phi(t)$ in rad

Damping	Phase Step θ_0(rad)	Frequency Step Ω_0(rad/sec)	Frequency Ramp Λ_0(rad/sec^2)
$\zeta < 1$	$\theta_0\left(\cos\sqrt{1-\zeta^2}\,\omega_n t\right.$ $\left.-\dfrac{\zeta}{\sqrt{1-\zeta^2}}\sin\sqrt{1-\zeta^2}\,\omega_n t\right)e^{-\zeta\omega_n t}$	$\dfrac{\Omega_0}{\omega_n}\left(\dfrac{1}{\sqrt{1-\zeta^2}}\sin\sqrt{1-\zeta^2}\,\omega_n t\right)e^{-\zeta\omega_n t}$	$\dfrac{\Lambda_0}{\omega_n^2}-\dfrac{\Lambda_0}{\omega_n^2}\left(\cos\sqrt{1-\zeta^2}\,\omega_n t\right.$ $\left.+\dfrac{\zeta}{\sqrt{1-\zeta^2}}\sin\sqrt{1-\zeta^2}\,\omega_n t\right)e^{-\zeta\omega_n t}$
$\zeta = 1$	$\theta_0(1-\omega_n t)e^{-\omega_n t}$	$\dfrac{\Omega_0}{\omega_n}(\omega_n t)e^{-\omega_n t}$	$\dfrac{\Lambda_0}{\omega_n^2}-\dfrac{\Lambda_0}{\omega_n^2}(1+\omega_n t)e^{-\omega_n t}$
$\zeta > 1$	$\theta_0\left(\cosh\sqrt{\zeta^2-1}\,\omega_n t\right.$ $\left.-\dfrac{\zeta}{\sqrt{\zeta^2-1}}\sinh\sqrt{\zeta^2-1}\,\omega_n t\right)e^{-\zeta\omega_n t}$	$\dfrac{\Omega_0}{\omega_n}\left(\dfrac{1}{\sqrt{\zeta^2-1}}\sinh\sqrt{\zeta^2-1}\,\omega_n t\right)e^{-\zeta\omega_n t}$	$\dfrac{\Lambda_0}{\omega_n^2}-\dfrac{\Lambda_0}{\omega_n^2}\left(\cosh\sqrt{\zeta^2-1}\,\omega_n t\right.$ $\left.+\dfrac{\zeta}{\sqrt{\zeta^2-1}}\sinh\sqrt{\zeta^2-1}\,\omega_n t\right)e^{-\zeta\omega_n t}$

input, the steady state response is given by

$$\phi = \frac{\Omega_0}{AKF(0)} \qquad \text{rad} \qquad (4.3\text{-}58)$$

Based on the work of Hoffman [15, 16], the curves for the ideal second-order loop are plotted in Figures 4.18a through 4.18c, versus $\omega_n t$, normalized time, with ω_n the loop natural frequency.

Now we consider the transient response of the Tausworthe-Crow third-order loop. By obtaining the inverse Laplace transform of $[1 - H(s)]\Theta_d(s)$, where $H(s)$ is given in (4.3-47) and $\Theta_d(s)$ is the Laplace transform of the phase step, etc., Tausworthe and Crow [10] obtained the transient response to a step phase (Figure 4.19a), frequency step (Figure 4.19b), and a frequency rate step (parabolic phase) (Figure 4.19c). Goldman [17] extended their results to the case of a jerk input, in which the input phase function is cubic.* All the third-order loop responses are plotted against $2B_L t$. Figure 4.19d illustrates the phase error due to an infinite duration jerk input at $t = 0$. Figure 4.19e illustrates the transient phase error for different finite jerk inputs. It must be cautioned again that phase errors

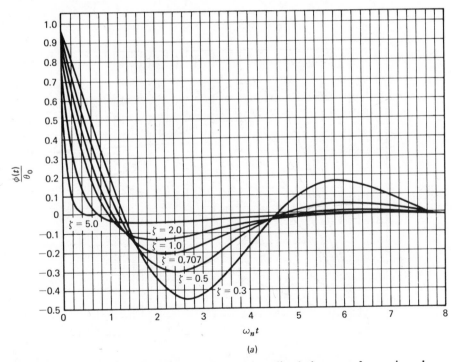

Figure 4.18 Transient phase error versus normalized time $\omega_n t$ for various loop damping factors due to (a) a step in phase (from Gardner [16] with permission).

*That is, the input phase function is $Jt^3/6$ starting at $t = 0$.

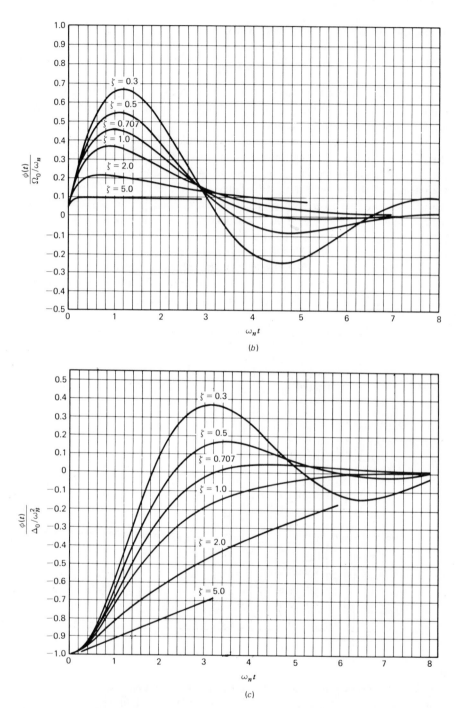

Figure 4.18 Continued. (*b*) a step in frequency, (*c*) a ramp in frequency (from Gardner [16] with permission).

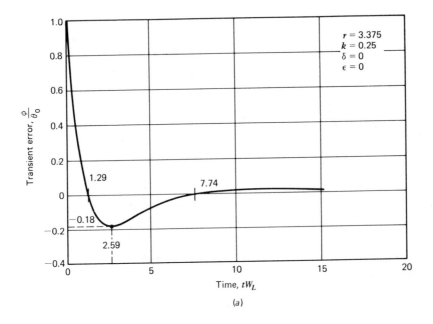

(a)

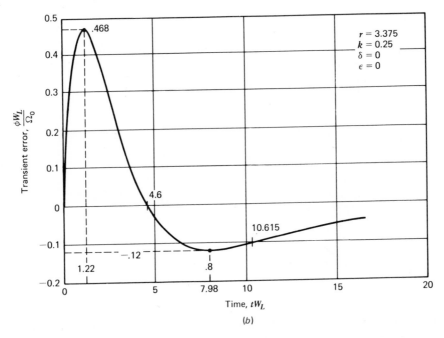

(b)

Figure 4.19 In-lock transient response of third-order loop to (a) a phase step, (b) a frequency step. $W_L = 2B_L$.

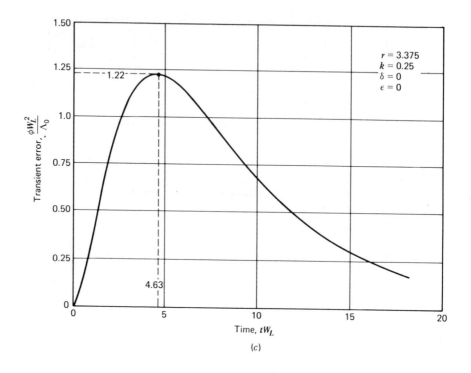

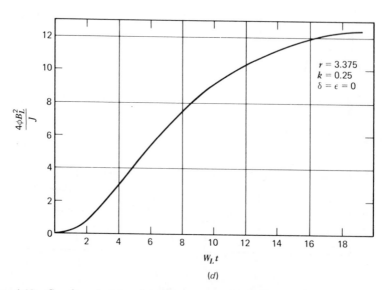

Figure 4.19 Continued. (*c*) a frequency ramp (from Tausworthe and Crow [10], with permission). (*d*) response of third-order PLL to jerk pulse amplitude *J* (jerk pulse duration $T = \infty$). From Goldman [31] with permission. $W_L = 2B_L$.

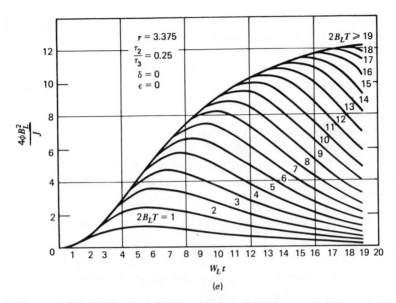

Figure 4.19 Continued. (e) phase error response of third-order PLL to jerk pulse amplitude J (Hz/sec^2) and duration T (sec). W_L is two-sided noise bandwidth of PLL (from Goldman [31] with permission).

exceeding 30° move the loop out of the linear region and hence are no longer applicable. All the curves for transient phase error are based on a linear model.

4.3.5 The Effects of Carrier Phase Noise

In this section we briefly consider the effects of carrier and voltage controlled oscillator phase instabilities (phase noise) on the operation of a PLL. Consider a received signal of the form

$$y(t) = \sqrt{2}A \sin[\omega_0 t + \theta(t) + \theta_0] + n(t) \qquad (4.3-59)$$

where A is the signal amplitude, $\theta(t)$ is the phase noise instability process, θ_0 is an unknown phase modeled as a uniform random variable on $(0, 2\pi)$, and $n(t)$ is WGN. From (4.3-20) we have

$$H(s) = [1 - H(s)]\theta(t) - H(s)\frac{n'(t)}{A} \qquad (4.3-60)$$

where $\theta(t)$ is any modulation process on the carrier, and $H(s)$ is the closed-loop transfer function.

There are two common ways to specify phase noise. The first method is to specify the single sideband (SSB) phase noise spectral density and the second is to specify the Allen variance [18]-[20]. We will discuss both methods briefly. First consider the frequency domain specification.

A typical voltage controlled oscillator phase noise spectral density is shown in Figure 4.20. The ordinate is normally specified by the oscillator manufacturer in dBc, which means dBs below the carrier or the equivalent SSB rad^2/Hz. Then twice the integral of the SSB density is the total phase noise variance in radians^2. The reason for this equivalence will be shown below.

Device manufacturers, who specify their oscillators in the frequency domain, usually plot the dBc value versus frequency since they measure the SSB phase noise spectral density by use of a narrowband wave analyzer. Now consider the following heuristic argument of why dBc and rad^2/Hz are equivalent numeric representations. Expanding (4.3-59) and neglecting thermal noise, we obtain

$$y(t) = \sqrt{2}A \sin(\omega_0 t + \theta_0) \cos \theta(t) + \sqrt{2}A \cos(\omega_0 t + \theta_0) \sin[\theta(t)]$$

$$(4.3-61)$$

Now assuming that $\theta(t)$ is a stationary random process* and further assuming with probability 1 that it is always small compared to 1, yields

$$y(t) \cong \sqrt{2}A \sin(\omega_0 t + \theta_0) + \sqrt{2}A\theta(t) \cos(\omega_0 t + \theta_0) \qquad (4.3-62)$$

The autocorrelation function is given by

$$R_y(\tau) = A^2 \cos(\omega_0 \tau) + A^2 R_\theta(\tau) \cos \omega_0 \tau$$

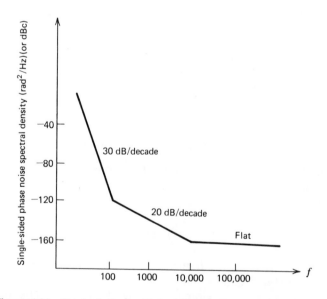

Figure 4.20 Typical single sideband phase noise spectral density.

*It is not actually clear that $\theta(t)$ is a stationary process (see [18] or [19]).

so that the power spectral density function is given by

$$\mathcal{S}_y(f) = \frac{A^2}{2}[\delta(f + f_0) + \delta(f - f_0)] + \frac{A^2}{2}[\mathcal{S}_\theta(f + f_0) + \mathcal{S}_\theta(f - f_0)]$$

(4.3-63)

The total mean squared phase noise is given by (accounting for the single sideband value by multiplying by 2)

$$\sigma_\theta^2 = 2\int_0^\infty \mathcal{S}_\theta(f)df \qquad \text{rad}^2$$

(4.3-64)

where $\mathcal{S}_\theta(f)$ is the single sideband phase noise spectral density. But the measured value (single sideband) is

$$A^2\mathcal{S}_\theta(f)$$

(4.3-65)

Hence twice the (one-sided) integrated value of (4.3-65) normalized by the carrier power yields

$$\int_0^\infty \frac{2A^2\mathcal{S}_\theta(f)df}{A^2} = 2\int_0^\infty \mathcal{S}_\theta(f)df = \sigma_\theta^2 \text{ rad}^2$$

(4.3-66)

which is what we desired to establish.

The alternative method of describing oscillator stability is the two-sample Allen variance. In order to describe the Allen variance we require a few definitions. Let

$$V(t) = V_0 \sin[2\pi ft + \varphi(t)]$$

(4.3-67)

Then the instantaneous frequency is $f + \dot{\varphi}(t)$, assuming that this quantity exists. Now define

$$y(t) = \frac{\dot{\varphi}(t)}{2\pi f}$$

(4.3-68)

as the instantaneous fractional frequency deviation from the nominal frequency f. Then define

$$y_k = \frac{1}{\tau}\int_{t_k}^{t_k+\tau} y(t)dt = \frac{\varphi(t_k + \tau) - \varphi(t_k)}{\tau}$$

(4.3-69)

where $t_{k+1} = t_k + T$, where $T \geq \tau$. Hence y_k is the τ-sec time average of the instantaneous fractional frequency offset during the kth measurement interval.

The generalized Allen variance frequency stability measure is defined by

$$\langle \sigma_y^2(N, T, \tau) \rangle = \left\langle \frac{1}{N-1}\sum_{n=1}^N \left(y_n - \frac{1}{N}\sum_{k=1}^N y_k\right)^2 \right\rangle$$

(4.3-70)

where $\langle x \rangle$ denotes the infinite time average of x. Some comments on this definition are in order. First, it is not necessarily true that as $N \to \infty$ $\langle \sigma_y^2(T, T, \tau) \rangle$ converges. Second, the preferred definition proposed in the literature is the Allen variance defined by

$$\langle \sigma_y^2(2, \tau, \tau) \rangle = \left\langle \sum_{n=1}^{2} \left(y_n - \frac{1}{2} \sum_{k=1}^{2} y_k \right)^2 \right\rangle \tag{4.3-71}$$

or equivalently [denoting $\langle \sigma_y^2(2, \tau, \tau) \rangle$ by $\langle \sigma_y^2(\tau) \rangle$]

$$\langle \sigma_y^2(\tau) \rangle = \left\langle \frac{(y_{k+1} - y_k)^2}{2} \right\rangle \tag{4.3-72}$$

Therefore, the Allen variance measurement is the time average of the difference of the τ-sec averaged frequency spaced apart by the averaging time (τ sec). Typical plots of the Allen variance are decreasing functions of τ. In Problem 11 it is shown that the Allen variance can be viewed as a highpass filtered version of the phase noise spectral density. In fact, knowing the phase noise spectral density allows one to compute the Allen variance [21]. Going from the Allen variance to σ_θ^2 is not a trivial matter and requires some assumptions about the spectral density near $f = 0$.

PROBLEM 11

Show that if the spectral density of $\dot{\theta}$ is known, then the ensemble average of the Allen variance is given by

$$E[\hat{\sigma}^2] = 2 \int_{-\infty}^{\infty} \mathcal{S}_{\dot{\theta}}(\omega) \left[\frac{\sin^2(\omega\tau/2)}{(\omega\tau/2)^2} - \frac{\sin^2(2\pi f\tau)}{(2\pi f\tau)^2} \right] \frac{d\omega}{2\pi}$$

Further show that when $\mathcal{S}_{\dot{\theta}}(\omega) = \omega^2 \mathcal{S}_\theta(\omega)$, that

$$E[\hat{\sigma}^2] = 2(2\pi)^2 \int_{-\infty}^{\infty} f^2 \mathcal{S}_\theta(f) \left[\frac{\sin^2(\pi f\tau)}{(\pi f\tau)^2} - \frac{\sin^2(2\pi f\tau)}{(2\pi f\tau)^2} \right] df$$

4.3.6 Optimization of the PLL Loop Filter

The problem we consider here is that of obtaining the optimum loop filter for the case when the loop has both noise and a transient modulation as the input. Consider the loop model shown in Figure 4.21. This model can be simplified to the linear filtering problem shown in Figure 4.22. The question arises, what is the optimum closed-loop transfer function $H(s)$ that in turn will provide the optimum loop filter $F(s)$? The answer depends upon the definition of optimum. First we would like the optimum transfer function to be causal. By *causal* we mean that the filter acts upon only the present and past inputs, that is,

$$\hat{\theta}(t) = \int_0^{\infty} h(u) \left[\theta(t-u) + \frac{n'(t-u)}{A} \right] du \tag{4.3-73}$$

where $h(t)$ is the Fourier transform of $H(\omega)$ and $h(t) = 0$ for $t < 0$. Secondly, partly because of mathematical convenience, we desire the filter that minimizes the mean squared phase error. We know that reducing the loop

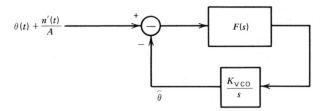

Figure 4.21 Linear PLL model.

bandwidth decreases the mean squared error due to thermal noise. Also, increasing the loop bandwidth decreases the transient error as well as oscillator noise. Hence it seems reasonable that an optimum exists if the transient error and oscillator noise are considered along with thermal receiver noise. Unfortunately, the optimum filter subjected to a transient input and thermal noise is time dependent. Consequently, Jaffee and Rechtin [22], who first performed the optimization, considered the following problem:

$$\min_{H(s)} (\sigma_m^2 + \sigma^2) \qquad \text{rad}^2$$

subject to $\epsilon_T^2 = \int_0^\infty \mu^2(t)\,dt = \text{constant}$, where $\mu(t)$ is the transient error due to doppler terms. This is equivalent to minimizing

$$\sigma_m^2 + \sigma^2 + \lambda^2 \epsilon_T^2 \qquad (4.3\text{-}74)$$

in which λ^2 is a Lagrange multiplier related to loop bandwidth, σ_m^2 is the phase error variance due to stochastic modulation, and σ^2 is the phase error due to thermal noise. In Section 4.3.3 we have shown that the spectra of the modulation plus transient distortion could be written as

$$\sigma_m^2 + \lambda^2 \epsilon_T^2 = \frac{1}{2\pi i} \int_{-i\infty}^{i\infty} |1 - H(s)|^2 \{\mathscr{S}_m(s) + E\lambda^2 [\Theta_d(s)\Theta_d^*(s)]\} ds$$

$$(4.3\text{-}75)$$

and the mean square error due to thermal noise is written as

$$\sigma^2 = \frac{1}{2\pi i} \int_{-i\infty}^{i\infty} |H(s)|^2 \frac{N_0}{2A^2}\,ds = \frac{1}{2\pi} \int_{-\infty}^{\infty} |H(i\omega)|^2 \frac{N_0}{2A^2}\,d\omega \qquad (4.3\text{-}76)$$

so that the spectrum at the input of Figure 4.21 is given by

$$\mathscr{S}(s) = \mathscr{S}_m(s) + \lambda^2 E\Theta_d(s)\Theta_d^*(s) + \frac{N_0}{2A^2} \qquad (4.3\text{-}77)$$

Figure 4.22 Simplified linear PLL model viewed as a filter.

where $E(\cdot)$ is the expected value of the quantity following it, $\Theta_d(s)$ is the Laplace transform of the doppler terms, and $\mathcal{S}_m(s)$ is the modulation spectral density. Yovits and Jackson [23] have obtained the optimum causal filter in this case when the noise is WGN. The result is given by

$$H_0(s) = 1 - \frac{\dfrac{\sqrt{N_0/2}}{A}}{[\mathcal{S}(s)]^+} \qquad (4.3\text{-}78)$$

where $[\mathcal{S}(s)]^+$ denotes a special type of square root (a factorization) of $\mathcal{S}(s)$ that has all poles and zeros of s in the left half plane. Any singularities on the real axis are equally divided between [ ]$^+$ and its mirror image [ ]$^-$, which has all its poles and zeros in the right half plane. Note that it follows that $\mathcal{S}(s) = [\mathcal{S}(s)]^+[\mathcal{S}(s)]^-$. Consider a simple example:

$$\left[A\frac{-s^2+a^2}{-s^2+b^2}\right]^+ = \left[\sqrt{A}\left(\frac{s+a}{s+b}\right)\sqrt{A}\left(\frac{-s+a}{-s+b}\right)\right]^+ = \sqrt{A}\frac{a+s}{b+s} \qquad a,b>0$$

$$(4.3\text{-}79)$$

where the pole is at $s = -b$ and the zero is at $s = -a$, both in the left half plane.

Following Tausworthe [8], we allowed $\Theta_d(s)$ to have a random phase. The doppler term $\theta_d(t)$ has the Laplace transform

$$\Theta_d(s) = \frac{\theta_0}{s} + \frac{\Omega_0}{s^2} + \frac{\Lambda_0}{s^3} + \cdots = \frac{Q(s)}{s^N} \qquad (4.3\text{-}80)$$

where the degree of $Q(s)$ is less than that of N and θ_0 is a uniform random variable defined on $(-\pi, \pi)$. Averaging over θ_0 and using the Yovits-Jackson result produces the optimum closed-loop transfer $H(s)$ given by

$$H_0(s) = 1 - \frac{\sqrt{N_0/2}/A}{[\lambda^2 E\Theta_d(s)\Theta_d^*(s) + \mathcal{S}_m(s) + N_0/2A^2]^+} \qquad (4.3\text{-}81)$$

Since we are considering only thermal noise and transient effects we obtain from (4.3-80)

$$H_0(\omega) = 1 - \frac{(s)^N}{\{(s)^N(-s)^N + [(2\lambda^2 A^2)/N_0]E[Q(s)Q(-s)]\}^+} \qquad (4.3\text{-}82)$$

We now apply (4.3-82) to the case of a random amplitude step in phase. We model the input phase function as $\theta_0 U(t)$, where θ_0 is a uniform random variable on $(-\pi, \pi)$, and $U(t)$ is the unit step function having a value of unity at $t \geq 0$ and zero elsewhere. Since $N = 1$, the denominator in our optimum closed-loop filter transfer function (4.3-82) becomes

$$\text{denominator} = \left[-s^2 + \frac{2\lambda^2 A^2}{N_0} E(\theta_0^2)\right]^+ \qquad (4.3\text{-}83)$$

Since $E(\theta_0^{[2]}) = \pi^2/3$, we have

$$\text{denominator} = -s^2 + \frac{\lambda A \pi}{\sqrt{3N_0/2}} \qquad (4.3\text{-}84)$$

Therefore, factoring (4.3-83) and taking the part with a left half pole produces

$$H_0(s) = \frac{\lambda A \pi}{\sqrt{3N_0/2}} \left(s + \frac{\lambda A \pi}{\sqrt{3N_0/2}} \right)^{-1} \qquad (4.3\text{-}85)$$

which is recognized as a single-pole lowpass filter. This implies that the loop filter is a constant and

$$2W_L = \frac{\lambda A \pi}{\sqrt{3N_0/2}} \qquad (4.3\text{-}86)$$

so that

$$H_0(s) = \frac{2W_L}{s + 2W_L} \qquad (4.3\text{-}87)$$

Notice that $\phi_{ss} = 0$ for this input and the resulting optimum filter is a direct consequence of the fact that $\int_0^\infty \mu^2(t)dt$ is finite. In the following filters it will be noted that $\phi_{ss} = 0$, due to the same reason.

As a <u>second</u> application we consider the optimum loop filter for a frequency offset plus a step phase input. In this case we model the input doppler signal as $\theta_d(t) = (\theta_0 + \Omega_0 t)U(t)$. Then

$$\Theta_d(s) = \frac{\theta_0}{s} + \frac{\Omega_0}{(s)^2}$$

so that the denominator becomes

$$\left\{ (s)^2(-s)^2 + \frac{2\lambda^2 A^2}{N_0} E[(\theta_0 s + \Omega_0)(-\theta_0 s + \Omega_0)] \right\}^+ \qquad (4.3\text{-}88)$$

Finding the root that lies in the left half s plane allows us to write

$$\text{denominator} = (s)^2 + \sqrt{\frac{2A^2\lambda^2\pi^2}{3N_0} + \frac{2\lambda A \Omega_0}{\sqrt{N_0/2}}}\,(s) + \frac{\lambda A \Omega_0}{\sqrt{N_0/2}} \qquad (4.3\text{-}89)$$

Using a simpler form, we have

$$\text{denominator} = s^2 + 2\zeta\omega_n s + \omega_n^2 \qquad (4.3\text{-}90)$$

where

$$\omega_n = \sqrt{\frac{\lambda A \Omega_0}{\sqrt{N_0/2}}}$$

$$2\zeta\omega_n = \sqrt{\frac{2A^2\lambda^2\pi^2}{3N_0} + \frac{2\lambda A \Omega_0}{\sqrt{N_0/2}}}$$

so that

$$H_0(s) = \frac{2\zeta\omega_n s + \omega_n^2}{s^2 + 2\zeta\omega_n s + \omega_n^2} \tag{4.3-91}$$

Solving for $F_0(s)$, via

$$F_0(s) = \frac{sH_0(s)}{AK[1 - H_0(s)]} \tag{4.3-92}$$

we obtain the optimum filter form (proportional plus integrator)

$$F_0(s) = \frac{\omega_n^2}{AK} + \frac{2\zeta\omega_n}{AKs} \tag{4.3-93}$$

As our final case we consider the optimum closed-loop filter to track a doppler rate. In this case we model the input doppler signal as

$$\theta_d(t) = \tfrac{1}{2}\Lambda_0 t^2 \quad \text{so that} \quad \Theta_d(s) = \frac{\tfrac{1}{2}\Lambda_0}{(s)^3} \tag{4.3-94}$$

The denominator can now be written [22] as

$$\text{denominator} = s^3 + 2\omega_n s^2 + 2\omega_n^2 s + \omega_n^3 \tag{4.3-95}$$

so that the optimum loop filter becomes

$$H_0(s) = \frac{2\omega_n s^2 + 2\omega_n^2 s + \omega_n^3}{s^3 + 2\omega_n s^2 + 2\omega_n^2 s + \omega_n^3}$$

or

$$F_0(s) = \frac{2\omega_n}{AK} + \frac{2\omega_n^2}{AKs} + \frac{\omega_n^3}{AKs^2} \tag{4.3-96}$$

which is a third-order PLL. The parameter ω_n is not the same as the second-order loop parameter.

Notice that in each case the optimum filter removes the static phase error completely as $t \to \infty$.

4.4 NONLINEAR MODEL RESULTS

At this point we have obtained many of the results that can be obtained using the linearized model. Acquisition and cycle slipping performance, for example, cannot be obtained from the linearized model. We now consider some relevant nonlinear results.

4.4.1 Mean Time to Cycle Slipping

One parameter that perhaps is most useful in defining threshold in a phase locked loop is the mean time to cycle slip. By this we mean the mean time it takes for the phase error to increase or decrease 2π rad. It can be shown that when thermal noise is present the probability of a slip approaches unity as $t \to \infty$.

In Chapter 10, Section 10.11, it is shown that the mean time (for a first-order Markov process driven tracking loop) to reach the level ϕ_L is given by

$$\bar{T}W_L = \frac{\pi^2}{\sigma_\phi^2} \int_0^{\phi_L}\int_0^{\phi_L} e^{-G(\phi)/\sigma_\phi^2} e^{G(\phi')/\sigma_\phi^2} d\phi/d\phi' \tag{4.4-1}$$

where in the PLL case $\phi_L = 2\pi$. Evaluating (4.4-1) with $\phi_L = 2\pi$ produces

$$\bar{T}W_L = \frac{\pi^2}{\sigma^2} I_0^2\left(\frac{1}{\sigma_\phi^2}\right) \tag{4.4-2}$$

where $I_0(x)$ is the modified Bessel function of zeroth order. This result (4.4-2) for the first-order PLL was first derived by Viterbi [4]. Notice from (4.4-2) that since $I_0(x) \sim e^\rho/\sqrt{2\pi\rho}$, we have asymptotically

$$W_L\bar{T} \cong \frac{\pi}{2} e^{2/\sigma^2} \qquad \text{for small } \sigma^2 \tag{4.4-3}$$

Lindsey has generalized Viterbi's result to include static phase error offsets [6]. His result yields

$$W_L\bar{T} = \frac{\pi \tanh(\pi\gamma/\sigma^2)}{\gamma} \left[I_0^2\left(\frac{1}{\sigma^2}\right) + 2\sum_{n=1}^\infty (-1)^n \frac{I_n^2(1/\sigma^2)}{1 + (n\sigma^2/\gamma)^2} \right] \tag{4.4-4}$$

with $\gamma = \Omega_0/AK$ and $I_n(x)$ the modified Bessel function of the nth order. Viterbi [4] has shown that the cycle slipping rate is the inverse of the mean slip time for the first order loop and approximately so for the second order loop, so that

$$\bar{s} = \frac{1}{\bar{T}} \qquad \text{slips per second} \tag{4.4-5}$$

Furthermore, Smith [24] and Lindsey have shown that the slip process is approximately Poisson distributed, so that the <u>probability of not slipping in t sec</u> is given by

$$p_{\bar{SL}} = e^{-\bar{s}t} \tag{4.4-6}$$

and the probability of K slips in t sec is

$$p_{SL}(K, t) = \frac{(\bar{s}t)^K e^{-\bar{s}t}}{K!} \tag{4.4-7}$$

So far all the above results were based on a first-order PLL. The second-order passive filter mean slip time has not been solved for exactly; however, simulation has produced useable results. Holmes [25] has made a simulation for $\rho = 1/\sigma^2$ between 0 and 5 dB, and the parameter $\gamma = \Omega_0/AK$ between 0 and 1 is shown below in Figure 4.23.

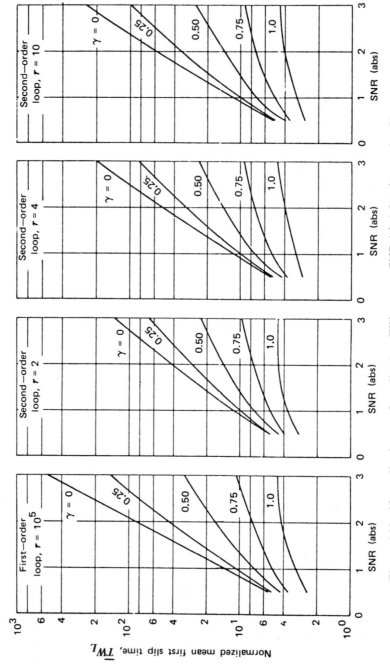

Figure 4.23 Normalized mean first slip time TW_L versus SNR (abs) for the passive filter second-order loop [25].

Based on the curves of Figure 4.23, it has been estimated that a second-order loop is about 1 dB inferior (ρ is reduced 1 dB), so that we have, as an approximation for the second-order loop,* the asymptotic result

$$W_L \bar{T} \cong \frac{\pi}{2} e^{1.59\rho} \qquad \rho = \frac{1}{\sigma^2} \qquad \zeta = 0.707 \qquad (4.4-8)$$

Sanneman and Rowbotham [28] also obtained computer simulation results that agreed quite closely to (4.4-8).

Tausworthe [29] has also considered the cycle slipping problem [27], although we will not discuss the details here.

4.4.2 Acquisition

There are three basic approaches to loop acquisition. The first one is natural pull-in, which is a property of any well-designed loop in which dc offsets are made small. Viterbi [4] has shown by considering the noise-free differential equation of the loop that the pull-in time is well approximated by ($|\Omega_0| \gg 2\pi B_L$)

$$T_{ACQ} = \frac{\Omega_0^2}{2\zeta\omega_n^3} \qquad \text{sec} \qquad (4.4-9)$$

where Ω_0 is the initial frequency offset in rad/sec, ζ is the damping factor, and ω_n is the loop natural frequency. However, it should be noted that this result applies to an idealized second-order PLL in that it has no dc offsets. This result can be in gross error if phase detector or loop filter offsets are not negligible. Furthermore, since the acquisition times can be very large, natural pull-in when $|\Omega_0| \gg 2\pi B_L$ is seldom used in practice.

When the initial frequency offset is on the order of a loop bandwidth, pull-in is quite fast. Nilsen [30] has studied this problem by means of simulation for an ideal operation amplifier loop filter with the results plotted in Figures 4.24 and 4.25. As can be seen from the plots, when the frequency offset is less than $\frac{1}{2}B_L$, pull-in is quite rapid. Figures 4.26 and 4.27 illustrate pull-in for the third-order loop. As is clear from the curves, pull-in is much less reliable for a third-order loop. Goldman [31] also has made acquisition time simulations. Hummels [32] has made simulations for acquisition time including the effects of a lock detector.

A second-order method for acquisition is to sweep the local oscillator across the range of carrier frequency uncertainty. Frazier and Page [33] have, by means of simulation, deduced the following empirical expression for the allowable sweep rate for a probability of acquisition of .9:

$$R_{90} = \frac{1 - \sqrt{2}/\sqrt{\rho}}{2\pi(1+d)}\omega_n^2 \qquad Hz/\text{sec} \qquad (4.4-10)$$

*It is expected that the passive filter and active filter loops have about the same mean slip time.

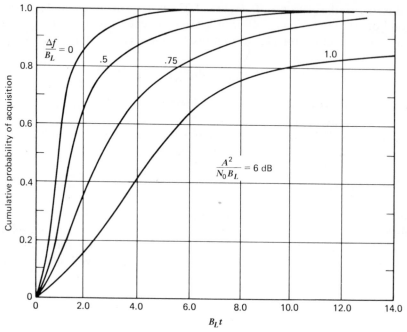

Figure 4.24 Cumulative probability of acquisition as a function of time (normalized) for an ideal second-order phase locked loop [30]: $A^2/N_0 B_L = 6$ dB (from Nilsen [30] with permission).

where

R_{90} = the sweep rate to achieve an acquisition probability of .9

$$\rho = \frac{A^2}{N_0 B_L} \geq 7 \text{ dB for the equation to be valid}$$

$$d = \begin{cases} \exp\left(\dfrac{-\pi\zeta}{\sqrt{1-\zeta^2}}\right) & \zeta < 1 \\ 0 & \zeta \geq 1 \end{cases}$$

ζ = loop damping factor $\quad \zeta \geq 0.5$

$$\omega_n = \begin{cases} \text{loop natural frequency (rad/sec)} \\ \dfrac{2B}{\left(\zeta + \dfrac{1}{4\zeta}\right)} \end{cases}$$

PROBLEM 12

(a) Show that when $\zeta = 0.707$ the 90% acquisition sweep rate becomes

$$R_{90} = 0.5431\left(1 - \frac{\sqrt{2}}{\sqrt{\rho}}\right) B_L^2 \qquad \text{Hz/sec}$$

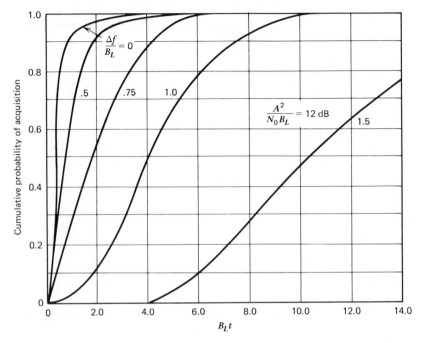

Figure 4.25 Cumulative probability of acquisition as a function of time (normalized) for an ideal second-order phase locked loop [30]: $A^2/N_0B_L = 12$ dB (from Nilsen [30] with permission).

(b) Determine the optimum B_L to maximize R_{90} for a given A^2/N_0. Is this value ≥ 7 dB?

(c) Let $\zeta = 0.707$, $B_L = 10$ Hz, $A^2/N_0 = 37$ dB-Hz, and determine the time it takes to sweep an 8-kHz doppler uncertainty.

Sweeping faster than (4.4-10) reduces the probability of acquisition below 90%, and sweeping more slowly increases the probability above 90%.

The <u>third</u> method of acquisition, to use a frequency aid such as a discriminator to provide the frequency acquisition, is probably the most common method used to prevent false lock.

A number of different strategies are possible, such as: (1) using a phase detector and frequency detector and adding these signals, which in turn control the VCO; or (2) using a phase frequency detector such as the one manufactured by Motorola, which is a detector that produces an output voltage in proportion to frequency when there is a frequency error and when in lock produces the signal proportional to phase error. Combinations of option (1) include turning off the frequency detector when the signal is acquired. Figure 4.28 illustrates two possible implementations. In the first implementation it is possible to switch out the frequency discriminator when the loop has acquired the signal, thereby rejecting unnecessary noise.

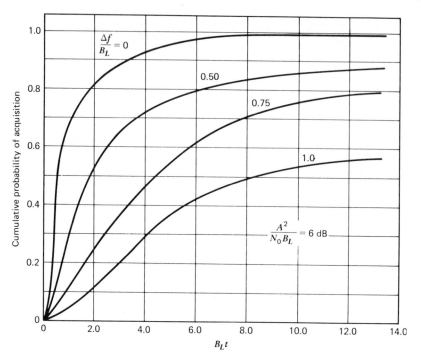

Figure 4.26 Cumulative probability of acquisition as a function of time (normalized) for third-order phase locked loop [30]: $A^2/N_0B_L = 6$ dB (from Nilsen [30] with permission).

4.4.3 A Lock Indicator

Acquisition is, of course, an important aspect of a coherent system. Perhaps equally important is the ability to determine when the loop has achieved lock. This process is accomplished by the use of a lock detector shown in Figure 4.29. The lock detector is formed by the addition of an in-phase phase detector (coherent amplitude detector) and a smoothing lowpass filter (LPF) along with a threshold detector. When the signal is not in lock (a frequency error exists), the lock detector output outputs an essentially zero mean noise process. However, when the signal is present, a dc value proportional to cos ϕ is generated out of the quadrature phase detector that is integrated by the LPF to produce a voltage that would be designed to exceed the detector threshold. Lock indication is not instantaneous obviously, since the LPF has some delay associated with its risetime.

We model the coherent amplitude detector (CAD) output, after the loop has locked, as

$$y(t) = A \cos \phi(t) + n(t) \qquad (4.4-11)$$

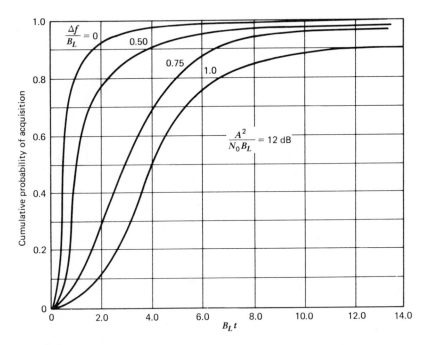

Figure 4.27 Cumulative probability of acquisition as a function of time (normalized) for third-order phase locked loop [30]: $A^2/N_0B_L = 12$ dB (from Nilsen [30] with permission).

where

$\phi(t)$ = the phase error process

A = signal voltage

$n(t)$ = CAD noise modeled as filtered WGN

Due to the sweep and possible dynamics, there will be a static phase error component plus a component due to noise; i.e.,

$$\phi(t) = \phi_{ss} + \phi_n(t) \qquad (4.4\text{-}12)$$

Assuming that the LPF is considerably narrower than the loop bandwidth and σ_ϕ^2 is small, the output of the filter will be approximately

$$Y_{LP}(t) = A(1 - \tfrac{1}{2}\sigma_\phi^2)\cos\phi_{ss}e(t) + N(t) \qquad (4.4\text{-}13)$$

where

$e(t)$ = step response of the LPF

$N(t)$ = lowpass filtered noise assumed to be Gaussian with variance σ_{LP}^2

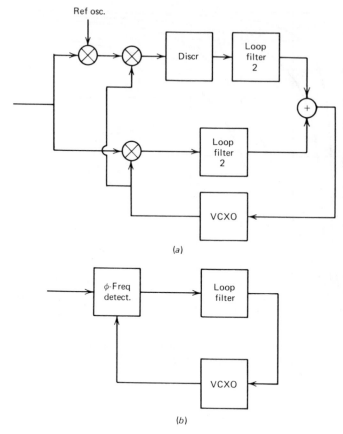

(a)

(b)

Figure 4.28 Some possible acquisition aiding systems. (a) Hybrid loop for acquisition aiding. (b) Phase-frequency detector acquisition aiding approach.

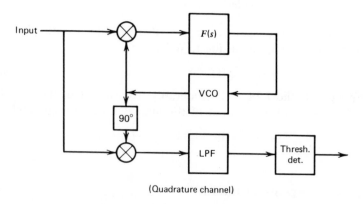

(Quadrature channel)

Figure 4.29 Coherent amplitude lock detector.

The variance of the noise out of the filter is given by

$$\sigma_{LP}^2 = \frac{N_0}{2} \int_{-\infty}^{\infty} |H(f)|^2 \, df \qquad (4.4\text{-}14)$$

with $H(f)$ being the transfer function of the LPF.

When the signal is not present, the level crossing rate [34] is given by

$$\lambda_{FA} = Cf_c e^{-(Th)^2/2} \qquad \text{false alarms per second} \qquad (4.4\text{-}15)$$

where Th is the threshold expressed in units of σ_{LP}, f_c is the 3-dB cutoff frequency of the filter, and C is dependent on the type of filter used,* a few cases being indicated in Table 4.3:

PROBLEM 13.

(a) Verify the threshold crossing formula.

(b) Verify the values of C given in the Table 4.3.

(c) Show that the mean rate to indicate loss of lock when the signal is present is $\lambda_{MD} = Cf_c \exp[-(A - Th)^2/2]$ after the signal charges the filter up.

Given a knowledge of the level crossing rate λ, it is possible to write down the probability of false alarm, P_{FA}, or the probability of missed detection. Modeling the false alarm or missed detection occurrences as Poisson events, we find that the probability of at least one false alarm or missed detection in T sec is given by

$$P_{FA} \cong 1 - \exp(-\lambda_{FA} T)$$
$$P_{MD} \cong 1 - \exp(-\lambda_{MD} T) \qquad (4.4\text{-}16)$$

with

$$\lambda_{FA} = Cf_c \exp[-(Th)^2/2]$$
$$\lambda_{MD} = Cf_c \exp[-(A - Th)^2/2] \qquad (4.4\text{-}17)$$
when the signal has charged the filter

Table 4.3 Threshold crossing parameters

C	Filter
1.15	Ideal lowpass
1.286	4-pole Butterworth
2.0	2-pole Butterworth

*In fact $C^2 = \int_{-\infty}^{\infty} \omega^2 \mathscr{S}_n(\omega) d\omega / \int_{-\infty}^{\infty} \mathscr{S}_n(\omega) d\omega$ with $\mathscr{S}(\omega)$ the power spectral density of the noise.

Typically, when the loop has acquired and indicated phase lock, an algorithm is used to require, for example, five successive out-of-lock indications before declaring that the loop is unlocked. This topic is discussed further in Chapter 9.

4.4.4 Phase Error Distribution and Moments

Based on the Fokker-Plank method of obtaining the probability density function from the stochastic PE of loop operation, Tikhonov [36] and Viterbi [4] obtained the probability density function in steady state for the first-order phase locked loop with the phase error reduced mod 2π. (See the Appendix for the derivation.) The result, for zero detuning, was the Tikhonov density [36]

$$P(\phi) = \frac{\exp(\rho \cos \phi)}{2\pi I_0(\rho)} \qquad |\phi| \le \pi \tag{4.4-18}$$

where

$$\rho = \frac{A^2}{N_0 B_L} \qquad \text{the loop SNR in } B_L \tag{4.4-19}$$

$$I_0(x) = \text{zeroth order modified Bessel function}$$

for large ρ

$$I_0(\rho) \sim \frac{e^\rho}{\sqrt{2\pi\rho}} \tag{4.4-20}$$

PROBLEM 14

Show that for large ρ that the Tikhonov phase error density function approaches a Gaussian density function.

The Tikhonov density function was also used as an approximation to the passive filter second-order loop by Charles and Lindsey [37]. They found by comparison to their experimental data that the Tikhonov density function was a reasonably good approximation. The variance, using the Tikhonov model, is given by (since the mean is zero),

$$\sigma^2 = \int_{-\pi}^{\pi} \frac{\phi^2}{2\pi I_0(\rho)} e^{(\rho \cos \phi)} d\phi \qquad \text{rad}^2 \tag{4.4-21}$$

Using the Jacobi-Anger expansion for $e^{(\rho \cos \phi)}$, we have

$$\sigma^2 = \frac{1}{2\pi I_0(\rho)} \int_{-\pi}^{\pi} \phi^2 \left[I_0(\rho) + 2 \sum_{n=1}^{\infty} I_n(\rho) \cos n\phi \right] d\phi \qquad \text{rad}^2 \tag{4.4-22}$$

or

$$\sigma^2 = \frac{\pi^2}{3} + 4 \sum_{n=1}^{\infty} \frac{(-1)^n I_n(\rho)}{n^2 I_0(\rho)} \qquad \text{rad}^2 \tag{4.4-23}$$

This result converges quite rapidly and is plotted in Figure 4.30. When $\rho = 10$ the difference between the linear and nonlinear models is negligible. Notice as $\rho \to 0$, $\sigma^2 \to \pi^2/3$, the variance of a uniformly distributed random variable on $(-\pi, \pi)$.

4.4.5 Effect of a Limiter Preceding a Second-Order PLL

In many systems employing phase lock receivers it is desirable to utilize a limiter preceding the loop in order to keep the input signal range down to a reasonable level and also to protect the phase detector from damage under high signal and/or noise levels.

The results given here basically follow that of references 6, 38, and 39. Consider the PLL preceded by a bandpass limiter shown in Figure 4.31. The input signal waveform is given by

$$x(t) = \sqrt{2}A \cos(\omega_0 t + \theta_0) + n'(t) \qquad (4.4\text{-}24)$$

where

$$n'(t) = \sqrt{2}n_s'(t) \sin(\omega_0 t + \theta_0) + \sqrt{2}n_c'(t) \cos(\omega_0 t + \theta_0) \qquad (4.4\text{-}25)$$

and $n'(t)$ is a broad band WGN. Out of the IF bandpass filter we have $y(t)$:

$$y(t) = \sqrt{2}\, A \cos(\omega_0 t + \theta_0) + n(t) \qquad (4.4\text{-}26)$$

$$n(t) = \sqrt{2}n_s(t) \sin(\omega_0 t + \theta_0) + \sqrt{2}n_c(t) \cos(\omega_0 t + \theta_0) \qquad (4.4\text{-}27)$$

where $n_s(t)$ and $n_c(t)$ are sample functions of baseband Gaussian noise processes having spectral densities $N_0/2$ and variances $N_0 B_i/2$. Combining the bandpass filtered signal with the noise, we have

$$y(t) = \sqrt{2} \sqrt{n_s(t) + [A + n_c(t)]^2} \cos[\omega_0 t + \theta_0 - \gamma(t)] \qquad (4.4\text{-}28)$$

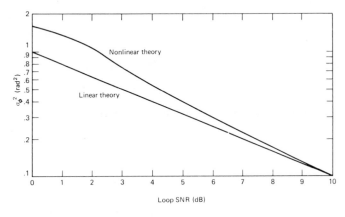

Figure 4.30 Linear and nonlinear phase error variance versus loop SNR.

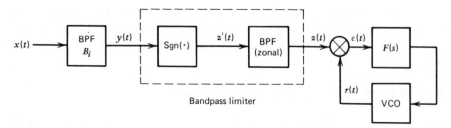

Figure 4.31 PLL preceded by a bandpass limiter.

where

$$\gamma(t) = \tan^{-1}\left(\frac{n_s}{A + n_c}\right) \tag{4.4-29}$$

The limiter, modeled as a signum function, removes the amplitude variations, so that

$$z'(t) = \text{sgn}\{\cos[\omega_0 t + \theta_0 - \gamma(t)]\} \tag{4.4-30}$$

Letting $\Theta = \omega_0 t + \theta_0 - \gamma(t)$, we see that $z(t) = \text{sgn}(\cos \Theta)$ is a square wave of unit amplitude and approximate period $2\pi/\omega_0$, so that using the Fourier expansion for $\text{sgn}(\cos \Theta)$, we have

$$z'(t) = \frac{4}{\pi} \sum_{m=0}^{\infty} (-1)^m \frac{\cos[(2m + 1)\Theta]}{2m + 1} \tag{4.4-31}$$

The function of the zonal filter is to remove all but the fundamental component of $z'(t)$, so that* (letting $m = 0$)

$$z(t) = \frac{4}{\pi} \cos[\omega_0 t + \theta_0 - \gamma(t)] \tag{4.4-32}$$

Notice the power in the fundamental is $P_L = 8/\pi^2$. We see that the limiter has removed the amplitude variations but retained the phase variations. Expanding $z(t)$, we have

$$z(t) = \frac{4}{\pi} \cos(\omega_0 t + \theta_0) \cos[\gamma(t)] + \frac{4}{\pi} \sin(\omega_0 t + \theta_0) \sin[\gamma(t)] \tag{4.4-33}$$

Adding and subtracting the average of $\cos \gamma(t)$ and $\sin \gamma(t)$, we have (dropping the t dependence of γ for convenience)

$$z(t) = \frac{4}{\pi} \overline{[\cos \gamma} \cos(\omega_0 t + \theta_0) + \overline{\sin \gamma} \sin(\omega_0 t + \theta_0)]$$

$$+ \frac{4}{\pi} [(\cos \gamma - \overline{\cos \gamma}) \cos(\omega_0 t + \theta_0)$$

$$+ (\sin \gamma - \overline{\sin \gamma}) \sin(\omega_0 t + \theta_0) \tag{4.4-34}$$

---

*We have assumed the limiter amplitude in unity; otherwise, the amplitude of $z(t)$ would be $4\alpha_L/\pi$, where α_L is the output amplitude of the limiter.

We now make the following definitions:

$$\alpha = \overline{\cos \gamma} \tag{4.4-35}$$

$$n_c(t) = \frac{2\sqrt{2}}{\pi} (\cos \gamma - \overline{\cos \gamma}) \tag{4.4-36}$$

$$n_s(t) = \frac{2\sqrt{2}}{\pi} (\sin \gamma - \overline{\sin \gamma}) \tag{4.4-37}$$

so that $z(t)$ can be written as

$$z(t) = \sqrt{2P_L}\, \alpha \cos(\omega_0 t + \theta_0) + \sqrt{2}\, n_c(t) \cos(\omega_0 t + \theta_0)$$
$$+ \sqrt{2}\, n_s(t) \sin(\omega_0 t + \theta_0) \tag{4.4-38}$$

Now consider $\sin \gamma$ and $\cos \gamma$. Using the obvious trigonometric identities, we have

$$\sin \gamma = \frac{n_s}{\sqrt{n_s^2 + (A + n_c)^2}} \tag{4.4-39}$$

$$\cos \gamma = \frac{A + n_c}{\sqrt{n_s^2 + (A + n_c)^2}} \tag{4.4-40}$$

clearly $\overline{\sin \gamma} = 0$ since $\sin \gamma$ is an odd function of n_s. Notice at very high SNR (in B_i) that

$$\sin \gamma \cong \frac{n_s}{A} \tag{4.4-41}$$

$$\cos \gamma \cong 1 \tag{4.4-42}$$

and

$$z(t) \cong \frac{4}{\pi} \cos(\omega_0 t + \theta_0) + \frac{4}{\pi} \frac{n_s}{A} \sin(\omega_0 t + \theta_0) \tag{4.4-43}$$

so that the limiter has removed the in-phase noise component. At this point one might conjecture that the noise power is halved, and therefore the loop is 3 dB better when a limiter precedes the loop. This conjecture, which for a number of years was thought to be true, was shown by Forney [40] not to be true, as deduced in Problem 15.

PROBLEM 15

Using the results of the limiter analysis, show that at high SNR (in B_i) that the loop operates with the same loop SNR as it does without the limiter if the loop gain is adjusted to be equal to the value when the limiter was not present. (This gain adjustment keeps the bandwidth the same.)

We will now derive the suppression factor from (4.4-40). Let

$$u = A + n_c \tag{4.4-44}$$

$$v = n_s \tag{4.4-45}$$

so that

$$\cos \gamma = \frac{u}{\sqrt{u^2 + v^2}} \tag{4.4-46}$$

Averaging, we obtain $[\sigma^2 = \text{var}(n_c) = \text{var}(n_s)]$

$$\overline{\cos \gamma} = \int_{-\infty}^{\infty}\int_{-\infty}^{\infty} \frac{u}{\sqrt{u^2 + v^2}} \frac{1}{2\pi\sigma^2} \exp\left[-\frac{1}{2\sigma^2}(u - A)^2 + v^2\right] du\, dv \tag{4.4-47}$$

Now let $u = r \cos \theta$ and $v = r \sin \theta$, so that

$$\overline{\cos \gamma} = \int_0^{2\pi}\int_0^{\infty} r \cos \theta \frac{1}{2\pi\sigma^2} \exp\left[-\frac{(A^2 + r^2)}{2\sigma^2} + \frac{Ar \cos \theta}{\sigma^2}\right] dr\, d\theta \tag{4.4-48}$$

or

$$\overline{\cos \gamma} = \int_0^{\infty} \frac{r}{\sigma^2} \exp\left(-\frac{A^2 + r^2}{2\sigma^2}\right) \int_0^{2\pi} \frac{\cos \theta}{2\pi} \exp\left(\frac{Ar \cos \theta}{\sigma^2}\right) d\theta\, dr \tag{4.4-49}$$

$$\overline{\cos \gamma} = \int_0^{\infty} \frac{r}{\sigma^2} \exp\left(-\frac{A^2}{2\sigma^2} - \frac{r^2}{2\sigma^2}\right) I_1\left(\frac{Ar}{\sigma^2}\right) dr \tag{4.4-50}$$

Since the following is true:

$$\int_0^{\infty} I_1(ay) \exp\left(-\frac{y^2}{2}\right) y\, dy = \frac{a\sqrt{\pi}}{(2\sqrt{2})} \exp\left(\frac{a^2}{4}\right)\left[I_0\left(\frac{a^2}{4}\right) + I_1\left(\frac{a^2}{4}\right)\right] \tag{4.4-51}$$

we obtain, using $a = \sqrt{2\rho_i}$ and the fact that $\rho_i = A^2/2\sigma^2$ (the input SNR), the result that the *signal amplitude suppression factor* is given by

$$\alpha = \sqrt{\frac{\pi}{2}} \sqrt{\frac{\rho_i}{2}} \exp\left(-\frac{\rho_i}{2}\right)\left[I_0\left(\frac{\rho_i}{2}\right) + I_1\left(\frac{\rho_i}{2}\right)\right] \tag{4.4-52}$$

This result, first derived by Davenport [38], has been very well approximated by Tausworthe [8] in the form

$$\alpha \cong \sqrt{\frac{0.7854\rho_i + 0.4768\rho_i^2}{1 + 1.024\rho_i + 0.4768\rho_i^2}} \tag{4.4-53}$$

The α derived by Davenport is shown in Figure 4.32. Note for large ρ_i, $\alpha \to 1$, and small $\rho_i \to \sqrt{\pi/4\rho_i}$.

Now the output of the phase detector of Figure 4.31 is the term

$$\begin{aligned}\epsilon(t) = \{&\sqrt{2P_L}\, \alpha \cos[\omega_0 t + \theta_0] + \sqrt{2}N_c(t) \cos(\omega_0 t + \theta_0) \\ &+ \sqrt{2}N_s(t) \sin(\omega_0 t + \theta_0)\}\{-\sqrt{2} \sin[\omega_0 t + \hat{\theta}(t)]\}\end{aligned} \tag{4.4-54}$$

where the reference signal is modeled by

$$r(t) = -\sqrt{2} \sin[\omega_0 t + \hat{\theta}(t)] \tag{4.4-55}$$

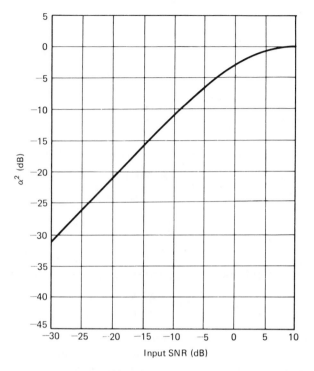

Figure 4.32 Signal amplitude suppression factor α^2 versus SNR (from Tausworthe [8], with permission).

The minus sign accounts for the fact that the reference signal is delayed by $\pi/2$ rad from the input signal. We have, for $\epsilon(t)$,

$$\epsilon(t) = \alpha\sqrt{P_L}\sin\phi(t) + N_c(t)\sin\phi(t) - N_s(t)\cos\phi(t) \qquad (4.4\text{-}56)$$

where $\phi(t) = \theta_0 - \hat{\theta}(t)$, and where we have neglected $O(2\omega_0)$ terms since they will be removed by the loop filter and oscillator. For small phase errors

$$\epsilon(t) \cong \alpha\sqrt{P_L}\sin\phi(t) - N_s(t)\cos\phi(t) \qquad (4.4\text{-}57)$$

Hence, we can define an equivalent coherent SNR at the output of the bandpass limiter considering only $N_s(t)$, since at high SNR in B_L that is the only contributor to the phase error.

$$\rho_0 = \frac{\alpha^2 P_L}{\sigma_s^2} \qquad (4.4\text{-}58)$$

where σ_s^2 is the power of the $\sqrt{2}\,N_s\sin(\omega_0 t + \theta_0)$ noise term. Springett and Simon [39] have evaluated σ_s^2 and found it to be given by

$$\sigma_s^2 = \frac{4}{\pi^2}\left(\frac{1 - e^{-\rho_i}}{\rho_i}\right) \qquad (4.4\text{-}59)$$

PROBLEM 16

(a) Show that $\sigma_s^2 = \overline{N_s^2}$.

(b) Show that $\sigma_s^2 = \dfrac{4}{\pi^2}\left(\dfrac{1 - e^{\rho_i}}{\rho_i}\right)$

so that ρ_0, the coherent SNR out of the limiter, is related to the input SNR ρ_i into the limiter by

$$\rho_0 = \frac{\alpha^2 P_L}{\sigma_s^2} = \frac{\dfrac{\pi}{2}\left(\dfrac{\rho_i}{2}\right)e^{-\rho_i}[I_0(\rho_i/2) + I_1(\rho_i/2)]^2 P_L}{(P_L/2)[(1 - e^{-\rho_i})/\rho_i]} \tag{4.4-60}$$

or

$$\rho_0 = \frac{\pi(\rho_i^2/2)e^{-\rho_i}[I_0(\rho_i/2) + I(\rho_i/2)]^2}{1 - e^{-\rho_i}} \tag{4.4-61}$$

Since he PLL performance depends on the ratio P/N_0B_L, we can define, following Lindsey [6], the limiter performance factor Γ_P defined by

$$\Gamma_P = \frac{A^2/N_0}{(P/N_0)_L} \tag{4.4-62}$$

where A^2/N_0 is the carrier power to one-sided noise spectral density into the limiter and $(P/N_0)_L$ is the same ratio out of the limiter disregarding $n_c(t)$ since it does not influence loop performance at high loop SNR. The limiter performance factor can be written in terms of ρ_0 and ρ_i in the form

$$\Gamma_P = \frac{\rho_i}{\rho_0}\frac{B_i}{B_0} \tag{4.4-63}$$

The ratio ρ_i/ρ_0 is given above; however, the ratio of input to output noise bandwidths B_i/B_0 has to be obtained. Springett and Simon [39] have approximated this ratio for the case of variously shaped IF filters. They found that

$$B_0 = B_i\left\{1 + \left(\frac{4}{\pi\Gamma_0} - 1\right)\exp[-\rho_i(1 - \pi/4)]\right\} \tag{4.4-64}$$

so that

$$\Gamma_P = \frac{1 - e^{-\rho_i}}{(\pi\rho_i/4)\exp(\rho_i)[I_0(\rho_i/2) + I_1(\rho_i/2)]^2\{1 + (4/\Gamma_0\pi - 1)\exp[-\rho_i(1 - \pi/4)]\}} \tag{4.4-65}$$

The parameter Γ_0 depends upon the IF filter shape. For an ideal bandpass filter (BPF), $\Gamma_0 = 1.16$. In addition, $\Gamma_0 = 1.12$ for a Gaussian shaped filter and $\Gamma_0 = 1.059$ for a single-pole filter. Now for the ideal BPF case

$$\Gamma_P \to \Gamma_0 \qquad \text{as } \rho_i \to 0$$

$$\Gamma_P \to 1 \qquad \text{as } \rho_i \to \infty \tag{4.4-66}$$

Γ_P is a measure of the phase error variance increase over the case when no limiter precedes the loop. To see this, note that

$$\sigma_\phi^2 = \left(\frac{N_0}{P_0}\right)_L B_L = \frac{N_0}{A^2} \Gamma_P B_L = \frac{N_0 B_L}{A^2} \Gamma_P \qquad \text{rad}^2 \qquad (4.4\text{-}67)$$

Figure 4.33 illustrates the value of $1/\Gamma_\rho$ versus ρ_i, the input SNR for the ideal filter case (worst case degradation).

For low value of ρ_i the Gaussian shaped filter produces a 0.50-dB decrease in loop SNR, and the one-pole filter produces a 0.25-dB decrease in loop SNR. In conclusion, the bandpass limiter has a very slight deleterious effect on loop operation.

For the second-order passive filter loop (by direct comparison to the closed-loop transfer function of the linear loop) we have, including the limiter effect,

$$H(s) = \frac{1 + \tau_2 s}{1 + (\tau_2 + 1/\alpha\sqrt{P_L}K)s + (\tau_1/\alpha\sqrt{P_L}K)s^2} \qquad (4.4\text{-}68)$$

The corresponding two-sided loop bandwidth is given by

$$W_L = \frac{1 + r}{2\tau_2(1 + \tau_2/r\tau_1)} \qquad (4.4\text{-}69)$$

and

$$r = \frac{\alpha\sqrt{P_L}K\tau^2}{\tau_1} \qquad (4.4\text{-}70)$$

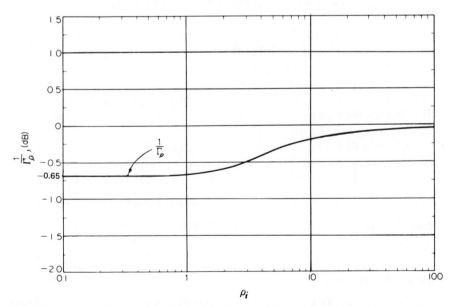

Figure 4.33 Variation of P_c/N_0 as a function of the input SNR, ρ_i (after Lindsey [38]).

The same modification to first- and third-order loops can be made as was done for the second-order loop above. The main difference is in the effect of the limiter on the closed-loop transfer function and the loop bandwidth.

PROBLEM 17

(a) If we let the "design point" be denoted by r_0, the suppression factor by α_0, and the loop noise bandwidth $(2B_{L_0})$ by W_{L_0}, show that the closed-loop transfer function, above the design point, is given by

$$H\alpha(s) = \frac{1 + \dfrac{r_0 + 1}{2W_{L_0}}s}{1 + \dfrac{r_0 + 1}{2W_{L_0}}s + \dfrac{\mu}{r_0}\left(\dfrac{r_0 + 1}{2W_{L_0}}\right)^2 s^2}$$

where $\mu = \alpha_0/\alpha$.

(b) Show that the two-sided bandwidth becomes

$$W_L = W_{L_0}\left(\frac{1 + r_0/\mu}{1 + r_0}\right)$$

(c) Compute σ_ϕ^2 using linear PLL theory when $N_0 = -168$ dBW-Hz, $C = -148$ dBW, $r_0 = 2$, $W_{L_0} = 20$ Hz, and $\rho_i = 3$ dB.

(d) Do the same calculation assuming a bandpass limiter precedes the PLL.

4.5 CARRIER TRACKING WITH A CW INTERFERENCE

This section assesses the impact of a continuous wave (cw) interference in a phase lock loop. We will find that the cw interferer can degrade or actually break lock if the interfering signal level is high enough. In 1973, Bruno [47] and Blanchard [48] independently performed similar analyses of the effects of a cw interferer on a coherently operating phase locked loop. Later Levitt [49] extended their analyses to the entire lock region. The presentation that follows is based on Levitt's work.

Consider a PLL that is initially locked to a carrier with amplitude A and frequency ω_0. In the presence of a cw interferer at offset frequency $\Delta\omega$, with $\alpha^2 = P_I/A^2$, with P_I the interference power, the PLL input is

$$y(t) = \sqrt{2}A[\sin \omega_0 t + \alpha \sin(\omega_0 + \Delta\omega)t] \tag{4.5-1}$$

Neglecting the $2\omega_0$ term, since the filter will remove it, the loop error signal is given by

$$\epsilon(t) = K_1\sqrt{2}y(t) \cos[\omega_0 t - \phi(t)] \tag{4.5-2}$$

with K_1 the phase detector gain. So

$$\epsilon(t) = K_1 A[(1 + \alpha \cos \Delta\omega t) \sin \phi(t) + \alpha \sin \Delta\omega t \cos \phi(t)] \tag{4.5-3}$$

where $\phi = -\hat{\theta}(t)$. Using the Heaviside operator $s = d/dt$, the loop phase error is defined by

$$\phi(t) = -K_1 K_{VCO} \left[\frac{F(s)}{s} \right] \epsilon(t) \tag{4.5-4}$$

where $F(s)$ is the loop filter transfer function and K_{VCO} is the gain of the voltage controlled oscillator (VCO).

Equations 4.5-3 and 4.5-4 cannot be solved analytically for $\phi(t)$ for all values of α and $\Delta\omega$. However, based on experimental evidence, Bruno and Blanchard considered the solution

$$\phi(t) = \theta_0 + \sigma \sin(\Delta\omega t + \nu) \qquad \text{rad} \tag{4.5-5}$$

where θ_0 is constrained to lie in the interval $(-\pi/2, \pi/2)$. Equation (4.5-5) implies that the effect of the interferer is to induce a static phase error along with a beat note at the difference frequency.

Substituting (4.5-5) into (4.5-2) and (4.5-3), and equating the dc, $\sin \Delta\omega t$ $\cos \Delta\omega t$ coefficients, produces

$$\sin \theta_0 = -\frac{\sigma^2 \delta \cos \Psi}{2J_0(\sigma)} \tag{4.5-6}$$

$$\sin(\theta_0 - \nu) = -\frac{\sigma^2 \delta \cos \Psi}{2\alpha J_1(\sigma)} \tag{4.5-7}$$

$$\left[\frac{\sigma\delta \sin \Psi + 2J_1(\sigma) \cos \theta_0}{J_0(\sigma) - J_2(\sigma)} \right]^2 + \left[\frac{\sigma^2 \delta \cos \Psi}{2J_1(\sigma)} \right]^2 = \alpha^2 \tag{4.5-8}$$

where Ψ is the phase angle of $F(j\Delta\omega)$ and δ is the normalized offset frequency:

$$\delta \equiv \frac{\Delta\omega}{AK_1 K_{VCO} |F(j\Delta\omega)|} \tag{4.5-9}$$

The Bessel functions in (4.5-6)–(4.5-8) are due to the Bessel function expansions of $\sin[\sigma \sin(\Delta\omega t + \nu)]$ and $\cos[\sigma \sin(\Delta\omega t + \nu)]$. This is accurate only if components at $2\Delta\omega$ and higher are negligible, which requires that σ be of the order of unity or less. Then terms of the form $J_n(\sigma)(n \geq 2)$ will be negligible compared to $J_0(\sigma)$ and $J_1(\sigma)$. This is not a severe limitation since larger values of σ correspond to values of θ_0 near $\pi/2$, where it is clear the loop will not remain in lock.

In the absence of noise, both Bruno and Blanchard determined that $\theta_0 = \pi/2$ is the point at which break lock occurs. Hence, for a given $\Delta\omega$, the loop remains in lock for $|\theta_0| \leq \pi/2$, $\sigma \leq \sigma_0$, or $\alpha \leq \alpha_0$. At $|\theta_0| = \pi/2$ we have from (4.5-6) that

$$\frac{\sigma_0^2}{J_0(\sigma_0)} = \frac{2}{|\delta \cos \psi|}] \Rightarrow \sigma_0 \tag{4.5-10}$$

Then using the solution of σ_0 from (4.5-10) in (4.5-8) with $|\theta_0| = \pi/2$ yields

$$\alpha_0^2 = \left[\frac{J_0(\sigma_0)}{J_1(\sigma_0)} \right]^2 \left\{ 1 + \left(\frac{2 \tan \psi}{\sigma_0} \right)^2 \left[\frac{J_1(\sigma_0)}{J_0(\sigma_0) - J_2(\sigma_0)} \right]^2 \right\} \tag{4.5-11}$$

Levitt [46] has shown that when

$$\frac{\sigma_0^2}{2}\left| \sin^2 \psi - \frac{\sigma_0^2}{48} \right| \ll 1 \tag{4.5-12}$$

then α_0^2 is given by

$$\alpha_0^2 \cong 2\left| \frac{\delta}{\cos \psi} \right| \tag{4.5-13}$$

PROBLEM 18

Establish the condition (4.5-12) by combining (4.5-10) and (4.5-11) to yield

$$\frac{\alpha_0^2}{2\left| \dfrac{\delta}{\cos \psi} \right|} = J_0(\sigma_0)\left\{ \left[\frac{\cos \psi}{2J_1(\sigma_0)/\sigma_0} \right]^2 + \left[\frac{\sin \psi}{J_0(\sigma_0) - J_2(\sigma_0)} \right]^2 \right\}$$

and then make the appropriate approximation for the Bessel functions.

In conclusion, solving for (4.5-10) produces σ_0, which allows one to determine the value of α_0 implicitly from (4.5-11). However, if σ_0 satisfies (4.5-12), then (4.5-13) can be used for the explicit solution of $\alpha_0^2 = P_I/A^2$, that is, the interference-to-signal power when break lock occurs in the absence of noise. Break lock would occur at a smaller value of α_0 when noise is present.

APPENDIX   STEADY-STATE SOLUTION OF THE FIRST-ORDER LOOP PHASE ERROR

The purpose of this appendix is to present a derivation of the Fokker-Planck equation (FPE) and its solution in the steady-state case. Before we present the FPE we briefly review Markov processes.

Markov Processes

A Markov process has the property that the probability density function of the process at any future instant, conditional on the past and present values of the process, is actually independent of the past values.

More specifically, let $x(t)$ be a random process. Let

$$x(t_0) = x_0 \qquad x(t_{-1}) = x_{-1} \qquad x(t_{-2}) = x_{-2}, \ldots \tag{A-1}$$

Then when $x(t)$ is a *Markov process* it is true that

$$p(x_1/x_0, x_{-1}, x_{-2}, x_{-3}, \ldots) = p(x_1/x_0) \tag{A-2}$$

Fokker-Planck Equation

The development of the FPE and of the theory of stochastic processes in general began in the nineteenth century when physicists were attempting to prove that heat in a medium is essentially a random motion of the constituent molecules. Rayleigh in the late 1890s obtained a partial differential equation for the probability density of a random walk problem. In addition, he applied similar techniques to the theory of gases to develop what was later called the FPE. Bachelier made a mathematical model of the French Stock Exchange and obtained a simple case of the FPE. The work of Rayleigh and Bachelier seems to have been relatively unnoticed by contempory physicists, who subsequently proceeded to develop the FPE for themselves. Einstein obtained a partial differential equation for the probability density function for a one-dimensional FPE. The theory of Brownian motion, named after an English botantist named Robert Brown, was subsequently generalized by many physicists, notably by, Smoluchowski, Langevin, Fokker, Planck, Ornstein, Burger, Fürth, and Kramers.

The FPE is a partial differential equation (derived from a stochastic differential equation) whose solution, at any time t, is the probability density function of the dynamical process described by the stochastic differential equation. The FPE [50, 51] for a first-order stochastic differential equation describing a Markov process is given by

$$\frac{\partial p(y, t)}{\partial t} = -\frac{\partial}{\partial y} [a_1(y)p(y, t)] + \frac{1}{2} \frac{\partial^2}{\partial y^2} [a_2(y)p(y, t)] \qquad \text{(A-3)}$$

with the initial condition

$$p(y, 0) = \delta(y - y_0) \qquad \text{(A-4)}$$

The coefficients $a_2(y)$ and $a_1(y)$ are defined by

$$a_n(y) = \lim_{\Delta t \to 0} \frac{E[(\Delta y)^n / y]}{\Delta t} \qquad \text{(A-5)}$$

where y is the Markov process under study and Δy is the change in the value of y at time t to the value of y at time $t + \Delta t$.

Steady-State Phase Error Density Function, First-Order PLL

We now apply the FPE to the stochastic differential equation of the loop [4]

$$\theta(t) = \phi(t) + \frac{F(s)}{s} [AK \sin \phi + Kn'(t)] \qquad \text{(A-6)}$$

Rearranging, we have

$$\dot{\theta}(t) = \phi(t) + AK \sin \phi + Kn'(t) \qquad \text{(A-7)}$$

where we have used the fact that for a first-order PLL, $F(s) = 1$.

First we establish that (A-7) defines a Markov process in the variable $\phi(t)$. Rewriting (A-6) as

$$\dot{\phi}(t) = \dot{\theta}(t) - AK \sin \phi - AKn'(t) \qquad \text{(A-8)}$$

and integrating (A-8) formally, we have

$$\phi(t) - \phi(0) = \theta(t) - \theta(0) - AK \int_0^t \sin \phi(u) \, du - K \int_0^t n'(u) \, du \quad \text{(A-9)}$$

Therefore, any future value depends only on the present value $\phi(0)$, when the noise process is white. The integrated white noise forms an "independent increment" process, so that knowledge of the past does not enhance the statistical knowledge of the future. We therefore conclude that $\phi(t)$ is a Markov process.

Now we can evaluate the moments. Let $\theta(t) = (\omega - \omega_0)t + \theta_0$, so that $\dot{\theta}(t) = \omega - \omega_0$. Then integrating (A-8), we have

$$\Delta\phi = \phi(t + \Delta t) - \phi(t) = [(\omega - \omega_0) - AK \sin \phi]\Delta t - K \int_t^{t+\Delta t} n'(u) \, du$$

$$\text{(A-10)}$$

Hence

$$a_1(\phi) = \lim_{\Delta t \to 0} \frac{E_\phi[(\omega - \omega_0) - AK \sin \phi]\Delta t - K \int_t^{t+\Delta t} n'(u) \, du}{\Delta t} \quad \text{(A-11)}$$

with $E_\phi(\,\cdot\,)$ denoting $E[(\,\cdot\,)/\phi]$. Hence we have

$$a_1(\phi) = \omega - \omega_0 - AK \sin \phi \quad \text{(A-12)}$$

Now

$$a_2(\phi) = \lim_{\Delta t \to 0} \frac{E[(\Delta\phi)^2/\phi]}{\Delta t} = \lim_{\Delta t \to 0} \frac{K^2}{\Delta t} \int_t^{t+\Delta t} \int_t^{t+\Delta t} E[n'(u)n'(v)] \, du \, dv$$

$$\text{(A-13)}$$

Since

$$E[n'(u)n'(v)] = \frac{N_0}{2} \delta(u - v)$$

we have

$$a_2(\phi) = \frac{K^2 N_0}{2} \quad \text{(A-14)}$$

It may be shown that $a_n(\phi) = 0$ for $n > 2$. Using $a_1(\phi)$ and $a_2(\phi)$ in our FPE (A-13), we have

$$\frac{\partial p(\phi)}{\partial t} = -\frac{\partial}{\partial \phi} [(\omega - \omega_0 - AK \sin \phi)p(\phi)] + \frac{K^2 N_0}{4} \frac{\partial^2 P(\phi, t)}{\partial \phi^2} \quad \text{(A-15)}$$

This FPE has not been solved analytically. However, La Frieda [52] has used a separation of variables technique to obtain the solution with the aid of the computer.

However, since our interest here is to obtain the steady-state solution, we let $\partial p(\phi)/\partial t = 0$. In addition we choose to obtain a solution mod 2π so that we, in effect, reduce the phase error to the region $-\pi$ to π. To do this we let

$$P(\phi, t) = \sum_{n=-\infty}^{\infty} p(\phi + 2\pi n, t) \tag{A-16}$$

Since $p(\phi, t)$ is a solution to the FPE, so is $P(\phi, t)$. Hence we solve

$$\frac{\partial P(\phi, t)}{\partial t} = \frac{\partial}{\partial \phi} (\omega - \omega_0 + AK \sin \phi) P(\phi, t) + \frac{K^2 N_0}{4} \frac{\partial^2 P(\phi, t)}{\partial \phi^2} \tag{A-17}$$

for $\phi \in (-\pi, \pi)$, with

$$P(\phi) = \delta(\phi)$$
$$P(\pi, t) = P(-\pi, t) \tag{A-18}$$

and in addition we require

$$\int_{-\pi}^{\pi} P(\phi, t) d\phi = 1 \qquad \forall t \tag{A-19}$$

In the steady state we have

$$P(\phi, t) = P(\phi) \tag{A-20}$$

Our partial differential equation reduces to

$$0 = \frac{d}{d\phi} [(\omega_0 - \omega + AK \sin \phi) P(\phi)] + \frac{K^2 N_0}{4} \frac{d^2 P(\phi)}{d\phi^2} \tag{A-21}$$

If the initial frequency detuning $(\omega_0 - \omega)$ is assumed to be zero, we have (integrating once)

$$0 = AK \sin \phi P(\phi) + \frac{K^2 N_0}{4} \frac{dP(\phi)}{d\phi} \tag{A-22}$$

Rearranging and integrating yields

$$P(\phi) = C \exp(\rho \cos \phi) \qquad \rho = \frac{4A}{KN_0} \tag{A-23}$$

Since

$$\int_{-\pi}^{\pi} P(\phi) d\phi = 1$$

we have that

$$P(\phi) = \frac{\exp(\rho \cos \phi)}{2\pi I_0(\rho)} \qquad -\pi \leq \phi < \pi \tag{A-24}$$

where

$$\rho = \frac{4A}{KN_0} = \frac{A^2}{N_0(AK/4)} = \frac{A^2}{N_0 B_L} \tag{A-25}$$

and is the loop SNR in B_L Hz. The above density function is usually called the *Tikhonov density function* in honor of the Russian scientist who obtained the first-order PLL solution.

PROBLEM A.1

Show that when ω-ω_0 is not zero the probability density function becomes

$$P(\phi) = \frac{\exp(\rho \cos \phi + B\phi)}{2\pi I_0(\rho)}\left[1 + D \int_{-\pi}^{\phi} \exp(-\rho \cos x - Bx)dx\right] \quad \text{(A-26)}$$

with

$$D = \frac{\exp(-2\pi B) - 1}{\displaystyle\int_{-\pi}^{\pi} \exp[-(\rho \cos x + Bx)]dx} \quad \text{(A-27)}$$

Higher-Order Loops

FPEs can be written for nth-order PLLs. However, exact solutions have not been found. Yet, Lindsey and Charles [37] have experimentally verified that the second-order PLL has a density function very close to the Tikhonov density function. For more details on nth-order loops, reference 6 should be consulted.

REFERENCES

1   DeBellescize, H., "La Reception Synchrone," *H. Onde Electrique*, June 11, 1932, pp. 225–240.

2   Wendt, K. R., and Fredendall, G. L., "Automatic Frequency and Phase Control of Synchronization in Television Receivers," *Proc. IRE*, Vol. 31, pp. 7–15, January 1943.

3   Cordell, R. R., "Integrated Circuit Phase-Locked Loops," National Telemetry Conference, Philadelphia, June 1976.

4   Viterbi, A. J., *Principles of Coherent Communication*, McGraw-Hill, New York, 1966.

5   Sakrison, D. J., *Communication Theory: Transmission of Waveforms and Digital Information*, John Wiley, New York, 1968.

6   Lindsey, W. C., *Synchronization Systems in Communication and Control*, Prentice-Hall, Englewood Cliffs, N.J., 1972.

7   Holmes, J. K., "On a Solution to the Second-Order Phase-Locked Loop," *IEEE Trans. Communications Tech.*, Vol. 18, No. 2, April 1970.

8   Tausworthe, R. C., "Theory and Design of Phase Locked Receivers," Vol. I, Tech. Report No. 32-819, February 15, 1966.

9   Blanchard, A., *Phase Locked Loops: Application to Coherent Receiver Design*, Wiley-Interscience, New York, 1976.

10 Tausworthe, R. C., and Crow, R. B., "Practical Design of Third-Order Tracking Loops," Interim Report 900-450, April 27, 1971.

11 Ghais, A., Myers, M., and Meer, S., "Phase-Locked Loops for the Apollo Vehicle Re-entry Tracking System," AD/COM Tech. Report No. G-63-7, May 3, 1976.

12 Savant, C. J., *Basic Feedback Control System Design*, McGraw-Hill, New York 1958, Chap. 4.

13 Tausworthe, R. C., and Crow, R. B., "Improvements in Deep Space Tracking by the use of Third-Order Loops", ITC 72 Conference Record, pp. 577–583, 1972.

14 Newton, Gould, and Kaiser, *Analytical Design of Linear Feedback Controls*, Wiley-Interscience, New York, 1967.

15 Hoffman, L. A., "Receiver Design and the Phase Lock Loop," Aerospace Corporation, El Segundo, Cal, May 1963.

16 Gardner, F. M., *Phaselock Techniques*, Wiley, New York, 1966.

17 Goldman, S. L., "Jerk Response of a Third-Order Phase Lock Loop," *IEEE Trans. Aerospace and Electronic Systems*, March 1976.

18 Barnes, J. A., and Allen, D. W., "A Statistical Model of Flicker Noise," *Proc. IEEE*, Vol. 54, No. 3, February 1966.

19 Baghdady, E. J., Lincoln, R. N., and Nelin, B. D., "Short Term Frequency Stability: Characterization Theory and Measurement", NASA Symposium on Stability, November 1964.

20 Barnes, J., Chi, A., *et al.*, "Characterization of Frequency Stability," *IEEE Trans. Instrumentation and Measurement*, Vol. IM-20, No. 2, May 1971.

21 Gray, R. M., and Tausworthe, R. C., "Frequency-Counted Measurements and Phase Locking to Noisy Oscillators", *IEEE Trans. Communications.*, Vol. COM-19, No. 1, February 1971.

22 Jaffee, R. M., and Rechtin, E., "Design and Performance of Phase-Locked Circuits Capable of Near-Optimum Performance over a Wide Range of Input Signal and Noise Levels," *IRE Trans. Information Theory*, Vol. IT-1, pp. 66–76, March 1955.

23 Yovits, M. C., and Jackson, J. L., "Linear Filter Optimization with Some Game-Theoretic Considerations," 1955 IRE National Convention Record, Part 4, pp. 193–199.

24 Smith, B. M., "The Phase Lock Loop with Filter: Frequency of Skipping Cycles", Letters to the Editor, *Proc. IEEE*, pp. 295, 1966.

25 Holmes, J. K., "First Slip Times vs Static Phase Error Offset for the First- and Second-Order Phase Locked Loop," *IEEE Communications. Tech.*, April 1971.

26 Baghdady, E. J., "Advanced Threshold Reduction Techniques Study," First Quarter Report, NASA Report N65-30842, Adcom. Inc., Cambridge, Mass., 1964.

27 Klapper J., and Frankle, J. T., *Phase-Locked and Frequency-Feedback Systems*, Academic Press, New York, 1972, Chap. 5.

28 Sanneman, R. W., and Rowbotham, J. R., "Unlock Characteristics of the Optimum Type II Phase Locked Loop," *IEEE Trans.*, Vol. ANE-11, March 1964.

29 Tausworthe, R. C., "Simplified Formula for Mean Cycle Slip Time of Phase-Locked Loops with Steady-State Phase Error," *IEEE Trans. Communications*, Vol. COM-30, No. 3, June 1972.

30 Nilsen, P. W., "Acquisition Performance of Second- and Third-Order Phase Lock Loops," Magnavox Research Lab Report No. MX-TM-3105-21, April 1971.

31 Goldman, S. L., "Second-Order Phase-Lock-Loop Acquisition Time in the Presence of Narrowband Gaussian Noise," *IEEE Trans Communications*, April 1973.

32 Hummels, D. R. "Some Simulation Results for the Time to Indicate Phase Lock," *IEEE Trans. Communications*, Vol COM-20, No. 1, February 1972.

33 Frazier, J. P., and Page, J., "Phase Lock Loop Frequency Acquisition Study," Trans. IRE, SET-8, pp. 210–227, September 1962.

34 Papoulis, A. *Probability, Random Variables, and Stochastic Processes*, McGraw-Hill, New York, 1965, Chap. 14.

35 Henderson, K. W., and Kautz, W. H., "Transient Responses of Conventional Filters", *IRE Trans. Circuit Theory*, December 1958.

36 Tikhonov, V. I., "The Operation of Phase and Automatic Frequency Control in the Presence of Noise," *Automation and Remote Control*, Vol. 21, No. 3, pp. 209–214, 1960.

37 Charles, F., and Lindsey, W. C., "Some Analytical and Experimental Phase-Locked Loop Results for Low Signal to Noise Ratios," *Proc. IEEE*, Vol. 55, No. 9, pp. 1152–1166, September 1966.

38 Lindsey, W. C., and Simon, M., *Telecommunication Systems Engineering*, Prentice-Hall, Englewood Cliffs, N. J. 1973.

39 Springett, J. D., and Simon, M. K., "An Analysis of the Phase Coherent/Incoherent Output of the Bandpass Limiter", *IEEE Trans. Communications Tech.*, February 1971.

40 Forney, G. D., "Coding and Coherent Deep Space Telemetry", McDonnel-Douglas Astronautics Corp., Santa Monica, Cal. March 17, 1967.

41 Davenport, W. B., "Signal-to-Noise Ratios in Bandpass Limiters," *J. Applied Physics*, Vol. 24, No. 6, June 1953.

42 Greenhall, C. A., "Signal and Noise in Nonlinear Devices," *JPL SPS* 37–44, Vol. IV, April 1967.

43 Tausworthe, R. C., "Analysis of Narrowband Signals Through the Bandpass Soft Limiter," *JPL SPS* 37–53, Vol. III, October 31, 1968.

44 Galejs, J., "Signal-to-Noise Ratios in Smooth Limiters," *IEEE Trans. Information Theory*, Vol. IT-1, 79–85. June 1959.

45 Baum, R. F., "The Correlation Function of Smoothly Limited Gaussian Noise," *IEEE Trans. Information Theory*, Vol. IT-3, pp. 193–197, September 1959.

46 Springett, J. C., "A Note on Signal-to-Noise and Signal-to-Noise Spectral Density Ratios at the Output of a Filter-Limiter Combination," *JPL SPS* 37–36, Vol. IV, December 31, 1965.

47 Bruno, F., "Tracking Performance and Loss of Lock of a Carrier Loop Due to the Presence of a Spoofed Spread Spectrum Signal," Proc. 1973 Symposium on Spread Spectrum Communications, Vol. I, M. L. Schiff, Ed., Naval Electronics Laboratory Center, San Diego, Cal., pp. 71–75, March 13 to 16, 1973.

48 Blanchard, A., "Interferences in Phase-Locked Loops," *IEEE Trans. Aerospace and Electronic Systems*, Vol. AES-10, pp. 686–697, September 1974.

49 Levitt, B., "Carrier Tracking Loop Performance in the Presence of Strong CW Interference," *JPL DSN* Progress Report, Vol. 42–51, 1979.

50 Fuller, A. T., "Analysis of Nonlinear Stochastic Systems by Means of the Fokker-Planck Equation", *Int. J. Control* Vol. 9, No. 6, pp. 603–655, 1969.

51 Middleton, D., *Introduction to Statistical Communication*, McGraw-Hill, New York, 1960, Chap. 10.

52 La Frieda, J. R., "Transient Analysis of Nonlinear Tracking Systems," University of Southern California, Dept. of EE, Los Angeles, June 1970.

5

TRACKING WITH SUPPRESSED CARRIER LOOPS

We have discussed the various aspects of tracking and acquisition performance of residual carrier loops in the last chapter. In this chapter, we consider tracking of suppressed carrier systems, that is, modulation schemes that have no residual carrier component to which a cw loop can track. By using suppressed carrier modulation, all of the power is put into the data, and thereby no energy is wasted on a carrier component.

By utilizing a loop in the system design designed to operate on a two-phase or a four-phase signal, successful tracking can be achieved. In this chapter, we will present and analyze various types of loops designed to track two-phase and four-phase signals.

A suppressed carrier signal is described by (using NRZ-L baseband symbols)

$$s(t) = \sqrt{2} \, Am(t) \sin(\omega_0 t + \theta_0)$$

where $m(t) = \pm 1$ during the symbol time and A is the rms amplitude, ω_0 the carrier frequency (rad/sec), and θ_0 is the unknown phase. The power spectral density of a suppressed carrier signal is shown in Figure 5.1, which illustrates the fact that there is no discrete residual carrier component left to lock up an ordinary phase lock loop. We will see that the suppressed carrier loops "reconstruct" the carrier by nonlinearly operating on the received signal.

First we consider the squaring loop.

5.1 SQUARING LOOP

Conceptually the simplest method by which to obtain a discrete spectral component is to square the input signal, to create a line spectrum at twice the carrier frequency. Numerous authors have analyzed the squaring loop [1–5, 8, 9].

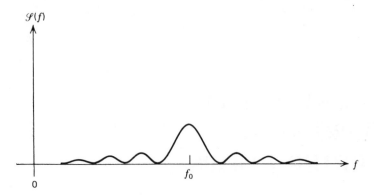

Figure 5.1   Typical power spectral density with no discrete residual component.

First we compute the stochastic differential equation (SDE) governing the operation of the squaring loop. Then, from the SDE, we obtain the phase error variance. Consider the model of the squaring loop shown in Figure 5.2*a* and Figure 5.2*b*. The bandpass filter preceding the squaring device has a positive frequency noise bandwidth of W-Hz. Following the squaring operation, a zonal filter is used to remove the baseband signals generated out of the squarer. The loop filter has Laplace transform $F(s)$. Let the input $U(t)$ be described by

$$U(t) = \sqrt{2}\, Am(t) \sin(\omega_0 t + \theta_0) + n(t) \qquad (5.1\text{-}1)$$

where A^2 is the total signal power, θ_0 is a constant, and $m(t)$ is the data sequence, where

$$m(t) = \sum_{k=-\infty}^{\infty} C_k p(t - kT) \qquad (5.1\text{-}2)$$

where C_k is a statistically independent random sequence taking the values of ± 1 with probability 1/2. The term $p(t)$ is the basic baseband symbol of duration T sec. We now consider the loop of Figure 5.2*a*. The noise $n(t)$ is modeled as white Gaussian noise (WGN) having a two-sided noise spectral density $N_0/2$. We express the noise out of the bandpass filter in the usual way as

$$n(t) = \sqrt{2}\, n_c(t) \cos(\omega_0 t + \theta_0) + \sqrt{2}\, n_s(t) \sin(\omega_0 t + \theta_0) \qquad (5.1\text{-}3a)$$

and the filtered version is given by

$$\tilde{n}(t) = \sqrt{2}\, \tilde{n}_c(t) \cos(\omega_0 t + \theta) + \sqrt{2}\, \tilde{n}_s(t) \sin(\omega_0 t + \theta_0) \qquad (5.1\text{-}3b)$$

where $\tilde{n}_c(t)$ and $\tilde{n}_s(t)$ are the filtered versions of $n_c(t)$ and $n_s(t)$, respectively. The filtered signal plus filtered noise is modeled as

$$x(t) = \sqrt{2}\, A\tilde{m}(t) \sin(\omega_0 t + \theta_0) + \tilde{n}(t) \qquad (5.1\text{-}4)$$

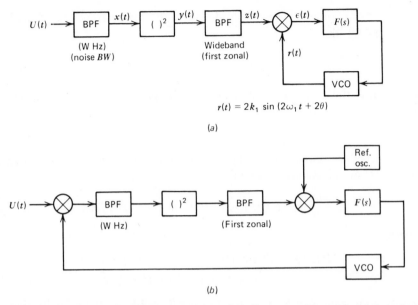

$$r(t) = 2k_1 \sin(2\omega_1 t + 2\theta)$$

(a)

(b)

Figure 5.2 Model for squaring loop; (a) short-loop version and (b) long-loop version.

where $\tilde{m}(t)$ is the filtered version of the baseband modulation sequence due to a baseband equivalent filter.

Following the squaring operation, we obtain

$$
\begin{aligned}
y(t) = &-[A\tilde{m}(t)]^2 \cos(2\omega_0 t + 2\theta_0) + 2A\tilde{m}(t)\tilde{n}_c(t)\sin(2\omega_0 t + 2\theta_0) \\
&- 2A\tilde{m}(t)\tilde{n}_s(t)\cos(2\omega_0 t + 2\theta_0) + \tilde{n}_c^2(t)\cos(2\omega_0 t + 2\theta_0) \\
&- \tilde{n}_s^2(t)\cos(2\omega_0 t + 2\theta_0) + 2\tilde{n}_s(t)\tilde{n}_c(t)\sin(2\omega_0 t + 2\theta_0) \\
&+ \text{baseband signals}
\end{aligned}
\tag{5.1-5}
$$

and thereby reconstruct a line spectrum at $f = 2f_0$.

The wideband filter following the squaring operation is assumed to remove the baseband components. Denote $\omega_0 t + \theta_0$ by Φ and $\omega_1 t + \hat{\theta}$ by $\hat{\Phi}$. Then the reference signal $r(t)$ out of the VCO is given by

$$r(t) = 2K_1 \sin(2\hat{\Phi}) \tag{5.1-6}$$

where $\sqrt{2}K_1$ is the rms value of the oscillator output voltage. The loop error signal out of the phase detector is given by

$$\epsilon(t) = K_m z(t) r(t) \tag{5.1-7}$$

where K_m is the phase detector gain. Hence the error signal $\epsilon(t)$ becomes

$$
\begin{aligned}
\epsilon(t) = K_m K_1 (\{A^2[\tilde{m}(t)]^2 &+ 2A\tilde{m}(t)\tilde{n}_s(t) - \tilde{n}_c^2(t) + \tilde{n}_s^2(t)\} \sin 2\phi \\
&+ [2A\tilde{m}(t)\tilde{n}_c(t) + 2\tilde{n}_c(t)\tilde{n}_s(t)] \cos 2\phi)
\end{aligned}
\tag{5.1-8}
$$

where $\phi = \Phi - \hat{\Phi}$ and is the instantaneous phase error at time t.

The instantaneous frequency relative to the center frequency $2\omega_0$ of the loop is given by

$$2\frac{d\hat{\Phi}}{dt} - 2\omega_1 = K_{VCO}F(s)\epsilon(t) \qquad (5.1\text{-}9)$$

where ω_1 is the zero input VCO nominal frequency. When $\omega_1 = \omega_0$, (5.1-9) becomes

$$2\dot{\phi} = -K_{VCO}F(s)\epsilon(t) \qquad (5.1\text{-}10)$$

From (5.1-9) we have, assuming $\omega_1 = \omega_0$ (which is equivalent to assuming that the static phase error is negligible)

$$2\frac{d\phi}{dt} = -KF(s)\{A^2[\tilde{m}(t)]^2 \sin 2\phi + N(t, 2\phi)\} \qquad (5.1\text{-}11)$$

where

$$N(t, 2\phi) = [2A\tilde{m}(t)\tilde{n}_s(t) - \tilde{n}_c^2(t) + \tilde{n}_s^2(t)] \sin 2\phi$$
$$+ 2[A\tilde{m}(t)\tilde{n}_c(t) + \tilde{n}_c(t)\tilde{n}_s(t)] \cos 2\phi \qquad (5.1\text{-}12)$$

and the loop gain $K = K_m K_{VCO} K_1$. Letting $\Phi_e = 2\phi$, we have finally that

$$\frac{d\Phi_e}{dt} + KPF(s)[\tilde{m}(t)]^2 \sin \Phi_e = -KF(s)N(t, \Phi_e) \qquad (5.1\text{-}13)$$

We shall see that the SDE for the squaring loop is the same for the Costas loop if we assume that the baseband data filter of the Costas loop is equivalent to a frequency translate of the presquaring filter of the squaring loop. This is normally an excellent assumption.

In order to obtain the phase error variance, it is necessary to separate the signal components from the noise components. Assuming that the symbol rate $R_s (R_s = 1/T)$ is large compared to B_L, it is possible to compute the complete spectra of $[\tilde{m}(t)]^2$, which are derived in the Appendix. However, here we will use the time and ensemble average of $[\tilde{m}(t)]^2$. First we add and subtract the time and ensemble average (which we donote by an overbar) from both sides of (5.1-13) to yield

$$\dot{\Phi}_e + KPF(s)\overline{[\tilde{m}(t)]^2} \sin \Phi_e = -KF(s)(N(t, \Phi_e) + \{(\tilde{m}(t)]^2 - \overline{[\tilde{m}(t)]^2}\} \sin \Phi_e)$$

$$(5.1\text{-}14)$$

Now we evaluate $\tilde{m}(t)$ by noting first from (5.1-2), that the filtered version of $m(t)$ is given by

$$\tilde{m}(t) = \sum_{k=-\infty}^{\infty} C_k q(t - kT) \qquad (5.1\text{-}15)$$

with C_k being a random data sequence independent from sample to sample and taking on the value 1 with probability 1/2 and the value -1 with probability 1/2. The signal term $q(t)$ is a baseband (equivalent) filtered version of $p(t)$. That is, $q(t) = 1/2\pi \int_{-\infty}^{\infty} P(\omega)H_{LP}(\omega)e^{j\omega t}d\omega$ with $P(\omega)$ given by $P(\omega) =$

$F\{p(t)\}$ and $H_{LP}(\omega)$ the low pass equivalent of the bandpass filter prior to the squarer.

Squaring and taking the ensemble and time average, we obtain

$$\overline{[\tilde{m}(t)]^2} = \frac{1}{T} \int_0^T \sum_{-\infty}^{\infty} q^2(t - kT)dt \tag{5.1-16}$$

or

$$\overline{[\tilde{m}(t)]^2} = \frac{1}{T} \int_0^{\infty} q^2(t)dt \tag{5.1-17}$$

with the lower limit being zero since $q(t)$ is zero for $t \le 0$. Using Parseval's theorem, we can rewrite (5.1-17) as

$$\overline{[\tilde{m}(t)]^2} = \frac{1}{T} \int_{-\infty}^{\infty} |P(f)|^2 |H_{LP}(f)|^2 df = \alpha \tag{5.1-18}$$

where $P(f)$ is the Fourier transform of $p(t)$ and $H_{LP}(f)$ is the baseband equivalent filter (of the bandpass) transfer function.

Since we recognize $1/T|P(f)|^2 = \mathcal{S}_m(f)$ as the power spectral density of the $m(t)$ sequence, we can write (5.1-18) as

$$\alpha = \overline{[\tilde{m}(t)]^2} = \int_{-\infty}^{\infty} \mathcal{S}_m(f)|H_{LP}(f)|^2 df \tag{5.1-19}$$

In order to determine the linearized loop tracking performance, we must linearize (5.1-14) to yield

$$\Phi_e = \frac{\alpha PKF(s)}{S + \alpha PKF(s)} \left[\frac{N(t, \Phi_e)}{\alpha P} + \frac{D(t, \Phi_e)}{\alpha P} \right] \tag{5.1-20}$$

where $D(t, \Phi_e)$, the self-noise distortion due to the filtering effect on $p(t)$, is given by

$$D(t, \Phi_e) = \{[\tilde{m}(t)]^2 - \overline{[\tilde{m}(t)]^2}\}\Phi_e \tag{5.1-21}$$

To complete our analysis we must determine the power spectral density of $N(t, \Phi_e)$ and $D(t, \Phi_e)$ near $f = 0$ under the assumption that $W \gg B_L$ (so that the spectral density is nearly constant about $f = 0$). First consider the autocorrection of the noise

$$R_N(\tau) = E[N(t)N(t + \tau)] \tag{5.1-22}$$

Assume that $\phi(t)$ is very narrowband, so that it is constant relative to $N(t)$. First, we have that

$$R_N(\tau) = R_{N_1}(\tau) \sin^2 2\phi + 4R_{N_2}(\tau) \cos^2 2\phi + 4R_{N_1 N_2}(\tau) \sin 2\phi \cos 2\phi \tag{5.1-23}$$

where

$$\begin{aligned} N_1(t) &= -\tilde{n}_c^2 + \tilde{n}_s^2 + 2A\tilde{m}\tilde{n}_s \\ N_2(t) &= A\tilde{m}\tilde{n}_c + \tilde{n}_c\tilde{n}_s \end{aligned} \tag{5.1-24}$$

Now, since the average of an odd number of zero mean of Gaussian random variables is zero and since $\tilde{n}_c(t)$ and $\tilde{n}_s(t)$ are uncorrelated, we obtain

$$R_{N_1 N_2}(\tau) = 0 \tag{5.1-25}$$

It is not difficult to show that

$$R_{N_1} = 4A^2 R_{\tilde{m}}(\tau) R_{\tilde{n}_c}(\tau) + 4R_{\tilde{n}_c}^2(\tau)$$
$$R_{N_2}(\tau) = 4R_{\tilde{n}_c}^2(\tau) + 4A^2 R_{\tilde{m}}(\tau) R_{\tilde{n}_c}(\tau) \tag{5.1-26}$$

Therefore, we conclude that the autocorrelation of $N(t)$ is given by*

$$R_N(\tau) = 4[R_{\tilde{n}_c}^2(\tau) + PR_{\tilde{n}_c}(\tau) R_{\tilde{m}}(\tau)] \tag{5.1-27}$$

The spectral density at $f = 0$ is given by

$$\mathscr{S}_n(0) = 4 \int_{-\infty}^{\infty} R_{\tilde{n}_c}^2(\tau) d\tau + 4A^2 \int_{-\infty}^{\infty} R_{\tilde{n}_c}(\tau) R_{\tilde{m}}(\tau) d\tau \tag{5.1-28}$$

Using Parseval's theorem, we obtain

$$\mathscr{S}_N(0) = N_0^2 \int_{-\infty}^{\infty} |H_{LP}(f)|^4 df + 2A^2 N_0 \int_{-\infty}^{\infty} \mathscr{S}_m(f) |H_{LP}(f)|^4 df \tag{5.1-29}$$

Now consider the self-noise or the distortion term:

$$\{[\tilde{m}(t)]^2 - \overline{[\tilde{m}(t)]^2}\} \sin \Phi_e$$

The spectra are obtained in the Appendix and can be shown to contain both line spectra at $f_K = K/T$ $(K = 1, 2, \ldots)$ and a continuous part that acts as self-noise within the loop. The self-noise distortion spectra at $\omega = 0$ is approximated by

$$\sigma_{\Phi_e}^2 \mathscr{S}_c(0) = 4T \sum_{l=1}^{\infty} \left[\int_{-\infty}^{\infty} \mathscr{S}_m(\omega) |H_{LP}(\omega)|^2 e^{j\omega lT} \frac{d\omega}{2\pi} \right]^2 \sigma_{\Phi_e}^2 \tag{5.1-30}$$

or, equivalently,

$$\mathscr{S}_c(0) = 4T \sum_{l=1}^{\infty} R_d^2(lT) \qquad R_d(\tau) = \int_{-\infty}^{\infty} \mathscr{S}_m(\omega) |H_{LP}(\omega)|^2 e^{j\omega\tau} \frac{d\omega}{2\pi} \tag{5.1-31}$$

where $\sigma_{\Phi_e}^2$ is the variance of the phase error $\Phi_e = 2\phi$, which is approximately equal to the variance of $\sin[\Phi_e(t)]$ for small phase errors, and $R_d(\tau)$ is the autocorrelation function of the filtered data. We have assumed $B_L \ll W$. We have

$$\sigma_{\Phi_e}^2 = \frac{\{\mathscr{S}_N(0) + \mathscr{S}_c(0)\sigma_{\Phi_e}^2\}2B_L}{(\alpha P)^2} \tag{5.1-32}$$

where B_L is the one-sided, closed-loop noise bandwidth. Therefore, we have, from (5.1-29), (5.1-30), and (5.1-32), the result

$$\sigma_{\Phi_e}^2 = \frac{\dfrac{4N_0 B_L}{\alpha P}\left(\dfrac{\alpha'}{\alpha} + \dfrac{N_0 B_{LP}'}{\alpha P}\right)}{1 - \dfrac{8TB_L}{(\alpha P)^2} \displaystyle\sum_{l=1}^{\infty} R_d^2(lT)} \qquad \text{rad}^2 \tag{5.1-33}$$

---

*Actually, we have approximated a cyclostationary process by a stationary process in the signal times noise term.

where

$$2B'_{LP} = \int_{-\infty}^{\infty} |H_{LP}(f)|^4 df \qquad (5.1\text{-}34)$$

and

$$R_d^2(lT) = \int_{-\infty}^{\infty} \mathscr{S}_m(f)|H_{LP}(f)|^2 e^{j2\pi lT} df \qquad (5.1\text{-}35)$$

The denominator of (5.1-33) is essentially unity under normal design conditions. We now consider an n-pole Butterworth filter as the baseband equivalent of the bandpass filter having a transfer function satisfying

$$|H_{LP}(f)|^2 = \frac{1}{1 + (f/f_0)^{2n}} \qquad (5.1\text{-}36)$$

Using the fact that the denominator is near unity, we obtain from Problem 1

$$\sigma_{\Phi_e}^2 = \frac{4N_0 B_L}{\alpha P} \left[\frac{\alpha'}{\alpha} + \frac{N_0 W}{2\alpha P} \left(1 - \frac{1}{2n}\right) \right] \qquad \text{rad}^2 \qquad (5.1\text{-}37)$$

where

$$\alpha = \int_{-\infty}^{\infty} \mathscr{S}_m(\omega)|H_{LP}(\omega)|^2 \frac{d\omega}{2\pi}$$

$$\alpha' = \int_{-\infty}^{\infty} \mathscr{S}_m(\omega)|H_{LP}(\omega)|^4 \frac{d\omega}{2\pi} \qquad n \geq 2 \qquad (5.1\text{-}38)$$

$$W = 2B_{LP} \qquad \alpha' \leq \alpha$$

It is to be noted that $\sigma_{\Phi_e}^2$ is the actual phase error variance in the cw loop of Figure 5.2a; however, this is relative to $2\omega_0$. As far as data demodulation is concerned, the carrier phase error must be divided by 2, so that σ_ϕ^2 is the important parameter for bit error rate estimates.

PROBLEM 1

Show that, for an n-pole Butterworth filter, that

$$B'_{LP} = B_{LP}\left(1 - \frac{1}{2n}\right) = \frac{W}{2}\left(1 - \frac{1}{2n}\right)$$

so that (5.1-37) follows.

To obtain the phase error variance of ϕ, we let $\phi = \Phi_e/2$, so that

$$\sigma_\phi^2 = \frac{N_0 B_L}{\alpha P} \left[\frac{\alpha'}{\alpha} + \frac{N_0 W}{2\alpha P} \left(1 - \frac{1}{2n}\right) \right] \qquad \text{rad}^2 \qquad (5.1\text{-}39)$$

for the n-pole Butterworth filter case where

$$2B_{LP} = \int_{-\infty}^{\infty} |H_{LP}(f)|^2 df \qquad (5.1\text{-}40)$$

In the general case,

$$\sigma_\phi^2 = \frac{N_0 B_L}{\alpha P} \left(\frac{\alpha'}{\alpha} + \frac{N_0 B'_{LP}}{\alpha P} \right) \qquad \text{rad}^2 \qquad (5.1\text{-}41)$$

with B'_{LP} defined in (5.1-34). We have neglected the denominator of (5.1-33) since it is normally essentially unity. Notice from (5.1-39) for a one-pole squaring loop, we have

$$\sigma_\phi^2 = \frac{N_0 B_L}{\alpha P} \left(\frac{\alpha'}{\alpha} + \frac{N_0 W}{4\alpha P} \right) \qquad \text{rad}^2 \qquad (5.1\text{-}42)$$

where W is the bandpass filter noise bandwidth. For a multipole Butterworth filter (baseband equivalent), we have approximatley

$$\sigma_\phi^2 = \frac{N_0 B_L}{\alpha P} \left(1 + \frac{N_0 W}{2\alpha P} \right) \qquad \text{rad}^2 \qquad (5.1\text{-}43)$$

In fact, (5.1-43) applies to any multipole filter that has sharp skirts. In this case, <u>W is the 3 dB bandwidth.</u>

PROBLEM 2

Show that α can be evaluated for a one-pole Butterworth filter and NRZ data to be

$$\alpha = 1 - \frac{1}{2\pi R} (1 - e^{-2\pi R})$$

with

$$R = f_0 T$$

where T is the symbol duration and f_0 is the 3 dB bandwidth.

It is to be stressed that we have obtained the linearized phase error variance, which is stationary. The actual phase error variance is unbounded, so that it is not a useful measure of tracking performance and is the reason why the linearized variance is normally the one of interest in tracking.

It should also be pointed out that the squaring process doubles any frequency uncertainties as well as doppler rates, etc. Therefore, the loop is stressed more in a squaring loop than a cw loop for the same doppler dynamics. It is possible to use an envelope detector rather than a squarer, which can actually produce better performance at higher SNRs and worse at lower SNRs [17].

5.1.1 Optimum Presquaring Filter for the Squaring Loop

An optimum filter exists to minimize the value of the phase error variance $\sigma_{\phi_e}^2$. The optimum filter is a function of the baseband modulation format. When $m(t)$, the symbol sequence, is NRZ, Layland [4] has shown that the optimum filter is well approximated by a rectangular bandpass filter with a bandwidth between $1/T$ and $2/T$, where $1/T$ is the symbol rate. In addition, Layland showed that if the modulating spectra is narrow with respect to the carrier frequency and the loop bandwidth is much narrower than the modulating spectrum, then the optimum (maximum SNR at $f = 2f_0$) presquaring filter $H_0(f)$ satisfies

$$|H_0(f)|^2 = K\,\frac{\mathcal{S}_m(f)}{\mathcal{S}_m(f) + N_0/2} \qquad (5.1\text{-}44)$$

where $\mathcal{S}_m(f)$ is the two-sided spectrum of the modulated carrier and $N_0/2$ is the two-sided spectral density. Since this can be factored as

$$|H_0(f)|^2 = [H_0(f)]^+[H_0(f)]^- \qquad (5.1\text{-}45)$$

a realizable optimum filter is of the form

$$H_0(f) = \sqrt{K}\left[\frac{\mathcal{S}_m(f)}{\mathcal{S}_m(f) + N_0/2}\right]^+ \qquad (5.1\text{-}46)$$

using the factorization of Chapter 4. We note that this always represents a realizable filter .At low SNR in the data bandwidth, the optimum filter approaches a matched filter, that is,

$$H_0(f) = K'[\mathcal{S}_m(f)]^+ \qquad (5.1\text{-}47)$$

with K' an arbitrary constant.

In the Costas loop section, the optimum data arm filter will be derived by a method different from that used by Layland.

5.2 COSTAS OR I-Q LOOP

The Costas or I-Q loop, as we shall see, turns out to be equivalent to the squaring loop under certain reasonable assumptions. However, the implementation is quite different. Among other differences, the Costas loop has potential phase detector imbalance problems, whereas the squaring loop can have "ideal squarer" problems, at least at the higher IF frequencies.

Riter [6] has shown that the optimum device for tracking suppressed carrier signals at low SNRs is either a Costas loop or a squaring loop since, as we will shown, they are equivalent. Actually, Riter showed that the in-phase arm should be weighted by $\tanh[KI(t)]$, where K is a known constant dependent on SNR and $I(t)$ is the matched in-phase output. It is easy to show that, when the SNR is low in the matched filter output, the hyperbolic nonlinearity becomes a linear function, with the weighting factor being unimportant. At high SNR, the hyperbolic tangent becomes a limiter function where again the weighting factor is unimportant. We will discuss both approximations to the optimum loop.

The Costas loop effectively squares the input signal and thereby forms a discrete line from which to track the incoming signal. The "linear" form of the Costas loop, shown in Figure 5.3, is the loop we now consider [7].

The bandwidth of the two lowpass filters must be large enough to pass the data. Again we consider the case when we have a suppressed carrier signal plus noise:

$$y(t) = \sqrt{2}\, Am(t) \sin(\omega_1 t + \theta_0) + n(t) \qquad (5.2\text{-}1)$$

where $A^2 = P$ is the signal power, $n(t)$ is modeled as WGN, and θ_0 is the unknown rf phase of the signal. The term $m(t)$ is a baseband, unit amplitude modulation sequence having symbols of duration of T sec [see (5.1-2)]. Let the wideband WGN be modeled as

$$n(t) = \sqrt{2}\, n_c(t) \cos(\omega_1 t + \theta_0) + \sqrt{2}\, n_s(t) \sin(\omega_1 t + \theta_0) \qquad (5.2\text{-}2)$$

where the spectral density of $n_c(t)$ and $n_s(t)$ is equal to $N_0/2$. At the filtered phase detector output (upper arm) after the lowpass filter, we have

$$y_c(t) = A\tilde{m}(t) \sin \phi + \tilde{n}_c(t) \cos \phi + \tilde{n}_s(t) \sin \phi \qquad (5.2\text{-}3)$$

with ϕ the phase error between θ_0 and $\hat{\theta}$, and $\tilde{m}(t)$ the filtered version of the data sequence. We have assumed that the lowpass filter removes the double-frequency terms and both $\tilde{n}_c(t)$ and $\tilde{n}_s(t)$ are band-limited sample functions of lowpass Gaussian processes with a noise bandwidth of B_{LP} Hz (one-sided). Out of the <u>coherent amplitude detector (CAD)</u>, after filtering by the lowpass filter, we have

$$y_s(t) = A\tilde{m}(t) \cos \phi - \tilde{n}_c(t) \sin \phi + \tilde{n}_s(t) \cos \phi \qquad (5.2\text{-}4)$$

where, as before, $\tilde{m}(t)$ is the filtered version of $m(t)$ and ϕ is the phase

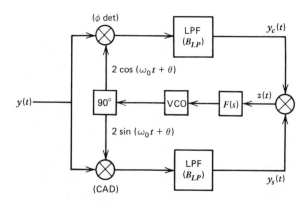

Figure 5.3 Costas loop model.

error, $\theta_0 - \hat{\theta}$. The error signal $z(t)$ is given by

$$z(t) = \frac{A^2}{2}(\tilde{m})^2 \sin 2\phi + A\tilde{m}n_s \sin 2\phi + \left(\frac{\tilde{n}_s^2}{2} - \frac{\tilde{n}_c^2}{2}\right) \sin 2\phi$$
$$+ (A\tilde{m} + \tilde{n}_s)\tilde{n}_c \cos 2\phi \qquad (5.2\text{-}5)$$

where the t dependence has been dropped for convenience. Since $\phi = \theta_0 - \theta + (\omega_1 - \omega_0)t$, we can write

$$\hat{\theta} = \frac{K_{\text{VCO}}F(s)}{s} z(t) \qquad (5.2\text{-}6)$$

or

$$\dot{\phi} + \frac{A^2 KF(s)}{2}(\tilde{m})^2 \sin 2\phi = KF(s)\left\{\left[\frac{\tilde{n}_c^2}{2} - \frac{\tilde{n}_s^2}{2} - A\tilde{m}\tilde{n}_s\right] \sin 2\phi \right.$$
$$\left. - [(A\tilde{m}\tilde{n}_c + \tilde{n}_s\tilde{n}_c) \cos 2\phi]\right\} + (\omega_1 - \omega_0)$$
$$(5.2\text{-}7)$$

where $K = K_{\text{VCO}}$ times any multiplier or filter gains in the loop and $\omega_1 - \omega_0$ is the initial frequency difference between the incoming signal and the VCO frequency.

Let $2\phi = \Phi_e$ and $P = A^2$, so that we have

$$\dot{\Phi}_e + KPF(s)(\tilde{m})^2 \sin \Phi_e = - KF(s)[(- \tilde{n}_c^2 + \tilde{n}_s^2 + 2A\tilde{m}\tilde{n}_s) \sin \Phi_e$$
$$+ 2(A\tilde{m}\tilde{n}_c + \tilde{n}_c\tilde{n}_s) \cos \Phi_e] + 2(\omega_i - \omega_0)$$
$$(5.2\text{-}8)$$

as the SDE of the Costas loop.

Upon comparing the SDE of the Costas loop with the SDE of the squaring loop [(5.1-13)], we see that they are in fact identical. We have assumed that the bandpass filter of the squaring loop, when translated to baseband, is the same as the lowpass filters in the Costas loop. The choice between the two loops is based on the ease of implementation at the frequency of interest.

Since the SDE is the same for both the Costas loop and the squaring loop, the phase error variance is given by ($W = 2B_{LP}$ and is the rf presquaring filter bandwidth):

$$\sigma^2_{\Phi_e} = \frac{\dfrac{4N_0 B_L}{\alpha P}\left(\dfrac{\alpha'}{\alpha} + \dfrac{N_0 B'_{LP}}{\alpha P}\right)}{\left[1 - \dfrac{8TB_L}{(\alpha P)^2}\displaystyle\sum_{l=1}^{\infty}\left(\displaystyle\int_{-\infty}^{\infty}\mathscr{S}_m(\omega)|H_{LP}(\omega)|^2 e^{j\omega lT}\,\dfrac{d\omega}{2\pi}\right)^2\right]} \quad \text{rad}^2 \quad (5.2\text{-}9)$$

where

$$2B'_{LP} = \int_{-\infty}^{\infty} |H(f)|^4 df \qquad (5.2\text{-}10)$$

$$2B_{LP} = \int_{-\infty}^{\infty} |H(f)|^2 df \qquad (5.2\text{-}11)$$

with $H(f)$ the lowpass arm filter transfer functions. It should be pointed out that $\sigma^2_{\Phi_e}$ is an equivalent variance useful for predicting cycle slipping; however σ^2_ϕ applies to the actual phase error and bit error rate degradations.

Letting $2\phi = \Phi_e$ and assuming that the denominator is essentially unity,* we obtain

$$\sigma^2_\phi = \frac{N_0 B_L}{\alpha P}\left(\frac{\alpha'}{\alpha} + \frac{N_0 B'_{LP}}{\alpha P}\right) \qquad \text{rad}^2 \qquad (5.2\text{-}12)$$

For an n-pole Butterworth lowpass filter, (5.2-12) becomes

$$\sigma^2_\phi = \frac{N_0 B_L}{\alpha P}\left[\frac{\alpha'}{\alpha} + \frac{N_0 B_{LP}}{\alpha P}\left(1 - \frac{1}{2n}\right)\right] \qquad \text{rad}^2 \qquad (5.2\text{-}13)$$

where B_{LP} is the arm filter noise bandwidth. For the often-used one-pole arm

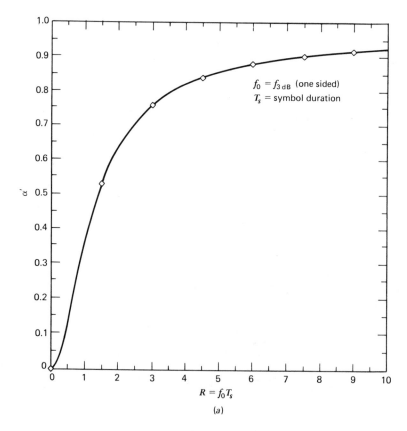

Figure 5.4   (a) α' versus f_0 for a one-pole RC filter (Manchester data).

*Which is the assumption made in the rest of the text unless noted otherwise.

filter loop, we obtain $[B_{LP} = (\pi/2)f_0]$

$$\sigma_\phi^2 = \frac{N_0 B_L}{\alpha P}\left(\frac{\alpha'}{\alpha} + \frac{N_0 B_{LP}}{2\alpha P}\right) \quad \text{rad}^2 \quad (5.2\text{-}14)$$

As before,

$$\alpha = \int_{-\infty}^{\infty} \mathcal{S}_m(f)|H(f)|^2 df$$

$$(5.2\text{-}15)$$

$$\alpha' = \int_{-\infty}^{\infty} \mathcal{S}_m(f)|H(f)|^4 df$$

In Figures 5.4a–5.4d, α and α' are plotted for NRZ and Manchester data, which are useful in estimating tracking performance for Costas and squaring loops.

It is not difficult to optimize the lowpass filter bandwidth based on

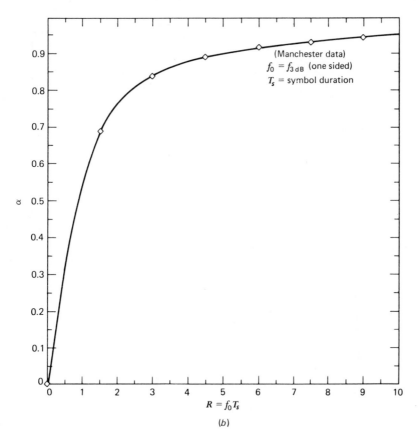

(b)

Figure 5.4 Continued. (b) α versus f_0 for a one-pole RC filter (Manchester data).

(5.2-13). Figures 5.5a and 5.5b illustrate the optimum values of the 3-dB bandwidth as a function of the E_s/N_0 ratio, with T the symbol duration. For NRZ data, the optimum value is about $f_{3\,dB} \cong 0.65/T$; for Manchester data, the optimum is around $f_{3\,dB} \cong 1.4/T$ over a broad range of E_s/N_0. The actual value of $f_{3\,dB}$ is not very critical in terms of minimizing σ_ϕ^2 since the optimum is quite broad for both modulation formats. In fact, 20%–30% changes in $f_{3\,dB}$ cause only a few percent increase in tracking error variance.

PROBLEM 3

Show that when a Costas loop has an imperfect hybrid (90° phase shifter) a dc bias due to the signal will exist at the output of the third multiplier, which induces a static phase error offset. Show that this bias is proportional to the sine of the angle off 90°.

PROBLEM 4

For the Costas loop preceded by a bandpass filter shown below,

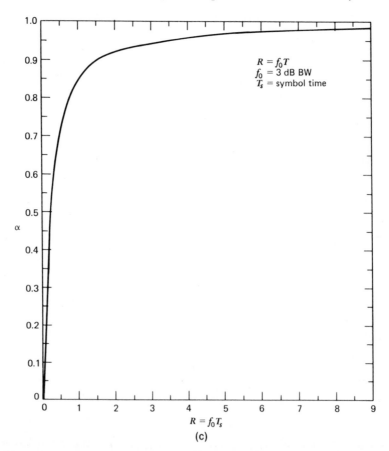

$$R = f_0 T$$
$$f_0 = 3 \text{ dB BW}$$
$$T_s = \text{symbol time}$$

$$\alpha$$

$$R = f_0 T_s$$

(c)

Figure 5.4   Continued. (c) α for NRZ data for one-pole RC lowpass filter.

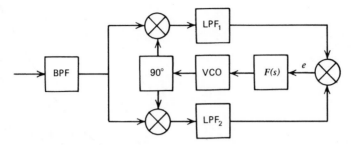

(a) Show that a dc bias due to noise may exist at the output of the third multiplier given by

$$\bar{e} = \frac{1}{2i} \int_{-\infty}^{\infty} [H_1^*(\omega)H_2(\omega) - H_2^*(\omega)H_1(\omega)]\mathscr{S}_L(\omega)\frac{d\omega}{2\pi}$$

where $H_1(\omega)$ is the transfer function of the upper arm filter, $H_2(\omega)$

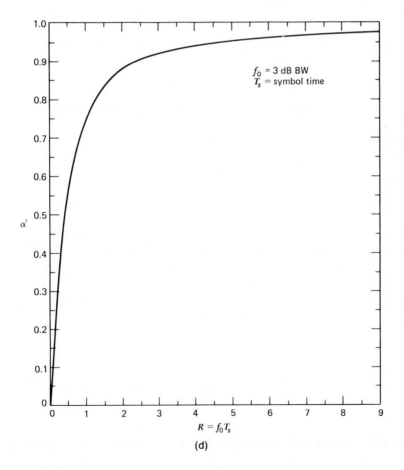

Figure 5.4 Continued. (*d*) α' for NRZ data for one-pole RC lowpass filter.

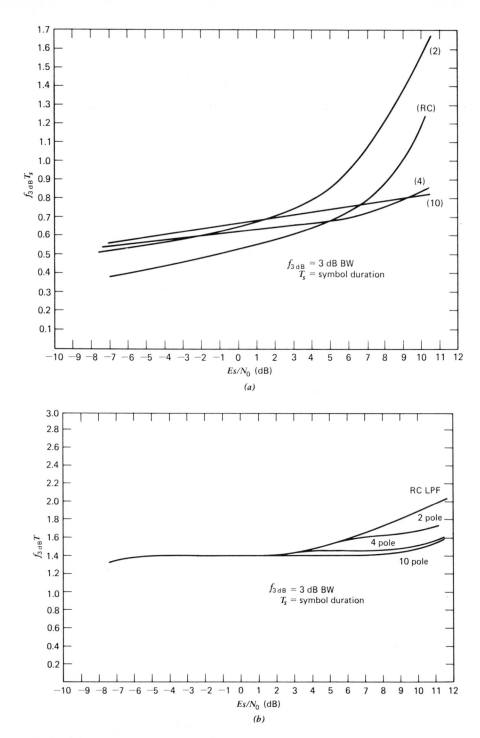

Figure 5.5 Optimum 3-dB data filter bandwidth for (*a*) NRZ data and (*b*) Manchester data.

is the transfer function of the lower arm, and $H^*(\omega)$ is the complex conjugate of $H(\omega)$. Also, $\mathcal{S}_L(\omega)$ is the baseband translation of the BPF noise power spectral density.

(b) For the case of lowpass one-pole arm filters, show that the mean output is given by

$$\bar{e} = \int_{-\infty}^{\infty} \frac{\left(\dfrac{\omega}{\omega_1} - \dfrac{\omega}{\omega_2}\right)\mathcal{S}_L(\omega)\dfrac{d\omega}{2\pi}}{\left[1 + \left(\dfrac{\omega}{\omega_1}\right)^2\right]\left[1 + \left(\dfrac{\omega}{\omega_2}\right)^2\right]}$$

(c) Show that $\bar{e} = 0$ when $\mathcal{S}_L(\omega)$ is symmetric.

This bias can, under normal operating conditions, range from a few tenths of a degree to a few degrees. Its effect increases as the loop SNR decreases and as the noise spectrum becomes asymmetric due to such effects as bandpass filter tilt, etc.

5.2.1 Optimum Data Filters for the Costas Loop

Now we consider the optimum data filter, based on linear theory (small phase errors) for the Costas loop. From (5.1-27), we have

$$R_N(\tau) = 4[R_{n_c}^2(\tau) + PR_{\tilde{m}}(\tau)R_{n_c}(\tau)] \tag{5.2-16}$$

Hence, the noise spectral density at $f = 0$ is given by

$$\mathcal{S}_N(0) = 4\int_{-\infty}^{\infty} \mathcal{S}_{n_c}(\omega)\mathcal{S}_{n_c}(-\omega)\frac{d\omega}{2\pi} + 4P\int_{-\infty}^{\infty} \mathcal{S}_{\tilde{m}}(\omega)\mathcal{S}_{n_c}(-\omega)\frac{d\omega}{2\pi} \tag{5.2-17}$$

Denoting the data filter by $H(\omega)$, we obtain

$$\mathcal{S}_N(0) = 4\int_{-\infty}^{\infty} \left(\frac{N_0}{2}\right)^2 |H(\omega)|^4 \frac{d\omega}{2\pi} + 4P\left(\frac{N_0}{2}\right)\int_{-\infty}^{\infty} \mathcal{S}_m(\omega)|H(\omega)|^4 \frac{d\omega}{2\pi} \tag{5.2-18}$$

We have shown that, when $B_L \ll B_{LP}$,

$$\sigma_{\Phi_e}^2 = \frac{2\mathcal{S}_N(0)B_L}{(\alpha P)^2} \tag{5.2-19}$$

or

$$\sigma_{\Phi_e}^2 = \frac{4N_0 B_L}{P^2}\left[\frac{\left(\dfrac{N_0}{2}\right)\displaystyle\int_{-\infty}^{\infty}|H(f)|^4\,df + P\displaystyle\int_{-\infty}^{\infty}\mathcal{S}_m(f)|H(f)|^4\,df}{\left\{\displaystyle\int_{-\infty}^{\infty}\mathcal{S}_m(f)|H(f)|^2\,df\right\}^2}\right] \tag{5.2-20}$$

The optimum filter is one that minimizes $\sigma_{\Phi_e}^2$ over all possible $H(f)$. (The self-noise term has been approximated as zero with negligible error.) Following Stiffler [8], we note that minimizing $\sigma_{\Phi_e}^2$ over all possible $H(f)$ is equivalent to minimizing the numerator subject to the denominator (or, more conveniently, the square root) being fixed. Thus, we desire to minimize the following functional:

$$f[H(f)] = \int_{-\infty}^{\infty} \left\{ \left[P\mathscr{S}_m(f) + \frac{N_0}{2} \right] |H(f)|^4 - \lambda P \mathscr{S}_m(f) |H(f)|^2 \right\} df \quad (5.2\text{-}21)$$

By completing the square, this can be written as

$$f(H_{LP}) = \int_{-\infty}^{\infty} \left\{ \left[P\mathscr{S}_m(f) + \frac{N_0}{2} \right] \left(|H(f)|^2 - \frac{\lambda P \mathscr{S}_m(f)}{2P\mathscr{S}_m(f) + N_0} \right)^2 - \frac{\lambda^2 P^2 \mathscr{S}_m(f)}{(4P\mathscr{S}_m(f) + 2N_0)} \right\}$$

$$(5.2\text{-}22)$$

Clearly, since the last term is independent of $H(f)$, the functional is minimized when

$$|H_0(f)|^2 = \frac{\lambda P \mathscr{S}_m(f)}{2\left(P\mathscr{S}_m(f) + \frac{N_0}{2} \right)} \quad (5.2\text{-}23)$$

therefore, the optimum filter satisfies

$$H_0(f) = \left[\frac{\lambda P \mathscr{S}_m(f)}{2(P\mathscr{S}_m(f) + N_0/2)} \right]^+ \quad (5.2\text{-}24)$$

This is the optimum filter under the approximations made. Notice that λ is arbitrary, since it does not affect the value of $\sigma_{\phi_e}^2$. Further, since only the magnitude squared of the optimum filter is specified, a realizable factorization is always possible, producing a causal filter. Notice at low SNR we find $[H_0(f)] = 1/N_0[\mathscr{S}_m(f)]^+$, the optimum data filter at <u>low SNR, which is a matched filter.</u>

PROBLEM 5

Consider the design of an optimum data filter when the signal spectrum is of the form

$$\mathscr{S}_m(f) = \frac{2}{\omega_0} \left[\frac{1}{1 + (\omega/\omega_0)^2} \right]$$

Find the functional form of the optimum data filter.

PROBLEM 6

Show that one bandpass filter placed in front of the Costas loop (now with only lowpass double-frequency rejection arm filters) has the same tracking performance as an ordinary Costas loop as long as the baseband equivalent of the bandpass filter is identical to the Costas arm filters. On a short-loop version, however, the internal loop delay is reduced!

5.2.2 Steady State Tracking for Costas and Squaring Loops

In this section, we point out that in both the Costas loop and squaring loop the "effective" carrier dynamics are doubled relative to a cw carrier tracking loop.

First consider a squaring loop. Since the squarer doubles the carrier frequency, the carrier frequency and all its derivatives are doubled. Therefore, the steady state tracking in an ideal second-order loop following the squarer would be given by

$$\phi_{ss_e}(t) = \frac{2\Omega_1}{\omega_n^2} + \frac{2\Omega_0}{AK} \qquad (5.2\text{-}25)$$

where Ω_0 is the frequency difference and Ω_1 is the frequency rate difference between the input signal and the reference oscillator.

In the Costas loop, the steady state error is one-half the value given in (5.2-25). However, since the S-curve is compressed to one-half the width, (5.2-25) is the effective (as far as cycle slipping) steady state phase error. In other words, the phase error control signal for a cw loop is given by $\sin \phi$, whereas, for a Costas loop, it is given by $\sin(2\phi)/2$. Hence, for small errors, the actual static phase error for both loops is the same but since the Costas loop has the $\sin(2\phi)/2$ error charactersitic, the effective error is 2ϕ as far as the loop losing lock.

To summarize, the effective steady state phase error for both a Costas loop and a squaring loop with an ideal second-order loop subject to a frequency and frequency rate error is given by

$$\phi_{ss_e}(t) = \frac{2\Omega_0}{AK} + \frac{2\Omega_1}{\omega_n^2} \qquad (5.2\text{-}26)$$

whereas the actual phase in the Costas loop and divided-by-2 phase error of the squaring loop for the same conditions are given by

$$\phi_{ss}(t) = \frac{\Omega_0}{AK} + \frac{\Omega_1}{\omega_n^2} \qquad (5.2\text{-}27)$$

In conclusion, for any type of loop filter, the effective dynamics, as far as cycle slipping, are doubled for either a Costas loop or a squaring loop.

5.3   THE INTEGRATE-AND-DUMP COSTAS LOOP

The Costas loop analyzed here is shown in Figure 5.6. The theory is based on Manchester symbols although the results hold for any symbols that are match-filtered.

Let the received carrier be denoted by

$$y(t) = \sqrt{2}\, Am(t) \sin(\omega_0 t + \theta) + n(t) \qquad (5.3\text{-}1)$$

with $m(t) = \pm 1$ the data symbol sequence with symbol duration of T sec, A is the rms signal amplitude, ω_0 the unknown center frequency, and θ the unknown rf phase. The Manchester PCM data sequence is assumed to have period T sec. Either PCM signal is assumed to occur with probability $1/2$. The wideband Gaussian noise process $n(t)$, with two-sided noise spectral density $N_0/2$, is represented by

$$n(t) = \sqrt{2}\, n_1(t)\cos(\omega_0 t + \theta) + \sqrt{2}\, n_2(t)\sin(\omega_0 t + \theta) \qquad (5.3\text{-}2)$$

where the spectral density of $n_1(t)$ and $n_2(t)$ is equal to $N_0/2$ over a bandwidth of $(-B_i, B_i)$ Hz, centered at $\omega_0/2\pi$ such that $\omega_0/2\pi \gg B_i \gg 1/T$.

At the phase detector output, we have

$$\begin{aligned}
Z_c(t) = {} & Am(t)\sin\phi(t) + Am(t)\sin(2\omega_0 t + \theta + \hat{\theta}) \\
& + n_1(t)\cos\phi(t) + n_1(t)\cos(2\omega_0 t + \theta + \hat{\theta}) \\
& + n_2(t)\sin\phi(t) + n_2(t)\sin(2\omega_0 t + \theta + \hat{\theta}) \qquad (5.3\text{-}3)
\end{aligned}$$

where $\phi \overset{\Delta}{=} \theta - \hat{\theta}$ represents the phase error.

Out of the matched integrate-and-dump filter, we have (neglecting $2\omega_0$ terms)

$$Y_c(T) = Ad(T')T\sin\phi + N_1\cos\phi + N_2\sin\phi \qquad (5.3\text{-}4)$$

where N_1 and N_2 are independent Gaussian random variables with zero mean and variance of $N_0 T/2$ and $d(T') = \pm 1$ according to the data polarity.* We assume that ϕ is essentially constant over one symbol time.

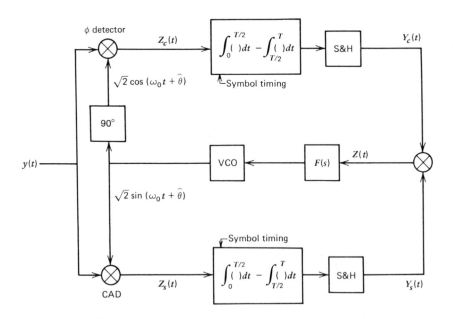

Figure 5.6 Costas loop with I & D data filters (Manchester symbols). (S & H symbolizes sample and hold for T sec.)

*It is assumed that $d(T')$ denotes the value of $d(t)$ at $t = T'$ just before a possible transition.

Now the output of the CAD is given by

$$Z_s(t) = Am(t) \cos \phi(t) - Am(t) \cos(2\omega_0 t + \theta + \hat{\theta})$$
$$- n_1(t) \sin \phi(t) + n_1(t) \sin(2\omega_0 t + \hat{\theta} + \theta)$$
$$+ n_2(t) \cos \phi(t) - n_2(t) \cos(2\omega_0 t + \hat{\theta} + \theta) \quad (5.3\text{-}5)$$

After the integrate-and-dump filter, we have (neglecting $2\omega_0$ terms)

$$Y_s(T) = Ad(T')T \cos \phi(T') - N_1 \cos \phi(T') + N_2 \sin \phi(T') \quad (5.3\text{-}6)$$

Using the equations for $Y_c(t)$ and $Y_s(t)$, we have that the control voltage is given by (suppressing the explicit time dependence)

$$Z = \frac{A^2 T^2}{2} \sin 2\phi + ATmN_2 \sin 2\phi + \frac{(N_2)^2}{2} \sin 2\phi$$
$$- \frac{(N_1)^2}{2} \sin 2\phi + (ATm + N_2)N_1 \cos 2\phi \quad (5.3\text{-}7)$$

since

$$\hat{\theta} = \frac{K_{VCO}K_m F(s)}{s} Z \quad (5.3\text{-}8)$$

We have for $\dot{\theta} = 0$ and $K = K_{VCO}K_m$ (K_m is the multiplier gain), the following SDE:

$$\dot{\phi} + \frac{A^2 T^2 KF(s)}{2} \sin 2\phi = KF(s)\left\{ \left(\frac{N_1^2}{2} - \frac{N_2^2}{2} - ATmN_2 \right) \sin 2\phi \right.$$
$$\left. - [(ATmN_1 + N_1 N_2) \cos 2\phi] \right\} \quad (5.3\text{-}9)$$

where $s = d/dt$ is the Heaviside operator. Letting $2\phi = \Phi$, we have

$$s\Phi + A^2 T^2 KF(s) \sin \Phi = KF(s)[(N_1^2 - N_2^2 - 2ATmN_2) \sin \Phi$$
$$- 2(ATmN_1 + N_1 N_2) \cos \Phi] \quad (5.3\text{-}10)$$

If we call all the noise terms to the right of the equal sign $N(t)$, then after linearizing we have

$$\Phi(t) = \left[\frac{A^2 T^2 KF(s)}{s + A^2 T^2 KF(s)} \right] \left[\frac{N}{(AT)^2} \right] \quad (5.3\text{-}11)$$

hence,

$$\sigma_\Phi^2 = \frac{N_0' B_L}{(AT)^4} \quad (5.3\text{-}12)$$

where

$$2B_L = \int_{-i\infty}^{i\infty} \left| \frac{(AT)^2 KF(s)}{s + A^2 T^2 KF(s)} \right|^2 \frac{ds}{2\pi i} \quad (5.3\text{-}13)$$

and N_0' is the one-sided spectral density at $f = 0$ assuming $B_L T \ll 1$. Hence, the problem reduces to computing the autocorrelation function of the term N, where N is given by

$$N = (N_1^2 - N_2^2 - 2ATmN_2) \sin 2\phi - 2(ATmN_1 + N_1N_2) \cos 2\phi$$

$$(5.3\text{-}14)$$

which is constant over one symbol time (T sec), since we are assuming a sample-and-hold data filter that holds the output for T sec. As long as the loop bandwidth B_L is much less than the data bandwidth, the delay will not affect the loop performance.

The autocorrelation function, as shown in Problem 7, is given by

$$R_N(\tau) = \begin{cases} \sigma^2\left(1 - \dfrac{|\tau|}{T}\right) & |\tau| \le T \\ 0 & |\tau| > t \end{cases}$$

$$(5.3\text{-}15)$$

where

$$\sigma^2 = E[N^2(t)]$$
$$= \sin^2 \phi (\overline{N_1^4} + \overline{N_2^4} + 4A^2T^2\overline{N_2^2} - 2\overline{N_2^2N_1^2})$$
$$+ 4 \cos^2 \phi (A^2T^2\overline{N_1^2} + \overline{N_1^2N_2^2})$$

$$(5.3\text{-}16)$$

PROBLEM 7

Show that the autocorrelation function of a random amplitude, fixed duration, rectangular, statistically independent sequence is given by

$$R_N(\tau) = \begin{cases} \sigma^2\left(1 - \dfrac{|\tau|}{T}\right) & |\tau| \le T \\ 0 & |\tau| > T \end{cases}$$

where σ^2 is the variance of the amplitude distribution.

Since

$$\overline{N_1^4} = \overline{N_2^4} = 3(\overline{N_1^2})^2 \qquad \text{and} \qquad \overline{N_1^2} = \overline{N_2^2} \qquad (5.3\text{-}17)$$

we have

$$\sigma^2 = 4[A^2T^2\overline{N_1^2} + (\overline{N_1^2})^2] \qquad (5.3\text{-}18)$$

and

$$\overline{N_1^2} = \overline{\left(\int_0^{T/2} n(t)dt - \int_{T/2}^{T} n(t)dt\right)^2} \qquad (5.3\text{-}19)$$

or

$$\overline{N_1^2} = \frac{N_0}{2} T \qquad (5.3\text{-}20)$$

We finally obtain

$$R_N(\tau) = 4\left(\frac{A^2N_0T^3}{2} + \frac{N_0^2T^2}{4}\right)\left(1 - \frac{|\tau|}{T}\right) \qquad (5.3\text{-}21)$$

The power spectral density is given by

$$\mathscr{S}(\omega) = 4T\left(\frac{A^2 N_0 T^3}{2} + \frac{N_0^2 T^2}{4}\right)\left[\frac{\sin(\omega T/2)}{\omega T/2}\right]^2 \tag{5.3-22}$$

If we assume that the loop is narrowband, that is, $B_L \ll B$, where B is the one-sided (baseband) data bandwidth, then the performance of the loop will depend essentially only on the spectrum at the origin. Hence, we have that

$$\mathscr{S}(0) = \frac{N_0'}{2} = 2A^2 N_0 T^4 + N_0^2 T^3 \tag{5.3-23}$$

so the linearized phase error variance of Φ is

$$\sigma_\Phi^2 = \frac{4N_0 B_L}{A^2}\left(1 + \frac{N_0}{2PT}\right) \tag{5.3-24}$$

or, since $\Phi = 2\phi$ and $P = A^2$, we have our final result:

$$\sigma_\phi^2 = \frac{N_0 B_L}{P}\left(1 + \frac{N_0}{2PT}\right) \tag{5.3-25}$$

By comparing the Costas loop with passive Butterworth filters, it is clear that the matched filter version is superior in the sense of smaller phase error variance. On the surface, it would appear that, since symbol synchronization is needed for the integrate-and-dump (I & D) filters, the I & D Costas loop is not suitable for the acquisition mode. However, Oldenwalder [10] has shown via simulation that by stepping the VCO through discrete frequencies the combination of carrier and bit synchronization loop could acquire the signal in reasonable acquisition times and, in addition, provide superior tracking performance. It is also possible to use passive filters for acquisition and switch in I & D matched filters for tracking (after the bit synchronizer is locked up).

5.4 MODIFIED COSTAS LOOP PERFORMANCE

Cahn [11] recently suggested removing one of the lowpass filter arms of an analog Costas loop in order to improve frequency acquisition and aid in removing false lock (which will be discussed later in this chapter). Basically, by removing the quadrature channel filter, Cahn has shown that the loop acts like a hybrid AFC-Costas* loop combination.

In Problem 8 this modified Costas loop is analyzed, with the result that the linearized tracking error variance with a one-pole RC lowpass filter, due

---

*AFC denotes automatic frequency control.

to thermal noise, is given by

$$\sigma_\phi^2 = \frac{N_0 B_L}{\alpha P} \left(1 + \frac{N_0 B_{LP}}{\alpha P}\right) \qquad \text{rad}^2 \qquad (5.4\text{-}1)$$

where

$$2B_{LP} = \int_{-\infty}^{\infty} |H(f)|^2 df \qquad (5.4\text{-}2)$$

Comparing (5.4-1) with (5.2-14), we see that both the signal-times-noise term ($\alpha' < \alpha$) and the noise-times-noise term is larger in the modified Costas loop. In fact, at very low SNR, σ_ϕ^2 is essentially twice as large for the modified loop.

PROBLEM 8

Derive (5.4-1) for the tracking performance of a one-pole RC arm filter Costas loop by showing that

(a)

$$\overline{mm} = \int_{-\infty}^{\infty} \mathscr{S}_m(f) H^*(f) df$$

$$= \int_{-\infty}^{\infty} \mathscr{S}_m(f) |H(f)|^2 df$$

(b)

$$\mathscr{S}_{\tilde{n} \times \tilde{n}}(0) = \int_{-\infty}^{\infty} \left(\frac{N_0}{2}\right)^2 |H(f)|^2 df$$

(c)

$$\sigma_{\Phi_e}^2 = \frac{4N_0 B_L}{\alpha P} \left(1 + \frac{N_0 B_{LP}}{\alpha P}\right)$$

Letting $2\phi = \Phi_e$ in (c) produces the final result. Assume that the inphase arm has noise bandwidth B_{LP} and the other arm is very wideband but removes the double frequency term.

5.5   COSTAS LOOP WITH RESIDUAL CARRIER

Normally, a Costas loop operates in the presence of a suppressed carrier signal. There are applications, however, where the carrier component is not completely suppressed. One example occurs when both ranging and data are transmitted simultaneously on the same carrier. We shall find that the modulation index of the data is quite critical, there being a value for which the loop will not operate, even with arbitrarily large SNRs. We follow reference 12 with a slight generalization. Simon also considered the problem [13], as well as Biederman [14].

Consider our Costas loop model shown in Figure 5.7, with Manchester data. The signal is described by

$$s(t) = \sqrt{2P} \, \sin[\omega_0 t + m_1 m(t) + m_2 \sin(\omega_{sc} t + \theta)] \qquad (5.5\text{-}1)$$

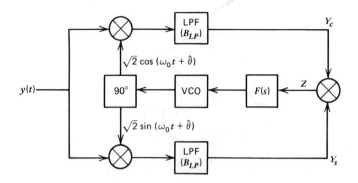

Figure 5.7   Costas loop model for data plus residual carrier signal.

with $m(t)$ being a Manchester data sequence and $\sin(\omega_{sc}t + \theta)$ being the ranging subcarrier. The signal can be expanded to the form

$$s(t) = \sqrt{2P} \cos m_1 \sin \omega_0 t \left\{ J_0(m_2) + \sum_{n=2,4,6,\ldots}^{\infty} 2J_n(m_2) \cos[n(\omega_{sc}t + \theta)] \right\}$$

$$+ \sqrt{2P} \sin m_1 \cos(\omega_0 t) m(t) \left\{ J_0(m_2) + \sum_{n=2,4,6,\ldots}^{\infty} 2J_n(m_2) \cos[n(\omega_{sc}t + \theta)] \right\}$$

$$+ \sqrt{2P} \cos m_1 \cos \omega_0 t \left\{ 2 \sum_{n=1,3,5,\ldots}^{\infty} J_n(m_2) \sin[n(\omega_{sc}t + \theta)] \right\}$$

$$- \sqrt{2P} \sin m_1 \sin(\omega_0 t) m(t) \left\{ 2 \sum_{n=1,3,5,\ldots}^{\infty} J_n(m_2) \sin[n(\omega_{sc}t + \theta)] \right\}$$

$$\tag{5.5-2}$$

The signal plus noise can be modeled as

$$y(t) = s(t) + \sqrt{2}\, n_s \sin(\omega_0 t) + \sqrt{2}\, n_c \cos(\omega_0 t) \tag{5.5-3}$$

The signal $Y_c(t)$ out of the upper lowpass filter is then

$$Y_c(t) = \sqrt{P} \cos m_1 J_0(m_2) \sin \phi + \sqrt{P} \sin m_1 J_0(m_2) \tilde{m}(t) \cos \phi$$

$$+ \tilde{n}_s(t) \sin \phi + \tilde{n}_c(t) \cos \phi + \text{heavily filtered terms} \tag{5.5-4}$$

Also, the lower arm filter output is given by

$$Y_s(t) = \sqrt{P} \cos m_1 J_0(m_2) \cos \phi - \sqrt{P} \sin m_1 J_0(m_2) \tilde{m}(t) \sin \phi$$

$$+ \tilde{n}_s(t) \cos \phi - \tilde{n}_c \sin \phi + \text{heavily filtered terms} \tag{5.5-5}$$

with $m(t)$ being the filtered version of $m(t)$ and $\tilde{n}_c(t)$ and $\tilde{n}_s(t)$ being the filtered versions of $n_c(t)$ and $n_s(t)$, respectively. It is assumed in what follows that no spectral components of the subcarrier modulation get into the error control signal. The error control signal $Z(t)$ becomes

$$Z = P \cos^2(m_1) J_0^2(m_2) \frac{\sin 2\phi}{2} + P \frac{\sin 2m_1}{2} J_0^2(m_2) \tilde{m}(t) \cos^2 \phi$$

$$+ \sqrt{P} \cos m_1 J_0(m_2) \tilde{n}_s \frac{\sin 2\phi}{2} + \sqrt{P} \cos m_1 J_0(m_2) \tilde{n}_c \cos^2 \phi$$

$$- P \frac{\sin 2m_1}{2} J_0^2(m_2) \tilde{m}(t) \sin^2 \phi - P \sin^2 m_1 J_0^2(m_2) (\tilde{m})^2 \frac{\sin 2\phi}{2}$$

$$- \sqrt{P} \sin m_1 J_0(m_2) \tilde{m} \sin^2 \phi - \sqrt{P} \sin m_1 J_0(m_2) \tilde{m} \frac{\sin 2\phi}{2}$$

$$+ \sqrt{P} \cos m_1 J_0(m_2) \frac{\sin 2\phi}{2} \tilde{n}_s + \sqrt{P} \sin m_1 J_0(m_2) \tilde{m} \tilde{n}_s \cos^2 \phi$$

$$+ \tilde{n}_s^2 \frac{\sin 2\phi}{2} + \tilde{n}_s \tilde{n}_c \cos^2 \phi - \sqrt{P} \cos m_1 J_0(m_2) \tilde{n}_c \sin^2 \phi$$

$$- \sqrt{P} \sin m_1 J_0(m_2) \tilde{m} \tilde{n}_c \frac{\sin 2\phi}{2} - \tilde{n}_c \tilde{n}_s \sin^2 \phi - \tilde{n}_c^2 \frac{\sin 2\phi}{2} \qquad (5.5\text{-}6)$$

Letting $\phi \to 0$ (high-loop SNR) for all terms except signal control terms, we have

$$Z(t) \cong P J_0^2(m_2) [\cos^2(m_1) - \alpha \sin^2(m_1)] \frac{\sin 2\phi}{2}$$

$$+ P \frac{\sin 2m_1}{2} J_0^2(m_2) \tilde{m}(t) + \sqrt{P} \cos m_1 J_0(m_2) \tilde{n}_c(t)$$

$$+ \sqrt{P} \sin m_1 J_0(m_2) \tilde{m}(t) \tilde{n}_s(t) + \tilde{n}_s(t) \tilde{n}_c(t) \qquad (5.5\text{-}7)$$

where

$$\alpha = \overline{(\tilde{m})^2} = \int_{-\infty}^{\infty} \mathscr{S}_m(\omega) |H_{LP}(\omega)|^2 \frac{d\omega}{2\pi} \qquad (5.5\text{-}8)$$

and it represents the power loss through the arm (data) filters. Notice that, if $\alpha = \text{ctn}^2(m_1)$, the loop will not operate at any loop SNR since the error control signal goes to zero.

Let

$$N(t) = \frac{P}{2} \sin 2m_1 J_0^2(m_2) \tilde{m}(t) + \sqrt{P} \cos m_1 J_0(m_2) \tilde{n}_c(t)$$

$$+ \sqrt{P} \sin m_1 J_0(m_2) \tilde{m}(t) \tilde{n}_s(t) + \tilde{n}_s(t) \tilde{n}_c(t) \qquad (5.5\text{-}9)$$

To close the loop, the spectral density must be obtained for the noise process $N(t)$. We have*

$$R_N(\tau) = \frac{P^2}{4} \sin^2 2m_1 J_0^4(m_2) R_{\tilde{m}}(\tau) + P \cos^2 m_1 J_0^2(m_2) R_{\tilde{n}}(\tau)$$

$$+ P \sin^2 m_1 J_0^2(m_2) R_{\tilde{m}}(\tau) R_{\tilde{n}}(\tau) + [R_n(\tau)]^2 \qquad (5.5\text{-}10)$$

---

*Again we model the cyclostationary SXN process as a stationary process.

where $R_{\tilde{n}}(\tau) = R_{\tilde{n}_c}(\tau) = R_{\tilde{n}_s}(\tau)$. Now

$$
\begin{aligned}
\mathscr{S}_N(0) = \frac{N_0'}{2} &= \int_{-\infty}^{\infty} R_N(\tau)\, d\tau \\
&= \frac{P^2}{4} \sin^2 2m_1 J_0^4(m_2) \int_{-\infty}^{\infty} R_{\tilde{m}}(\tau)\, d\tau \\
&\quad + P \cos^2 m_1 J_0^2(m_2) \int_{-\infty}^{\infty} R_{\tilde{n}}(\tau)\, d\tau \\
&\quad + P \sin^2 m_1 J_0^2(m_2) \int_{-\infty}^{\infty} R_{\tilde{m}}(\tau) R_{\tilde{n}}(\tau)\, d\tau \\
&\quad + \int_{-\infty}^{\infty} R_{\tilde{n}}^2(\tau)\, d\tau
\end{aligned}
\tag{5.5-11}
$$

Now we evaluate each component of $\mathscr{S}_N(0)$.

1 Consider

$$
I_1 = \int_{-\infty}^{\infty} R_{\tilde{m}}(\tau)\, d\tau
$$

or

$$
I_1 = \int_{-\infty}^{\infty} R_{\tilde{m}}(\tau)\, d\tau = \left[\frac{\sin^4 \omega\tau/2}{(\omega\tau/2)^2} |H(\omega)|^2 \right]\Bigg|_{\omega=0} = 0
\tag{5.5-12}
$$

2 Consider

$$
I_2 = \int_{-\infty}^{\infty} R_{\tilde{n}}(\tau)\, d\tau = \frac{N_0}{2}
\tag{5.5-13}
$$

3 Consider

$$
I_3 = \int_{-\infty}^{\infty} R_{\tilde{m}}(\tau) R_{\tilde{n}}(\tau)\, d\tau = \alpha' \frac{N_0}{2}
\tag{5.5-14}
$$

where

$$
\alpha' = \int_{-\infty}^{\infty} \mathscr{S}_m(f) |H(f)|^4\, df
\tag{5.5-15}
$$

4 Consider

$$
I_4 = \int_{-\infty}^{\infty} R_{\tilde{n}}^2(\tau)\, d\tau = \int_{-\infty}^{\infty} \left[\int_{-\infty}^{\infty} |H(f)|^2 e^{-i\omega\tau} \left(\frac{N_0}{2} \right) df \right]^2 d\tau
\tag{5.5-16}
$$

or

$$
I_4 = \left(\frac{N_0}{2} \right) \int_{-\infty}^{\infty} |H(f)|^4\, df
\tag{5.5-17}
$$

so that

$$
I_4 = \frac{B_{LP}' N_0^2}{2}
\tag{5.5-18}
$$

where

$$2B'_{LP} = \int_{-\infty}^{\infty} |H(f)|^4 \, df \qquad (5.5\text{-}19)$$

so that

$$\mathscr{S}_N(0) = P \cos^2 m_1 J_0^2(m_2) \frac{N_0}{2}$$

$$+ P \sin^2 m_1 J_0^2(m_2)\alpha' \frac{N_0}{2} + \frac{N_0^2}{2} B'_{LP} \qquad (5.5\text{-}20)$$

now

$$\hat{\theta} = \frac{K_{VCO}K_m}{s} F(s)Z(t) \qquad (5.5\text{-}21)$$

or

$$\dot{\phi} + PJ_0^2(m_1)(\cos^2 m_1 - \alpha \sin^2 m_1)KF(s)\frac{\sin 2\phi}{2}$$

$$= -P \frac{\sin^2 m_1}{2} J_0^2(m_2)\tilde{m}(t) - \sqrt{P} \cos m_1 J_0(m_2)\tilde{n}_c(t)$$

$$- \sqrt{P} \sin m_1 J_0(m_2)\tilde{m}(t)\tilde{n}_s(t) - \tilde{n}_s(t)\tilde{n}_c(t) \qquad (5.5\text{-}22)$$

let $2\phi = \Phi$

$$\dot{\Phi} + \overbrace{PJ_0^2(m_1)(\cos^2 m_1 - \alpha \sin^2 m_1)KF(s)}^{\gamma} \sin \Phi = KF(s)2N(t)$$

$$(5.5\text{-}23)$$

For small Φ, with γ defined in (5.5-23), we have

$$\Phi = \left[\frac{\gamma KF(s)}{s + \gamma KF(s)} \right]^2 \frac{2N(t)}{\gamma} \qquad (5.5\text{-}24)$$

where $H_L(f)$ is the closed loop response and

$$2B_L = \int_{-\infty}^{\infty} |H_L(f)|^2 \, df \qquad (5.5\text{-}25)$$

so that

$$\sigma_\Phi^2 = \frac{4\mathscr{S}_N(0)(2B_L)}{\gamma^2} \qquad (5.5\text{-}26)$$

Using the value of $\mathscr{S}_N(0)$ and γ, we have finally that (using $\phi = \Phi/2$)

$$\sigma_\phi^2 = \frac{1}{(\cos^2 m_1 - \alpha \sin^2 m_1)^2 J_0^2(m_2)P} \frac{N_0 B_L}{} \left[\cos^2 m + \alpha' \sin^2 m_1 + \frac{N_0 B'_{LP}}{J_0^2(m_2)P} \right]$$

$$(5.5\text{-}27)$$

which is our general result. Notice if $m_2 = 0$ and $m_1 = \pi/2$, we obtain

$$\sigma_\phi^2 = \frac{N_0 B_L}{\alpha P} \left(\frac{\alpha'}{\alpha} + \frac{N_0 B'_{LP}}{\alpha P} \right) \qquad (5.5\text{-}28)$$

which is the result derived in Section 5.2.

In the case where the tone is not present, we have $J_0(m_2) = 1$, so that (5.5-27) becomes

$$\sigma_\phi^2 = \frac{1}{(\cos^2 m_1 - \alpha \sin^2 m_1)^2} \frac{N_0 B_L}{P} \left(\cos^2 m_1 + \alpha' \sin^2 m_1 + \frac{N_0 B'_{LP}}{P} \right)$$

(5.5-29)

In conclusion, we note that the modulation index of the data must be such that it is not near the value $\operatorname{ctn}^2(m_1) = \alpha$ or the loop will be seriously degraded in tracking performance. The NRZ case can be worked out in the same manner.

PROBLEM 9

Let the input to a Costas loop be jammed by a coherent tone interferer. The total input is given by

$$y(t) = \sqrt{2P}\, d(t) \sin(\omega_0 t + \theta_0) + \sqrt{2P_I} \cos(\omega_0 t + \theta_0) + n(t)$$

where P is the signal power, $d(t)$ the binary valued data sequence with duration T sec., P_I the interfering signal power, and $n(t)$ is modeled as WGN. Show that the linearized phase error is given by

$$\sigma_\phi^2 = \frac{N_0 B_L}{(\alpha P - P_I)^2} \left[(\alpha' P + P_I) + \frac{2\alpha P_I P T}{N_0} + N_0 B'_{LP} \right]$$

where

$$2B'_{LP} = \int_{-\infty}^{\infty} |H(f)|^4\, df$$

$$\alpha' = \int_{-\infty}^{\infty} \mathscr{S}_d(f) |H(f)|^4\, df$$

$$\alpha = \int_{-\infty}^{\infty} \mathscr{S}_d(f) |H(f)|^2\, df \qquad (P_I < \alpha P)$$

Notice, when $P_I = 0$, the result reduces to (5.2-12), as it should.

5.6 COSTAS LOOPS WITH HARD-LIMITED IN-PHASE CHANNELS

Since the optimum nonlinearity in the in-phase arm of a Costas loop, at high arm filter SNR, is a limiter, it is worthwhile considering what performance gains are possible employing one. Figure 5.8 illustrates the loop under discussion. Simon [15] has obtained high and low SNR approximations for the hard-limited Costas loop. The basic approach was first to obtain the slope of the S-curve, which now includes a signal suppression due to the limiter. Next, an approximation for the equivalent noise spectral density was determined for both low and high arm SNR. Finally, the

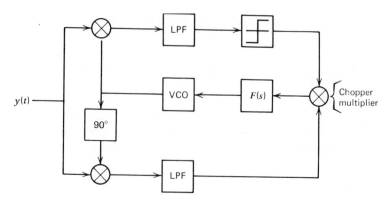

Figure 5.8 Costas loop model with hard-limited in-phase channel.

tracking variance was determined to be of the form

$$\sigma_\phi^2 = \frac{N_0 B_L}{P} \mathscr{S}_L^{-1} \tag{5.6-1}$$

Numerical evaluation of $\mathscr{S}_L^{-1}$ is plotted in Figures 5.9a and 5.9b; the squaring loss $\mathscr{S}_L$ is plotted versus $2B_L T$ for NRZ data and a one-pole RC lowpass filter. Figure 5.9a uses the high SNR approximation. Simon [15] has shown, using the results for low SNR, that the squaring loss (and therefore the tracking error variance) is at most only 1.08 dB larger for the hard-limited loop than the conventional loop. From Figure 5.9a, it is clear that there is only a small loss at low SNR ($R_d \leq -2$ dB), and from Figure 5.9b there appears to be an advantage at higher SNR ($R_d \geq -3.5$ dB). The approximation of Figure 5.9b indicates about a 0.5-dB decrease in tracking error variance for nominal arm filter output SNRs (6 dB). In conclusion, using a hard-limited in-phase channel Costas loop will, under most system specifications, improve tracking performance and allow a chopper multiplier to be used for the third multiplier, thereby potentially reducing dc offsets.

5.7 ACQUISITION BY SWEEPING FOR SQUARING AND COSTAS LOOPS

When false lock is not a problem, carrier sweeping is an efficient acquisition aid. The results of Frazier and Page [16] can be modified to account for the squaring loss. Recall from Chapter 4 that Frazier and Page showed that the sweep rate that would result in the cw loop acquiring 90% of the time is given, to a good approximation, by

$$R_{90} = \frac{(1 - \sqrt{2}/\sqrt{\rho})\omega_n^2}{2\pi(1 + d)} \qquad \text{Hz/sec} \tag{5.7-1}$$

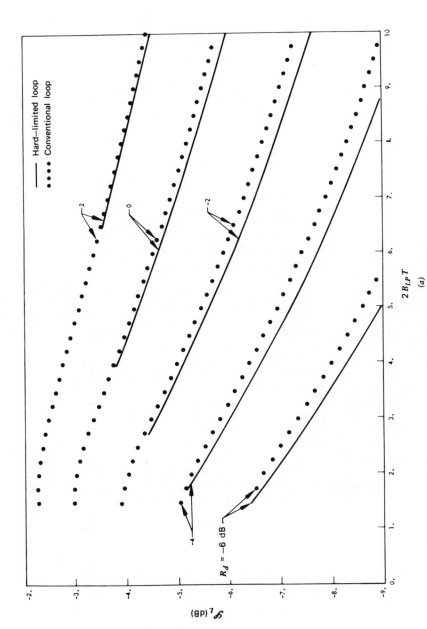

Figure 5.9 Squaring loss versus $2B_{LP}T$ with $R_d = PT/N_0$ as a parameter, RC filter (NRZ data) using (*a*) high SNR approximation.

151

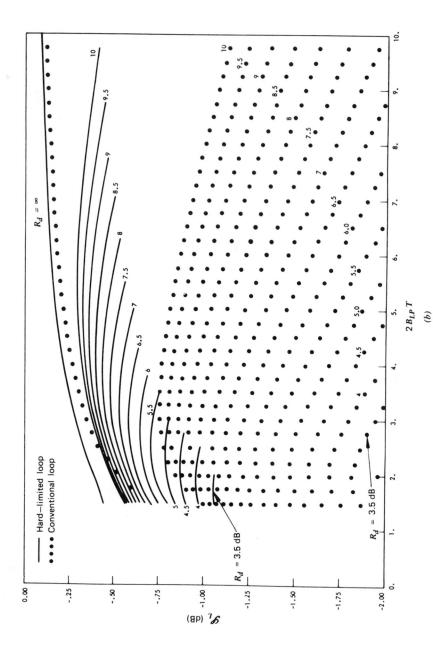

Figure 5.9 Continued. (*b*) low SNR approximation (from Simon [15] with permission).

where

ρ = loop SNR in B_L

ω_n = natural frequency of the loop in rad/sec

$d = \exp(-\pi\zeta/\sqrt{1-\zeta^2})$, when $\zeta \leq 1$ and zero otherwise

ζ = loop damping factor

For $\zeta = 0.707$, it can be shown that

$$R_{90} = 0.543\left(1 - \frac{\sqrt{2}}{\sqrt{\rho}}\right)B_L^2$$

We first consider the squaring loop. In order to modify this result, an effective ρ must be defined. It was shown that [see (5.1-37)]

$$(\rho_e)^{-1} = \sigma_{\Phi_e}^2 = \frac{4N_0B_L}{\alpha P}\left[\frac{\alpha'}{\alpha} + \frac{N_0W}{2\alpha P}\left(1 - \frac{1}{2n}\right)\right] \qquad \text{rad}^2 \qquad (5.7\text{-}2)$$

for the n-pole baseband equivalent Butterworth filter, squaring loop, phase error variance. The term ρ_e denotes the effective SNR. As before, W is the positive frequency bandwidth, and α is the relative power passed by the presquaring filter of bandwidth W. A model is shown in Figure 5.10.

The variance expression, of course, depends upon the type of filter used.

Acquisition time is arbitrarily defined here as the time to sweep the total uncertainty including doppler and oscillator instabilities. Denoting this frequency range by f_d and letting $\dot{f}_d$ denote the maximum doppler rate, it follows that

$$T_{acq} = \frac{f_d}{R_{90}/2 - |\dot{f}_d|} \qquad \text{sec} \qquad (5.7\text{-}3)$$

is the time to make one complete sweep under the worst-case doppler rate environment. Notice, due to the squaring, the nominal effective rate of sweeping relative to the original carrier is $R_{90}/2$. The sweep rate (R_{90}) plus $2|\dot{f}_d|$ is the maximum allowable effective sweep rate to achieve 90% probability of acquisition. The minimum sweep rate is equal to $R_{90} - 2|\dot{f}_d|$.

For the Costas loop, the same results apply since the same SDE applies,

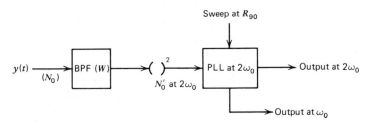

Figure 5.10 Model for squaring loop acquisition.

except that the phase error variance expression used in the allowable sweep rate expression must be for the Costas loop. The doubling effect of the doppler and doppler rate also effectively occurs in the Costas loop, so T_{acq} of (5.7-3) also applies to the Costas loop case. The only essential difference between the Costas loop and the squaring loop is that the Costas operates at a nominal frequency of ω_0 rad/sec, whereas the squaring loop operates at $2\omega_0$ rad/sec and the Costas loop is swept at $R_{90}/2 - |\dot{f}_d|$ Hz/sec. The reason that the Costas loop "sees" effectively twice the doppler is because the Costas loop S-curve is compressed to one-half that of a cw loop and, hence, it can tolerate only one-half the static phase error.

5.8 LOCK DETECTOR FOR COSTAS AND SQUARING LOOPS

Closely allied with the subject of acquisition is the topic of lock indication or detection. The lock detectors for Costas loops and squaring loops are different, and we will discuss them separately. The lock detector's function is to indicate the presence or absence of the signal.

The squaring loop lock detector that is commonly used is basically the same type as that used on the cw loop. It operates off a coherent amplitude detector (CAD). A block diagram is shown in Figure 5.11.

After the signal is squared, the resulting line spectra are handled just as they are in the cw loop case; that is, a CAD is used to develop a signal to drive the lowpass filter (LPF) and, in turn, control the AGC lock detection circuitry and, possibly, an acquisition to tracking bandwidth change.

In order to obtain statistics of the lock detector, it is necessary to describe the stochastic process at the output of the BPF. We model the

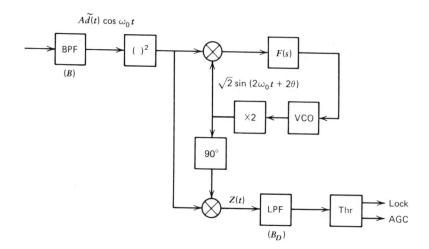

Figure 5.11 Squaring loop lock detector.

filtered WGN in the usual way:

$$\tilde{n}(t) = \sqrt{2}\,\tilde{n}_c(t)\cos\omega_0 t + \sqrt{2}\,\tilde{n}_s(t)\sin\omega_0 t \tag{5.8-1}$$

and the filtered signal as

$$s(t) = \sqrt{2}\,A\tilde{d}(t)\cos(\omega_0 t) \tag{5.8-2}$$

with $\tilde{d}(t)$ being the baseband equivalent filtered version of $d(t)$, the binary-valued data sequence. The square of the signal plus noise is given by

$$y^2(t) = A^2\tilde{d}^2\cos 2\omega_0 t + \tilde{n}_c^2\cos\omega_0 t - \tilde{n}_s^2\cos 2\omega_0 t$$
$$+ 2a\tilde{d}\tilde{n}_c\cos 2\omega_0 t + 2A\tilde{d}\tilde{n}_s\sin 2\omega_0 t + 2\tilde{n}_c\tilde{n}_s\sin 2\omega_0 t$$
$$+ \text{dc terms} \tag{5.8-3}$$

where $\tilde{n}_c(t)$ and $\tilde{n}_s(t)$ are the filtered versions of $n_c(t)$ and $n_s(t)$. After multiplying by an appropriate form of the reference signal, we obtain*

$$z(t) = A^2\tilde{d}^2 + \tilde{n}_c^2 - \tilde{n}_s^2 + 2A\tilde{d}\tilde{n}_c + O(2\omega_0) \tag{5.8-4}$$

For moderate false alarm probabilities and probabilities of detector, a Gaussian assumption can be invoked at the output of the LPF to determine the lock detector performance, as was done in Chapter 4 if $B/B_D \gg 1$. A more accurate description of acquisition includes both the loop and the lock detector as a system. However, this analysis has not been reported in the open literature. Typically, an algorithm would be used to increase the lock detector probability of detection after acquisition.

There is more than one type of lock detector that can be incorporated into the system when a Costas loop is used. Two of the more common ways are shown in Figures 5.12a and 5.12b.

The square law detectors can be replaced with envelope detectors with negligible change in performance.

First consider the symmetric lock detector used in Figure 5.12a and assume $\phi = 0$. Let the signal plus noise be described by

$$y(t) = \sqrt{2}\,Ad(t)\cos\omega_0 t + \sqrt{2}\,n_s\sin\omega_0 t + \sqrt{2}\,n_c\cos\omega_0 t \tag{5.8-5}$$

so that the filtered quadrature signal $Q(t)$ is given by

$$Q(t) = \tilde{n}_s + O(2\omega_0) \tag{5.8-6}$$

The input signal $I(t)$ is given by

$$I(t) = A\tilde{d}(t) + \tilde{n}_c + O(2\omega_0) \tag{5.8-7}$$

where $O(2\omega_0)$ denotes terms at $2\omega_0$ and are assumed to be negligible.

The lock detector produces, discarding negligible terms,

$$z(t) = I^2(t) - Q^2(t) = A^2\tilde{d}^2 + 2A\tilde{d}\tilde{n}_c + \tilde{n}_c^2 - \tilde{n}_s^2 \tag{5.8-8}$$

---

*We have assumed that $\phi \approx 0$ in this analysis. It is a simple matter to generalize the result to the nonzero case.

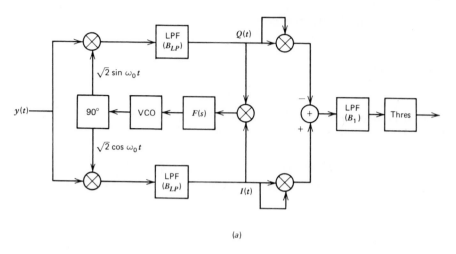

(a)

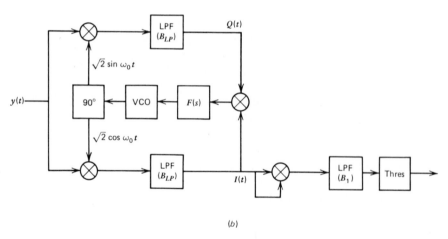

(b)

Figure 5.12 (a) Coherent lock detector. (b) Noncoherent lock detector.

which is exactly the same result obtained for the squaring loop. This signal can be used to provide a coherent AGC also.

The second lock detector of Figure 5.12b has an input to the LPF of the form

$$z(t) = A^2\tilde{d}^2 + 2A\tilde{d}\tilde{n}_c + \tilde{n}_c^2 \qquad (5.8\text{-}9)$$

Notice that since this signal is not zero mean, it can be used for a noncoherent AGC control.

The results obtained above assume that $\phi = 0$ but can be easily generalized to the nonzero case. Again the method of Chapter 4 can be used to approximate the probability of detection and false alarm.

5.9   FALSE LOCK IN COSTAS AND SQUARING LOOPS

This section deals with false lock in Costas and squaring loops, with emphasis on Costas loops. All the basic results apply only to long-loop squaring loops* since both loops have the same SDE. The false lock theory developed here applies to other data symbol formatting as well as to Manchester symbols.

In Costas loops, there are four principal mechanisms that are identifiable and can lead to false lock during acquisition. One mechanism is due to spurious oscillations (resulting from mixers, for example), which generate a sufficient error voltage to provide a stable lock to that spurious oscillation. Another mechanism can be attributed to accumulated delay in a long-loop implementation. A method of analyzing this effect has been reported by Leonhardt and Fleischmann [18], Lindsey [19], Develet [20], Tausworthe [21], Gardner [22], and others. A third mechanism is amplitude and phase modulation of the carrier due to periodic antenna movement. The fourth false lock mechanism is data related and is the subject of this section. The principal reason for data related false lock is the distortion due to filtering of the frequency offset signal.

As we have seen earlier in this chapter, optimizing the tracking performance of a Costas loop specifies the arm filter bandwidth. However, false lock protection is improved by increasing the arm filter bandwidth and thereby decreasing loop performance.

Early work on this data related false lock problem was attempted by Olson [23]; unfortunately, he neglected the arm filtering effect in his analysis and then concluded that an I^2-Q^2 lock would not respond to false locks. As we will see in the following, when the arm filtering is included, this is not the case. The results given here closely follow an original study by Hedin, Holmes, Lindsey, and Woo [24] that was later published as a paper [25]. Further works in this area are cited [26–29].

We consider false lock for both periodic and random data and then consider the lock detector output and show that it is represented by a function very similar to the error signal.

In the calculations to follow, it will be convenient to utilize complex notation. Those readers who are not familiar with it should refer to an appropriate text, such as the one by Stein and Jones [30].

To obtain a feeling for the false lock phenomenon, consider the Costas loop of Figure 5.13 and the waveforms of Figure 5.14 depicted for the case $\Delta = 2\pi/T$, where Δ is the false lock frequency, that is, the difference between the input and reference frequencies. The first figure on the left shows the beat note $\sin(\Delta t + \phi)$ ($\phi > 0$). In the second figure, the function $H(s)d(t)\sin(\Delta t + \phi)$ is shown assuming a single-pole RC arm filter with nominal distortion. The notation $H(s)f(t)$ denotes the Heaviside operator

---

*See Figure 5.2*b* for a block diagram of a long-loop version of a squaring loop.

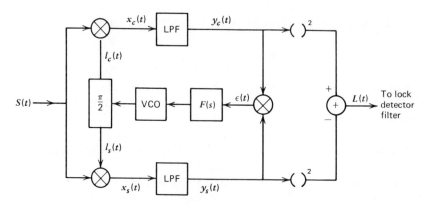

Figure 5.13 Block diagram of a Costas loop.

operating on the time function $f(t)$. The third figure depicts $H(s)d(t)\cos(\Delta t + \phi)$, and the fourth shows the product signal that drives the loop filter and VCO. For the value of ϕ shown ($\simeq 45°$), a "slice" is removed from the resultant sine wave so that the dc value is negative. Further, the loop will only respond to the dc value of the product plus any noise present (assuming $B_L \ll$ arm bandwidth). When ϕ is smaller, say, 22.5°, the last figure shows that the "slice" diminishes in size, thus providing proportional control.

The figures on the right depict the same type of waveforms for the case when $\phi < 0$. As can be seen, the dc value is now positive and, thus, the false lock point is stable. It can be demonstrated that if the data bits change sign, the slice remains pointed in the same direction, which indicates the false lock is a stable lock phenomenon.

5.9.1 Costas Loop Error Signal in the General Case

Figure 5.13 shows the block diagram of a typical Costas loop implementation in which the lock detector is also illustrated. The incoming signal with unity power can be written as

$$S(t) = \sqrt{2}\, d(t) \cos(\omega_c t + \theta) \qquad (5.9\text{-}1)$$

where ω_c is the incoming carrier frequency, θ the incoming signal phase, and $d(t)$ the Manchester symbol data sequence. Let the local oscillator be at frequency ω'_c and phase $\hat{\theta}$ and let

$$\Delta \equiv \omega_c - \omega'_c \qquad \text{rad/sec} \qquad (5.9\text{-}2)$$

be the frequency difference between the incoming carrier and the local oscillator, that is,

$$l_c(t) = \sqrt{2} \cos(\omega'_c t + \hat{\theta})$$

$$l_s(t) = -\sqrt{2} \sin(\omega'_c t + \hat{\theta}) \qquad (5.9\text{-}3)$$

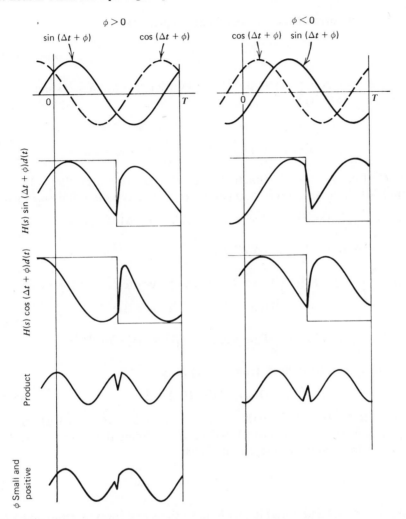

Figure 5.14 Error waveforms for the Costas loop during false lock showing the origin of the dc component ($\Delta t = 1/T$).

Neglecting double frequency signals, $x_c(t)$ and $x_s(t)$ can be written as

$$x_c(t) = d(t) \cos(\Delta t + \phi) = \text{Re}[d(t)e^{j\phi}e^{j\Delta t}]$$

$$x_s(t) = d(t) \sin(\Delta t + \phi) = \text{Re}\left[\frac{1}{j} d(t) e^{j\phi} e^{j\Delta t}\right]$$

(5.9-4)

where Re stands for the real part of a complex number, and where ϕ is the phase error. The signals written in the form (5.9-4) are bandpass signals with complex envelopes $d(t)$ and $(1/j) d(t)$, respectively. Let $D(\omega)$ be the

Fourier transform of the baseband data signal $d(t)$:

$$D(\omega) = \int_{-\infty}^{\infty} d(t)e^{-j\omega t}\, dt \tag{5.9-5}$$

Then (see, for example, [30, p. 77]) the Fourier spectra of $x_c(t)$ and $x_s(t)$ are given by, respectively,

$$x_c(\omega) = \tfrac{1}{2}D(\omega - \Delta)e^{j\phi} + \tfrac{1}{2}D^*(-\omega - \Delta)e^{-j\phi}$$
$$x_s(\omega) = \frac{1}{2j} D(\omega - \Delta)e^{j\phi} - \frac{1}{2j} D^*(-\omega - \Delta)e^{-j\phi} \tag{5.9-6}$$

Let $H(\omega)$ be the transfer function of the Costas loop arm filters. Then the outputs of these arm filters, $y_c(t)$ and $y_s(t)$, will have the following spectra:

$$Y_c(\omega) = \tfrac{1}{2}D(\omega - \Delta)e^{j\phi}H(\omega) + \tfrac{1}{2}D^*(-\omega - \Delta)e^{-j\phi}H(\omega)$$
$$Y_s(\omega) = \frac{1}{2j}D(\omega - \Delta)e^{j\phi}H(\omega) - \frac{1}{2j}D^*(-\omega - \Delta)e^{-j\phi}H(\omega) \tag{5.9-7}$$

Since $H(\omega) = H^*(-\omega)$, (5.9-7) can also be written in the following form:

$$Y_c(\omega) = \tfrac{1}{2}\tilde{D}(\omega - \Delta)e^{j\phi} + \tfrac{1}{2}\tilde{D}^*(-\omega - \Delta)e^{-j\phi}$$
$$Y_s(\omega) = \frac{1}{2j}\tilde{D}(\omega - \Delta)e^{j\phi} - \frac{1}{2j}\tilde{D}^*(-\omega - \Delta)e^{-j\phi} \tag{5.9-8}$$

where $\tilde{D}(\omega)$ can be defined from (5.9-7) to be

$$\tilde{D}(\omega) = D(\omega)H(\omega + \Delta) \tag{5.9-9}$$

Notice here $\tilde{D}(\omega) \neq \tilde{D}(-\omega)^*$, since $H(\omega + \Delta) \neq H^*(-\omega + \Delta)$ if the frequency error Δ is not zero. However, if we define the complex envelope $\tilde{d}(t)$ to be the inverse transform of $\tilde{D}(\omega)$:

$$\tilde{d}(t) = \frac{1}{2\pi} \int_{-\infty}^{\infty} \tilde{D}(\omega)e^{j\omega t}\, d\omega \tag{5.9-10}$$

then since $Y_c(\omega)$ and $Y_s(\omega)$ in (5.9-8) have the same form as $x_c(\omega)$ and $x_s(\omega)$ in (5.9-6), they are also bandpass signals and can be written analogously as

$$y_c(t) = \text{Re}[\tilde{d}(t)e^{j(\Delta t + \phi)}]$$
$$y_s(t) = \text{Re}\left[\frac{1}{j} \tilde{d}(t)e^{j(\Delta t + \phi)}\right] \tag{5.9-11}$$

Using the complex identity

$$\text{Re}\, Z_1 \cdot \text{Re}\, Z_2 = \tfrac{1}{2}[\text{Re}(Z_1 Z_2) + \text{Re}(Z_1 Z_2^*)]$$

the Costas error signal $\epsilon(t)$ can be computed from (5.9-11) to be the following:

$$\epsilon(t) = y_c(t)y_s(t) = \tfrac{1}{2}\text{Re}\left\{\frac{1}{j}[\tilde{d}(t)]^2 e^{j2(\Delta t + \phi)}\right\} \tag{5.9-12}$$

The above equation can also be written in the alternate form

$$\epsilon(t) = \tfrac{1}{2} \operatorname{Im}\{[\tilde{d}(t)]^2 e^{j2(\Delta t + \phi)}\} \qquad (5.9\text{-}13)$$

where Im( ) stands for the imaginary part of the following complex number. Since $[\tilde{d}(t)]^2$ is in general complex (except when $\Delta = 0$), the Costas error signal $\epsilon(t)$ can be written as

$$\epsilon(t) = \tfrac{1}{2}|[\tilde{d}(t)]^2| \sin[2(\Delta t + \phi) + \zeta(t)] \qquad (5.9\text{-}14)$$

where

$$[\tilde{d}(t)]^2 = |[\tilde{d}(t)]^2| e^{j\zeta(t)}$$

Notice from (5.9-9) that when $\Delta = 0$, $\tilde{d}(t)$ is real and thus $\zeta(t)$ is zero. Thus in the case of true lock the error signal is of the well-known form $\tfrac{1}{2}\alpha \sin 2\phi$, with $\alpha = \tilde{d}(t)^2$, the signal degradation due to filter distortion. However, in the case of false lock ($\Delta \neq 0$), the error signal has the phase variation $\zeta(t)$ in addition to the amplitude variation, which gives another dimension of complexity to the dynamic behavior of the loop in the case of false lock.

The major driving force in the loop error signal that drives the loop into lock during sweep acquisition is the dc component in $\epsilon(t)$. When there is filter distortion, $[\tilde{d}(t)]^2 \neq 1$ in (5.9-14). In fact, $[\tilde{d}(t)]^2$ will have dc plus higher harmonics, and it is the higher harmonics in $[\tilde{d}(t)]^2$ that, when multiplied by the term $e^{j(2\Delta t + 2\phi)}$ in (5.9-13), give rise to a dc component in the error signal $\epsilon(t)$ when $\Delta \neq 0$. This component drives the loop into false lock. To evaluate the relative magnitudes of these dc components at various Δ, the spectrum of $[\tilde{d}(t)]^2$ has to be evaluated at various Δ. We consider first periodic sequences and then treat the case of random data.

5.9.2 Periodic Data

So far the results are general, in that they can include both aperiodic data and periodic data. For a periodic data sequence with period p the input baseband data signal has the following line spectrum:

$$D_p(\omega) = 2\pi \sum_{k=-\infty}^{\infty} C_k \delta(\omega - k\omega_0) \qquad (5.9\text{-}15)$$

where the subscript p stands for periodic signals, and where $\omega_0 = 2\pi/pT$, with T equal to the Manchester symbol duration, $\delta(\omega)$ the Dirac δ function, and C_k the Fourier coefficients of $d(t)$:

$$C_k = \frac{1}{pT} \int_0^{pT} d(t) e^{-jk\omega_0 t} \, dt \qquad (5.9\text{-}16)$$

With $D_p(\omega)$ given by (5.9-15), the signal $\tilde{d}(t)$ can be computed from (5.9-10) as follows:

$$\tilde{d}_p(t) = \sum_{k=-\infty}^{\infty} C_k H(k\omega_0 + \Delta) e^{jk\omega_0 t} \qquad (5.9\text{-}17)$$

For Manchester encoded binary periodic sequences with period p the function $[\tilde{d}(t)]^2$ can be represented via

$$[\tilde{d}_p(t)]^2 = \sum_{k=-\infty}^{\infty} \sum_{l=-\infty}^{\infty} C_k C_l H(k\omega_0 + \Delta) H(l\omega_0 + \Delta) e^{j(k+l)\omega_0 t} \qquad (5.9\text{-}18)$$

where $\omega_0 = 2\pi/(pT)$, and $\{C_k\}$ are the Fourier coefficients of the periodic data signal $d(t)$, that is,

$$C_k = \frac{1}{pT} \int_0^{pT} d(t) e^{-jk\omega_0 t}\, dt \qquad (5.9\text{-}19)$$

For a Manchester encoded binary sequence with period p it is easily shown that C_k can be evaluated to be the following [for the interval $(0, T)$]:

$$C_k = \frac{1}{pT} \sum_{n=0}^{p-1} d_n e^{-jkn\omega_0 T} \cdot \frac{2 - 2\cos \omega T/2}{j\omega} e^{-j\omega T/2}\Big|_{\omega = k\omega_0} \qquad (5.9\text{-}20)$$

where $d_0, d_1, \ldots, d_{p-1}$ represents the ± 1 data sequence. Since the harmonics in $[\tilde{d}(t)]^2$ in (5.9-18) occur at multiples of

$$\frac{\omega_0}{2\pi} = \frac{1}{pT}$$

the frequency offset Δ can be at values [see (5.9-13)]

$$\Delta_f = \frac{\Delta}{2\pi} = \frac{m}{2pT} = \frac{m}{2p} R_s \qquad m = 0, \pm 1, \pm 2, \ldots \qquad (5.9\text{-}21)$$

to produce a dc component in $\epsilon(t)$. It is straightforward to see from (5.9-13) and (5.9-18) that this dc component is of the form

$$\epsilon(\text{dc}) = \tfrac{1}{2}|Q| \sin(2\phi + \psi) \qquad (5.9\text{-}22)$$

where Q is observed from (5.9-18) to be the complex quantity

$$Q = |Q| e^{j\psi} \equiv \sum_{\substack{k=-\infty \\ (k+l)\omega_0 = -2\Delta}}^{\infty} \sum_{l=-\infty}^{\infty} C_k C_l H(k\omega_0 + \Delta) H(l\omega_0 + \Delta) \qquad (5.9\text{-}23)$$

Q is a function of the frequency offset Δ, and the magnitude of Q gives the maximum amplitude of $\epsilon(\text{dc})$ (at $2\phi + \psi = \pm \pi/2$). When $\Delta \neq 0$, ψ may not be zero for an arbitrary periodic sequence. But for sequences with equal numbers of ± 1's it is observed that ψ is zero for the various values of Δ given in (5.9-21).

More generally the Costas error signal $\epsilon_p(t)$ for periodic input signals is given from (5.9-13) by

$$\epsilon_p(t) = \frac{1}{2} \text{Im} \left[\sum_{k=-\infty}^{\infty} \sum_{l=-\infty}^{\infty} C_k C_l H(k\omega_0 + \Delta) H(l\omega_0 + \Delta) e^{j(k+l)\omega_0 t} e^{j2(\Delta t + \phi)} \right]$$

$$(5.9\text{-}24)$$

For several periodic sequences and for one-pole RC data arm filters with transfer function

$$H(\omega) = \frac{1}{1 + j(\omega/\omega_i)} \tag{5.9-25}$$

the relative magnitudes of the dc components in the error signal, that is, in (5.9-22), are computed for various values of $\Delta_f T$, where Δ_f is the frequency detuning (Hz) and T is the symbol duration (sec). The periodic sequences selected are shown in Table 5.1 for Manchester data.

The sequences (C) and (D) are actually rate 1/3 convolutionally encoded sequences of (A) and (B).

For the product $(\omega_i/2\pi) \times T = 1.4$ (the minimum phase variance values) and using Manchester data, the computer results of $|Q|$ for these sentences at various values of Δ_f are shown in Figure 5.14, where the $|Q|$'s are given in terms of dBs from unity. For example, consider sequence (D) in Figure 5.15. The difference between the dc components in the error signal $\epsilon(t)$ when $\Delta_f T = 0$ and when $\Delta f_T = 1.0$ is about 3.5 dB. Since, in both cases $\epsilon(dc)$ has a sine dependence on 2ϕ, the loop can be driven into lock at $\Delta_f T = 1$ as well as at $\Delta_f T = 0$. We note from Figure 5.15 that there is a countable number of false lock frequencies, the number depending on the period of the data sequence.

Periodic PN sequencies of periods 7 and 31 were also investigated. Their resulting dc components in the Costas error signal at various frequency offsets are shown in Figures 5.16a and 5.16b. Note that the situations of PN sequences approach that of a pure random sequence, which is shown in Figure 5.16c and will be discussed below.

5.9.3 Random Data Sequences

When the data is random and values ± 1 are taken with equal probability, the results obtained in the previous sections can be further simplified. From (5.9-13) it is observed that the Costas error signals $\epsilon(t)$ can be written

Table 5.1 Periodic sequence examples*

No.	Period	Data sequence
A	2	$-1, 1$
B	4	$-1, -1, 1, 1$
C	6	$-1, -1, -1, -1, 1, 1$
D	12	$-1, 1, 1, 1, -1, -1, -1, -1, -1, 1, 1, 1$

*These sequences were picked based upon an idle pattern that occurred in the space shuttle S-band receiver built at TRW DSSG.

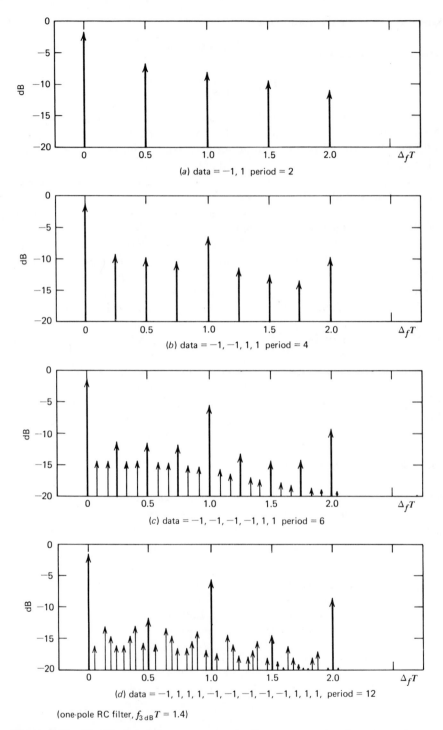

(a) data = −1, 1 period = 2

(b) data = −1, −1, 1, 1 period = 4

(c) data = −1, −1, −1, −1, 1, 1 period = 6

(d) data = −1, 1, 1, 1, −1, −1, −1, −1, −1, 1, 1, 1, period = 12

(one-pole RC filter, $f_{3\,dB}T = 1.4$)

Figure 5.15 Amplitude of dc component in Costas error signal (or lock detector signal), Manchester data.

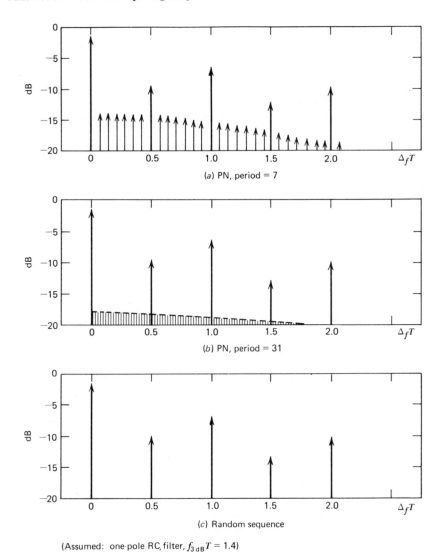

(a) PN, period = 7

(b) PN, period = 31

(c) Random sequence

(Assumed: one-pole RC, filter, $f_{3\,dB}T = 1.4$)

Figure 5.16 Amplitude of dc component in Costas error signal (or lock detector signal), Manchester data.

as:

$$\epsilon(t) = \tfrac{1}{2}\,\mathrm{Im}\{\tilde{d}(t)^2 e^{j2(\Delta t + \phi)}\} \tag{5.9-26}$$

where $\tilde{d}(t)$ is a complex function with transform (equation 5.9-9)

$$\tilde{D}(\omega) = D(\omega)H(\omega + \Delta) \tag{5.9-27}$$

where $D(\omega)$, $H(\omega)$ are, respectively, the Fourier transforms of the data sequence and the transfer function of the arm filters. The data sequence $d(t)$ can be written as

$$d(t) = \sum_{k=-\infty}^{\infty} d_k m(t - kT) \tag{5.9-28}$$

where $m(t)$ is the Manchester symbol with duration T and d_k are random ± 1 data. Let $M(\omega)$ be the transform of $m(t)$. Then the transform of $d(t)$ is given by

$$D(\omega) = \sum_{k=-\infty}^{\infty} d_k M(\omega) e^{-jk\omega T} \tag{5.9-29}$$

and $\tilde{D}(\omega)$ is thus equal to

$$\tilde{D}(\omega) = \sum_{k=-\infty}^{\infty} d_k M(\omega) H(\omega + \Delta) e^{-jk\omega T} \tag{5.9-30}$$

From (5.9-30) it is obvious that $d(t)$ is actually the following complex waveform:

$$\tilde{d}(t) = \sum_{k=-\infty}^{\infty} d_k q(t - kT) \tag{5.9-31}$$

where $q(t)$ is the inverse transform of $M(\omega) H(\omega + \Delta)$, that is,

$$q(t) = \frac{1}{2\pi} \int_{-\infty}^{\infty} H(\omega + \Delta) M(\omega) e^{j\omega t} \, d\omega \tag{5.9-32}$$

Since $H(\omega + \Delta)$ is not symmetric with respect to zero when $\Delta \neq 0$, $q(t)$ is in general not real. With $\tilde{d}(t)$ given above, the term $[\tilde{d}(t)]^2$ in the loop error signal $\epsilon(t)$ can be written as

$$[\tilde{d}(t)]^2 = \sum_{k=-\infty}^{\infty} q^2(t - kT) + \sum_{\substack{k=-\infty \\ k \neq l}}^{\infty} \sum_{l=-\infty}^{\infty} d_k d_l q(t - kt) q(t - lt) \tag{5.9-33}$$

The first term in (5.9-33) is a periodic (complex) function with period T, which will give rise to line spectra with harmonics at the symbol rate. For random data the second term in (5.9-33) only gives rise to continuous spectra, and, in the case that d_k are ± 1 with equal probability and statistically independent this term has zero mean. Thus for random data the loop error signal is equal to

$$\epsilon(t) = \tfrac{1}{2} \operatorname{Im}\left\{ \left[\sum_{k=-\infty}^{\infty} q^2(t - kT) \right] e^{j2(\Delta t + \phi)} \right\} \tag{5.9-34}$$

Since the function $\sum_k q^2(t - kT)$ is periodic, it can be represented by a Fourier series:

$$\sum_k q^2(t - kT) = \sum_{n=-\infty}^{\infty} C_n \exp^{(j[2n\pi/T]t)} \tag{5.9-35}$$

when the Fourier coefficients C_n are given by

$$C_n = \frac{1}{T} \mathscr{F}[q^2(t)]_{\omega = n(2\pi/T)} \tag{5.9-36}$$

where $\mathcal{F}[q^2(t)]$ stands for the Fourier transform of $q^2(t)$. It is easier to compute $\mathcal{F}[q^2(t)]$ by first computing the time domain function $q^2(t)$. We only carry out the computations here for the case of an RC filter with transfer function

$$H(\omega) = \frac{1}{1 + j(\omega/\omega_i)} \tag{5.9-37}$$

where $\omega_i = 1/RC$ is the 3-dB cutoff frequency in rad/sec. We assume a Manchester symbol and leave the case of NRZ symbols for a problem.

Writing the Manchester symbol waveform $m(t)$ in terms of three unit step functions, we immediately obtain

$$Q(\omega) = \mathcal{F}[q(t)] = M(\omega)H(\omega + \Delta)$$

$$= (1 - 2e^{-j\omega(T/2)} + e^{-j\omega T}) \frac{1}{j\omega} \cdot \frac{1}{1 + j[(\omega + \Delta)/\omega_i]} \tag{5.9-38}$$

From $Q(\omega)$ it follows directly that $q(t)$ is given by

$$q(t) = \frac{1}{1 + j(\Delta/\omega_0)} \left(m(t) - \{ e^{-(\omega_i + j\Delta)t} u(t) \right.$$

$$- 2e^{-(\omega_i + j\Delta)(t - T/2)} u\left(t - \frac{T}{2} \right)$$

$$+ e^{-(\omega_i + j\Delta)(t - T)} u(t - T) \} \bigg) \tag{5.9-39}$$

where $u(t)$ is the unit step function and $m(t)$ is the undistorted Manchester symbol waveform. From (5.9-39) it can be shown by first squaring $q(t)$ and then taking its Fourier transform that

$$\frac{1}{T} \mathcal{F}[q^2(t)] = \left(\frac{1}{1 + j(\Delta/\omega_i)} \right)^2 \left\{ e^{-j\omega T/2} \left(\frac{\sin \omega T/2}{\omega T/2} \right) \right.$$

$$- \frac{2}{(j\omega + \alpha)T} [1 + 2e^{-j\omega T/2}(1 - e^{-\alpha T/2}) + e^{-j\omega T}(e^{-\alpha T} - 2e^{-\alpha T/2}]$$

$$+ \frac{1}{(j\omega + 2\alpha)T} [1 + 4e^{-j\omega T/2}(1 - e^{-\alpha T/2})$$

$$+ e^{-j\omega T}(1 + 2e^{-\alpha T} - 4e^{-\alpha T/2})] \bigg\} \tag{5.9-40}$$

where the complex number α is defined to be

$$\alpha \equiv \omega_0 + j\Delta = \frac{1}{R_1 C} + j\Delta \tag{5.9-41}$$

Substituting (5.9-40) into (5.9-36), we can obtain the Fourier coefficients C_n of the periodic function $\Sigma_k q^2(t - kt)$. These Fourier coefficients, for a given frequency offset Δ, are the following:

$$
\begin{aligned}
C_n = &\left(\frac{1}{1+j\dfrac{\Delta T}{2\pi R}}\right)^2 \left[(-1)^n \frac{\sin n\pi}{n\pi} - \frac{1}{2\pi R}\left(\frac{1+j\dfrac{\Delta T}{2\pi R}}{1+j\dfrac{2n\pi+\Delta T}{2\pi R}}\right) \right. \\[2ex]
&\times \left(\frac{1}{1+j\dfrac{n\pi+\Delta T}{2\pi R}}\right)\{1+2(-1)^n - 2e^{-(\pi R+j\Delta(T/2))}[1+(-1)^n] \\[2ex]
&\left. + e^{-(2\pi R+j\Delta T)}\} \vphantom{\Big]}\right]
\end{aligned}
\tag{5.9-42}
$$

Where R is defined to be the ratio between the 3-dB bandwidth of the RC filter and the symbol rate, that is,

$$
R = \frac{\omega_i T}{2\pi} = f_i T
\tag{5.9-43}
$$

Since the harmonics $\{C_n\}$ are separated by $1/T$, the possible values of Δ, which will give rise to a dc component in the error signal, thus creating an error voltage which drives the VCO into lock, are at multiples of half the symbol rate, that is,

$$
\Delta = \frac{k\pi}{T} \qquad k = 0, \pm 1, \pm 2, \ldots
\tag{5.9-44}
$$

The case of $k = 0$ will be the case of true lock, while the other cases will be possible false lock offset frequencies. At those values of Δ, see (5.9-44), the Fourier coefficients C_n can be computed from (5.9-42) to be the following:

$$
\begin{aligned}
C_n = &\left(\frac{1}{1+j\dfrac{k}{2R}}\right)^2 \left((-1)^n \frac{\sin n\pi}{n\pi} \right. \\[2ex]
&- \frac{1}{2\pi R} \frac{1+j\dfrac{k}{2R}}{1+j\left(\dfrac{n+k}{2R}\right)} \frac{1}{1+j\left(\dfrac{2n+k}{2R}\right)} \\[2ex]
&\left. \times \{1+2(-1)^n - 2[1+(-1)^n]e^{-(\pi R+jk(\pi/2))} + (-1)^k e^{-2R}\}\right)
\end{aligned}
\tag{5.9-45}
$$

The magnitude of the dc component in $\epsilon(t)$, for $\Delta = k\pi/T$, is then given by C_{-k}, which is computed by substituting $n = -k$ in (5.9-45) to be the following:

$$C_{-k} = \text{dc component in } \epsilon(t) \text{ when } \Delta = k\pi/T$$

$$= \left(\frac{1}{1+j\dfrac{k}{2R}}\right)^2 \left\{(-1)^k \frac{\sin k\pi}{k\pi} - \frac{1}{2\pi R}\left(\frac{1+j\dfrac{k}{2R}}{1-j\dfrac{k}{2R}}\right)\right.$$

$$\left. \times[1 + 2(-1)^k + 4e^{-\pi R}\cos\frac{k\pi}{2} + (-1)^k e^{-2\pi R}]\right\} \qquad (5.9\text{-}46)$$

The dc component (when $\Delta = k\pi/T$) is always real. Hence C_k is actually the magnitude of the loop error signal:

$$\epsilon(t)_{\text{dc}} = \tfrac{1}{2}C_{-k}\sin 2\phi \qquad (5.9\text{-}47)$$

For $k = 0, 1, 2$ the following values of C_{-k} are obtained:

$$k = 0, \qquad C_0 = 1 - \frac{1}{2\pi R}(3 - 4e^{-\pi R} + e^{-2\pi R})$$

$$k = 1, \qquad C_{-1} = \frac{1 + e^{-2\pi R}}{2\pi R[1 + (1/2R)^2]} \qquad (5.9\text{-}48)$$

$$k = 2, \qquad C_{-2} = -\frac{3 + 4e^{-\pi R} + e^{-2\pi R}}{2\pi R[1 + (1/R)^2]}$$

where R is defined in (5.9-43). For arbitrary $k \geq 1$, the following C_{-k}'s gives the magnitudes of the dc components in the loop error signal corresponding to $\Delta = k\pi/T$ in Table 5.2.
Notice that the magnitude of this dc component is the same for $\Delta =$

Table 5.2 Random data—Manchester data—magnitudes of dc component in $\overline{\epsilon(t)}$

Frequency Offset Δ		C_n
$\Delta_f = 0$	true lock	$1 - \dfrac{1}{2\pi R}(3 - 4e^{-\pi R} + e^{-2\pi R})$
$\Delta_f = \dfrac{n}{2T}$	n odd $(\pm 1, \pm 3, \pm 5, \ldots)$	$\dfrac{1 + e^{-2\pi R}}{2\pi R[1 + (n/2R)^2]}$
$\Delta_f = \dfrac{n}{2T}$	n even $(\pm 2, \pm 4, \pm 6, \ldots)$	$-\dfrac{3 - (-1)^{n/2}4e^{-\pi R} + e^{-2\pi R}}{2\pi R[1 + (n/2R)^2]}$

$\pm k\pi/T$:

$$M_{dc}\left(\Delta = \frac{k\pi}{T}\right) = M_{dc}\left(\Delta = -\frac{k\pi}{T}\right) \tag{5.9-49}$$

Hence, the signal level of false lock on the carrier plus a multiple of half the symbol rate is identical to that of locking on the carrier minus the same multiple of half the symbol rate.

In Figure 5.17 $|C_0|$ through $|C_{-4}|$ are plotted as a function of $R = f_{3\,dB}T = f_iT$ for the case of Manchester data with one-pole lowpass arm filters. In order to obtain a feel of the false lock protection level as a function of bandwidth, $|C_0/C_{-k}|$, $k = 1, \ldots, 5$ are plotted again for the case of one-pole lowpass filters with Manchester data in Figure 5.18.

The above false lock results for Manchester data have been compared

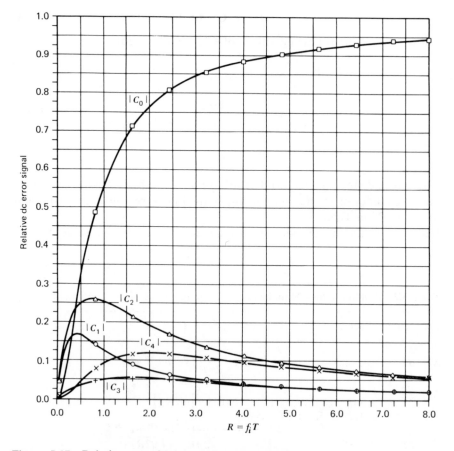

Figure 5.17 Relative magnitudes of Costas loop true lock and false lock error signals versus arm filter time bandwidth product, Manchester data.

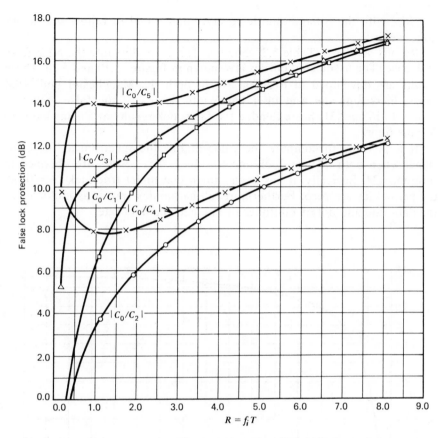

Figure 5.18 False lock protection versus arm filter BT product, Manchester data.

Table 5.3 Random data (NRZ)—magnitudes of dc components in $\overline{\epsilon(t)}$

Frequency Offset Δ_f		C_n
$\Delta_f = 0$	true lock	$1 - \dfrac{1}{2\pi R}[1 - e^{-2\pi R}]$
$\Delta_f = \dfrac{n}{2T}$	n odd $(\pm 1, \pm 3, \pm 5, \ldots)$	$-\dfrac{1 + e^{-2\pi R}}{2\pi R[1 + (n/2R)^2]}$
$\Delta_f = \dfrac{n}{2T}$	n even $(\pm 2, \pm 4, \pm 6, \ldots)$	$-\dfrac{1 - e^{-2\pi R}}{2\pi R[1 + n/2R)^2]}$

at TRW with the S-band shuttle receiver and found to follow the theory quite well.

Results similar to Table 5.2, shown in Table 5.3, illustrate the NRZ false lock case. The associated curves are shown in Figures 5.19 and 5.20. These results are developed in Problem 10. In both the Manchester and NRZ cases, increasing the arm filter bandwidth decreases the false lock tendency. LaFlame [31] has extended the false lock results of Costas loops to N-pole filters.

It is possible to compute the value of the line spectrum of the error signal when both in false lock and true lock. The results are obtained in reference 24. Basically, it has been shown that under true lock ($\phi = 0$) all the harmonics are zero. Furthermore, all harmonics appear as multiples of the data rate and not as multiples of one-half the data rate.

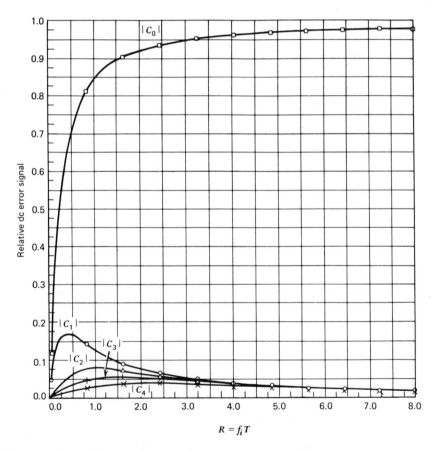

Figure 5.19 Relative magnitudes of Costas loop true lock and false lock error signals versus arm filter time bandwidth product, NRZ data.

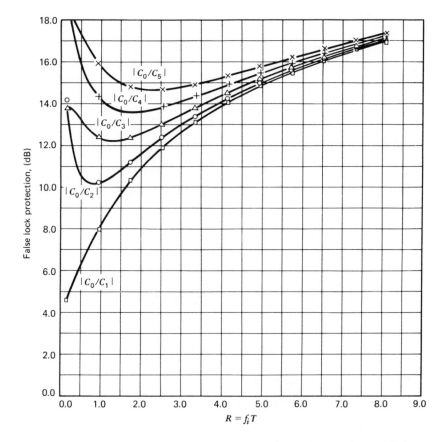

Figure 5.20 False lock protection versus arm filter BT product, NRZ data.

PROBLEM 10

Using the same methods as in the text, show for NRZ data that

$$C_{-k} = \left[\frac{1}{1 + j(k/2R)}\right]^2 \left\{ (-1)^k \frac{\sin \pi k}{\pi k} - \frac{1}{2\pi R - jk\pi} [-2e^{-2\pi R}(-1)^k + 2] \right.$$
$$\left. + \frac{1}{4\pi R} [2 - 2e^{-2\pi R}(-1)^k] \right\}$$

It therefore follows that

$$C_0 = 1 - \frac{1}{2\pi R} (1 - e^{-2\pi R})$$

$$C_{-1} = \frac{-1}{[1 + (1/2R)^2]} \frac{1}{2\pi R} (1 + e^{-2\pi R})$$

$$C_{-2} = \frac{-(1 - e^{-2\pi R})}{[1 + (1/R)^2]2\pi R}$$

and in general

$$C_{-k} = \frac{-(1-e^{-2\pi R})}{2\pi R[1+(k/2R)^2]} \qquad k \text{ odd}$$

$$C_{-k} = \frac{-(1+e^{-2\pi R})}{2\pi R[1+(k/2R)^2]} \qquad k \text{ even}$$

PROBLEM 11

By considering the in-phase and quadrature phase outputs in an integrate-and-dump type arm filter Costas loop show that

(a) For NRZ waveforms that the mean product (third multiplier output) is given by

$$\overline{\epsilon(t)} = -\frac{1}{2}\sum_{k=-\infty}^{\infty} p(t-kT)\frac{1}{\Delta^2}\sin 2\phi$$

$$= 0 \qquad m \text{ even}$$

when $\Delta = m\pi/T$ rad/sec $(m \neq 0)$. Hint: $\sum_{k=-\infty}^{\infty} p(t-kT)$ is a constant.

(b) For true lock $(m = 0)$

$$\overline{\epsilon(t)} = \frac{T^2}{8}\sin 2\phi$$

(c) The false lock protection is

$$FL_{\text{prot}} = 10 \log[\tfrac{2}{8}(m\pi)^2] \qquad dB \qquad m \text{ odd}$$

and there is no false lock when m is even. Note for $m = 1$, $FL_{\text{prot}} = 3.9$ dB at $\Delta_f = 1/2T$.

PROBLEM 12

Do Problem 11 for the case of Manchester symbols.

PROBLEM 13

The false lock theory presented here considers only the signal $\times$ signal term in determining the false lock levels. However, in reality, the signal $\times$ noise terms also change. Show that the more complete comparison of false lock in a Costas loop is specified by a comparison of $\mathscr{S}_L^{-1}$, where (see [27])

$$\sigma_\phi^2 = \frac{4N_0B_L}{A^2}\mathscr{S}_L^{-1}$$

and where

$$\mathscr{S}_L^{-1} = \frac{\left[\frac{1}{T}\int_{-\infty}^{\infty}|H(f)|^4 M\left(-\omega-\frac{n\pi}{T}\right)M\left(\omega+\frac{n\pi}{T}\right)df + \frac{N_0}{2P}\int_{-\infty}^{\infty}|H(f)|^4\,df\right]}{\left[\frac{1}{T}\int_{-\infty}^{\infty}|H(f)|^2 M\left(-\omega+\frac{n\pi}{T}\right)M\left(\omega+\frac{n\pi}{T}\right)df\right]^2}$$

with $M(\omega)$ the Fourier transform of the data pulse and $H(f)$ the transfer function of the arm filters.

Although the theory presented was derived for Costas loops, it applies to long-loop squaring loops when the lowpass filter equivalent of the presquaring filter is the same as the Costas arm filter. Chang and Kleinberg [32] have extended the results here to the case of short-loop squaring loops. They have found that the false lock margin can be lower than that shown in Figure 5.20 for the long loop.

5.9.4 Lock Detector—Costas Loop

First we consider periodic data. From Figure 5.12a the signal going into the lock detector filter is given by

$$\mathcal{L}(t) = y_c^2(t) - y_s^2(t) \qquad (5.9\text{-}50)$$

From 5.9-11 we have

$$\mathcal{L}_c(t) = \text{Re}[\tilde{d}(t)e^{j(\Delta t + \phi)}]$$
$$\mathcal{L}_s(t) = \text{Im}[\tilde{d}(t)e^{j(\Delta t + \phi)}] \qquad (5.9\text{-}51)$$

Since $\text{Re}^2(Z) - \text{Im}^2(Z) = \text{Re}(Z^2)$ for any complex number Z, $\mathcal{L}(t)$ is thus given by

$$\mathcal{L}(t) = \text{Re}[[\tilde{d}(t)]^2 e^{j2(\Delta t + \phi)}] \qquad (5.9\text{-}52)$$

which can also be written as

$$\mathcal{L}(t) = |[\tilde{d}(t)]^2| \cos[2(\Delta t + \phi) + \zeta(t)] \qquad (5.9\text{-}53)$$

where

$$[\tilde{d}(t)]^2 \equiv |[\tilde{d}(t)]^2| e^{j\zeta(t)}$$

Following the same arguments as in Section 5.9.2 for periodic signals the lock detector signal $\mathcal{L}_p(t)$ is given by

$$\mathcal{L}_p(t) = \text{Re}\left[\sum_{k=-\infty}^{\infty} \sum_{l=-\infty}^{\infty} C_k C_l H(k\omega_0 + \Delta) H(l\omega_0 + \Delta) e^{j(k+l)\omega_0 t} \, e^{j2(\Delta t + \phi)} \right]$$
$$\omega_0 = 2\pi/pT \qquad (5.9\text{-}54)$$

where $\{C_k\}$ are the Fourier coefficients of the baseband data sequence $d(t)$, Δ is the frequency error in rad/sec, ϕ is the phase error, $\omega_0 = 2\pi/pT$, with p the period of the sequence, and T the Manchester symbol duration. The dc term in this signal is given by

$$\mathcal{L}_p(\text{dc}) = |Q| \cos(2\phi + \psi) \qquad (5.9\text{-}55)$$

where $Q = |Q|e^{j\psi}$ is defined in (5.9-23). This indicates that the dc component of the lock detector signal (which will be integrated up to exceed threshold by the lock detector filters and thus will falsely indicate lock when $\Delta \neq 0$) does have a cosine dependence on $2\phi + \psi$.

Next we compute the rf components in the lock detector signal $\mathcal{L}(t)$. From (5.9-24) we observe for the possible false lock frequencies

$$\Delta = \frac{m2\pi}{p}\frac{R_s}{2}, \qquad m = \pm 1, \pm 2, \ldots$$

the rf component in $\mathcal{L}_p(t)$ is given by

$$\mathcal{L}_p(\text{rf at } \omega_r) = \text{Re}(Q^+ e^{j2\phi} e^{j\omega_r t} + Q^- e^{j2\phi} e^{-j\omega_r t}) \tag{5.9-56}$$

where

$$Q^+ = \sum_{k=-\infty}^{\infty} \sum_{l=-\infty}^{\infty} C_k C_l H(k\omega_0 + \Delta) H(l\omega_0 + \Delta)$$

$$(k+l)\omega_0 + 2\Delta = \omega_r \tag{5.9-57}$$

$$Q^- = \sum_{k=-\infty}^{\infty} \sum_{l=-\infty}^{\infty} C_k C_l H(k\omega_0 + \Delta) H(l\omega_0 + \Delta)$$

$$(k+l)\omega_0 + 2\Delta = -\omega_r \tag{5.9-58}$$

From 5.9-56 the lock detector rf component at ω_r can be simplified to be the following:

$$\begin{aligned}
\mathcal{L}_p(\text{rf at } \omega_r) &= \text{Re}(Q^+ e^{j2\phi} e^{j\omega_r t}) + \text{Re}(Q^{-*} e^{-j2\phi} e^{j\omega_r t}) \\
&= \text{Re}[(Q^+ e^{j2\phi} + Q^{-*} e^{-j2\phi}) e^{j\omega_r t}] \\
&= |Z| \cos(\omega_r t + \eta)
\end{aligned} \tag{5.9-59}$$

where the asterisk denotes the complex conjugate and where

$$Z \equiv |Z| e^{j\eta} \equiv Q^+ e^{j2\phi} + Q^{-*} e^{-j2\phi} \tag{5.9-60}$$

It can be shown that the same rf component at ω_r in the loop error signal $\epsilon(t)$ is given by (see Problem 14)

$$\begin{aligned}
\epsilon_p(\text{rf at } \omega_r) &= \tfrac{1}{2}\text{Im}[(Q^+ e^{j2\phi} - Q^{-*} e^{-j2\phi}) e^{j\omega_r t}] \\
&= \tfrac{1}{2}|Y| \sin(\omega_r t + \zeta)
\end{aligned} \tag{5.9-61}$$

where

$$Y = |Y| e^{j\zeta} = Q^+ e^{j2\phi} - Q^{-*} e^{-j2\phi} \tag{5.9-62}$$

Comparing $\epsilon_p(\text{rf})$ and $\mathcal{L}_p(\text{rf})$ from the above equations, we see that they are related by phase rotations of each other, except for the constant factor $\tfrac{1}{2}$.

PROBLEM 14

Show that the rf component of the error signal at ω_r can be written as

$$\epsilon_p(\omega_r) = \tfrac{1}{2}|Y| \sin(\omega_r t + \zeta)$$

where

$$Y = |Y| e^{j\zeta} = Q^+ e^{j2\phi} - Q^{-*} e^{-j2\phi}$$

We now consider the case of random data for the lock detector. Using the same techniques as used on the error signal, it is not difficult to show that the dc component of the lock detector with Manchester data is given by

$$\mathcal{L}_{dc}(t) = \begin{cases} \left[1 - \dfrac{1}{2\pi R}(3 - 4e^{-\pi R} + e^{-2\pi R}) \right] \cos 2\phi & k = 0 \quad \Delta = \pm\dfrac{k\pi}{T} \\[3mm] -\dfrac{1 + 2(-1)^k - 4e^{-\pi R}\cos(k\pi/2) + (-1)^k e^{-2\pi R}}{2\pi R[1 + (k/2R)^2]}\cos 2\phi & k \geq 1 \end{cases}$$

(5.9-63)

The rf components, when $\Delta = k\pi/T$, are

$$\mathcal{L}_{rf}(t)\big|_{\omega = 2\pi l/T} = |Q_{kl}(\phi)| \cos\left(\frac{2\pi l}{T} t + \eta\right)$$

(5.9-64)

where

$$Q_{kl}(\phi) \equiv C_{-k+l}e^{j2\phi} + C^{*}_{-k+l}e^{-j2\phi} = |Q_{kl}(\phi)|e^{j\eta}$$

(5.9-65)

and C_n is given in (5.9-45).

Notice even in the case of true lock ($\phi = 0$) the rf components of the lock detector output are not zero (though they are smaller than in the case of false lock) while in the phase detector output the same rf component is zero when ϕ is zero in the case of true lock.

PROBLEM 15

Establish the above result for the lock detector dc output with Manchester data when $\Delta = \pm k\pi/T$ with random data.

5.9.5 False Lock Detectors

There are many ways to detect false lock. One way is to wait until the doppler data are decoded on the ground and then note whether a large dis-

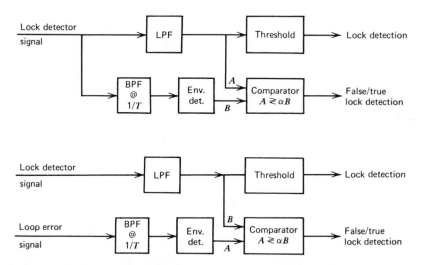

Figure 5.21 False lock detection schemes for $\Delta = 1/2T$, which is relatively insensitive to input signal level.

crepancy is observed. Another way, which is more practical, is to utilize the bit synchronizer that if designed properly will not recognize frame synchronization when the carrier loop is in false lock.

A more direct approach is to construct a false lock detector, two types of which are discussed below, under the assumption that false lock can only occur at ± one-half the data rate. See Figure 5.21a. This detector is

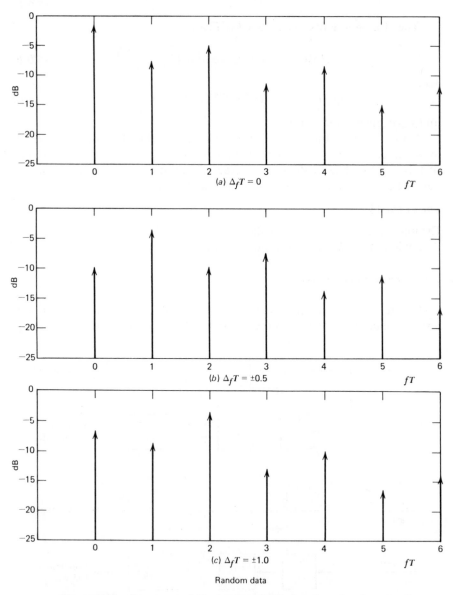

Random data

Figure 5.22 Spectrum of Costas loop lock detector signal at $\phi = 0$.

based on the fact that the dc component at the lock detector output is greater in level than any of the higher harmonics at multiples of the symbol rate when the loop is in true lock. Further, under false lock conditions the output component at the data rate is greater in level than the dc component. Thus by comparing the level of the detector output at the $f = 1/T$ frequency to the dc output of the detector the true state of lock can be ascertained. See Figure 5.22, in which the lock detector levels are shown for the case $f_{3\,dB}T = 1.4$, assuming Manchester data.

Another scheme, shown in Figure 5.21b, compares the data rate harmonic in the loop error signal with the dc component of the lock detector. This scheme appears to have a greater potential at the cost of greater complexity.

5.10   DECISION-DIRECTED FEEDBACK LOOPS

This section briefly considers decision-directed loops. These loops are not as well-understood as the Costas loop or the squaring loop, but tracking performance is one performance parameter that can be estimated.

5.10.1   Decision-Directed Feedback Loop with Delay

One coherent tracking device that will track a suppressed carrier signal is the decision-directed feedback loop (DDFL), which estimates the modulation and multiplies it by a delayed version of the input. The loop is shown in Figure 5.23. If we model the input signal plus noise by

$$y(t) = \sqrt{2P}\, m(t) \sin(\omega_0 t) + \sqrt{2}\, n_s \sin \omega_0 t + \sqrt{2}\, n_c \cos \omega_0 t \quad (5.10\text{-}1)$$

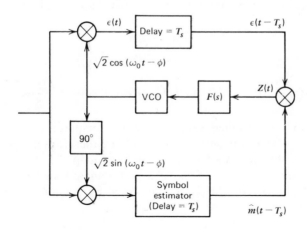

Figure 5.23   Decision directed feedback loop with delay.

where P is the signal power, $m(t)$ is the unit amplitude data sequence, and $n_s(t)$ and $n_c(t)$ are quadrature noise terms, then the error signal $\epsilon(t)$ is given by

$$\epsilon(t) = \sqrt{P}\, m(t) \sin \phi + n_c(t) \sin \phi + n_s(t) \cos \phi \qquad (5.10\text{-}2)$$

with ϕ the phase error.

The symbol estimator (usually a matched filter) produces $\hat{m}(t - T_s)$, which is multiplied by a delayed version of $\epsilon(t)$ to produce

$$z(t) = \sqrt{P}\, \hat{m}(t - T_s) m(t - T_s) \sin \phi + \hat{m}(t - T_s)$$
$$\times [n_c(t - T_s) \cos \phi + n_s(t - T_s) \sin \phi] \qquad (5.10\text{-}3)$$

When the loop bandwidth is narrow compared to the data rate $\mathscr{R} = 1/T_s$ the delay can be neglected (as far as tracking performance is concerned) so that we have

$$z(t) = \sqrt{P}\, \hat{m}(t - T_s) m(t - T_s) \sin \phi + \hat{m}(t - T_s) n(t - T_s) \qquad (5.10\text{-}4)$$

where, as in Chapter 4, $n(t)$ is modeled as a sample function of a WGN process. Now consider the product $\hat{m}(t)m(t)$. Since we have assumed that the loop is narrow, we can approximate $\hat{m}(t)m(t)$ with its time average or ensemble average, since the loop, when narrow, integrates over many symbols. Thus we can approximate the error signal by

$$z(t) = \sqrt{P}\,[1 - 2\text{PE}(\phi)] \sin \phi + \hat{m}(t - T_s) n(t - T_s) \qquad (5.10\text{-}5)$$

where we have used

$$E[m(t)\hat{m}(t)\,|\,\phi] = 1 - 2\text{PE}(\phi) \qquad (5.10\text{-}6)$$

with $\text{PE}(\phi)$ being the probability of a symbol error and with ϕ, the phase error, assumed constant over a symbol time. We neglect the self noise induced by the variation around the mean, $1 - 2\text{PE}(\phi)$.

To close the loop we note, letting the oscillator gain be k, that

$$\hat{\theta}(t) = \frac{k}{s}\, z(t) \qquad (5.10\text{-}7)$$

so that

$$\dot{\phi}(t) \cong \dot{\theta}(t) - k\sqrt{P}\, F(s) \left\{ [1 - 2\text{PE}(\phi)] \sin \phi + \frac{n(t - T_s)}{\sqrt{P}} \right\} \qquad (5.10\text{-}8)$$

since the noise is assumed very broadband compared to $\hat{m}(t - T_s)$. Note that $[1 - 2\text{PE}(\phi)] \sin \phi$ is periodic in π rad and therefore the DDFL has the same π-rad ambiguity as a Costas or squaring loop. To analyze a loop that has a nonlinear error control signal it is a standard analysis technique to linearize around $\phi = 0$. This is accomplished by computing the equivalent gain at $\phi = 0$:

$$A_e = \frac{d}{d\phi}\{[1 - 2\text{PE}(\phi)] \sin \phi\}\big|_{\phi=0} \qquad (5.10\text{-}9)$$

The so called S-curve is $E[z(t)|\phi]$ and A_e is just the slope of the S-curve at the origin. For the case of matched filter detection of PSK data in WGN it is well-known that

$$PE(\phi) = \frac{1}{\sqrt{2\pi}} \int_{\sqrt{2R}\cos\phi}^{\infty} \exp(-t^2/2)dt = Q(\sqrt{2R}\cos\phi) \quad (5.10\text{-}10)$$

Calculation of A_e for this case leads us to

$$A_e = [1 - 2Q(\sqrt{2R})] \quad (5.10\text{-}11)$$

The linearized SDE becomes (in Heaviside operator notation)

$$s\phi + A_e k\sqrt{P}F(s)\phi = kF(s)n(t - T_s) \quad (5.10\text{-}12)$$

or

$$\phi = \frac{A_e\sqrt{P}\,kF(s)}{s + A_e k\sqrt{P}\,F(s)}\frac{n(t - T_s)}{A_e\sqrt{P}} \quad (5.10\text{-}13)$$

Therefore, making the usual narrowband assumption, we obtain for the linearized tracking error variance, the expression

$$\sigma_\phi^2 = \frac{N_0 B_L}{P}\left[\frac{1}{[1 - 2Q(\sqrt{2R})]^2}\right] \quad (5.10\text{-}14)$$

In Problem 16 it is shown that at low SNR (5.10-14) yields

$$\sigma_\phi^2 = \frac{N_0 B_L}{P}\left(\frac{\pi}{4}\frac{N_0}{PT_s}\right) = \frac{\pi}{4}\frac{B_L T_s}{R^2} \quad (5.10\text{-}15)$$

with $R = PT_s/N_0$.

PROBLEM 16

Show that at low SNR the DDFL has a phase error variance given by

$$\sigma_\phi^2 = \frac{\pi}{4}\frac{B_L T_s}{R^2}$$

Notice that the integrate-and-dump Costas loop has a linearized tracking error variance (5.3-25) at low SNR of

$$\sigma_\phi^2 \cong \frac{B_L T_s}{2R^2} \quad (5.10\text{-}16)$$

From the phase error variance equations at low SNR it appears that the DDFL is not as good as the integrate-and-dump Costas loop. Furthermore, the analog delay required for the DDFL is useable at one data rate, so that different delays must be switched in if the system has a multiple error rate requirement. In addition, the DDFL must have bit timing in order to sample the data filter at the proper point in time, which requires bit synchronizer to supply the timing for the carrier tracking loop just as in the I and D Costas loop.

5.10.2   Decision-Directed Feedback Loop Without Delay

In order to aviod the delay problems of the above DDFL the symbol estimator can be replaced with an integrator-limiter combination shown in Figure 5.24.

In Figure 5.24, an integrator is placed in the data path to perform a real time integration. Hence in the early portion of the symbol the error probability is near $\frac{1}{2}$, whereas in the latter portion of the symbol it is near $Q(\sqrt{2R})$. In Problem 17 it is shown that the phase error variance is approximately given by

$$\sigma_\phi^2 = \frac{N_0 B_L}{P}\left[1 - 2Q(\sqrt{2R}) - \frac{1}{2R} + \frac{Q(\sqrt{2R})}{R} + \frac{e^{-R}}{\sqrt{\pi R}}\right]^{-2} \quad (5.10\text{-}17)$$

As can be seen by direction comparison with the results of Problem 16, this loop without delay is inferior to the DDFL. The results in Problem 17 are based on time-averaging $1 - 2\text{PE}(\phi)$ over one symbol time to determine the error probability.

PROBLEM 17

(a)   For the loop of Figure 5.24, show that the average probability of error can be written as

$$\overline{\text{PE}}\Big|_{\phi=0} = \frac{1}{2} - \frac{1}{T}\frac{1}{\sqrt{2\pi}}\int_0^T \int_0^{\sqrt{2Rt/T}} e^{-x^2/2}\,dx$$

(b)   By considering horizontal differential areas in the x (ordinate) t (absicca) plane, deduce that

$$\overline{\text{PE}} = Q(\sqrt{2R}) - \frac{e^{-R}}{2\sqrt{\pi R}} + \frac{1}{4R} - \frac{1}{2R}Q(\sqrt{2R})$$

(c)   Close the loop and deduce that (5.10-17) holds.

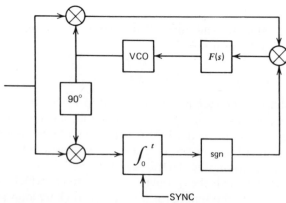

Figure 5.24   Decision feedback loop without delay.

5.10.3 Data Aided Loop

Another loop, the data aided loop [5], can be utilized when a square-wave subcarrier us used to "carry" the data. Consider a received signal of the form

$$s(t) = \sqrt{2} A \cos[\omega_0 t + \theta S(t)d(t)] \qquad (5.10\text{-}18)$$

where

A is the rms signal amplitude,

ω_0 is the radian center frequency,

θ is the modulation index of data,

$s(t)$ is the squarewave subcarrier (unit amplitude),

$d(t)$ is a unit amplitude binary-valued data sequence.

Expanding, we have

$$s(t) = \sqrt{2} A \cos(\omega_0 t) \cos \theta - \sqrt{2} A \sin(\omega_0 t) S(t)d(t) \sin \theta$$

and therefore the power in the carrier is $A^2 \cos \theta^2$ and the power in the data is $A^2 \sin \theta^2$. The data aided loop is shown in Figure (5.25) along with the subcarrier tracker and the data detector.

The loop works as follows. The upper branch acts as an ordinary PLL,

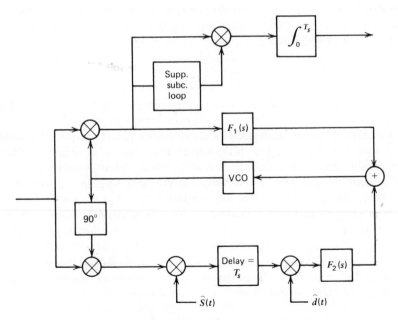

Figure 5.25 Data aided loop for subcarrier modulation.

whereas the lower branch acts like a data and subcarrier wipe-off loop, with the filtered sum driving the VCO.

It is shown in Problem 18 that the error control signal is the sum of two signals proportional to $\sin \phi$, with ϕ being the phase error. The upper loop filter passes some data modulated subcarrier energy that can get into the VCO. Normally, however, this interference would be negligible when $B_L T_s \ll 1$. The delay of T_s sec from the data symbol is assumed to be much less than $1/B_L$. It is to be noted that this loop can also be used when there is no subcarrier employed in the modulation process.

PROBLEM 18

Determine the performance of the data aided loop by

(a) Showing that the SDE can be modeled approximately as the sum of two independent noise terms and two signal components.
(b) Assume PSK matched filter detection and $F_1(s) = F_2(s)$; determine the linear error variance.

5.11 QUADRIPHASE AND MULTIPHASE TRACKING LOOPS

We now consider quadriphase tracking loops since quadriphase signals are now quite commonly used in present-day communication systems. In this section, we discuss various loops that will track quadriphase signals. We will discuss in some detail only the modified four-phase Costas loop.

5.11.1 Nth Power Loop

If a multiphase signal of the form

$$s(t) = \sqrt{2P} \sin\left[\omega_0 t + (n-1)\frac{2\pi}{N} + \theta_0\right] \qquad n = 1, \ldots, N \qquad (5.11\text{-}1)$$

is transmitted and it is desired to lock on the carrier phase, the simplest loop that will work is the Nth power loop shown in Figure 5.26. The filtered signal is then raised to the Nth power, which provides a spectral component at $N\omega_0$. This line component is tracked by an ordinary PLL. A divider circuit provides a data demodulation signal at ω_0 (rad/sec).

It is easy to show that instead of two possible lock points for binary signals there are now N possible lock points. One method to deal with this problem is to transmit a known sequence and then adjust the carrier phase until the signal is correctly detected. It is easy to show that the modulation is removed by taking the Nth power of an N-phase signal. Consider a quadriphase signal of the form

$$s(t) = \sqrt{2P} \sin\left[\omega_0 t + (n-1)\frac{\pi}{2} + \theta_0\right] \qquad n = 1, 2, 3, 4 \qquad (5.11\text{-}2)$$

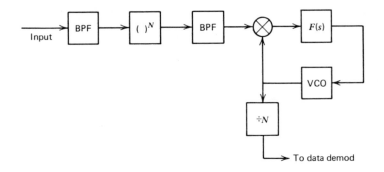

Figure 5.26 Nth power loop.

and let

$$z(t) = s^4(t) = \frac{P^2}{2} \cos[4\omega_0 t + (n-1)2\pi + 4\theta_0]$$ (5.11-3)

Clearly, for any n, we have

$$z(t) = \frac{P^2}{2} \cos(4\omega_0 t + 4\theta_0)$$ (5.11-4)

which has all the modulation removed and can easily be tracked by a PLL centered at $4\omega_0$.

We now consider a baseband equivalent of the N-phase squaring loop.

5.11.2 N-Phase Costas Loop

The N-phase Costas loop is a generalization of the biphase Costas loop. First we consider a four-phase Costas loop, shown in a conceptual block diagram in Figure 5.27.

There are $N+1$ multipliers used in the loop, one for each of the N phases, plus the multiplier of these signals. The error signal generated in the loop is proportional to $\sin N\phi$, where ϕ is the phase error in the loop. The data filters are picked to pass most of the signal but to reject unwanted noise. For NRZ data, one-pole RC filters should have their 3-dB frequencies around $0.8\,\mathcal{R}_s$, where $\mathcal{R}_s$ is the data rate.

PROBLEM 19

Determine the phase error variance of the linearized quadriphase Costas loop under the assumption that the phase error is small.

The N-phase Costas Loop is a generalization of the quadriphase Costas Loop and is shown in Figure 5.28. The N-phase Costas loop works in a manner similar to that of the quadriphase Costas loop; however, there are now N phase ambiguities instead of just 4.

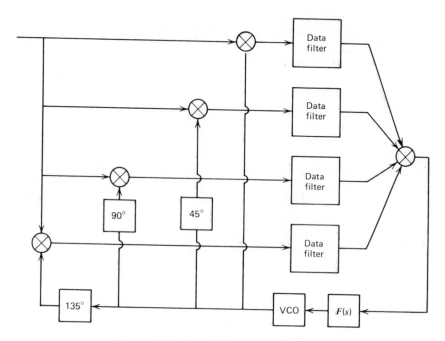

Figure 5.27 Quadriphase Costas loop.

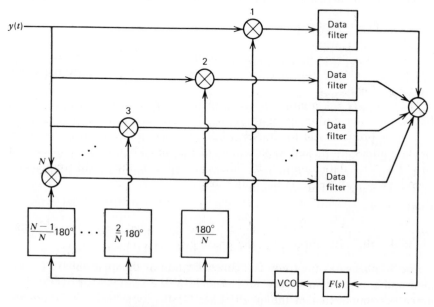

Figure 5.28 *N*-phase Costas loop.

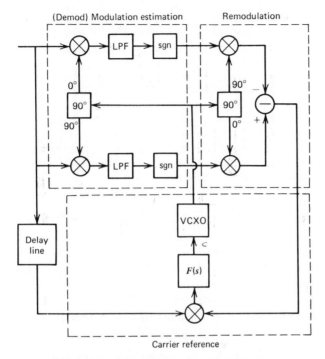

Figure 5.29 Demod-remod quadriphase loop.

5.11.3 Demod-Remod Quadriphase Tracking Loop

The demod-remod quadriphase tracking loop is another type of loop that will successfully track a four-phase signal. Figure 5.29 illustrates the demod-remod loop [33]. Basically, the data signals on both phases of the quadriphase signal are estimated in the demod section, then remodulated onto the locally generated carrier reference signal. This reference signal, when multiplied by a delayed version of the input, produces an appropriate signal for phase tracking.

We will not analyze this loop; however, it can be shown that its performance is identical to that of the modified four-phase Costas loop discussed in the next section.

5.11.4 Modified Four-Phase Costas Loop—SQPSK* Modulation

We now discuss a loop used on the TDRSS multiple-access ground (MAG) receiver for tracking both a balanced and an unbalanced quadriphase signal. Ths signal structure is staggered quadriphase signaling. Braun and

*SQPSK denotes staggered QPSK (one-half symbol offset).

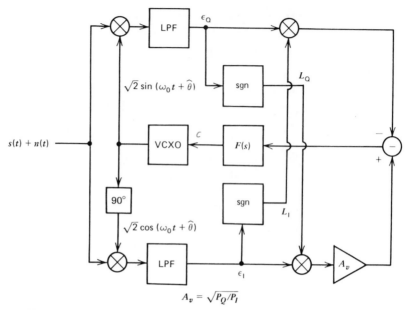

$$A_v = \sqrt{P_Q/P_I}$$

Figure 5.30 Modified four-phase Costas loop arbitrary P_Q/P_I ratio.

Lindsey [34] have shown that the closed-loop approximation to the maximum *a posteriori* probability (MAP) density estimate leads to a carrier demodulator at high E_b/N_0 shown in Figure 5.30 for both balanced and unbalanced signals, respectively. We assume that the power in the Q channel is no less than that in the I channel, that is $P_Q \geq P_I$.

To understand the operation of the loop, it is convenient to compute the S-curve with noise present for the case of NRZ data. The received quadriphase signal plus noise is written as

$$s(t) = \sqrt{2(1-\alpha)S}\,m_Q(t)\,\sin(\omega_0 t + \theta) + \sqrt{2\alpha S}\,m_I(t)\,\cos(\omega_0 t + \theta)$$

$$(5.11\text{-}5)$$

$$n(t) = \sqrt{2}\,n_c(t)\,\cos(\omega_0 t + \theta) + \sqrt{2}\,n_s(t)\,\sin(\omega_0 t + \theta)$$

where

αS is the power in the I channel.

$(1 - \alpha)S$ is the power in the Q channel
($\alpha = \frac{1}{5}$ if the power ratios are 4:1).

$m_Q(t)$ is the Q channel data stream (± 1).

$m_I(t)$ is the I channel data stream (± 1).

$n_c(t)$ and $n_s(t)$ are independent Gaussian random noise processes with two-sided noise density of $N_0/2$.

R_Q is the Q channel symbol rate ($= 1/T_Q$).

R_I is the I channel symbol rate ($= 1/T_I$).

Notice that the voltage gain A_v increases the gain of the lower powered channel (I channel). When the power levels are equal, $A_v = 1$, but in general $A_v = \sqrt{P_Q/P_I}$. The reference signals are

$$r(t) = \sqrt{2}\sin(\omega_0 t + \hat{\theta}) \tag{5.11-6}$$

$$r'(t) = \sqrt{2}\cos(\omega_0 t + \hat{\theta}) \tag{5.11-7}$$

The output of the LPFs (assumed to be one-pole RC filters) are given by

$$\epsilon_Q(t) = \sqrt{(1-\alpha)S}\,\tilde{m}_Q(t)\cos\phi - \sqrt{\alpha S}\,\tilde{m}_1(t)\sin\phi - \tilde{n}_c(t)\sin\phi + \tilde{n}_s(t)\cos\phi \tag{5.11-8}$$

$$\epsilon_1(t) = \sqrt{(1-\alpha)S}\,\tilde{m}_Q(t)\sin\phi + \sqrt{\alpha S}\,\tilde{m}_1(t)\cos\phi + \tilde{n}_c(t)\cos\phi + \tilde{n}_s(t)\sin\phi \tag{5.11-9}$$

where

$\tilde{n}_c(t)$ is the filtered version of $n_c(t)(\sigma_{\tilde{n}_c}^2 = N_0 B)$.

$\tilde{n}_s(t)$ is the filtered version of $n_s(t)(\sigma_{\tilde{n}_s}^2 = N_0 B)$.

$\tilde{m}_Q(t)$ is the filtered version of $m_Q(t)$.

$\tilde{m}_1(t)$ is the filtered version of $m_1(t)$.

$\phi = \theta - \hat{\theta}$ is the instantaneous phase error (rad).

B is the arm filter noise bandwidth (Hz).

Out of the hard limiters we have the signals

$$L_Q(t) = \text{sgn}[\epsilon_Q(t)] \tag{5.11-10}$$

$$L_1(t) = \text{sgn}[\epsilon_1(t)] \tag{5.11-11}$$

Without noise and with small ϕ we have

$$L_Q(t) \cong m_Q(t) \tag{5.11-12}$$

$$L_1(t) \cong m_1(t) \tag{5.11-13}$$

The control signal C, which, after filtering, is used to drive the VCO, is formed by

$$C = A_v\epsilon_1(t)L_Q(t) - \epsilon_Q(t)L_1(t) \tag{5.11-14}$$

Evaluating (5.11-14), using (5.11-8), (5.11-9), (5.11-12), and (5.11-13), we obtain

$$\begin{aligned}C(\phi) = {} & A_v\sqrt{(1-\alpha)S}\,\tilde{m}_Q(t)\,\text{sgn}[\epsilon_Q(t)]\sin\phi + A_v\sqrt{\alpha S}\,\tilde{m}_1(t)\,\text{sgn}[\epsilon_Q(t)]\cos\phi \\ & - \sqrt{(1-\alpha)S}\,\tilde{m}_Q(t)\,\text{sgn}[\epsilon_1(t)]\cos\phi \\ & + \sqrt{\alpha S}\,\tilde{m}_1(t)\,\text{sgn}[\epsilon_1(t)]\sin\phi + n'(t)\end{aligned} \tag{5.11-15}$$

where $n'(t)$ is given by

$$\begin{aligned}n'(t) = {} & \tilde{n}_c(t)\{A_v\,\text{sgn}[\epsilon_Q(t)]\cos\phi + \text{sgn}[\epsilon_1(t)]\sin\phi\} \\ & + \tilde{n}_s(t)\{A_v\,\text{sgn}[\epsilon_Q(t)]\sin\phi - \text{sgn}[\epsilon_1(t)]\cos\phi\}\end{aligned} \tag{5.11-16}$$

Having the equations for the error control signal C, we can obtain the closed-loop tracking performance by determining S-curve gain slope at $\phi = 0$ and the standard deviation of the error control signal. Therefore, the next step is to determine the S-curve, which is the mean value of the control signal C conditioned on ϕ.

Denote the S-curve by $\bar{C}(\phi)$. Then

$$\bar{C}(\phi) = E[C(\phi)|\phi] \qquad (5.11\text{-}17)$$

where $E(\cdot)$ denotes a statistical expectation. In order to proceed without making the problem nearly impossible to solve, we make the approximation that

$$\tilde{m}_Q(t) \cong \sqrt{\gamma}\, m_Q(t) \qquad (5.11\text{-}18)$$

$$\tilde{m}_I(t) \cong \sqrt{\gamma}\, m_I(t) \qquad (5.11\text{-}19)$$

that is, we approximate the filtering effect on the data streams as reducing the amplitude by $\sqrt{\gamma}$ and leaving the symbols undistorted! The parameter γ is the relative power loss defined by

$$\gamma = \int_{-\infty}^{\infty} \mathscr{S}_d(f)|H(f)|^2\, dt \qquad (5.11\text{-}20)$$

with $\mathscr{S}_d(f)$ the spectral density of the data stream and $H(f)$ the transfer function of the lowpass filters (assumed to be identical). This approximation allows us to use conditional expectations, with the conditioning being on random variables rather than random processes. With this assumption, it can be shown that [35], [36]

$$\begin{aligned}
\bar{C}(\phi) = &\ [(A_v)\sqrt{(1-\alpha)\gamma_Q S} + \sqrt{\alpha\gamma_I S}\,]\sin\phi \\
&+ (A_v)[-\sqrt{(1-\alpha)\gamma_Q S}\,\sin\phi + \sqrt{\alpha\gamma_I S}\,\cos\phi]Q(F) \\
&- (A_v)[\sqrt{(1-\alpha)\gamma_Q S}\,\sin\phi + \sqrt{\alpha\gamma_I S}\,\cos\phi]Q(G) \\
&+ [\sqrt{(1-\alpha)\gamma_Q S}\,\cos\phi - \sqrt{\alpha\gamma_I S}\,\sin\phi]Q(A) \\
&- [\sqrt{(1-\alpha)\gamma QS}\,\cos\phi + \sqrt{\alpha\gamma_I S}\,\sin\phi]Q(B) \qquad (5.11\text{-}21)
\end{aligned}$$

where

$$\begin{aligned}
F &= \sqrt{2R_Q}\sqrt{1-\alpha}\,\cos\phi + \sqrt{2R_I}\sqrt{\alpha}\,\sin\phi \\
G &= \sqrt{2R_Q}\sqrt{1-\alpha}\,\cos\phi - \sqrt{2R_I}\sqrt{\alpha}\,\sin\phi
\end{aligned} \qquad (5.11\text{-}22)$$

and

$$\begin{aligned}
A &= \sqrt{2R_I\alpha}\,\cos\phi + \sqrt{2R_Q(1-\alpha)}\,\sin\phi \\
B &= \sqrt{2R_I\alpha}\,\cos\phi - \sqrt{2R_Q(1-\alpha)}\,\sin\phi
\end{aligned} \qquad (5.11\text{-}23)$$

and

$$Q(u) = \int_u^\infty \frac{e^{-t^2/2}}{\sqrt{2\pi}} \, dt \qquad (5.11\text{-}24)$$

For the two-arm filter SNR parameters we obtain

$$R_Q = \frac{\gamma_Q S}{2\sigma^2} = \frac{\gamma_Q S}{2N_0 B}$$

$$R_1 = \frac{\gamma_1 S}{2\sigma^2} = \frac{\gamma_1 S}{2N_0 B} \qquad (5.11\text{-}25)$$

where $B[B = (\pi/2)f_{3\,dB}]$ is the one-sided noise bandwidth of the arm filters and $N_0/2$ is the two-sided noise spectral density. The S-curves, based on (5.11-21) are plotted in Figures 5.31 through 5.34 for the case of NRZ and Manchester data (the shape is the same but the filtering losses are slightly different in each case).

The range of R in the balanced case (equal data rate and power)is 0–5 dB on Figure 5.31 and 10–30 dB on Figure 5.32. Notice that stable lock points for the balanced case occur at 0°, 90°, 180° and 270°, so that the loop may lock onto one of four allowable phases. This ambiguity must be resolved for correct demodulation of the data.

In Figures 5.33 and 5.34, a 4:1 power ratio unbalanced case is illustrated. When $R \le 5$ dB, the only stable lock points occur at 0° and 180° hence, only two phases are allowable. However, when $R \ge 10$ dB, a lock point at ±90° can also occur!

We now consider the tracking error variance. To obtain the tracking

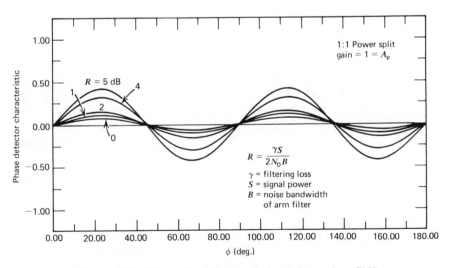

Figure 5.31 Normalized S-curve balanced case low SNR.

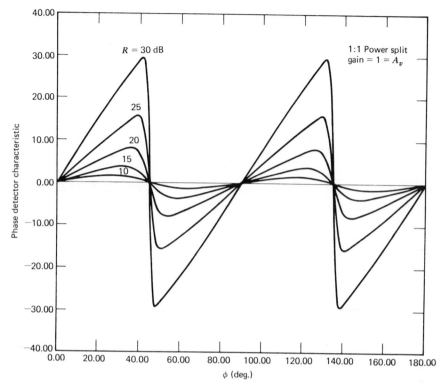

Figure 5.32 Normalized S-curve balanced case high SNR.

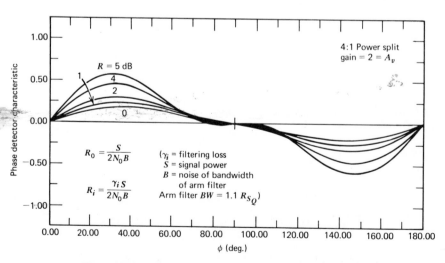

Figure 5.33 Normalized S-curve for unbalanced case.

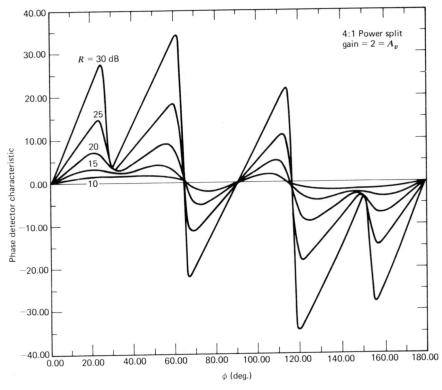

Figure 5.34   S-curve for unbalanced case.

error variance it is necessary to obtain the slope of the S-curve at the origin, which we call $\bar{C}'(0)$. In Problem 20, it is shown that, for the balanced case

$$\bar{C}'(0) = \sqrt{2\gamma S}\left[1 - 2Q(\sqrt{R}) - \sqrt{\frac{2}{\pi}}\,\sqrt{R}\,\exp(-R/2)\right] \qquad (5.11\text{-}26)$$

where γ is defined in (5.11-20).

PROBLEM 20

Evaluate

$$\frac{\partial \bar{C}(\phi)}{\partial \phi}\Big|_{\phi=0}$$

in (5.11-20) to establish for the balanced power case that

$$\bar{C}'(0) = \sqrt{2\gamma S}\left[1 - 2Q(\sqrt{R}) - \sqrt{\frac{2}{\pi}}\,\sqrt{R}\,e^{-R/2}\right]$$

In order to compute the tracking error variance it is necessary to obtain the SDE of operation. We write the control signal $C(\phi)$ as

$$C(\phi) = \bar{C}(\phi) + [C(\phi) - \bar{C}(\phi)] \qquad (5.11\text{-}27)$$

The first term is the average control signal. The second term, the error or noise on the control signal, will be denoted $N(t, \phi)$. It is convenient to express the error signal in terms of unit gains at the origin, so let us define $g_n(\phi)$ as the S-curve with unit slope at the origin

$$\bar{C}(\phi) = \bar{C}'(0)g_n(\phi) \qquad (5.11\text{-}28)$$

so that

$$g_n'(0) = 1 \qquad (5.11\text{-}29)$$

Let K_v be the gain of the VCO; then

$$\hat{\theta} = \frac{C'(0)K_vF(s)}{s}\left[g_n(\phi) + \frac{N(\phi, t)}{C'(0)}\right] \qquad (5.11\text{-}30)$$

Since

$$\phi = \theta - \hat{\theta}$$

we obtain

$$\theta - \phi = \frac{C'(0)K_vF(s)}{s}\left[g_n(\phi) + \frac{N(\phi, t)}{C'(0)}\right] \qquad (5.11\text{-}31)$$

To proceed, it is convenient to linearize the loop equation so that we obtain

$$\phi = -[H(s)]\frac{N(t, \phi)}{C'(0)} \qquad (5.11\text{-}32)$$

where

$$H(s) = \frac{k_vC'(0)F(s)}{s + k_vC'(0)F(s)} \qquad (5.11\text{-}33)$$

is the closed-loop transfer function and we have used $g_n(\phi) \simeq \phi$ for small ϕ. It then follows that

$$\sigma_\phi^2 = \int_{-\infty}^{\infty} \frac{|H(f)|^2 S_N(f)\, df}{[C'(0)]^2} \qquad (5.11\text{-}34)$$

When $S_N(f)$ is essentially constant near $f = 0$, then (following the usual procedure), we have

$$\sigma_\phi^2 = \frac{\dfrac{N_0'}{2}(2B_L)}{[C'(0)]^2} \qquad (5.11\text{-}35)$$

where

$$2B_L = \int_{-\infty}^{\infty} |H(f)|^2\, df \qquad (5.11\text{-}36)$$

with $H(f)$ the closed-loop transfer function of the loop. In order to compute the tracking error variance, it is necessary to compute $R_n(\tau, \phi)$, where*

$$R_N(\tau, \phi) = E\{[C(\phi, t) - \overline{C(\phi, t)}][C(\phi, t + \tau) - \overline{C(\phi, t + \tau)}]\,|\,\phi\}$$
(5.11-37)

Weber *et al.* [33] have shown for an equivalent loop structure that at low values of R ($R \leq 5$ dB), $R_N(\tau, \phi) \cong R_N(\tau, 0)$.

In the following calculations we assume $\phi = 0$ in the determination of $R_N(\tau, 0)$. Since $R_N(\tau, 0) = R_N(\tau)$, using (5.11-15) $R_N(\tau)$ is given by [($A_v = 1$, $\alpha = \frac{1}{2}$, and $\tilde{m}(t) = \sqrt{\gamma}\, m(t)$]

$$R_N(\tau) = E\left(\left\{\sqrt{\frac{\gamma S}{2}}\, m_{\rm I}(t)\, {\rm sgn}[\epsilon_Q(t)] - \sqrt{\frac{\gamma S}{2}}\, m_Q(t)\, {\rm sgn}[\epsilon_{\rm I}(t)]\right.\right.$$

$$\left.+ \tilde{n}_c(t)\, {\rm sgn}[\epsilon_{\rm I}(t)] - \tilde{n}_s(t)\, {\rm sgn}[\epsilon_Q(t)]\right\} \cdot$$

$$\left\{\sqrt{\frac{\gamma S}{2}}\, m_{\rm I}(t + \tau)\, {\rm sgn}[\epsilon_Q(t + \tau)] - \sqrt{\frac{\gamma S}{2}}\, m_Q(t + \tau)\, {\rm sgn}[\epsilon_Q(t + \tau)]\right.$$

$$\left.\left.+ \tilde{n}_c(t + \tau)\, {\rm sgn}[\epsilon_{\rm I}(t + \tau)] + \tilde{n}_s(t + \tau)\, {\rm sgn}[\epsilon_Q(t + \tau)]\right\}\right)$$
(5.11-38)

with γ being defined in (5.11-20) and where, with $\phi = 0$, we have

$$\epsilon_{\rm I}(t) = \sqrt{\frac{\gamma S}{2}}\, m_{\rm I}(t) + \tilde{n}_c(t)$$
(5.11-39)

$$\epsilon_Q(t) = \sqrt{\frac{\gamma S}{2}}\, m_Q(t) + \tilde{n}_s(t)$$
(5.11-40)

After considerable algebra and some approximations [35], we obtain

$$R_N(\tau) = \frac{4}{\pi}\, R\Delta(\tau)\, e^{-R} R_{\tilde{n}}(\tau) + \frac{2}{\pi}\, R^2 e^{-R} \Lambda^2(\tau) \frac{R_{\tilde{n}}^2(\tau)}{\sigma^2}$$

$$+ \frac{2}{\pi}\, R e^{-R} \Lambda(\tau) \frac{R_{\tilde{n}}^3(\tau)}{3\sigma^4}\, (R^2 - 2R + 1)$$

$$- \frac{4\sqrt{2}\sqrt{R}}{\sqrt{\pi}}\, \Lambda(\tau) R_{\tilde{n}}(\tau) e^{-R/2}\, {\rm erf}\left(\sqrt{\frac{R}{2}}\right)$$

$$+ 2R_{\tilde{n}}(\tau)\Lambda(\tau)\, {\rm erf}^2\left(\sqrt{\frac{R}{2}}\right) + \frac{2R_{\tilde{n}}(\tau)}{\pi}\, e^{-R}\rho^2(\tau)\Lambda(\tau)$$

$$+ \frac{2R_{\tilde{n}}(\tau)}{\pi}\, e^{-R}\, \frac{\rho^3(\tau)}{3}\, (R^2 - 2R + 1)$$
(5.11-41)

---

*Again we approximate all cyclostationary processes by stationary processes.

where

$$\Lambda(\tau) = \begin{cases} \left(1 - \dfrac{|\tau|}{T}\right) & |\tau| \le T \\ 0 & |\tau| > T \end{cases} \tag{5.11-42}$$

$$\sigma^2 = N_0 B \tag{5.11-43}$$

$$\text{erf}(x) = \frac{2}{\sqrt{\pi}} \int_0^x e^{-t^2}\, dt \tag{5.11-44}$$

and

$$R_{\tilde{n}}(\tau) = \sigma^2 \rho(\tau) \tag{5.11-45}$$

is the (unnormalized) autocorrelation function of $\tilde{n}_s(t)$ or $\tilde{n}_c(t)$. Now

$$S_N(0) = \mathscr{F}[R_N(\tau)] = \frac{N_0'}{2} \tag{5.11-46}$$

where $F[R_N(\tau)]$ is the Fourier transform of $R_N(\tau)$. Hence from (5.11-41) we obtain

$$\frac{N_0'}{2} = \left[\frac{4}{\pi} R e^{-R} + 2\,\text{erf}^2\left(\sqrt{\frac{R}{2}}\right) - \frac{4\sqrt{2}\sqrt{R}}{\sqrt{\pi}} e^{-R/2}\right.$$
$$\left. \times \text{erf}\left(\sqrt{\frac{R}{2}}\right)\right]\left[\int_{-\infty}^{\infty} S_d(f) \frac{N_0}{2} |H(f)|^2\, df\right]$$
$$+ \left[\frac{2}{3\pi} R e^{-R}(R^2 - 2R + 1) + 2\frac{R}{\pi} e^{-R}\right]\left[\frac{1}{\sigma^4} \int_{-\infty}^{\infty} \Lambda(\tau) R_{\tilde{n}}^3(\tau)\, d\tau\right]$$
$$+ \left[\frac{2}{\pi} R^2 e^{-R}\right]\left[\frac{1}{\sigma^2} \int_{-\infty}^{\infty} \Lambda^2(\tau) R_{\tilde{n}}^2(\tau)\, d\tau\right]$$
$$+ \left[\frac{2}{3\pi} e^{-R}(R^2 - 2R + 1)\right]\left[\frac{1}{\sigma^6} \int_{-\infty}^{\infty} R_{\tilde{n}}^4(\tau)\, d\tau\right] \tag{5.11-47}$$

where

$$R_{\tilde{n}}(\tau) = N_0 B e^{-4B|\tau|} \tag{5.11-48}$$

Evaluating the integrals in (5.11-47) yields (T = symbol duration)

$$\frac{N_0'}{2} = \left[\frac{4}{\pi} R e^{-R} + 2\,\text{erf}^2\left(\sqrt{\frac{R}{2}}\right) - \frac{4\sqrt{2}\sqrt{R}}{\sqrt{\pi}} e^{-R/2}\,\text{erf}\left(\sqrt{\frac{R}{2}}\right)\right]\frac{\gamma N_0}{2}$$
$$+ \left[\frac{2}{3\pi} R e^{-R}(R^2 - 2R + 1) + \frac{2R}{\pi} e^{-R}\right]\left\{\frac{N_0}{6}[1 - e^{-12BT}]\right.$$
$$\left. - \frac{N}{72BT} + e^{-12BT}\left[\frac{N_0}{6} + \frac{N_0}{72BT}\right]\right\}$$
$$+ \left[\frac{2}{\pi} R^2 e^{-R}\right]\left(\frac{N_0}{4} - \frac{N_0}{16BT} + \frac{N_0}{128B^2T^2} - \frac{N_0 e^{-8BT}}{128B^2T^2}\right)$$
$$+ \left[\frac{2}{3\pi} e^{-R}(R^2 - 2R + 1)\right]\frac{N_0}{8} \tag{5.11-49}$$

Finally, using (5.11-49) and (5.11-26) in (5.11-35) we obtain

$$\sigma_\phi^2 \cong \frac{N_0 B_L}{\gamma S[1 - 2Q(\sqrt{R}) - \sqrt{2/\pi}\, R\, e^{-R/2}]^2} \left\{ \gamma \left[\frac{2}{\pi} R e^{-R} + \text{erf}^2\left(\sqrt{\frac{R}{2}}\right) \right.\right.$$

$$- \frac{2\sqrt{2}\sqrt{R}}{\sqrt{\pi}} e^{-R/2} \text{erf}\left(\sqrt{\frac{R}{2}}\right) \Bigg]$$

$$+ \left[\frac{1}{3\pi} R e^{-R}(R^2 - 2R + 1) + \frac{R}{\pi} e^{-R} \right]$$

$$\times \left[\frac{1}{3}(1 - e^{-12BT}) - \frac{1}{36BT} + e^{-12BT}\left(\frac{1}{3} + \frac{1}{36BT}\right) \right]$$

$$+ \left(\frac{1}{\pi} R^2 e^{-R}\right)\left(\frac{1}{2} - \frac{1}{8BT} + \frac{1}{64B^2 T^2} - \frac{e^{-8BT}}{64B^2 T^2}\right)$$

$$+ \left[\frac{1}{12\pi} e^{-R}(R^2 - 2R + 1) \right] \right\} \qquad \text{balanced case NRZ data}$$

$$(5.11\text{-}50)$$

which is the final result for the phase error varaince of the balanced, modified, four-phase offset Costas loop using NRZ data symbols with a staggered quadriphase signal input.

Notice as R becomes large the expression approaches

$$\sigma_\phi^2 \to \frac{N_0 B_L}{S} \qquad (5.11\text{-}51)$$

which is a well-known result for a cw phase locked loop.

Using the same basic methods for the Manchester data case, we obtain [35]:

$$\sigma_\phi^2 \cong \frac{N_0 B_L}{\gamma S[1 - 2Q(\sqrt{R}) - \sqrt{2/\pi}\sqrt{R} e^{-R/2}]^2}$$

$$\left\{ \gamma \left[\frac{2}{\pi} R e^{-R} + \text{erf}^2\left(\sqrt{\frac{R}{2}}\right) - \frac{2\sqrt{2}}{\sqrt{\pi}} \sqrt{R} e^{-R/2} \text{erf}\left(\sqrt{\frac{R}{2}}\right) \right] \right.$$

$$+ \left[\frac{1}{3\pi} R e^{-R}(R^2 - 2R + 1) + \frac{R}{\pi} e^{-R} \right]\left(\frac{1}{3} - \frac{1}{12BT} + \frac{e^{-6BT}}{9BT} - \frac{e^{-12BT}}{36BT}\right)$$

$$+ \left[\frac{1}{\pi} R^2 e^{-R}\right]\left[\frac{1}{2} - \frac{3}{8BT} + \frac{9}{64B^2 T^2} - e^{-4BT}\left(\frac{1}{4} + \frac{1}{8BT} + \frac{5}{32B^2 T^2}\right) + \frac{e^{-8BT}}{64B^2 T^2}\right]$$

$$+ \frac{1}{12\pi} e^{-R}(R^2 + 2R + 1) \right\} \qquad \text{balanced case Manchester data}$$

$$(5.11\text{-}52)$$

In the unbalanced case it can be shown [36] that the slope of the S-curve at the origin is given by

$$C'(0) = [(A_v)\sqrt{(1-\alpha)\gamma_Q S} + \sqrt{\alpha \gamma_I S}]$$

$$+ (A_v)\{-2\sqrt{(1-\alpha)\gamma_Q S} Q[\sqrt{2R_Q(1-\alpha)}]$$

$$+ 2\sqrt{\alpha \gamma_I S} Q'(\sqrt{2R_Q(1-\alpha)} \cdot \sqrt{2R_I \alpha}\}$$

$$+ [-2\sqrt{\alpha \gamma_I S} Q(\sqrt{2R_I \alpha}) + 2\sqrt{(1-\alpha)\gamma_Q S} Q'(\sqrt{2R_I \alpha})\sqrt{2R_Q(1-\alpha)}]$$

$$(5.11\text{-}53)$$

and the phase error variance for NRZ data is given by

$$
\begin{aligned}
\sigma_Q^2 \cong \frac{N_0 B_L}{[C'(0)]^2} \Bigg\{ &\frac{16}{5} \frac{R_Q}{\gamma_Q} \left[\sqrt{\gamma_I}\, \mathrm{erf}\left(\sqrt{\frac{4}{5} R_Q}\right) - \sqrt{\gamma_Q}\, \mathrm{erf}\left(\sqrt{\frac{R_2}{5}}\right) \right]^2 B T_Q \left(1 - \frac{1}{3}\frac{T_Q}{T_I}\right) \\
&+ \gamma_Q \left[\frac{16 R_Q}{5\pi} e^{-2R_I/5} - \frac{16\sqrt{R_Q}}{\sqrt{5\pi}}\, \mathrm{erf}\left(\sqrt{\frac{4}{5} R_Q}\right) e^{-R_I/5} + 4\, \mathrm{erf}^2\left(\sqrt{\frac{4}{5} R_Q}\right)\right] \\
&+ \gamma_I \left[\frac{16 R_I}{5\pi} e^{-8R_Q/5} - \frac{8\sqrt{R_I}}{\sqrt{5\pi}}\, \mathrm{erf}\left(\sqrt{\frac{R_I}{5}}\right) e^{-4R_Q/5} + \mathrm{erf}^2\left(\sqrt{\frac{R_I}{5}}\right)\right] \\
&+ \frac{64}{25\pi} R_I R_Q \left(e^{-2R_I/5} + 4e^{-8R_Q/5}\right) \left[\frac{1}{8} - \left(\frac{1}{T_Q} + \frac{1}{T_I}\right)\frac{1}{64B} + \frac{1}{256 B^2 T_I T_Q}\right] \\
&+ \left\{ \frac{4}{\pi} e^{-8R_Q/5} + \frac{1}{\pi} e^{-2R_I/5} - \frac{4}{\pi} \exp\left[-\left(\frac{R_I + 4R_Q}{5}\right)\right]\right\}\Bigg\} \\
&+ \left[\frac{32 R_Q}{5\pi} e^{-8R_Q/5} + \frac{16 R_Q}{30\pi} e^{-2R_I/5}\left(\frac{2}{5} R_I - 1\right)^2\right]\left(\frac{1}{3} - \frac{1}{36 B T_Q}\right) \\
&+ \left[\frac{2 R_I}{5\pi} e^{-2R_I/5} + \frac{16}{30\pi} R_I e^{-8R_Q/5}\left(\frac{8}{5} R_Q - 1\right)^2\right]\left(\frac{1}{3} - \frac{1}{36 B T_I}\right) \\
&+ \left[\frac{1}{12\pi} e^{-2R_I/5}\left(\frac{2}{5} R_I - 1\right)^2 + \frac{1}{3\pi} e^{-8R_Q/5}\left(\frac{8}{5} R_Q - 1\right)^2\right]
\end{aligned}
$$

$$\text{NRZ}_Q\text{–NRZ}_I \quad (4:1) \tag{5.11-54}$$

where $C'(0)$ is given in (5.11-53) and for the Manchester data case we obtain

$$
\begin{aligned}
\sigma_\phi^2 \cong \frac{N_0 B_L}{[C'(0)]^2} \Bigg(&\frac{4}{5} S \left[\sqrt{\gamma_2}\, \mathrm{erf}\left(\sqrt{\frac{4}{5} R_Q}\right) - \sqrt{\gamma_I}\, \mathrm{erf}\left(\sqrt{\frac{R_I}{5}}\right)\right]^2 \frac{1}{N_0(1/T_Q)}\left(\frac{T_Q}{T_I}\right) \\
&+ \gamma_I \left[\frac{16 R_Q}{5\pi} e^{-2R_I/5} - \frac{16\sqrt{R_Q}}{\sqrt{5\pi}}\, \mathrm{erf}\left(\sqrt{\frac{4}{5} R_Q}\right) e^{-R_I/5} + 4\, \mathrm{erf}^2\left(\sqrt{\frac{4}{5} R_Q}\right)\right] \\
&+ \gamma_2 \left[\frac{16 R_I}{5\pi} e^{-8R_Q/5} - \frac{8\sqrt{R_I}}{\sqrt{5\pi}}\, \mathrm{erf}\left(\sqrt{\frac{R_I}{5}}\right) e^{-4R_Q/5} + \mathrm{erf}^2\left(\sqrt{\frac{R_I}{5}}\right)\right] \\
&+ \left[\frac{64}{25\pi} R_I R_Q \left(e^{-2R_I/5} + 4e^{-8R_Q/5}\right)\right]\left[\frac{1}{8} - \frac{3}{64}\left(\frac{1}{T_Q B} + \frac{1}{T_I B}\right) + \frac{9}{256 T_I T_Q B^2}\right] \\
&+ \left\{ \frac{4}{\pi} e^{-8R_Q/5} + \frac{1}{\pi} e^{-2R_I/5} - \frac{4}{\pi} \exp\left[-\left(\frac{R_I + 4R_Q}{5}\right)\right]\right\}\Bigg\} \\
&+ \left[\frac{32 R_Q}{5\pi} e^{-8R_Q/5} + \frac{16}{30\pi} R_Q e^{-2R_I/5}\left(\frac{2}{5} R_I - 1\right)^2\right]\left(\frac{1}{3} - \frac{1}{12 B T_Q}\right) \\
&+ \left[\frac{2 R_I}{5\pi} e^{-2R_I/5} + \frac{16}{30\pi} R_I e^{-8R_Q/5}\left(\frac{8}{5} R_Q - 1\right)^2\right]\left(\frac{1}{3} - \frac{1}{12 B T_I}\right) \\
&+ \left[\frac{1}{12\pi} e^{-2R_I/5}\left(\frac{2}{5} R_I - 1\right)^2 + \frac{1}{3\pi} e^{-8R_Q/5}\left(\frac{8}{5} R_Q - 1\right)^2\right]\Bigg)
\end{aligned}
$$

$$\text{Manchester}_I\text{–Manchester}_Q \quad (4:1) \tag{5.11-55}$$

5.12   MEAN SLIP TIME FOR MULTIPHASE LOOPS

We now discuss the mean slip time for multiphase loops. From Chapter 10, the mean slip time for an n-phase first-order (flat noise spectral density) carrier tracking loop is given by

$$\bar{T}W_L = \frac{1}{4\sigma_\phi^2} \int_0^{\phi_u} \int_0^{\phi_u} \exp\left\{-\left[\frac{G(\phi)}{\sigma_\phi^2}\right]\right\} \exp\left[\frac{G(\phi')}{\sigma_\phi^2}\right] d\phi\, d\phi' \quad (5.12\text{-}1)$$

where ϕ_u is equal to $2\pi/n$ for the balanced case and $G(\phi)$ is the indefinite integral of the normalized (unit gain) S-curve and σ_ϕ^2 is the linearized phase error variance (relative to ϕ, not $n\phi$), and $W_L = 2B_L$ is the two-sided loop noise bandwidth, with B_L the one-sided loop noise bandwidth. To "convert" to a second-order loop, σ_ϕ^2 should be increased by 1 dB based on mean slip time simulations [37]. The above equation also applies (approximately) when an unbalanced quadriphase loop is used. For example, in an unbalanced say, (4:1), modified Costas loop, the next stable lock point, for $R \le 5$ dB, is at 180°, so the upper limits would be π. For the case when the noise spectral density varies with ϕ, see Problem 13, Chapter 10.

We now obtain some approximate mean slip time formulas for the Costas loop and the balanced four-phase modified Costas loop. The approximations are based on assuming a sinusoidal S-curve, which appears to be a good approximation at low SNRs.

First consider the Costas loop. In this case the S-curve normalized to unit slope is of the form

$$g_n(\phi) = \frac{\sin 2\phi}{2} \quad (5.12\text{-}2)$$

so that

$$\int^\phi \frac{\sin 2\phi'}{2} d\phi' = -\tfrac{1}{4}\cos 2\phi \quad (5.12\text{-}3)$$

Therefore ($n = 2$)

$$\bar{T}W_L = \frac{1}{4\sigma_\phi^2} \int_0^\pi \int_0^\pi \exp\left[-\left(\frac{\cos 2\phi}{4\sigma_\phi^2}\right)\right] \exp\left(\frac{\cos 2\phi'}{4\sigma_\phi^2}\right) d\phi\, d\phi' \quad (5.12\text{-}4)$$

Make the change of variables:

$$2\phi = y$$

$$2\phi' = x \quad (5.12\text{-}5)$$

so that we obtain

$$\bar{T}W_L = \frac{1}{4^2\sigma_\phi^2} \int_0^{2\pi} \int_0^{2\pi} \exp\left(\frac{\cos y}{4\sigma_\phi^2}\right) \exp\left[-\left(\frac{\cos x}{4\sigma_\phi^2}\right)\right] dy\, dx \quad (5.12\text{-}6)$$

Since

$$I_0(z) = \frac{1}{2\pi} \int_0^{2\pi} e^{\pm z \cos \theta} \, d\theta \qquad (5.12\text{-}7)$$

we have for the first-order Costas loop mean slip time:

$$\bar{T}W_L = \frac{\pi^2}{4\sigma_\phi^2} I_0^2\left(\frac{1}{4\sigma_\phi^2}\right) \sim \frac{\pi}{2} e^{2\rho_e} \qquad \rho_e = \frac{1}{4\sigma_\phi^2} \qquad (5.12\text{-}8)$$

with $\sim$ denoting asymptotic equality. Notice that the effective phase error variance is $4\sigma_\phi^2$, which is well-known, and the effective $\text{SNR}_e = 1/4\sigma_\phi^2$.

Now consider the four-phase balanced, modified, Costas loop. Since $n = 4$, we have

$$\bar{T}W_L = \frac{1}{4\sigma_\phi^2} \int_0^{\pi/2} \int_0^{\pi/2} \exp\left\{-\left[\frac{G(\phi)}{\sigma_\phi^2}\right]\right\} \exp\left[\frac{G(\phi')}{\sigma_\phi^2}\right] d\phi \, d\phi' \quad (5.12\text{-}9)$$

From Figure 5.32 it is seen that the normalized S-curve is well approximated by

$$g_n(\phi) = \frac{\sin 4\phi}{4} \qquad (5.12\text{-}10)$$

so that

$$G(\phi) = -\frac{1}{16} \cos 4\phi \qquad (5.12\text{-}11)$$

Let

$$4\phi = y$$
$$4\phi' = x$$

then

$$\bar{T}W_L = \frac{1}{4 \cdot 4^2 \sigma_\phi^2} \int_0^{2\pi} \int_0^{2\pi} \exp\left(\frac{\cos y}{16\sigma_\phi^2}\right) \exp\left[-\left(\frac{\cos x}{16\sigma_\phi^2}\right)\right] dx \, dy$$

$$(5.12\text{-}12)$$

Evaluating, we obtain for the first-order four-phase modified balanced Costas loop

$$\bar{T}W_L = \frac{\pi^2}{16\sigma_\phi^2} I_0^2\left(\frac{1}{16\sigma_\phi^2}\right) \sim \frac{\pi}{2} e^{2\rho_e} \qquad \rho_e = \frac{1}{16\sigma_\phi^2} \qquad (5.12\text{-}13)$$

In this case the effective phase error variance is $16\sigma_\phi^2$, since the S-curve has been shrunk by a factor of 4. The resulting effective SNR is $\text{SNR}_e = 1/16\sigma_\phi^2$. Hence the four-phase loop loses on the order of 12 dB more than a cw loop. In fact, the Costas loop thresholds about 6 dB above a cw loop.

The asymptotic expressions for the second-order Costas loop and four-phase modified Costas loop are given by

$$TW_L \sim \frac{\pi}{2} \exp\!\left(\frac{\pi}{2}\,\rho_e\right) \qquad \rho_e = \frac{1}{4\sigma_\phi^2} \tag{5.12-14}$$

for the Costas loop and

$$TW_L \sim \frac{\pi}{2} \exp\!\left(\frac{\pi}{2}\,\rho_e\right) \qquad \rho_e = \frac{1}{16\sigma_\phi^2} \tag{5.12-15}$$

for the four-phase modified Costas loops.

In general, it can be shown for balanced N-phase first-order loops that have sinusoidal S-curves that the mean slip time (essentially) satisfies

$$\bar{T}W_L = \frac{1}{N^2\sigma_\phi^2} \, I_0^2\!\left(\frac{1}{N^2\sigma_\phi^2}\right) \tag{5.12-16}$$

APPENDIX   SPECTRA OF THE SELF-NOISE TERM FOR SQUARING AND COSTAS LOOPS

In this appendix we compute the spectra of the term $[\tilde{m}(t)]^2$ to assess the noise performance of the squaring loop. The filtered sequence $\tilde{m}(t)$ is expressed as

$$\tilde{m}(t) = \sum_{k=-\infty}^{\infty} C_k q(t - kT) \tag{A-1}$$

where $q(t)$ is the filtered version of the basic pulse $p(t)$ with amplitude of unity and the C_k are independent random variables having the value ± 1 with probability .5.

To obtain the power spectra of $[\tilde{m}(t)]^2$ we use (from Chapter 2) the formula

$$\mathcal{S}(\omega) = \lim_{N \to \infty} \frac{|\mathcal{F}\{[\tilde{m}_N(t)]^2\}|^2}{(2N+1)T} \tag{A-2}$$

where $\tilde{m}_N(t)$ is the truncated sequence

$$\tilde{m}_N(t) = \sum_{k=-N}^{N} C_k q(t - kT) \tag{A-3}$$

and $\mathcal{F}[a(t)]$ is the Fourier transform of $a(t)$. Expanding $[\tilde{m}_N(t)]^2$, we have

$$[\tilde{m}_N(t)]^2 = \sum_{k=-N}^{N} q^2(t - kT) + \sum_{k=-N}^{N} \sum_{\substack{i=-N \\ i \neq k}}^{N} C_k C_i q(t - kT) q(t - iT) \tag{A-4}$$

$$\triangleq S_1(t) + S_2(t) \tag{A-5}$$

where the first term denotes the $2N + 1$ diagonal terms and the last term denotes the $4N^2 + 2N$ off-diagonal terms. Taking Fourier transforms and squaring, we have

$$\mathscr{S}(\omega) = \lim_{N \to \infty} \frac{|\overline{\mathscr{F}[\tilde{m}_N(t)]^2}|^2}{(2N+1)T} = \lim_{N \to \infty} \frac{\overline{\mathscr{F}(S_1)\mathscr{F}^*(S_1) + 2\mathscr{F}(S_1)\mathscr{F}^*(S_2) + \mathscr{F}(S_2)\mathscr{F}^*(S_2)}}{(2N+1)T}$$

(A-6)

The cross term has zero expectation since

$$2\mathrm{Re}[\overline{\mathscr{F}(S_1)\mathscr{F}(S_2)^*}] = 2\mathrm{Re}\,\mathscr{F}\left[\sum_{k=-N}^{N} q^2(t-kT)\right]$$

$$\times \mathscr{F}\left[\sum_{k=-N}^{N}\sum_{\substack{i=-N \\ i \neq k}}^{N} \overline{C_k C_i}\, q(t-kT)q(t-iT)\right]$$

$$= 0 \text{ since } \overline{C_k C_i} = 0 \qquad \text{when } i \neq k \qquad \text{(A-7)}$$

since $q(t)$ is real and $\mathrm{Re}(x)$ denotes the real part of x.

Therefore the spectra can be written as

$$\mathscr{S}(\omega) = \mathscr{S}_d(\omega) + \mathscr{S}_c(\omega) \tag{A-8}$$

where $\mathscr{S}_d(\omega)$ is the discrete line spectrum [the first term of (A-4)] and $\mathscr{S}_c(\omega)$ is the continuous portion of the spectra (the second term of A-4).

First consider the discrete line spectra due to the periodic component. Denote the first term of $[\tilde{m}_N(t)]^2$ as $D(t)$ for $N \to \infty$. Then

$$D(t) = \sum_{k=-\infty}^{\infty} q^2(t-kT) = q^2(t) \circledast \sum_{k=-\infty}^{\infty} \delta(t-kT) \tag{A-9}$$

The Fourier transform of $D(t)$ is (since the $\circledast$ denotes convolution) given by

$$\mathscr{F}[D(t)] = \mathscr{F}(q^2(t))\mathscr{F}\left[\sum_{K=-\infty}^{\infty} \delta(t-kT)\right] = \mathscr{F}[q^2(t)]\sum_{k=-\infty}^{\infty}\delta\left(\omega - \frac{2\pi K}{T}\right)$$

(A-10)

now

$$\mathscr{F}[q^2(t)] = \frac{1}{T}\int_{-\infty}^{\infty} q^2(t)e^{-j\omega t}\,dt \tag{A-11}$$

Therefore using (A-11) in (A-10), we obtain

$$\mathscr{S}_d(f) = \frac{1}{T^2}\left[\int_0^{\infty} q^2(t)\,dt\right]^2 + \frac{1}{T^2}\left[\int_0^{\infty} q^2(t)\exp\left(-j\frac{2\pi}{T}t\right)\right]^2$$

$$+ \frac{1}{T^2}\left[\int_0^{\infty} q^2(t)\exp\left(-j\frac{4\pi}{T}t\right)dt\right]^2 + \cdots \tag{A-12}$$

By Passeval's theorem

$$\frac{1}{T^2}\left[\int_0^{\infty} q(t)q(t)\exp\left(-j\frac{2\pi K}{T}t\right)dt\right]^2 = \frac{1}{T^2}\left[\int_{-\infty}^{\infty} Q^*(\omega)Q\left(\omega + \frac{2\pi K}{T}\right)\frac{d\omega}{2\pi}\right]^2$$

But

$$\frac{1}{T}\int_{-\infty}^{\infty} Q^*(\omega)Q\left(\omega + \frac{2\pi K}{T}\right)\frac{d\omega}{2\pi} = \frac{1}{T}\int_{-\infty}^{\infty} P^*(\omega)P\left(\omega + \frac{2\pi K}{T}\right)H^*_{LP}(\omega)$$

$$\times\, H_{LP}\left(\omega + \frac{2\pi K}{T}\right)\frac{d\omega}{2\pi} \qquad (A\text{-}13)$$

where $P(\omega) = \mathscr{F}[p(t)]$ the Fourier transform of the unfiltered pulse. $H_{LP}(\omega)$ is the low pass equivalent of the presquaring bandpass filter. At $K = 0$ (dc term)

$$\alpha = \frac{1}{T}\int_0^{\infty} q^2(t)\,dt = \frac{1}{T}\int_{-\infty}^{\infty} |P(\omega)|^2|H_{LP}(\omega)|^2\frac{d\omega}{2\pi} = \int_{-\infty}^{\infty} \mathscr{S}_m(\omega)|H_{LP}(\omega)|^2\frac{d\omega}{2\pi}$$
$$(A\text{-}14)$$

with $\mathscr{S}_m(\omega)$ being the power spectra of the baseband modulation (NRZ, Bi, ϕ-L, etc), and α being the fraction of total power passed by presquaring or data filter. From our assumption

$$B_L \ll \frac{1}{T} \qquad (A\text{-}15)$$

it is clear that only the dc term appears in the loop bandwidth. The remaining terms are removed by the loop filter and the VCO. The dc term is given by

$$\mathscr{S}_d^{dc}(f) = \frac{1}{T^2}\left[\int_0^{\infty} q^2(t)dt\right]^2 \delta(f) = \alpha^2\delta(f) \qquad (A\text{-}16)$$

Now we evaluate the continuous spectra. To do this we must evaluate

$$\mathscr{S}_c(f) = \lim_{N\to\infty}$$

$$\frac{\left[\mathscr{F}\sum_{K=-N}^{N}\sum_{i=-N}^{N} C_kC_i(q - KT)q(t - iT)\right]\left[\mathscr{F}\sum_{K=-N}^{N}\sum_{\substack{i=-N \\ i\neq K}}^{N} C_kC_iq(t - kT)q(t - iT)\right]^*}{(2N + 1)T}$$
$$(A\text{-}17)$$

Consider the Fourier transform of $q(t - kT)q(t - iT)$

$$\mathscr{F}\{q(t - kT)q(t - iT)\} = \int_{-\infty}^{\infty} q(t - kT)q(t - iT)e^{-j\omega t}\,dt$$

$$= Q(\omega)e^{-j\omega kT} \circledast Q(\omega)e^{-j\omega iT}$$
$$(A\text{-}18)$$

where the $\circledast$ denotes the convolution operation.
Then

$$\overline{\mathscr{F}\{\ \}\mathscr{F}^*\{\ \}} = \sum_{\substack{k=-N \\ i\neq k}}^{N}\sum_{\substack{k'=-N}}^{N}\sum_{\substack{i=-N \\ i'\neq k'}}^{N}\sum_{i'=-N}^{N} \overline{C_kC_{k'}C_iC_{i'}}|Q(\omega)e^{-j\omega kT} \circledast Q(\omega)e^{-j\omega iT}|$$

$$\times\, |Q(\omega)e^{-j\omega k'T} \circledast Q(\omega)e^{-j\omega i'T}| \qquad (A\text{-}19)$$

where $\{\ \}$ denotes $\{q(t - kT)q(t - iT)\}$.

The terms that are nonzero correspond to the case when

$$k = k' \text{ and } i = i' \text{ or } i' = k \text{ and } i = k'$$

Hence

$$\overline{\mathcal{F}\{\ \}F^*\{\ \}} = 2 \sum_{k=-N}^{N} \sum_{\substack{i=-N \\ i \neq k}}^{N} |Q(\omega)e^{-j\omega kT} \circledast Q(\omega)e^{-j\omega iT}|^2 \qquad \text{(A-20)}$$

Consider the convolution

$$Q(\omega)e^{-j\omega kT} \circledast Q(\omega)e^{-j\omega iT} = \int_{-\infty}^{\infty} Q(\omega')e^{-j\omega kT} Q(\omega - \omega')e^{-j(\omega-\omega')iT} \frac{d\omega'}{2\pi}$$
$$\text{at } \omega = 0 \qquad \text{(A-21)}$$

$$\text{convolution} = \int_{-\infty}^{\infty} |Q(\omega')|^2 e^{j\omega'(i-k)T} \frac{d\omega'}{2\pi} \qquad \text{(A-22)}$$

Hence

$$\overline{\mathcal{F}\{\ \}\mathcal{F}^*\{\ \}}\Big|_{\omega=0} = 2 \sum_{k=-N}^{N} \sum_{\substack{i=-N \\ i \neq K}}^{N} \left[\int_{-\infty}^{\infty} |Q(\omega')|^2 e^{j\omega(i-k)T} \frac{d\omega'}{2\pi} \right]^2 \qquad \text{(A-23)}$$

and

$$Q(\omega) = P(\omega)H_{LP}(\omega) \qquad \text{(A-24)}$$

with $H_{LP}(\omega)$ being the lowpass equivalent of the bandpass filter. So

$$\mathcal{S}_c(0) = \lim_{T \to \infty} \frac{2}{(2N+1)T} \sum_{\substack{i=-N \\ k=-N \\ i \neq k}}^{N} \sum^{N} \left[\int\int_{-\infty}^{\infty} |P(\omega)H_{LP}(\omega)|^2 e^{j\omega(i-k)T} \frac{d\omega}{2\pi} \right]^2$$

$$\text{(A-25)}$$

Define

$$R_d(kT) = \int_{-\infty}^{\infty} \mathcal{S}_p(\omega)|H_{LP}(\omega)|^2 e^{-j\omega lT} \frac{d\omega}{2\pi} \qquad \text{(A-26)}$$

But

$$\int_{-\infty}^{\infty} |Q(\omega)|^2 e^{j\omega kT} \frac{d\omega}{2\pi} = \int_{-\infty}^{\infty} |P(\omega)|^2 |H_{LP}(\omega)|^2 e^{j\omega kT} \frac{d\omega}{2\pi} \qquad \text{(A-27)}$$

Therefore

$$\mathcal{S}_c(0) = \lim_{T \to \infty} \frac{2T}{2N+1} \sum_{i=-N}^{N} \sum_{\substack{k=-N \\ i \neq k}}^{N} \{R_d(i-k)T\}^2 \qquad \text{(A-28)}$$

since

$$\mathcal{S}_p(\omega) = \frac{|P(\omega)|^2}{T} \qquad \text{(A-29)}$$

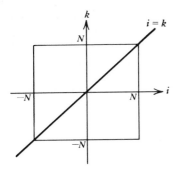

Figure 5.A.1 Diagram for double summation.

for random data $[P(1) = \frac{1}{2}$ and $P(-1) = \frac{1}{2}]$. Let S be denoted by

$$S = \sum_{k=-N}^{N} \sum_{\substack{i=-N \\ i \neq k}}^{N} \{R_d[(i - k)T]\}^2 \qquad \text{(A-30)}$$

Consider Figure 5.A.1

Consider the sum S_1. Let $l = i - k$; for each $l > 0$ there are $2N + 1 - l$ combinations for that given l. For $l < 0$ there are $2N + 1 + l$ combinations. Therefore

$$S = \sum_{\substack{l=-(2N-1) \\ l \neq 0}}^{2N-1} (2N + 1 - |l|)[R_d(lT)]^2 = 2 \sum_{l=1}^{2N-1} (2N + 1 - l)[R_d(lT)]^2$$

$$\text{(A-31)}$$

So if

$$\sum_{l=0}^{\infty} l[R_d(lT)]^2 < \infty \qquad \text{(A-32)}$$

we have that

$$\mathscr{S}_c(0) = \lim_{N \to \infty} \frac{2T}{2N + 1} \left\{ 2 \sum_{l=1}^{2N-1} (2N + 1 - l)[R_d(lT)]^2 \right\} \qquad \text{(A-33)}$$

and finally

$$\mathscr{S}_c(0) = 4T \sum_{l=1}^{\infty} [R_d(lT)]^2 \qquad \text{(A-34)}$$

This result was apparently first obtained by Weidner [38], although by an approach different from that used by the author.

REFERENCES

1 Van Trees, H. L., "Optimum Power Division in Coherent Communication Systems," Lincoln Lab., Lexington, Mass., Technical Report 301, February 1963.

2   Didday, R. L., and Lindsey, W. C., "Subcarrier Tracking Methods and Communication System Design," *IEEE Trans. Communications Tech.*, Vol. 16, August 1968.

3   Viterbi, A. J., *Principles of Coherent Communications*, McGraw-Hill, New York, 1966.

4   Layland, J. W., "An Optimum Squaring Loop Prefilter," *IEEE Trans. Communications Tech.*, October 1970.

5   Lindsey, W. C., and Simon, M., *Telecommunication Systems Engineering*, Prentice-Hall, Englewood Cliffs, N.J., 1973.

6   Riter, S., "An Optimum Phase Reference Detector for Fully Modulated Phase Shift Keyed Signals," *IEEE AES-5*, No. 4, July 1969.

7   Holmes, J. K., and Lindsey, W. C., "A Comparison of the Performance of the Costas Loop for Various Data Filters," TRW IOC 7130.60-005, February 1974.

8   Stiffler, J. J., *Theory of Synchronous Communications*, Prentice-Hall, Englewood Cliffs, N.J., 1971, Chap. 8.

9   Weidner, M. Y., "Costas Loop Tracking with Linear Time Invariant Detection Filtering," TRW IOC, May 12, 1975.

10   Oldenwalder, J. P., "Carrier Tracking, But Synchronization, and Coding for S-Band Communication Links," National Electronics Conference, December 1974, San Diego, Cal.

11   Cahn, C., "Improving Frequency Acquisition of a Costas Loop," *IEEE Communications*, Vol. COM-25, No. 12, December 1977.

12   Holmes, J. K., "Costas Loop Carrier Performance Calculations for Shuttle Orbiter for the Costas Loop Mode of Operation with a Residual Carrier Component," TRW IOC SCTE-50-75-107JKH, November 18, 1975.

13   Simon, M., "The Effects of Residual Carrier on Costas Loop Performance as Applied to the Space Shuttle Orbiter S-Band Uplink," *IEEE Trans. Communications*, Part I, Special Issue on Shuttle, November 1978.

14   Biederman, L., "Performance Analysis of a Costas Loop Under Different Than Normal Conditions," TRW IOD SCTE-50-76-267/LB, August 23, 1976.

15   Simon, M., "Tracking Performance of Costas Loops with Hartd-Limited In-Phase Channels," *IEEE Trans. Communications*, April 1978.

16   Frazier, J. P., and Page, J., "Phase-Lock Loop Frequency Acquisition Study," *Trans. IRE SET-8*, pp. 210–277, September 1962.

17   Lindsey, W. C., and Woo, K. T., "Analysis of Squaring Circuit Mechanizations in Costas and Squaring Loops," *IEEE Trans. AES*, September 1978.

18   Leonhardt, R., and Fleischmann, H. H., "Pull-In Range of Phase-Lock Circuits with Arbitrary Feedback Filter," *The Radio and Electronic Engineer*, August 1978.

19   Lindsey, W. C., *Synchronization Systems in Communication and Control*, Prentice-Hall, Englewood Cliffs, N.J., 1972.

20   Develet, J. A., Jr., "The Influence of Time Delay on Second-Order Phase-Lock Loop Acquisition Range," STL Report No. 9332.6-9, September 1962.

21   Tausworthe, R. C., "Acquisition and False-Lock Behavior of Phase-Locked Loops With Noisy Inputs," *JPL SPS* 37–46, Vol. IV.

22   Gardner, F. J., *Phase Lock Techniques*, Wiley, New York, 1966.

23   Olson, M. L., "False-Lock Detection in Costas Demodulators," *IEEE Trans. AES*, No. 2, March 1975.

24   Hedin, G., Holmes, J. K., Lindsey, W. C., and Woo, K. T., "Analysis of False Lock Phenomena In Costas Loops During Acquisition," TRW Systems Memo SCTE-50-76-249/KTW, July 15, 1976.

25   Hedin, G., Holmes, J. K., Lindsey, W. C., and Woo, K. T., "Theory of False Lock in

Costas Loops," *IEEE Trans. Communications*, Vol. COM-26, No. 1, pp. 1–12, January 1978.

26 Simon, M. K., "False Lock Performance of Costas Receivers," in *Integrated Source and Channel Encoded Digital Communication System Design Study*, Final Report, July 31, 1976, Axiomatix, Marina del Rey, Cal. Appendix J.

27 Hedin, G., Holmes, J. K., Lindsey, W. C., and Woo, K. T., "False Lock Phenomenon in Costas and Squaring Loops," *Proceedings of the 1977 National Telecommunications Conference*, Los Angeles, Cal., pp. 34:4-1 to 34:4-6.

28 Simon, M. K., "The False Lock Performance of Costas Loops with Hard-Limited In-Phase Channels," *IEEE Trans. Communications*, Vol. COM-26, No. 1, pp. 23–34, January 1978.

29 Simon, M. K., "On the False Lock Behavior of Polarity-Type Costas Loops with Manchester Coded Input," *Proceedings of the 1977 National Telecommunications Conference*, Los Angeles, Cal., pp. 30:1-1 to 30:1-5.

30 Stein, S., Jones, J. J., *Modern Communication Principles with Applications to Digital Signaling*, McGraw-Hill, New York, 1967.

31 LaFlame, D. T., "False Lock Performance of Costas Loops with N-Pole Filters," *IEEE Trans. Communications*, Volume COM-26, No. 11, Part II, November 1978.

32 Chang, H., and Kleinberg, S., "False Lock Performance of Squaring, Costas, Fourth-Power, and Quadriphase Phase-Locked Loops," Part I, *IEEE Trans. on Communications*, Vol. COM-18, August 1980.

33 Weber, C., Vandervoet, D., Salcedo, R., and Jarett, D., "Spacecraft Switched TDMA Analysis and Laboratory Demonstration," TRW Final Technical Report, Vol. II, No. 25588-6001-RV-00, February 1976.

34 Braun, W. R., and Lindsey, W. C., "Carrier Synchronization Techniques for Unbalanced QPSK Signals, Parts I and II," *IEEE Trans. Communications*, September 1978.

35 Holmes, J. K., and Osborne, H. C., "DG-2 Modified Costas Loop Tracking Performance, Part I: Balanced SQPSK," TRW Memo Number LRD-78-200-035, November 1978.

36 Holmes, J. K., and Osborne, H. C., "DG-2 Modified Costas Loop Tracking Performance, Part II: Unbalanced QPSK," TRW Memo Number LRD-78-200-036, November 1978.

37 Holmes, J. K., "First Slip Times versus Static Phase Error Offset for the First and Second-Order Phase Locked Loop," *IEEE Communications Tech.*, April 1971.

38 Weidner, M., "Costas Loop Tracking Performance with Linear Time Invariant Detection Filtering," TRW IOC, May 12, 1975.

6

COHERENT DETECTION
OF COHERENT
MODULATED SIGNALS

The purpose of this chapter is to present the basic coherent detection methods and bit error rate performance of the commonly used modulation schemes. We discuss PSK, MPSK, FSK, along with coherently demodulated differential encoding. Simple coding schemes, such as MPSK and MFSK, are discussed, and we also discuss Viterbi (maximum likelihood decoding) of convolutionally encoded data. We do not consider algebraic type coding or decoding since they are beyond the scope of this book. Noncoherent schemes are not discussed here. Finally, some imperfections in the channel are considered.

Figure 6.1 is a model of a communication system, typically with an analog source such as an analog voltage proportional to temperature, battery voltage, etc. The source could be digitalized in time also.

The function of the sampler is to convert the signal to a digital sequence. The source encoder could be an A/D converter that converts the sampled voltage to a digital word that is quantized to, say, K bits. More generally, the source encoder is a device for mapping the data samples to digital words.

A channel encoder (if used) has the function of encoding the source encoder sequence into a new sequence of length N symbols (having the same time duration as the K data bits) with usually a higher bandwidth requirement and the ability to tolerate symbol errors and still correctly associate the symbol sequence to the bit sequence. In other words, the channel encoder adds redundancy to the encoded sequence.

The function of the modulator is to convert the sequence of encoded symbols into waveforms suitable for transmission over the communication channel. Each signal waveform is modulated in amplitude, phase, frequency, or a combination of the above in accordance with the information transmitted onto a carrier. The channel may distort and attenuate the transmitted signal as well as add noise. This noise is actually the

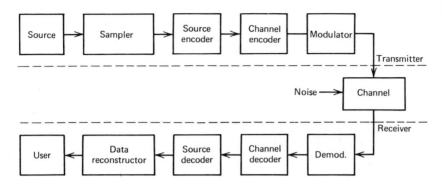

Figure 6.1 Coded digital communication system model.

receiver front end noise and is often Gaussianly distributed in amplitude. A very commonly modeled channel is that of the white Gaussian noise (WGN) channel, which assumes no signal distortion but simply a signal attenuation along with the addition of WGN.

The receiver simply reverses the respective operations of the transmitter (as well as it can), first to obtain a carrier demodulated signal that is (hopefully) the same as the transmitted baseband signal. The channel decoder interprets the received noise-corrupted code words as a bit sequence. The source decoder transforms the data sequence into a sequence of pulses of various amplitudes. The data reconstructor converts these pulses into an analog signal. This device may be a D/A converter or a more general device. The reconstructed signal is then sent to the signal destination.

We have described a general coded digital communication system. If the encoder and decoder are removed, then we have an uncoded digital communication system. Both coded and uncoded systems will be the subject of this chapter.

6.1 INFORMATION OR SOURCE RATE

We define the source or information rate R of a source measured in bits per second that produces one of a set of M equally likely messages in T sec, by

$$R = \frac{1}{T} \log_2 M \qquad \text{bits/sec} \qquad (6.1\text{-}1)$$

If, for example, we have an alphabet of four messages to be transmitted over a channel, then we could binary-encode the messages in the following way:

$$
\begin{aligned}
&1\text{--}00\\
&2\text{--}01\\
&3\text{--}10\\
&4\text{--}11
\end{aligned}
\qquad\qquad (6.1\text{-}2)
$$

If the messages are statistically independent and equally likely, then the binary bits will also be independent and equally likely, and clearly we will have

$$
R = \frac{2}{T} \qquad \text{bits/sec} \qquad\qquad (6.1\text{-}3)
$$

Moreover, the definition also yields

$$
R = \frac{1}{T}\log_2 4 = \frac{2}{T} \qquad\qquad (6.1\text{-}4)
$$

Suppose now two sources are connected simultaneously to a single transmitter. It must be able to specify $M = M_1 M_2$ messages in time T so that

$$
R = \frac{1}{T}\log_2 M = R_1 + R_2 \qquad\qquad (6.1\text{-}5)
$$

where

$$
R_i = \frac{1}{T}\log_2 M_i \qquad i = 1, 2 \qquad\qquad (6.1\text{-}6)
$$

Hence, we see that the rate of two sources combined is the sum of the individual rates.

6.2 SIGNAL STRUCTURES

As we mentioned in the introduction, the basic schemes for rf transmission are PSK, FSK, ASK, and combinations of them. First consider *phase shift keying* (PSK), which is modeled as

$$
x(t) = \sqrt{2}A \sin[\omega_0 t + \theta_i(t) + \theta_0] \qquad t \in (0, T) \qquad (6.2\text{-}1)
$$

where

$$
\begin{aligned}
\theta_i &= \pi \qquad \text{if} \quad d(t) = 0\\
&= 0 \qquad \text{if} \quad d(t) = 1
\end{aligned}
$$

and A^2 is the average transmitted power. Therefore, for T sec ($R = 1/T$ in this binary case), the data bit is either a "one" or a "zero," and the signal-dependent portion of the carrier phase is either π or 0 rad. Each bit, of duration T sec, produces a carrier phase of 0 or π rad. MPSK denotes

M possible carrier phases instead of 2. In this case, $R = (1/T) \log_2 M$, so that 8 PSK has symbols containing three bits of information each.

Frequency shift keying (FSK) is a modulation scheme is which one of two possible tones are transmitted according to whether a "zero" or "one" is the data bit. The signal structure is given by

$$x(t) = \sqrt{2}A \sin(\omega_i t + \theta_0) \tag{6.2-2}$$

$$\omega_i = \omega_1 \quad \text{when} \quad d(t) = 1$$
$$\omega_i = \omega_0 \quad \text{when} \quad d(t) = 0$$
$$(\omega_0 \neq \omega_1)$$

MFSK denotes *multiple frequency shift keying* in which for each of the M messages there corresponds a distinct carrier frequency.

Amplitude shift keying (ASK), in binary form, amplitude-modulates a carrier so that the transmitted signal is of the form

$$s(t) = \sqrt{2}A_i(t) \sin(\omega_0 t + \theta_0) \tag{6.2-3}$$

where

$$A_i(t) = 0 \quad \text{or} \quad A \text{ for } t \in (0, T)$$

Combined *amplitude and phase shift keying* is called APSK. All the above systems are digital modulation techniques, operating on amplitude discrete and time discrete data waveforms; that is, there are a discrete number of amplitudes, phases, or frequencies that can change only every T sec. We will not address other types of modulation methods, such as continuous amplitude schemes, in this text.

Some of the above schemes require coherent carrier demodulation. These systems are called *coherent communication systems* when the carrier and phase are required for demodulation, while those that do not require a phase estimate are termed *noncoherent (also incoherent) communication systems*. Furthermore, coherent systems require bit, or symbol, synchronization, which is normally provided by a bit (symbol) synchronizer. In addition, spread spectrum systems have to be "despread" before the carrier demodulation is performed.

6.3 OPTIMUM RECEIVERS FOR COHERENT RECEPTION

First let us consider a binary receiver structure where either $S_0(t)$ or $S_1(t)$ is received during $t \in (0, T)$. The reception is assumed to be corrupted by a Gaussian noise process with a positive definite covariance function. Consider basing our decisions on the *a posteriori* probabilities [4]:

$$P[i \mid y(t)] \quad \text{probability that } S_i(t)$$
$$i = 0, 1 \qquad \text{was transmitted given that} \tag{6.3-1}$$
$$y(t) \text{ was received for } t \in (0, T)$$

where

$$y(t) = S_i(t) + n(t) \qquad i = 0, 1 \qquad t \in (0, T) \tag{6.3-2}$$

We assume that $n(t)$, the noise process, has a positive definite covariance function. If $P[0 \mid y(t)] > P[1 \mid y(t)]$, then we decide that $S_0(t)$ was transmitted; otherwise, we decide that $S_1(t)$ was transmitted. Suppose

$$P[0 \mid y(t)] > P[1 \mid y(t)] \tag{6.3-3}$$

using the maximum *a posteriori* criterion, we must decide in favor of $S_0(t)$. The probability of error, PE, is the probability that $S_1(t)$ was transmitted:

$$\text{PE} = P[1 \mid y(t)] = 1 - P[0 \mid y(t)] \tag{6.3-4}$$

Suppose alternatively that we decided on $S_1(t)$ being transmitted. The probability of error in this case would be

$$\text{PE}' = P[0 \mid y(t)] = 1 - P[1 \mid y(t)] \tag{6.3-5}$$

We now show that PE < PE′. By hypothesis

$$P[0 \mid y(t)] > P[1 \mid y(t)] \tag{6.3-6}$$

or

$$1 - P[0 \mid y(t)] < 1 - P[1 \mid y(t)] \tag{6.3-7}$$

or

$$\text{PE} < \text{PE}' \tag{6.3-8}$$

We have just shown that in the binary case, maximizing the *a posteriori* probability corresponds to minimizing the probability of error and therefore is the optimum strategy.

6.4 RECEIVER STRUCTURE BASED ON MAXIMIZING $P[i \mid y(t)]$ [4, 7]

Let $P(S_i)$ be the *a priori* probability of reception of S_i. [Commonly, $P(S_i) = \frac{1}{2}$.] Let $y(t)$ be observed for T sec. Let us make the $y(t)$ process discrete:

Observe $y(t_k)$, $(t_k = \Delta t + t_{k-1})$

$$T - \Delta t < K\Delta t \le T$$

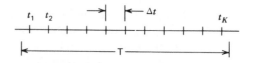

Figure 6.2 Observation interval.

Let

$$\mathbf{y} = [y(t_1), y(t_2), \ldots, y(t_k)]$$
$$\mathbf{n} = [n(t_1), n(t_2), \ldots, n(t_k)] \qquad (6.4\text{-}1)$$
$$\mathbf{S}_i = [S_i(t_1), S_i(t_2), \ldots, S_i(t_k)]$$

be K-dimensional vectors representing the respective K samples. Using Bayes' rule

$$P(i \mid \mathbf{y}) = \frac{p[\mathbf{y} \mid \mathbf{S}_i) P(\mathbf{S}_i)}{P(\mathbf{y})} \qquad (6.4\text{-}2)$$

Let the Gaussian noise have covariance function $R_n(\tau)$. Then

$$p(\mathbf{y} \mid \mathbf{S}_i) = p(\mathbf{y} - \mathbf{S}_i \mid \mathbf{n}) = \frac{\exp[-\frac{1}{2}(\mathbf{y} - \mathbf{S}_i) R_n^{-1}(\mathbf{y} - \mathbf{S}_i)^T]}{(2\pi)^{K/2} |R_n|^{1/2}} \qquad (6.4\text{-}3)$$

$[R_n]$ is a $K \times K$ matrix of the noise whose jkth component is $R_n(t_j - t_k)$. Taking natural logs yields

$$\ln P(i \mid \mathbf{y}) = \ln P(\mathbf{y} \mid \mathbf{S}_i) + \ln P(\mathbf{S}_i) - \ln P(\mathbf{y}) \qquad (6.4\text{-}4)$$

Since $P(\mathbf{y})$ does not depend on i and since $P(\mathbf{y} \mid \mathbf{S}_i)$ contains $(2\pi)^{K/2} |R_n|^{1/2}$, which does not depend on i, we can write

$$\ln P(i \mid \mathbf{y}) = C + \ln P_i - \frac{1}{2}(\mathbf{y} - \mathbf{S}_i)[R_n^{-1}](\mathbf{y} - \mathbf{S}_i)^T \qquad (6.4\text{-}5)$$

or

$$\ln P(i \mid \mathbf{y}) = D + \ln P_i - \frac{1}{2}\mathbf{S}_i R_n^{-1}\mathbf{S}_i^T + \mathbf{y}R_n^{-1}\mathbf{S}_i^T \qquad (6.4\text{-}6)$$

where $\mathbf{S}_i^T$ denotes the transpose of $\mathbf{S}_i$. Now define $\mathbf{q}_i$ by

$$\mathbf{q}_i \Delta t = \mathbf{S}_i R_n^{-1} \qquad \text{or} \qquad \mathbf{q}_i^T \Delta t = R_n^{-1}\mathbf{S}_i^T \qquad (6.4\text{-}7)$$

Postmultiplying by R_n, we obtain

$$S_i(t_k) = \sum_{m=1}^{K} R_n(t_k - t_m) q_i(t_m)\Delta t \qquad (6.4\text{-}8)$$

Using

$$\ln P[i \mid y(t)] = D + \ln P(\mathbf{S}_i) + \mathbf{y}\mathbf{q}_i^T \Delta t - \frac{1}{2}\mathbf{S}_i \mathbf{q}_i^T \Delta t$$
$$= D + \ln P(\mathbf{S}_i) + \sum_{k=1}^{k} y(t_k)q_i(t_k)\Delta t$$
$$- \frac{1}{2}\sum_{k=1}^{K} S_i(t_k)q_i(t_k)\Delta t \qquad (6.4\text{-}9)$$

Now let $\Delta t \to 0$ and $k \to \infty \ni K\Delta t = T$. Then we get

$$\ln P[i \mid y(t)] = D + \ln P(\mathbf{S}_i) + \int_0^T y(t)q(t)\,dt - \frac{1}{2}\int_0^T S(t)q(t)\,dt \qquad (6.4\text{-}10)$$

Also, we can show that

$$S_i(t) = \int_0^T R_n(t - \zeta) q_i(\zeta) \, d\zeta \tag{6.4-11}$$

Define

$$d_i^2 = \int_0^T S_i(t) q_i(t) \, dt \qquad i = 0, 1 \tag{6.4-12}$$

Hence the decision that $S_0(t)$ be transmitted is made if

$$\int_0^T y(t) q_0(t) \, dt - \frac{d_0^2}{2} + \ln P(S_0) > \int_0^T y(t) q_1(t) \, dt - \frac{d_1^2}{2} + \ln P(S_1) \tag{6.4-13}$$

otherwise $S_1(t)$ is presumed transmitted. The functions $q_i(t)$ are determined by (6.4-11). The receiver structure is shown in Figure 6.3a.

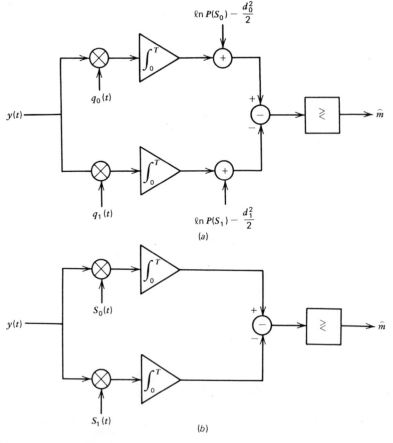

Figure 6.3 (a) Optimum receiver general case (coherent receiver). (b) Optimum detector for white noise case (coherent receiver).

6.4.1 White Noise Case with Equal Signal Energy

Assume now that $R_n(\tau) = (N_0/2)\delta(\tau)$, so that we assume WGN, and that

$$\int_0^T S_1^2(t) \, dt = \int_0^T S_2^2(t) \, dt$$

Then

$$S_i(t) = \frac{N_0}{2} \int_0^T S(t-x)q_i(x) \, dx = \frac{N_0}{2} q_i(t) \qquad (6.4\text{-}14)$$

and

$$d_0^2 = \int_0^T S_0(t)q_0(t) \, dt = \int_0^T S_1(t)q_1(t) \, dt = d_1^2 \qquad (6.4\text{-}15)$$

Then the decision rule is to choose $S_0(t)$ if

$$\ln P(S_0) + \int_0^T y(t)S_0(t) \, dt > \int_0^T y(t)S_1(t) \, dt + \ln P(S_1) \qquad (6.4\text{-}16)$$

and to choose $S_1(t)$ if

$$\ln P(S_0) + \int_0^T y(t)S_0(t) \, dt < \int_0^T y(t)S_1(t) \, dt + \ln P(S_1) \qquad (6.4\text{-}17)$$

Now, if $P(S_0) = P(S_1)$, then we have a correlation detector as shown in Figure 6.3b.

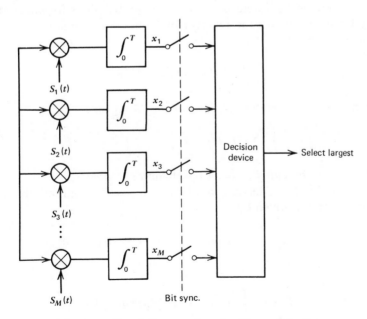

Figure 6.4 Correlation receiver for M-ary detection.

The above derivation can be done on a more rigorous basis using a Karhunen-Loeve expansion. The above results are based on reference 4. From (6.4-16) and (6.4-17), we can show, with some algebra, that the likelihood ratio test

$$L = \frac{p(y \mid i = 0)}{p(y \mid i = 1)} \gtrless \frac{P(S_i)}{P(S_0)} \tag{6.4-18}$$

is an equivalent statement of (6.4-16) and (6.4-17) for the case of equal signal energy. It is well-known [2, 4, 5, 6] that whether we use minimum error probability, minimum risk, minimax, or Neyman-Pearson tests, the likelihood ratio defines the functional form of the optimum decision rule.

We can generalize the binary case to the M-ary case by noting that for M signals that are of equal energy and equal *a priori* probability of occurrence, we can compare by pairs using the binary rule so that we can conclude that we should choose signal i as having been transmitted when

$$\int_0^T y(t)S_i(t)\,dt > \int_0^T y(t)S_j(t)\,dt \qquad \forall j \neq i \tag{6.4-19}$$

The optimum M-ary receiver is illustrated in Figure 6.4.

6.5   ERROR PROBABILITIES FOR BLOCK CODED SIGNALS [8]

6.5.1   Signal Correlation Description

In this section we shall determine the error probabilities for equal energy, equally likely signals for the receiver of Figure 6.4. Define the *normalized signal correlation* by

$$\lambda_{ij} = \frac{1}{ST} \int_0^T S_i(t)S_j(t)\,dt \qquad -1 \leq \lambda_{ij} \leq 1 \tag{6.5-1}$$

Where $E = ST$ is the energy of each of the M signals:

$$E = \int_0^T S_i^2(t)\,dt \qquad \forall i \qquad i = 1, 2, \ldots, M \tag{6.5-2}$$

Here $\{S_i(t)\}$ can be either a set of baseband or rf signals.

Assuming WGN we have, given that $S_1(t)$ was transmitted, that the ith correlator output is given by

$$x_i = \int_0^T y(t)S_i(t)\,dt = \int_0^T [S_1(t) + n(t)]S_i(t)\,dt \tag{6.5-3}$$

So

$$E(x_i) = \int_0^T \overline{[S_1(t) + n(t)]}S_i(t)\,dt = E\lambda_{1i} \tag{6.5-4}$$

The covariances are given by (the overbar denotes ensemble average)

$$E[(x_i - \bar{x}_i)(x_j - \bar{x}_j)] = \overline{\int_0^T n(t)S_i(t)\, dt \int_0^T n(u)S_i(u)\, du} \qquad (6.5\text{-}5)$$

Using the fact the noise autocorrelation satisfies

$$R_n(\tau) = \frac{N_0}{2}\delta(\tau) \qquad (6.5\text{-}6)$$

we obtain

$$E[(x_i - \bar{x}_i)(x_j - \bar{x}_j) = \frac{N_0}{2}\int_0^T S_i(t)S_j(t)\, dt = \frac{N_0 E}{2}\lambda_{ij} \qquad (6.5\text{-}7)$$

therefore

$$\text{var}(x_i) = \frac{N_0 E}{2} \qquad (6.5\text{-}8)$$

hence, the covariance matrix of the correlator outputs is

$$[Q] = \frac{N_0 E}{2}[\Lambda] \qquad (6.5\text{-}9)$$

where

$$[\Lambda] = [\lambda_{ij}] = \begin{bmatrix} \lambda_{11} & \lambda_{12} & \lambda_{13}\dots\lambda_{1N} \\ \lambda_{21} & \lambda_{22} & \lambda_{23}\dots\lambda_{2N} \\ \vdots & & \\ \lambda_{N1} & \lambda_{N2} & \lambda_{N3}\dots\lambda_{NN} \end{bmatrix} \qquad (6.5\text{-}10)$$

The joint probability density function of the correlator output vector **x**, where $\mathbf{x} = (x_1, x_2, \dots, x_m)$, given $S_1(t)$ was transmitted, is given by

$$p(\mathbf{x}\,|\,S_1) = \frac{1}{(2\pi)^{N/2}|Q|^{1/2}}\exp[-\tfrac{1}{2}(\mathbf{x} - \bar{\mathbf{x}})[Q^{-1}](\mathbf{x} - \bar{\mathbf{x}})'] \qquad (6.5\text{-}11)$$

where

$(\mathbf{x} - \bar{\mathbf{x}})^t$ is the transpose of $(\mathbf{x} - \bar{\mathbf{x}})$.

$|Q|$ is the determinate of the matrix Q. $\qquad (6.5\text{-}12)$

Q^{-1} is the inverse of Q.

This result applies whenever Q^{-1} exists; that is, Q is not singular.

As an example of *orthogonal signals*, consider the following signal set. Note the set of signals $\{S_i(t)\}$ is orthogonal if and only if

$$\begin{aligned} \lambda_{ij} &= 0 \qquad i \neq j \\ &= 1 \qquad i = j \end{aligned} \qquad (6.5\text{-}13)$$

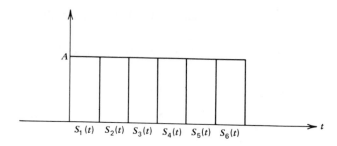

Figure 6.5 $M = 6$, orthogonal signal set.

In this case

$$[\Lambda] = [I] = \begin{bmatrix} 1 & 0 & 0 & \cdots & 0 \\ 0 & 1 & 0 & \cdots & 0 \\ \vdots & & & \ddots & \\ \cdot & \vdots & \cdot & 0 & 0 & 1 \end{bmatrix} \qquad (6.5\text{-}14)$$

For our example, pick $S_1(t)$ to be a pulse of height A and duration $T/6$ seconds in the interval $(0, T/6)$ as shown in Figure 6.5. Then $S_2(t)$ is of height A in the interval $(T/6, T/3)$, and so on. Clearly $\lambda_{ij} = 0$ for $j \neq i$.

In this case, the joint density function becomes simply the product of the individual densities.

When coded signals are BPSK modulated (one of two phases for each symbol) so that

$$x_0(t) = \sqrt{2P}\,\sin\left(\frac{K\pi n}{T}t\right) = -x_1(t) \qquad t \in \left(0, \frac{T}{n}\right) \qquad (6.5\text{-}15)$$

we can obtain a simplification in the signal correlation of coded words:

$$\lambda_{ij} = \frac{1}{E}\int_0^T s_i(t)s_j(t)\,dt \qquad (6.5\text{-}16)$$

The signals $s_i(t)$ are composed of n connected segments of the appropriately time shifted versions of $x_0(t)$ and $x_1(t)$ (that is, n symbols per word), so that

$$\lambda_{ij} = \frac{1}{E}\sum_{k=1}^{n}\int_{(k-1)T/n}^{kT/n} a_k x_0\left[t - \frac{(k-1)T}{n}\right]b_k x_0\left[t - \frac{(k-1)T}{n}\right]dt \qquad (6.5\text{-}17)$$

where $\{a_k\}$ and $\{b_k\}$ are n-dimension binary vectors representing the encoding sequences of $s_i(t)$ and $s_j(t)$, respectively. Evaluating (6.5-17), we obtain

$$\lambda_{ij} = \frac{\text{number of agreements} - \text{number of disagreements}}{n} \qquad (6.5\text{-}18)$$

where number of agreements means the number of agreements in $(a_k b_k)$ and the number of disagreements is defined as the number of disagreements in $(a_k b_k)$. Problem 1 illustrates the use of (6.5-18).

PROBLEM 1

(a) Let the binary two-component signal vectors

$$S_1 = (0, 0) \qquad H_1 = \begin{bmatrix} 0 & 0 \\ 0 & 1 \end{bmatrix} = \begin{bmatrix} S_1 \\ S_2 \end{bmatrix}$$
$$S_2 = (0, 1)$$

have a correlation defined by

$$\lambda_{S_1 S_2} = \frac{A - D}{M}$$

where A is the number of term-by-term agreements and D is the number of term-by-term disagreements, with M being the number of components in the signal vector. Show that

$$\lambda_{S_1 S_2} = 0 \qquad \text{orthogonal vectors}$$

(b) Generate a four-component vector that is orthogonal, by noting that (*Note*: The overbar denotes the binary complement.)

$$[H_2] = \begin{bmatrix} H_1 & H_1 \\ H_1 & \bar{H}_1 \end{bmatrix} = \begin{bmatrix} 0 & 0 & 0 & 0 \\ 0 & 1 & 0 & 1 \\ 0 & 0 & 1 & 1 \\ 0 & 1 & 1 & 0 \end{bmatrix} = \begin{Bmatrix} S_1 \\ S_2 \\ S_3 \\ S_4 \end{Bmatrix}$$

produces four mutually orthogonal vectors

$$\lambda_{S_i S_j} = 0 \qquad i \neq j$$

(c) Show that in general we can generate H_{l+1} via

$$[H_{l+1}] = \begin{bmatrix} H_l & H_l \\ H_l & \bar{H}_l \end{bmatrix}$$

to obtain an orthogonal set of vectors.

(d) Show that the transmission rate R is given by

$$R = \frac{n}{T} \qquad \text{bits/sec}$$

where n is the number of bits encoded into signal vectors of length 2^n code symbols. The actual transmitted signals would be transformed, for example, by $0 \leftrightarrow -1$, $1 \leftrightarrow 1$, to be of equal signal power.

A signal set related to the orthogonal set is the *biorthogonal set*. A biorthogonal code consists of an orthogonal code plus its negatives. As an example, for the case of one bit $(K = 1)$, we have

$$[H_1^0] = \begin{bmatrix} 0 & 0 \\ 0 & 1 \end{bmatrix} \qquad \text{orthogonal case}$$

$$\qquad \qquad \qquad \qquad \qquad \qquad \qquad \qquad \qquad \qquad (6.5\text{-}19)$$

$$[H_1^B] = \begin{bmatrix} 0 & 0 \\ 0 & 1 \\ 1 & 1 \\ 1 & 0 \end{bmatrix} \qquad \text{biorthogonal case}$$

Given $M/2$ signals that are orthogonal, a biorthogonal set can be obtained to produce M signals by properly augmenting the set with the negatives: The code correlation can be defined by

$$\lambda_{ij} = \begin{cases} 0 & 0 \neq |i - j| \neq M/2 \\ -1 & |i - j| = M/2 \\ 1 & i = j \end{cases} \qquad (6.5\text{-}20)$$

That is, set $S_1, \ldots, S_{M/2}$ to be orthogonal; then

$$S_{M/2+1} = -S_1, \quad S_{M/2+2} = -S_2, \ldots, S_M = -S_{M/2}$$

PROBLEM 2

(a) Show that the normalized code correlation for biorthogonal codes is given by

$$[\Lambda] = \begin{bmatrix} I & -I \\ -I & I \end{bmatrix}$$

where $[I]$ is an $N \times N$ unit matrix:

$$[I] = \begin{bmatrix} 1 & 0 & 0 & 0 & . & . & . \\ 0 & 1 & 0 & 0 & . & . & . \\ \vdots & & & & & & \\ 0 & . & . & . & 0 & 1 & 0 \\ 0 & . & . & . & 0 & 0 & 1 \end{bmatrix}$$

(b) Further show that the matrix is singular.

Another signal set is optimum in the sense that, for no bandwidth constraints, it minimizes the probability of error. This code is the *transorthogonal* signal set. All of the signal correlations are negative. Furthermore, the signal set is called a *regular simplex* set since

$$\lambda_{ij} = \frac{-1}{M - 1} \qquad i \neq j \qquad M \text{ even} \qquad (6.5\text{-}21)$$

where again M is the number of signals in the signal set. The correlation matrix $[\Lambda]$ is given by

$$[\Lambda] = \begin{bmatrix} 1 & -\dfrac{1}{M-1} & -\dfrac{1}{M-1} & \cdots & -\dfrac{1}{M-1} \\ -\dfrac{1}{M-1} & & & & \\ \vdots & & & & \\ -\dfrac{1}{M-1} & -\dfrac{1}{M-1} & & \cdots & -\dfrac{1}{M-1} \end{bmatrix} \qquad (6.5\text{-}22)$$

In this case, since the sum of the columns is zero, the correlation matrix is singular.

We now show that regular simplex codes are optimum by bounding the signal correlation elements λ_{ij} for signal structures that can be described by vectors with ± 1 elements.

Theorem 1   Define

$$\lambda_{\text{avg}} = \frac{1}{M(M-1)} \sum_i \sum_{\substack{j \\ i \neq j}} \lambda_{ij} \qquad (6.5\text{-}23)$$

Then

$$\min \lambda_{\text{avg}} \geq \begin{cases} -\dfrac{1}{M-1} & M \text{ even} \\[2mm] -\dfrac{1}{M} & M \text{ odd} \end{cases} \qquad (6.5\text{-}24)$$

Proof

$$\lambda_{\text{avg}} = \frac{1}{M(M-1)} \sum_{\substack{i=1 \\ i \neq j}}^{M} \sum_{j=1}^{M} \lambda_{ij} \qquad (6.5\text{-}25)$$

or

$$\lambda_{\text{avg}} = \frac{1}{M(M-1)} \left(\sum_{i=1}^{M} \sum_{j=1}^{M} \lambda_{ij} - \sum_{i=1}^{M} \lambda_{ii} \right) \qquad (6.5\text{-}26)$$

with L being the length of the code words, S_i being represented by

$$S_i = (a_{i1}, a_{i2}, a_{i3}, \ldots, a_{iL})$$

with

$$a_{ij} = \pm 1$$

Then

$$\lambda_{\text{avg}} = \frac{1}{M(M-1)L} \left(\sum_{i=1}^{M} \sum_{j=1}^{M} S_i \cdot S_j - \sum_{i=1}^{M} S_i S_i \right) \qquad (6.5\text{-}27)$$

or

$$\lambda_{\text{avg}} = \frac{1}{M(M-1)L} \left(\left\| \sum_{i=1}^{M} \mathbf{S}_i \right\|^2 - \sum_{i=1}^{M} \|\mathbf{S}_i\|^2 \right) \tag{6.5-28}$$

Since $\|\mathbf{S}_i\|^2 = L$, $\forall i$, and denoting

$$V^2 = \left\| \sum_{i=1}^{M} \mathbf{S}_i \right\|^2$$

we obtain

$$\lambda_{\text{avg}} = \frac{1}{M(M-1)L} (V^2 - ML) \tag{6.5-29}$$

Now we obtain a lower bound on V^2. Since V^2 is the sum of the squares of the column sums, that is,

$$V^2 = \sum_{i=1}^{M} \sum_{j=1}^{M} \left(\sum_{l=1}^{L} a_{il} a_{jl} \right)$$

$$V^2 = \sum_{l=1}^{L} \left(\sum_{j=1}^{M} a_{jl} \right) \left(\sum_{i=1}^{M} a_{il} \right) \tag{6.5-30}$$

clearly V^2 could be made zero if there are an even number of terms. For M odd, the smallest V^2 that can be made is L. Therefore we conclude that

$$\min \lambda_{\text{avg}} \geq -\frac{1}{M-1} \qquad M \text{ even}$$

$$\min \lambda_{\text{avg}} \geq -\frac{1}{M} \qquad M \text{ odd} \tag{6.5-31}$$

Note that if all $\lambda_{ij} = -1/(M-1)$ (M even), we have achieved the bound and therefore have an optimum code. The regular simplex codes achieve this bound.

As a final example, let us consider polyphase signal sets (MPSK). This signal set consists of M equally spaced signal vectors located on the unit circle. The code correlation matrix has elements

$$\lambda_{ij} = \cos\left[\frac{2\pi}{M} (i - j) \right] \qquad i, j = 1, \ldots, M \tag{6.5-32}$$

Hence the code correlation matrix is given by

$$[\Lambda] = \begin{bmatrix} 1 & \cos\dfrac{2\pi}{M} & \cos\dfrac{4\pi}{M} & \cdots & \cos\dfrac{2\pi(M-1)}{M} \\[2mm] \cos\dfrac{2\pi}{M} & 1 & \cos\dfrac{2\pi}{M} & \cdots & \\[2mm] \vdots & & & \ddots & \\[2mm] \cos\dfrac{2\pi(M-1)}{M} & \cos\dfrac{2\pi(M-2)}{M} & \cdots & & 1 \end{bmatrix}$$

$$\tag{6.5-33}$$

6.5.2 Performance of Phase Coherent Receivers with det $\Lambda \neq 0$

In this section we determine the probability of error for phase coherent receivers with det $\Lambda \neq 0$. The cases when det $\Lambda = 0$ will be treated separately. We assume that all signals have equal energy, are equally likely, and have equal time duration. For the case of equally likely signals the probability of correct detection will be given by [assuming signal $x_1(t)$ was sent],

$$PC_W = P[x_2 < x_1, x_3 < x_1, \ldots, x_M < x_1 \mid x_1(t)] \tag{6.5-34}$$

Therefore

$$PC_W = \int_{-\infty}^{\infty} \int_{-\infty}^{x_1} \int_{-\infty}^{x_1} \cdots \int_{-\infty}^{x_1} p_x[x_1, x_2, \ldots, x_M \mid x_1(t)] \, dx_1 \cdots dx_M \tag{6.5-35}$$

Equation 6.5-1 is the M-dimensional Gaussian density function used in (6.5-34) with mean and covariance matrix given by (6.5-4) and (6.5-9), respectively. Now let $u_i = x_i - \bar{x}_i$, then

$$PC_W = \int_{-\infty}^{\infty} \int_{-\infty}^{x_1 - \bar{x}_2} \int_{-\infty}^{x_1 - \bar{x}_3} \cdots \int_{-\infty}^{x_1 - \bar{x}_M} p_u[u_1, u_2, u_3, \ldots, u_M \mid x_1(t)] \, d\mathbf{u} \tag{6.5-36}$$

where

$$\mathbf{u} = (u_1, u_2, u_3, \ldots, u_M) \tag{6.5-37}$$

Now since $x_1 = u_1 + \bar{x}_1$, we can rearrange PC_W to yield

$$PC_W = \int_{-\infty}^{\infty} \int_{-\infty}^{u_1 + \bar{x}_1 - \bar{x}_2} \int_{-\infty}^{u_1 + \bar{x}_1 - x_3} \cdots \int_{-\infty}^{u_1 + \bar{x}_1 - \bar{x}_M} p_u[u_1, u_2, \ldots, u_m \mid x_1(t)] \, d\mathbf{u} \tag{6.5-38}$$

Now $p_u(u_1, u_2, \ldots, u_m)$ is an M-dimensional Gaussian density function with mean zero and covariance $(N_0 E_1/2)[\Lambda]$. Finally let $v_i = u_i/\sigma_x$. Then we have

$$PC_W = \int_{-\infty}^{\infty} \int_{-\infty}^{v_1 + (\bar{x}_1 - \bar{x}_2)/\sigma_x} \int_{-\infty}^{v_1 + (\bar{x}_1 - \bar{x}_2)/\sigma_x} \cdots \int_{-\infty}^{v_1 + (\bar{x}_1 - \bar{x}_M)/\sigma_x}$$
$$\times p_v[v_1, v_2, \ldots, v_M \mid x_1(t)] \, d\mathbf{v} \tag{6.5-39}$$

where $\mathbf{v} = (v_1, v_2, \ldots, v_m)$, and $p_v(\mathbf{v})$ is an M-dimensional Gaussian density function with zero mean and covariance matrix $[\Lambda]$. Since

$$\frac{\bar{x}_1 - \bar{x}_i}{\sigma_x} = \sqrt{\frac{2E}{N_0}} \, (1 - \lambda_{1i}) \tag{6.5-40}$$

we obtain, finally, our results:

$$PC_W = \int_{-\infty}^{\infty} \int_{-\infty}^{v_1 + \sqrt{2R_W}(1 - \lambda_{12})} \cdots \int_{-\infty}^{v_1 + \sqrt{2R_W}(1 - \lambda_{1M})} p_v[v_1, v_2, v_3, \ldots, v_M \mid x_1(t)] \, d\mathbf{v} \tag{6.5-41}$$

where

$$p_v(v_1, v_2, \ldots, v_M \mid s_1) = \frac{\exp(-\frac{1}{2}v\Lambda^{-1}v^T)}{(2\pi)^{M/2}|\Lambda|^{1/2}} \tag{6.5-42}$$

with $[\Lambda]^{-1}$ the inverse of $[\Lambda]$, $|\Lambda|$ the determinate of $[\Lambda]$, and R_W, which equals E/N_0 the word energy-to-noise spectral density ratio.

PROBLEM 3

Consider the case $M = 2$, with $P(S_i) = P(S_2) = \frac{1}{2}$ and $E_1 = E_2$, with the normalized signal correlation $\lambda_{12} = \lambda$. Show that

(a) $\qquad P_C = \dfrac{1}{2\pi\sqrt{1-\lambda^2}} \displaystyle\int_{-\infty}^{\infty} \int_{-\infty}^{v_1+\sqrt{2R_W}(1-\lambda)} \exp\left[\dfrac{(v_2-\lambda v_1)^2}{2(1-\lambda^2)}\right] dv_1\, dv_2$

(b) Using

$$\text{erfc}(x) = \frac{2}{\sqrt{\pi}} \int_x^{\infty} e^{-y^2}\, dy$$

and

$$\text{erfc}(-x) = 1 + \text{erf}(x)$$

that

$$PC = \frac{1}{2} + \frac{\sqrt{1-\lambda^2}}{2\sqrt{\pi(1-\lambda)}} \int_{-\infty}^{\infty} \exp\left\{-\frac{1}{2}\left[\frac{\sqrt{2(1-\lambda^2)}}{1-\lambda}z - \sqrt{2R_w}\right]^2\right\} \text{erf}(z)\, dz$$

(c) Using

$$\int_{-\infty}^{\infty} \exp(-a^2 z^2 + bz)\, \text{erf}(z)\, dz = \frac{\sqrt{\pi}}{a} e^{b^2/(4a^2)}\, \text{erf}\left(\frac{b}{2a\sqrt{1+a^2}}\right)$$

show that

$$PC = \tfrac{1}{2} + \tfrac{1}{2}\text{erf}\,\sqrt{R_W\left(\frac{1-\lambda}{2}\right)}$$

or

$$PE = \tfrac{1}{2}\text{erfc}\,\sqrt{R_W\left(\frac{1-\lambda}{2}\right)}$$

with PE being the probability of error. Note when $M = 2$, we have that $PE = PE_W = PE_B$ and $R_W = R_B$.

From the results of Problem 3 (in the binary case) it is clear that the optimum value of λ is -1, since

$$PE = \tfrac{1}{2}\text{erfc}\,\sqrt{R_W\left(\frac{1-\lambda}{2}\right)} \tag{6.5-43}$$

with $\text{erfc}(x)$ being monotonically decreasing with x.

PROBLEM 4

Show that in the binary case the probability of error can be written as

$$PE = \tfrac{1}{2}\text{erfc}\left[\frac{d}{\sqrt{2N_0}}\right]$$

where

$$d^2 = \int_{-\infty}^{\infty} [x_1(t) - x_2(t)]^2 \, dt$$

where d is the "distance" between signals.

6.5.3 Performance of Phase Coherent Receivers with Orthogonal Signals

Of the various schemes that will produce orthogonal signals, we consider a couple here. First, as an example, consider a set of tones that are mutually disjoint in time. Note this implies a receiver that has a broad bandwidth at the front end so as not to produce intersymbol interference. The signal structure is

$$x_i(t) = \sqrt{2MS} \, \sin\left(\frac{\pi k M t}{T}\right) \qquad \frac{i-1}{M} T \le t \le \frac{iT}{M} \qquad (6.5\text{-}44)$$

$$i = 1, 2, \ldots, M$$

This is a form of time division multiplexing. As another example, consider MFSK, that is, a set of coherent tones placed $\frac{1}{2}T$ Hz apart (see Section 6.5.7) where T is the tone duration. The signal set is given by

$$x_i(t) = \sqrt{2S} \, \sin\left[\frac{\pi(k+i)t}{T}\right] \qquad \begin{matrix} 0 \le t \le T \\ i = 1, \ldots, M \end{matrix} \qquad (6.5\text{-}45)$$

Now we shall evaluate the performance of orthogonal signals given $x_1(t)$ was sent, using (6.5-41) and (6.5-42). We again assume equal energy and equally likely signals. When the signals are orthogonal, $[\Lambda] = [I]$, so that

$$p_v(v_1, v_2, \ldots, v_M) = \frac{1}{(2\pi)^{M/2}} \prod_{i=1}^{M} e^{v_i^2/2} \qquad (6.5\text{-}46)$$

Let $v_1 = \sqrt{2}x$ and $v_i = \sqrt{2}y_i$, $i = 2, \ldots, M$, so that

$$PC_W = \int_{-\infty}^{\infty} \int_{-\infty}^{x+\sqrt{R_W}} \cdots \int_{-\infty}^{x+\sqrt{R_W}} \frac{1}{(2\pi)^{M/2}} \prod_{i=1}^{M} e^{-y_i^2} (2)^{M/2} \, dx \, dy_2 \ldots dy_M$$

$$(6.5\text{-}47)$$

Using the definition of erfc( ), we obtain

$$PC_W = \int_{-\infty}^{\infty} \frac{e^{-x^2}}{\sqrt{\pi}} [\tfrac{1}{2} \operatorname{erfc}(-x - \sqrt{R_W})]^{M-1} \, dx \qquad (6.5\text{-}48)$$

which is the probability of a correct word detection with a word containing $\log_2 n$ bits. The probability of a word error is simply

$$PE_W = 1 - PC_W \qquad (6.5\text{-}49)$$

Again, the probability of error is independent of which signal was transmitted. In Figure 6.6, PE_W is plotted versus R_B, the bit energy-to-noise spectral density ratio ($R_B = R_W/N$) for $M = 2^n$.

Now that we have obtained the word error probability, we consider the bit error probability. If a word is in error, one or more bits may be in error in general. If a word is incorrectly detected since $\lambda_{ij} = 0$ ($i \neq j$), the decision is equally likely to be made in favor of any one of the $M - 1$ incorrect signals.

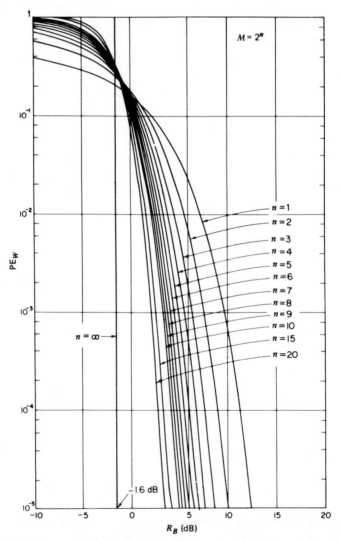

Figure 6.6 Word error probability performance for an orthogonal signal set (from Lindsey and Simon [3] with permission).

Assuming that each code word contains n bits ($M = 2^n$), the probability of K bits being in error is given by

$$p_k = \frac{\dfrac{n!}{k!(n-k)!}}{\left[\displaystyle\sum_{j=1}^{n} \dfrac{n!}{j!(n-j)!}\right]} = \frac{\left(\begin{array}{l}\text{number of ways of}\\ \text{arranging } k \text{ bits in}\\ n \text{ slots}\end{array}\right)}{\left(\begin{array}{l}\text{total number of ways}\\ \text{of arranging the bits in } n \text{ slots}\\ \text{including one at a time, etc.}\end{array}\right)} \qquad (6.5\text{-}50)$$

Therefore, the average number of decoded data bits per word in error is given by

$$\bar{k} = \sum_{k=1}^{n} k P_k = \frac{n 2^{n-1}}{2^n - 1} \qquad (6.5\text{-}51)$$

Further, it follows that the probability of a given bit being in error, given a word is in error, is expressed by

$$P(B \mid W) = \frac{1}{n} \frac{n 2^{n-1}}{2^n - 1} = \frac{2^{n-1}}{2^n - 1} \qquad (6.5\text{-}52)$$

Using Bayes' rule

$$PE_B = \frac{P(B \mid W) PE_W}{P(W \mid B)} \qquad (6.5\text{-}53)$$

where B denotes a bit error and W denotes a word error. Since $P(W/B) = 1$, we conclude that

$$PE_B = \frac{2^{n-1}}{2^n - 1} PE_W \qquad (6.5\text{-}54)$$

It is then a simple matter to evaluate the bit error rate using (6.5-48), (6.5-49), and (6.5-54). Figure 6.7 illustrates the bit error rate for orthogonal codes plotted against $R_B = R_W/\log_2 M$. Notice that $PE_B \simeq PE_W/2$ for $n \geq 3$.

6.5.4 Performance of Phase Coherent Receivers with Biorthogonal Signals

We have analyzed the orthogonal signal case and we now consider the case of Biorthogonal signals. As was mentioned previously, there are only $M/2$ independent correlator outputs since each of the $M/2$ correlator values has its negative in the remaining set of $M/2$ correlators. The signal correlation matrix Λ can be represented for biorthogonal signals by

$$[\Lambda] = \begin{bmatrix} I & -I \\ -I & I \end{bmatrix} \qquad (6.5\text{-}55)$$

Clearly the determinate is zero. To avoid the problem of singularity we eliminate the $M/2$ dependent variables.

The original decision rule is to pick the largest positive correlation. An

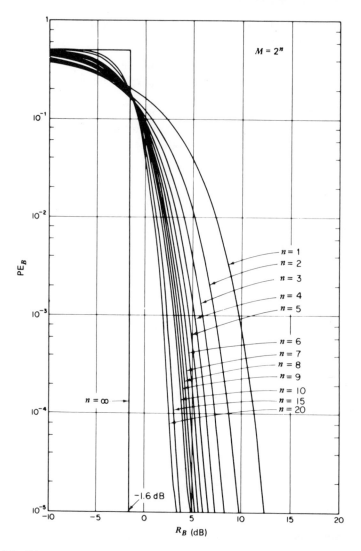

Figure 6.7 Bit error probability performance for an orthogonal signal set (from Lindsey and Simon [3] with permission).

equivalent rule is to correlate with only the first $M/2$ correlators using both magnitude and algebraic sign to determine which signal was transmitted. With only $M/2$ correlators used, the new signal correlation matrix is no longer singular. We assume equal energy and equally likely signals.

The correct word probability, based on the equivalent decision rule, is specified by

$$\text{PC}_W = P[x_1 > 0, |x_1| > |x_2|, |x_1| > |x_3|, \ldots, |x_1| > |x_{M/2}| \mid x_1(t)] \quad (6.5\text{-}56)$$

or

$$PC_W = \int_0^\infty \int_{-x_1}^{x_1} \int_{-x_1}^{x_1} \cdots \int_{-x_1}^{x_1} p(x_1, x_2, \ldots, x_{M/2} \mid x_1(t)) \, d\mathbf{x} \qquad (6.5\text{-}57)$$

where $p[\mathbf{x} \mid x_1(t)]$ is a Gaussian probability density with mean $E(1, 0, 0, \ldots, 0)$ and covariance $[Q] = (N_0 E/2)[I]$ and $[I]$ is an $M/2$ by $M/2$ identity matrix. Recall E is the word energy. Let $u_i = (x_i - \bar{x}_i)/\sigma_x$ so that we obtain

$$PC_W = \frac{1}{(2\pi)^{M/4}} \int_{-\sqrt{2R_W}}^\infty e^{-u_1^2/2} \left(\int_{-u_1 - \sqrt{2R_W}}^{u_1 + \sqrt{2R_W}} e^{-u^2/2} \, du \right)^{M/2-1} du_1 \qquad (6.5\text{-}58)$$

Now letting $x = u_1/\sqrt{2}$ and $y = u/\sqrt{2}$, we obtain $(R_W = E/N_0)$

$$PC_W = \int_{-\sqrt{R_W}}^\infty \frac{e^{-x^2}}{\sqrt{\pi}} [\operatorname{erf}(x + \sqrt{R_W}]^{M/2-1} \, dx \qquad (6.5\text{-}59)$$

The error probability is given by

$$PE_W = 1 - PC_W \qquad (6.5\text{-}60)$$

In Figure 6.8 the word error probability is plotted. The error probability is smaller (better) than orthogonal signals, but when $M = 32$ the performance is essentially the same, since the difference decreases as M increased $(M = 2^n)$.

PROBLEM 5

Show that biorthogonal codes have half the bandwidth requirement of orthogonal codes. *Hint*: Consider the ratio of bandwidth to data rate.

Now we shall obtain the bit error probability of biorthogonal codes. In a biorthogonal signal set two classes or types of error can occur. An error of the first kind occurs when $-x_1(t)$ is selected and $x_1(t)$ is transmitted. The probability of this event is denoted by P_1 and given by

$$P_1 = P[x_1 < 0, |x_1| > |x_2|, |x_1| > |x_3|, \ldots, x_1 | > |x_{M/2}| \mid x_1(t)] \qquad (6.5\text{-}61)$$

or

$$P_1 = \int_{-\infty}^0 \int_{-x_1}^{x_1} \cdots \int_{-x_1}^{x_1} p[\mathbf{x} \mid x_1(t)] \, d\mathbf{x}$$

where $p[\mathbf{x} \mid x_1(t)]$ is an $M/2$-dimensional Gaussian density function with mean $E(1, 0, 0, \ldots, 0)$ and covariance $[Q] = (N_0 E/2)[I]$, with I again being an $M/2$ by $M/2$ identity matrix. Using the same change of variables as before, we obtain

$$P_1 = \int_{-\infty}^{-\sqrt{R_W}} \frac{e^{-x^2}}{\sqrt{\pi}} [\operatorname{erf}(x + \sqrt{R_W})]^{M/2-1} \, dx \qquad (6.5\text{-}62)$$

The probability of selecting any one of the $M - 2$ signals that is

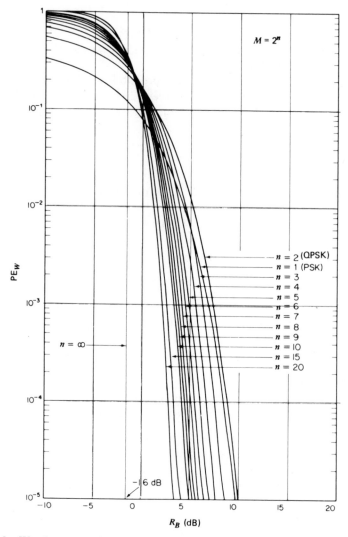

$M = 2^n$

$n = 2$ (QPSK)
$n = 1$ (PSK)
$n = 3$
$n = 4$
$n = 5$
$n = 6$
$n = 7$
$n = 8$
$n = 9$
$n = 10$
$n = 15$
$n = 20$

$n = \infty$

-6 dB

R_B (dB)

Figure 6.8 Word error probability for a biorthogonal signal set (from Lindsey and Simon [3] with permission).

orthogonal to the transmitted signal corresponds to an error of the second kind and is given numerically by

$$P_2 = PE_W - P_1 \qquad (6.5\text{-}63)$$

It is assumed that complementary data vectors (binary) are coded into complementary code words to minimize the probability that a word error will cause all bits to be in error.

Now if an error of the first kind is made, all n bits will be in error.

Therefore, the conditional bit error probability, given that an error of the first kind has occurred, is exactly 1. When an error of the second kind is made, the average number of data bits in error (NBE) is given by

$$E[\text{NBE} \mid E_2] = \frac{\sum\limits_{k=1}^{n-1} k \dfrac{n!}{k!(n-k)!}}{\sum\limits_{j=1}^{n-1} \dfrac{n!}{j!(n-j)!}} \tag{6.5-64}$$

This can be evaluated to yield the simple result

$$E(\text{NBE} \mid E_2) = \frac{n}{2} \tag{6.5-65}$$

The bit error probability, given that an error of the second kind is made, is then simply $\frac{1}{2}$.

Finally, the resulting error probability is given by

$$\text{PE}_B = P_1 + \tfrac{1}{2} P_2 \tag{6.5-66}$$

The bit error probability is plotted in Figure 6.9. Note that when $n = 1$ the PSK case, as well as the QPSK case, is obtained.

6.5.5 Performance of Phase Coherent Receivers with Transorthogonal Signals

In this section we will show that a transorthogonal signal set is equivalent to an orthogonal signal set that has reduced power requirements relative to an orthogonal code set. Let the signal set $x_i(t)$ be transorthogonal, then construct a new signal set $x_i^1(t)$ defined by

$$x_i^1(t) = \begin{cases} x_i(t) & 0 \le t < T \\ \sqrt{S} & T \le t < \dfrac{M}{M-1} T \end{cases} \tag{6.5-67}$$

Then the normalized signal correlations are given by $\{T' = [M/(M-1)]T\}$

$$\lambda_{ij}' = \frac{1}{E'} \int_0^{T'} x_i'(t) x_j'(t) \, dt \tag{6.5-68}$$

with

$$E' = ST + S\left(\frac{M}{M-1} - 1\right) T = ST \frac{M}{M-1} \tag{6.5-69}$$

Evaluating, we obtain

$$i \ne j: \quad \lambda_{ij}' = \frac{M-1}{MST} \left[\int_0^T x_i(t) x_j(t) \, dt + \int_T^{[M/(M-1)]T} s \, dt \right] = 0 \tag{6.5-70}$$

$$i = j: \quad \lambda_{ij}' = \frac{M-1}{MST} \left[ST + \left(\frac{M}{M-1} - 1\right) ST \right] = 1 \tag{6.5-71}$$

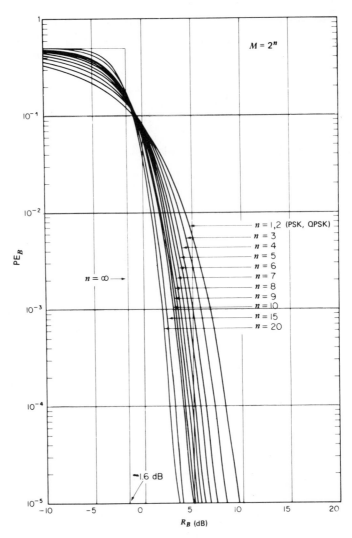

Figure 6.9 Bit error probability performance for a biorthogonal signal set $n = 1$ represents PSR and QPSK (from Lindsey and Simon [3] with permission).

Our new signal set $\{x_i^1(t)\}$ is an orthogonal signal set with energy $E' = [M/(M-1)]E$. Furthermore, the error rate of the old and new systems must be the same since adding a constant signal over the interval $(T, M/(M-1)T)$ does not change the detectability of the code. Therefore, from Section 6.5.3, we have the probability of detection is given by

$$PC_W = \int_{-\infty}^{\infty} \frac{e^{-x^2}}{\sqrt{\pi}} \left[\tfrac{1}{2} \operatorname{erfc}\left(-x - \sqrt{R_W} \cdot \frac{M}{M-1} \right) \right]^{M-1} dx \qquad (6.5\text{-}72)$$

Since a transorthogonal code set has the same uniform distant properties as an orthogonal signal set, the bit error probability is related to the word error probability in the same way as orthogonal signal sets:

$$PE_B = \frac{2^{n-1}}{2^n - 1} PE_W \qquad (6.5\text{-}73)$$

The orthogonal error probability curves can be used to obtain the word and bit error performance by increasing the specified $R_W = E/N_0$ by $M/(M-1)$.

PROBLEM 6

Show that for orthogonal signals that

$$\lim_{M\to\infty} PC_W = \begin{cases} 0 & \text{if } \dfrac{E}{N_0} > \ln 2 \\[2mm] 1 & \text{if } \dfrac{E}{N_0} < \ln 2 \end{cases}$$

Hence, orthogonal codes achieve the Shannon limit; that is, error-free transmission occurs when

$$R = \frac{1}{T} < \frac{S}{N_0 \ln 2} = C_\infty$$

6.5.6 Performance of Phase Coherent Receivers with Multiple Phase Shift Keying

We now assess the performance of multiple phase shift keying (MPSK). The MPSK signal is of the form

$$x_j(t) = \sqrt{2S} \sin\left[\omega_0 t + \frac{2\pi(j-1)}{M}\right] \qquad j = 1, \ldots, M \qquad (6.5\text{-}74)$$

When $M = 2$ we have PSK signaling, and when $M = 4$ we have quadriphase signaling.

Since a square matrix is nonsingular if and only if its columns are linearly independent [9], it follows that our MPSK code correlation matrix is singular, because each column sums to zero. To circumvent this problem, we deal with the statistics directly.

Since we have a two-dimensional problem, we can form an optimum detector by observing projections on an orthogonal basis such as

$$V_c = \int_0^T y(t)\sqrt{2s}\, \cos \omega t\, dt$$

$$V_s = \int_0^T y(t)\sqrt{2s}\, \sin \omega t\, dt \qquad (6.5\text{-}75)$$

The decision rule becomes: choose the hypothesis that

$$\theta_j = \frac{2\pi(j-1)}{M}$$

was sent when

$$\hat{\theta} = \tan^{-1}\left(\frac{V_c}{V_s}\right) \tag{6.5-76}$$

is closest to some θ_j. Now we compute the probability density function (pdf) of the amplitude and phase. Since V_c and V_s are independent Gaussian random variables with means, conditioned on the signal $x_j(t)$ having been transmitted, we have that

$$\bar{V}_s = 2S \int_0^T \sin(\omega t + \theta_j) \sin(\omega t) \, dt = ST \cos \theta_j$$

$$\tag{6.5-77}$$

$$\bar{V}_c = 2S \int_0^T \sin(\omega t + \theta_j) \cos(\omega t) \, dt = ST \sin \theta_j$$

We have neglected the double-frequency terms since they are negligible. The variances are given by

$$\sigma_{V_s}^2 = 2S \int_0^T \int_0^T \overline{n(t)n(u)} \sin \omega t \sin \omega u \, du \, dt = \frac{STN_0}{2}$$

$$\tag{6.5-78}$$

$$\sigma_{V_c}^2 = 2S \int_0^T \int_0^T \overline{n(u)n(t)} \cos \omega t \cos \omega u \, du \, dt = \frac{STN_0}{2}$$

Hence the joint p.d.f. is given by

$$p(V_c, V_s) = \frac{1}{STN_0\pi} \exp\left[-\frac{(V_s - ST \cos \theta_j)^2 + (V_c - ST \sin \theta_j)^2}{STN_0}\right]$$

$$\tag{6.5-79}$$

Define

$$V = \sqrt{V_c^2 + V_s^2}$$

$$\tag{6.5-80}$$

$$\hat{\theta} = \tan^{-1}\left(\frac{V_c}{V_s}\right)$$

Therefore, transforming to the variables V and $\hat{\theta}$ yields

$$p(V, \hat{\theta}) = \frac{p(V_c, V_s)}{\begin{vmatrix} \dfrac{\partial V}{\partial V_c} & \dfrac{\partial V}{\partial V_s} \\[2mm] \dfrac{\partial \hat{\theta}}{\partial V_c} & \dfrac{\partial \hat{\theta}}{\partial V_s} \end{vmatrix}} = Vp(V_c, V_s) \tag{6.5-81}$$

so that

$$p(V, \hat{\theta}) = \frac{V}{STN_0\pi} \exp\left[-\frac{V^2 - 2ST(V \cos \hat{\theta} \cos \theta_j + V \sin \hat{\theta} \sin \theta_j) + (ST)^2}{STN_0}\right]$$

$$\tag{6.5-82}$$

Simplifying, we obtain

$$p(V, \hat{\theta}) = \frac{V}{STN_0\pi} \exp\left[-\frac{V^2 - 2STV \cos(\hat{\theta} - \theta_j) + (ST)^2}{STN_0}\right]$$

$$0 \le V < \infty \qquad 0 \le \hat{\theta} \le 2\pi \qquad\qquad (6.5\text{-}83)$$

To obtain the p.d.f. of $\hat{\theta}$ we integrate out the variable V to arrive at

$$p(\hat{\theta}) = \frac{1}{STN_0\pi} \int_0^\infty \exp\left[-\frac{V^2 - 2STV \cos(\hat{\theta} - \theta_j) + (ST)^2}{STN_0}\right] dV$$

$$(6.5\text{-}84)$$

Now, if we let $r = V/\sqrt{STN_0}$ and $R_W = ST/N_0$, we obtain

$$p(\hat{\theta}) = \int_0^\infty \frac{r}{\pi} \exp[-r^2 - 2r\sqrt{R_W} \cos(\hat{\theta} - \theta_j) + R_W] \, dr$$

$$\hat{\theta} \in (0, 2\pi) \qquad \text{zero otherwise} \qquad (6.5\text{-}85)$$

It is now clear that $p(\hat{\theta})$ is maximum when $\hat{\theta} = \theta_j$. The probability of correct detection, given that signal j was transmitted, is given by

$$PC(j) = \int_{(j-3/2)2\pi/M}^{(j-1/2)2\pi/M} p(\hat{\theta}) \, d\hat{\theta} \qquad (6.5\text{-}86)$$

Now substituting (6.4-85) into (6.5-86) produces

$$PC(j) = \int_0^\infty \int_{(j-3/2)2\pi/M}^{(j-1/2)2\pi/M} \frac{r}{\pi} \exp\{-[r^2 - 2r\sqrt{R_W} \cos(\hat{\theta} - \theta_j) + R_W]\} \, dr \, d\hat{\theta}$$

$$(6.5\text{-}87)$$

Now let $\phi = \hat{\theta} - 2(j - 1)\pi/M$; then

$$PC(j) = \int_0^\infty \int_{-\pi/M}^{\pi/M} \frac{r}{\pi} \exp\{-[r^2 - 2r\sqrt{R_W} \cos(\phi) + R_W]\} \, d\hat{\theta} \, dr \quad (6.5\text{-}88)$$

where we have used the fact that

$$\theta_j = (j - 1)\frac{2\pi}{M} \qquad (6.5\text{-}89)$$

From (6.5-88) we see that $PC(j)$ is independent of j. Therefore, the average probability of error is equal to $PC(j)$. Using symmetry, we have

$$PC_W = \frac{2}{\pi} \exp(-R_d) \int_0^\infty r \exp(-r^2) \left[\int_0^{\pi/M} \exp(2r\sqrt{R_W} \cos \phi) \, d\phi\right] dr$$

$$(6.5\text{-}90)$$

To further simplify the form, let

$$x = r \cos \phi$$

$$y = r \sin \phi$$

then

$$dy\, dx = r\, dr\, d\phi$$

It is shown in Problem 7 that

$$PC_W = \frac{2}{\pi} \int_0^\infty \exp\{-(x - \sqrt{R_W})^2\} \left[\int_0^{x\,\tan(\pi/M)} \exp(-y^2)\, dy \right] dx$$

$$(6.5\text{-}91)$$

This is our final result for the word error probability for MPSK signals ($PE_W = 1 - PC_W$). The results are plotted in Figure 6.10 against $R_b = R_W/\log_2 M$.

PROBLEM 7

Show that (6.5-91) follows from (6.5-90), with the appropriate change of variables.

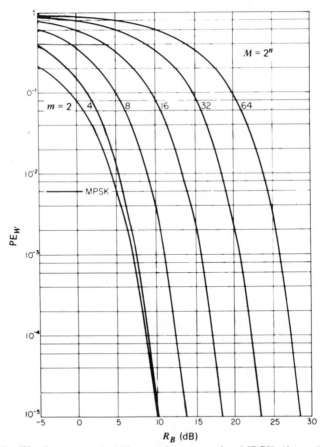

Figure 6.10 Word error probability performance for MPSK (from Lindsey and Simon [3] with permission).

PROBLEM 8

Show that a bound on MPSK can be obtained by determining the following probabilities:

$$P_1 = P\left[V_s > V_c \tan\frac{\pi}{M}\right]$$

$$P_2 = P\left[V_s < -V_c \tan\frac{\pi}{M}\right]$$

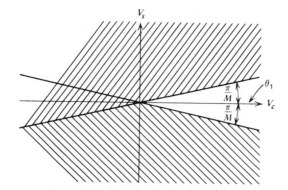

as shown in the Figure, P_1 and P_2 are the probabilities that the phase estimate $\hat{\theta}$ lies outside the zone of correct detection when θ_1 was transmitted, except that the zone of width $2\pi/M$ centered at π rad is included twice. Conclude therefore that

$$PE_W = P_1 + P_2 - P_3$$

where P_3 is the probability of lying in the intersecting area. Now neglect P_3 to obtain an upper bound on PE_W, finally showing that

$$PE_W \le \text{erfc}\left[\sqrt{R_W}\,\sin\left(\frac{\pi}{M}\right)\right]$$

This bound becomes tight as R_W increases and M increases. For BPSK ($M = 2$) for example, it is double the correct value.

From the curves we see that BPSK and quadriphase have essentially the same word error probability at the smaller values of PE_W. In fact, in Problem 8 it is shown that quadriphase and BPSK have the exact same *bit error rate*, but quadriphase signaling requires one-half the bandwidth! This is one of the reasons quadriphase signaling is so popular in communication system design.

PROBLEM 9

Show that quadriphase (QPSK) and BPSK modulation have the same bit error probability [11] when detected in an optimum manner. When the

QPSK signal is broken up into two binary PSK signals, the equivalence is obvious. The problem is to show the equivalence in the less obvious case of word detection (two bits at a time). *Hint*: Consider errors of types I and II as in Section 6.5.4.

Now consider the bit error rate PE_B as determined from the word error rate PE_W. Let $P(K)$ denote the probability that the phase $K(2\pi/M)$ rad away from zero is considered as being present when, in fact, $\theta = 0$ was transmitted. Thus the $P(K)$ can be viewed as transition probabilities [12]. Now to compute the bit error probabilities, assume that the mapping from bits to words (in this case two bits per word) is done via a Gray code. A Gray code has the property that there is exactly one bit difference when comparing adjacent words. In other words, deciding ϕ_1 was transmitted when really ϕ_2 was transmitted would produce one bit error.

In the case $M = 2$, the problem is trivial since a word contains one bit and therefore

$$PE_B(2) = PE_W(2) \qquad (6.5\text{-}92)$$

When $M = 4$, we have

$$PE_B(4) = \tfrac{1}{2}P(1) + \tfrac{1}{2}P(-1) + 1 \cdot p(2)$$

since with a single transition error one-half the bits are in error, etc. Using symmetry, we obtain

$$PE_B(4) = P(1) + P(2) \qquad (6.5\text{-}93)$$

For $M = 8$ we can obtain

$$PE_B(8) = 1 \cdot P(-3) + 1 \cdot P(3) + \tfrac{1}{3}P(1) + \tfrac{1}{3}P(1)$$
$$+ \tfrac{2}{3}[P(2) + P(-2)]$$

or from symmetry

$$PE_B(8) = \tfrac{2}{3}P(1) + \tfrac{4}{3}P(2) + 2P(3) \qquad (6.5\text{-}94)$$

These results are exact, but require computing the "transition probabilities" starting at (6.5-81) with the limits adjusted $K(2\pi/M)$ larger or smaller for a transition of K phases. For example

$$P(1) = \int_{(j-1/2)2\pi/M}^{(j+1/2)2\pi/M} p(\hat{\theta} \mid j) \, d\hat{\theta} \qquad (6.5\text{-}95)$$

where we now have used $p(\hat{\theta} \mid j)$ to denote the density function of $\hat{\theta}$ given θ_j was transmitted. An equation similar to (6.5-86) will be obtained.

As an approximation, however, it can be shown that the dominant terms in each bit error rate expression were the single phase error terms, so that for M symbols

$$PE_B \cong \frac{2}{\log_2(M)} P(1) \qquad (6.5\text{-}96)$$

Furthermore

$$PE_W \cong 2P(1)$$

so that

$$PE_B(M) \cong \frac{PE_W(M)}{\log_2(M)} \qquad (6.5\text{-}97)$$

Equation 6.5-97 only holds approximately, and then only when PE_W is small. Hence (6.5-97) and Figure 6.10 can be used to compute $PE_B(M)$.

PROBLEM 10

Determine the general expression for $P(K)$ assuming M possible phase levels following the method of this section.

6.5.7 Performance of Phase Coherent Receivers with Frequency Shift Keying

In the last section we looked at phase shift keying. Another method of communicating is frequency shift keying. We discuss coherent detection here as that is our objective for this chapter. Noncoherent detection performance is well documented elsewhere.

Consider the coherent MFSK receiver, shown in Figure 6.11. First we show that if we set each tone apart by $\frac{1}{2}T$ Hz, then the signal correlation is zero. Let

$$x_i(t) = \sqrt{2S}\,\sin(\omega_0 t + 2\pi i \Delta f t) \qquad (6.5\text{-}98)$$

Then λ_{ij} is given by

$$\lambda_{ij} = \frac{\int_0^T x_i(t)x_j(t)\,dt}{ST} \qquad (6.5\text{-}99)$$

Evaluating

$$\begin{aligned}
\lambda_{ij} &= \frac{2}{T}\int_0^T \sin(\omega_0 t + 2\pi i \Delta f t)\,\sin(\omega_0 t + 2\pi j \Delta f t)\,dt \\
&= \frac{1}{T}\int_0^T \cos[2\pi(i-j)\Delta f t]\,dt \\
&\quad -\frac{1}{T}\int_0^T \cos[2\omega_0 t + 2\pi(i+j)\Delta f t]\,dt \qquad (6.5\text{-}100)
\end{aligned}$$

Since the second term will be negligible if $\omega_0/2\pi \gg \Delta f$, we have

$$\lambda_{ij} = \frac{\sin[2\pi(i-j)\Delta f T]}{2\pi(i-j)\Delta f T} \qquad (6.5\text{-}101)$$

Now the first zero occurs for

$$2\pi\Delta f T = \pi \qquad (6.5\text{-}102)$$

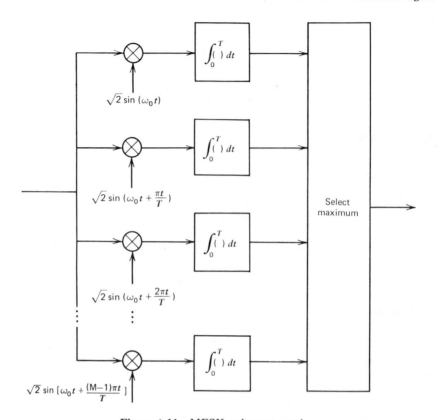

Figure 6.11   MFSK coherent receiver.

or

$$\Delta f = \frac{1}{2T} \tag{6.5-103}$$

Hence, when the angular frequency spacing of the MFSK signal set is π/T, the result is an orthogonal signal set.

Now consider binary FSK. We shall now obtain the optimum spacing Δf to minimize the probability of error. Using the results of Problem 3, we have that the probability of error is given by ($R = E_B/N_0$)

$$PE_B = \tfrac{1}{2}\,\text{erfc}\left[\sqrt{R\left(\frac{1-\lambda}{2}\right)}\right] \tag{6.5-104}$$

Clearly, PE_B is minimized when λ is minimized. From (6.5-101) we see that λ is given by

$$\lambda = \frac{\sin[2\pi\Delta fT]}{2\pi\Delta fT} \tag{6.5-105}$$

Hence, to find the minimum, we solve

$$f'(x) = 0 \qquad (6.5\text{-}106)$$

where

$$f(x) = \frac{\sin x}{x} \qquad (6.5\text{-}107)$$

Hence

$$f'(x) = \frac{\cos x}{x} - \frac{\sin x}{x^2} = 0 \qquad (6.5\text{-}108)$$

Therefore, the solution of $\tan x = x$ is given by $x = \pm 4.493$. The corresponding value of λ is

$$\lambda_0 = \frac{\sin(4.493)}{4.493} = -0.217 \qquad (6.5\text{-}109)$$

Therefore

$$\Delta f_0 = \frac{4.493}{2\pi T} = \frac{0.715}{T} \qquad (6.5\text{-}110)$$

which is a well-known result. The corresponding bit error rate is

$$\text{PE} = \tfrac{1}{2}\,\text{erfc}(\sqrt{0.61R}) \qquad (6.5\text{-}111)$$

Hence, since a PSK system has $\lambda = -1$, we see that coherent FSK is about 2.2 dB inferior to PSK, but 0.8 dB better than coherent orthogonal FSK ($\lambda = 0$).

6.5.8 Performance of Phase Coherent Receivers with Coherent Phase Frequency Shift Keying

Coherent phase frequency shift keying (CPFSK) is an attractive, narrow bandwidth, modulation format that can be detected either coherently or noncoherently. Numerous papers have been written on this subject, some under the name of minimum shift keying [MSK] [13–24]. It has been determined that this modulation scheme can lead to about 1-dB improvement over BPSK performance at $\text{PE} = 10^{-5}$.

In a binary FSK system, a tone at frequency f_1 is transmitted when a data bit "one" is input to the modulator, and another tone is transmitted at frequency f_0 when a data bit "zero" is input to the modulator.

In CPFSK, however, the initial phase of the tone corresponding to a particular bit depends on previous data bits (continuous phase), and it is possible to obtain a better bit error rate by observing the received signal over more than one bit interval. Forney [21] has discussed the use of the Viterbi algorithm for detection of coherent CPFSK when the modulation index is 0.5 (that is, MSK). (For this index a trellis diagram is a useful tool)

The tone spacing in MSK is one-half that employed in conventional orthogonal FSK; hence the name "minimum shift keying."

The CPFSK waveform during the first bit interval (of T sec) can be expressed as

$$x(t) = \sqrt{2P} \cos\left(\omega_0 t + \frac{d_1 \pi h t}{T} + \theta_1\right) \qquad 0 \le t \le T \qquad (6.5\text{-}112)$$

where d_1 is the first data bit having value of ± 1, θ_1 is the rf carrier phase at the beginning of the observation interval, and $\underline{h}$ is the modulation index (or sometimes the deviation ratio). Note that the maximum phase deviation is $\pm \pi h$ rad, and the frequency shift from the carrier frequency is $\pm(h/2T)$Hz.

During the ith bit period the signal can be written as

$$x(t) = \sqrt{2P} \cos\left(\omega_0 t + \frac{d_i \pi h(t - iT + T)}{T} + \sum_{j=1}^{i-1} d_j \pi h + \theta_1\right) \quad (6.5\text{-}113)$$

Notice that the phase is continuous across the bit boundaries. When a single bit period is used, the bit error rate performance is given by (6.5-111),

$$\text{PE} = \tfrac{1}{2} \text{erfc}[\sqrt{0.61R}] \qquad R = \frac{E_B}{N_0} \qquad (6.5\text{-}114)$$

We now establish that when we decode bit "one" by observing bit "one" plus later bits we can improve the performance over the single-bit observation case.

Consider the case of observing two bit periods to detect the first bit. Let $x_{11}(t)$ denote the rf signal corresponding to data sequence 11 and define $x_{10}(t)$, $x_{01}(t)$, and $x_{00}(t)$ in a similar manner, (all are assumed to have an *a priori* probability of $\tfrac{1}{4}$). Over the observation interval of two bits a maximum likelihood decision on the first bit can be obtained by correlating the input signal with the four references $x_{ij}(t)$ and deciding that the first bit i is determined by the reference having the highest correlation. We approximate (overbound) the probability of error by using the union bound [2, pp. 264], which states that for detection of M equally likely signals in WGN the probability of error, given signal k is transmitted, is bounded by

$$P(E \mid m_i) \le \sum_{\substack{k=0 \\ (k \ne i)}}^{M-1} P_2(x_i, x_k) \qquad (6.5\text{-}115)$$

where $P_2(x_i, x_k)$ is the probability that the received signal is closer to signal x_k than to x_i when x_i is transmitted. It turns out that $P_2(x_i, x_k)$ is just the probability of error for a system that uses the signals $x_i(t)$ and $x_k(t)$ as signals to communicate one of two likely messages. In this case, we have

$$P_2(x_i, x_k) = \tfrac{1}{2} \text{erfc}\left[\sqrt{R\left(\frac{1 - \lambda_{ik}}{2}\right)}\right] \qquad (6.5\text{-}116)$$

Table 6.1   Signal correlations

ij	kl	λ_{ijkl}
10	00	$\frac{1}{2}[\text{sinc}(2\pi h) + \cos(2\pi h)]$
10	01	$\text{sinc}(2\pi h)$
11	00	$\text{sinc}(4\pi h)$
11	01	$\frac{1}{2}[\text{sinc}(2\pi h) + \cos(2\pi h)]$

Hence, knowing the signal correlation allows one to bound the error probability. For our problem the average probability of error is given by

$$PE \leq \tfrac{1}{4}[P(11, 00) + P(11, 01) + P(10, 00) + P(10, 01)$$
$$+ P(00, 11) + P(00, 10) + P(01, 11) + P(01, 10)] \quad (6.5\text{-}117)$$

In this equation $P(11, 00)$ means the probability that the noisy input signal (corresponding to 11) correlates better with the signal corresponding to 00. Notice that only the quadruples listed correspond to differences in the first and third bit. Remember we are making a decision *only* on the first bit by considering all possible bit pairs! For our problem, we define λ_{ijkl} as the signal correlation between signal x_{ij} and x_{kl} with i, j, k, and l taking on the value "zero" or "one." These values are listed in Table 6.1 [$\sinh(x) = \sin x/x$]. Note that $\lambda_{ijkl} = \lambda_{klij}$ also. Let λ_0 be the maximum value of λ_{ijkl} in Table 6.1 so that we obtain

$$P_2(ij, kl) \leq \tfrac{1}{2}\text{erfc}\left[\sqrt{R_W\left(\frac{1-\lambda_0}{2}\right)} \right] \quad (6.5\text{-}118)$$

since $\tfrac{1}{2}\text{erfc}(x)$ is a monotonically increasing function of x. (Note $R_W = 2PT/N_0$.) Using the union bound for the average probability of error produces

$$PE \leq \tfrac{1}{4}(8)(\tfrac{1}{2})\, \text{erfc}\left[\sqrt{R_W\left(\frac{1-\lambda_0}{2}\right)} \right] \quad (6.5\text{-}119)$$

or

$$PE \leq \text{erfc}\left[\sqrt{R_W\left(\frac{1-\lambda_0}{2}\right)} \right] \quad (6.5\text{-}120)$$

By direct evaluation, when $h = 0.773$ we have

$$PE \leq \text{erfc}[\sqrt{R_B(1.0298)}] \quad (6.5\text{-}121)$$

which is about 0.127 dB better than PSK because of processing two bits at a time. It can be shown [15] that when $h = 0.5$ [this case is usually called minimum shift keying (MSK)], that

$$PE \leq \text{erfc}[\sqrt{R_B}] \quad (6.5\text{-}122)$$

so that the performance is essentially equal to PSK (where we are comparing asymptotic exponents only).

The same procedure can be used to bound the three-bit case [15], with the result that

$$PE \leq erfc[\sqrt{1.215 R_b}] \tag{6.5-123}$$

which is about 0.85 dB better than PSK when $h = 0.715$. (It should be pointed out that union bounds are tight at high SNR, or $\leq 10^{-4}$ BER.) Osborne and Luntz [16] have used these techniques and another to form composite upper bounds that are illustrated in Figure 6.12. It should be noted from Figure 6.12 there is about a 1-dB gain over BPSK with three bits and about

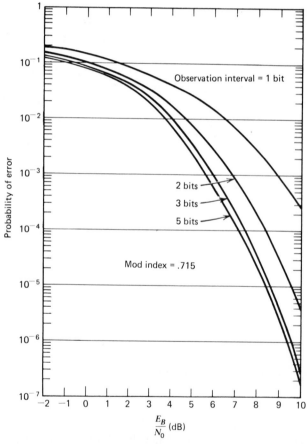

Figure 6.12 Upper bounds on CPFSK performance (from Osborne and Lutz [16] with permission).

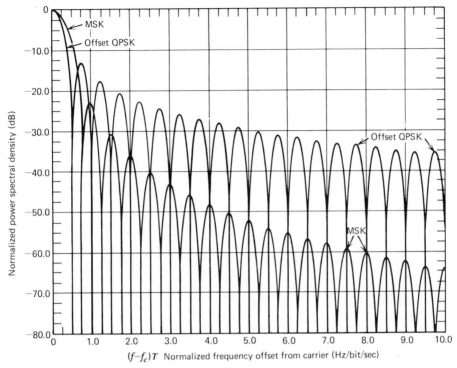

Figure 6.13 Power spectral densities for MSK and offset QPSK (from Grone-meyer and McBride [18] with permission).

a 1.2-dB gain with five bits (all at BER $= 10^{-5}$).* Notice after three bits of observation the improvement is very small.

Before we leave the topic of CPFSK we point out that, due to the continuous phase at the bit transitions, the spectral density is quite narrow. Gronemeyer and McBride [22] have computed the spectral density of MSK and offset QPSK (OQPSK) by means of a Markov process representation, which results in the following:

$$\mathcal{S}_{MSK}(f) = \frac{8PT(1 + \cos 4\pi fT)}{\pi^2(1 - 16T^2f^2)^2} \qquad f \ge 0 \qquad (6.5\text{-}124)$$

$$\mathcal{S}_{OQPSK}(f) = 2PT \left[\frac{\sin(2\pi fT)}{2\pi fT} \right]^2 \qquad f \ge 0 \qquad (6.5\text{-}125)$$

where f is the frequency offset from the carrier frequency, T is the data bit duration, and P is the power in the carrier. The power spectral densities are plotted in Figure 6.13. MSK modulation falls off as $(R/f)^4$, with $R = 1/T$

*These numbers do not agree exactly with the asymptotic exponent comparison of above.

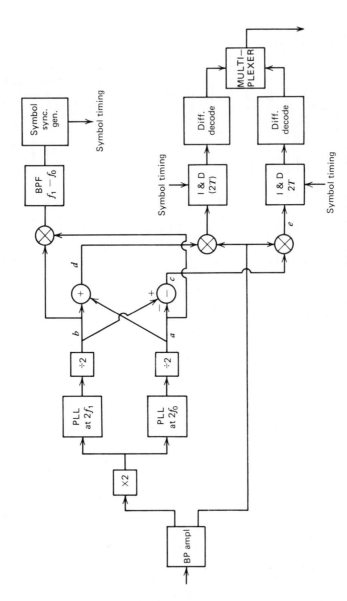

Figure 6.14 An MSK receiver with differential decoding.

while OQPSK falls off as $(R/f)^2$. The main lobe of the MSK spectrum is wider than that of OQPSK: the first null of MSK occurs at $f = \frac{3}{4}R$, whereas for OQPSK it occurs at $f = \frac{1}{2}R$. Simon [23] has demonstrated that even narrow spectra can be obtained by making the symbol pulse shape sinusoidal rather than rectangular (SFSK).

As a final note on MSK we indicate a possible MSK receiver system [17, 24, 25, 46] that could be used for coherent demodulation. The receiver is shown in Figure 6.14 for $M = 4$ (needs $2T$ sec of observation).

The MSK signal is passed through the bandpass amplifier and doubled in the squaring device, which provides spectral components at $2f_1$ and $2f_0$. The spectral lines are tracked by the respective PLLs to produce a clean reference. The divide-by-2 circuits produce $\pm\cos(\omega_0 t)$ at point a and $\pm\cos(\omega_1 t)$ at point b, with the $\pm$ sign being necessary because of the divide-by-2 process. Out of the summer, point c can be either $\pm\cos \omega_c t \cos(\pi t/2T)$ or $\pm\sin \omega_c t \sin(\pi t/2T)$, with the same type of signals appearing at point d, except that the order is inverted. Since the signal can be expressed as

$$y(t) = \pm \left[\sqrt{P} \cos \omega_c t \cos\left(\frac{\pi t}{2T}\right) \pm \sqrt{P} \sin \omega_c t \sin\left(\frac{\pi t}{2T}\right) \right] \quad (6.5\text{-}126)$$

we see that the outputs d and c can be multiplied by the MSK signal to produce the algebraic sign of each quadrature component. Secondly, because of the ambiguity in sign, a differential decoder (see next section), can be used when a differential encoder is used in the transmitter.

Although we have not discussed noncoherent detection of CPFSK, gains over BPSK of 1 dB are also possible [16].

6.6 PERFORMANCE OF COHERENT DETECTION OF DIFFERENTIAL MPSK SIGNALS

As we noticed in the last section, it is not always possible to know the absolute phase of the received signal. Commonly the phase can be resolved only to one of two possible phases in PSK, for example, or one of four possible phases in quadriphase transmission.

One solution is to use a short synchronization code which, when correlated at the ground, will resolve the ambiguity. Another method is to provide for a residual carrier component for synchronization purposes.

An alternative method is to employ differential encoding along with differential decoding. Since our goal is only to discuss coherent detection schemes, we only consider coherent detection of differentially encoded MPSK signals in what follows. Let

$$x_i(t) = \sqrt{2S} \sin(\omega_0 t + \theta^{(i)})$$

represent the signal to be transmitted in the ith interval $(i-1)T \leq t < iT$.

```
                                         ┌───────Arbitrary initial symbol
                                         ↓
Binary message                             1  1  1  0  0  1  0  1  1
Differentially encoded message        1  1  1  1  0  1  1  0  0  0
Transmitted phase (rad)               0  0  0  0  π  0  0  π  π  π
   Detected phase = transmitted phase + θ₀
   No change in encoded sequence ↔ 1 in data sequence
   A change in the encoded sequence ↔ 0 in the data sequence
```

Figure 6.15 Differential encoder for the binary case.

The phase $\theta^{(i)}$ takes on one of M possible phases:

$$\theta_i = \frac{2(i-1)\pi}{M} \qquad i = 1, 2, \ldots, M$$

The transmitted phase corresponds to the encoded sequence (bit) for which differences of phase determine the actual data sequence. The receiver then estimates each phase and takes differences that in turn determine the data sequence (bit). Because of the differencing in transmission and reception, absolute phase determination is not required.

Consider the binary case as illustrated in Figure 6.15, which will aid in understanding the procedure. As can be seen in the figure, the received phase has an unknown additional phase θ_0, which "cancels out" since differences in the correctly received phase provide the correctly decoded binary message.

Consider now the four-phase case for which an example is given in Figure 6.16. Let the initial phase be 0 rad, corresponding to 00 in the encoded sequence. Since the message pair 10 is inputted, the encoded message must reflect a change of 2. The encoded sequence must then be 10. The next message bit pair is 01. Again, using the present bit pair phase minus the last bit pair phase, we arrive at

$$\frac{3\pi}{2} - \pi = \frac{\pi}{2} \leftrightarrow 01 \qquad \text{decoded bit pair} \qquad (6.6\text{-}1)$$

The next received phase (assuming no errors) is $(3\pi/4)(11)$, so that the

```
                    00 ↔ 0 ↔ 0
                    01 ↔ 1 ↔ π/2
                    10 ↔ 2 ↔ π
                    11 ↔ 3 ↔ 3π/2
```

($M=4$) Message		10	01	00	01	11	10
Encoded message	00	10	11	11	00	11	01
Transmitted phase	0	π	$\dfrac{3\pi}{2}$	$\dfrac{3\pi}{2}$	0	$\dfrac{3\pi}{2}$	$\dfrac{\pi}{2}$

Figure 6.16 Differential encoder for the $M=4$ case.

decoded bit pair is

$$\frac{3\pi}{2} - \frac{3\pi}{2} = 0 \leftrightarrow 00 \qquad (6.6\text{-}2)$$

This procedure continues in this manner. Now let the present phase estimate be $\pi/2(01)$ and the last phase be $3\pi/2$. Taking the difference between present phase and the last phase, we obtain

$$\frac{\pi}{2} - \frac{3\pi}{2} = -\pi \qquad (6.6\text{-}3)$$

To decode this negative value we add 2π to the first (present) phase whenever it is smaller than the last phase, so that we now obtain

$$2\pi + \frac{\pi}{2} - \frac{3\pi}{2} = \pi \leftrightarrow 10 \qquad \text{decoded bit pair} \qquad (6.6\text{-}4)$$

In other words, we take differences mod 2π.

This procedure can be generalized to larger values of $M = 2^n$, of course. It should be clear at this point in our discussion that M-ary signals have M-fold ambiguity.

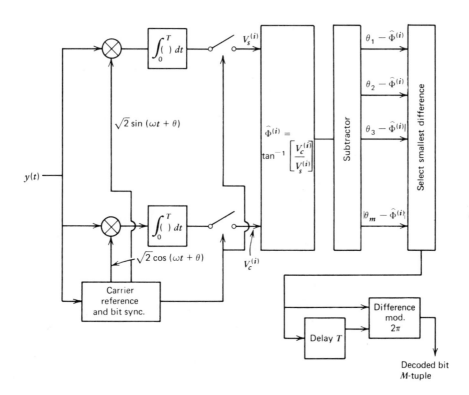

Figure 6.17 Coherent detection of MPSK signals that are differentially encoded.

We assume in the analysis to follow that an M-phase tracking loop is used to track a suppressed carrier signal. Let each M-ary phase estimate be denoted by $\hat{\theta}^{(i)}$. Then the demodulated data is obtained by subtracting $\hat{\theta}^{(i-1)}$ from $\hat{\theta}^{(i)}$ mod 2π as discussed in the example; that is, we consider

$$\hat{\theta}^{(i)} - \hat{\theta}^{(i-1)} \bmod 2\pi \qquad (6.6\text{-}5)$$

A receiver that provides coherent demodulation is shown in Figure 6.17. Now note that if both phase estimates are correct, then the transmitted data M-tuple will be decoded correctly. The probability of this event is PC_{MPSK}^2, where PC_{MPSK} is the probability of correct detection of an MPSK signal, and is given by (6.5-81). Now if both $\hat{\theta}^{(i)}$ and $\hat{\theta}^{(i-1)}$ are in error by a fixed but equal amount, the data bits are decoded correctly since the

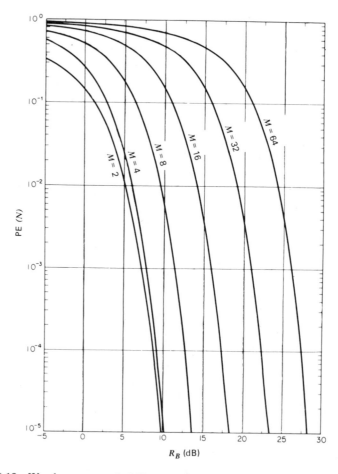

Figure 6.18 Word error probability performance for coherent detection of differentially encoded MPSK (from Lindsey and Simon [3] with permission).

difference of the two phases is the correct value. Denote by PE_K the value of being in error by $(2\pi K/M)$ $(K = 1, 2, 3, \ldots, M)$ rad from the true phase. Then the probability of correct decoding is given by

$$PC_W = PC_{MPSK}^2 + \sum_{k=1}^{M-1} PE_k^2 \qquad (6.6\text{-}6)$$

The general result is given in reference 3. Let us consider the binary case. The probability of correct detection is given by

$$PC_W = (1 - PE_{PSK})^2 + (PE_{PSK})^2 \qquad (6.6\text{-}7)$$

Hence the probability of error is given by

$$PE_W = 2PE_{PSK}(1 - PE_{PSK}) \qquad (6.6\text{-}8)$$

The general results, based on reference 3, are shown in Figure 6.18.

6.7 OPTIMAL DECODING OF CONVOLUTIONAL CODES

In 1967 Viterbi [26] defined a decoding scheme for decoding convolutional codes that he showed to be asymptotically optimum. Heller [47] showed that these codes were quite promising. Later, Omura [27] and Forney [28] proved the optimality of Viterbi's maximum likelihood procedure. In this section, we shall discuss the algorithm, its optimality, the trellis diagram, the state diagram, and other associated topics. Typical coding gains relative to PSK are in the range of 4 to 6 dB at error rates of 10^{-5}. We consider this encoding-decoding combination because many actual systems have utilized them. The Tracking Data Relay Station System (TDRSS), for example, has a coding gain on the low rate multiple access (MA) system of about 5 dB. This section is based primarily on reference 31.

6.7.1 Representations of Convolutional Codes

There are numerous methods [29] of describing convolutional codes, including the (1) polynomial matrix approach, (2) the scalar matrix approach, (3) the shift register approach, (4) the state diagram approach, (5) the trellis approach, and finally (6) the tree approach. We will only consider approaches (3), (4), and (5).

A convolutional encoder can be implemented with a K-stage shift register, n mod 2 adders, and one multiplexer, as shown in Figure 6.19 for the case $K = 3$ and $n = 2$. It is controlled by a clock. The binary input data, prefaced by three zeros to "load" the register, is fed into the encoder one bit at a time (m bits at a time in the general case). In Figure 6.19, one bit at a time is input to the encoder, and two code symbols are obtained at their respective outputs. When multiplexed (interleaved) onto the output line, a symbol appears at the output in this example at twice the rate as bits are

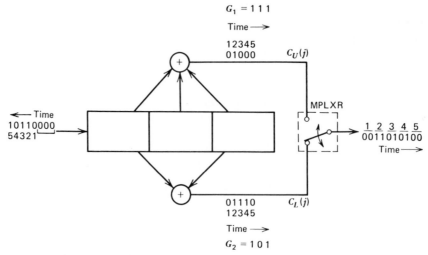

Figure 6.19   A $K = 3$, $n = 2$ convolutional encoder ($m = 1$).

input to the encoder. Therefore, this is a rate $\frac{1}{2}$ convolutional encoder. If one symbol in the pair is identical to the input bit, we have a *systematic code*. The two generator sequences G_1 and G_2 are a convenient way to describe the codes.

Notice from Figure 6.20 that each symbol output is described by

$$C_{Uj} = I_j \oplus I_{j-2}$$
$$C_{Lj} = I_j \oplus I_{j-1} \oplus I_{j-2} \tag{6.7-1}$$

and hence each symbol output can be written in the form

$$C_j = \sum_{k=0}^{2} W_k I_{j-k} \tag{6.7-2}$$

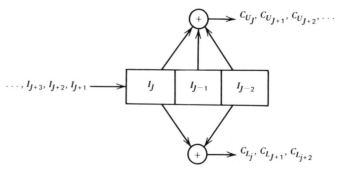

Figure 6.20   A $k = 3$, $n = 2$ convolutional encoder showing the input and the output symbols.

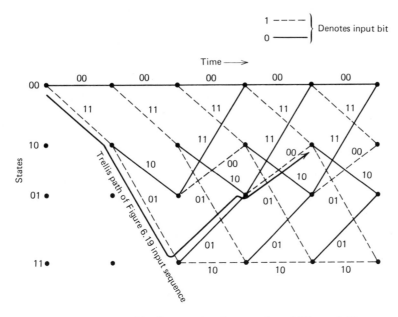

Figure 6.21 Trellis diagram for the encoder of Figure 6.19.

where $\{C_j\}$, the output sequence, is a *convolution* of the input sequence $\{I_j\}$ and the tap weights $\{W_k\}$. Hence the name *convolutional encoder*.

Another way to characterize convolutional codes is by the *trellis diagram*, one of which is constructed in Figure 6.21 for the encoder of Figure 6.19. First define the *state* of the encoder at a given instant as the $K - 1$ leftmost register cells. In our example the state is the left two cells. We use a dashed line to indicate a "one" just prior to the first cell and a solid line to indicate a "zero" just prior to the first cell. The pairs of numbers at the output are the encoded symbols that are transmitted over the channel. The first symbol corresponds to the top encoder output, and the second symbol corresponds to the lower encoder output.

To illustrate the process with an example, assume that the input bit sequence is

$$(\leftarrow \text{Time})$$

$$1\ 0\ 1\ 1\ 0\ \underline{0\ 0\ 0}$$

initial register loading

The first three bits, 000, load the register. Next a "zero" is inputted, which produces two output symbols 00, and the "state" remains at 00. Next a "one" is inputted, which produces the coded symbol pair 11, resulting in the coded state 10. The shift register contains 100 at this time. The next input is a "one" also, which produces the coded symbol pair 01. Now the

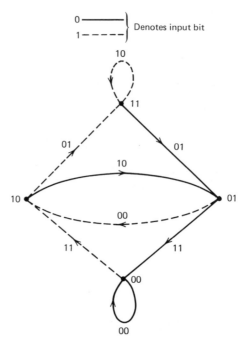

Figure 6.22   State diagram encoder representation for the encoder of Figure 6.19.

encoder state is 11. Next a "zero" placed in the input to the encoder produces the coded symbol pair 01, and the coder state is now 01. By this point, the reader should be able to determine what the next bit inputs will do. The trellis diagram relates the states and the input bits together. For an example, the "path" is shown in Figure 6.21.

Our next characterization of convolutional codes is the *state diagram*. In Figure 6.22 the state diagram for the encoder of Figure 6.19 is illustrated. This diagram can be constructed directly from the trellis diagram. Again the states are indicated at the nodes, and the coded symbol pairs are shown on the line segments between nodes. Dashed lines indicate a "one" was just fed into the encoder, and solid lines indicate a "zero" was fed into the encoder. A path can be drawn through this diagram showing the evolution of the states just as in the trellis diagram.

6.7.2   The Viterbi Decoding Algorithm for the Binary Symmetric Channel

In this section we discuss the Viterbi (maximum likelihood) procedure for decoding bits on a binary symmetric channel (BSC). A BSC channel is one in which errors convert a transmitted "one" (symbol) to received "zero" (symbol) and *vice versa*. All these errors are statistically independent and occur with probability p.

Given that $p < .5$, it is well known [32] that the optimum decoder of equally likely sequences is one that compares all possible allowable transmitted sequences with the received sequence and decides on the received sequence on the basis of the closest Hamming distance (that is, the transmitted sequence that differs from the received sequence in the minimum number of symbols), to a possible transmitted sequence.

Viterbi noted that it is not necessary to consider the entire received symbol sequence at one time in determining which is the most likely (minimum Hamming distance) transmitted sequence. Rather, reductions·in possible bit patterns could be made in a bit-by-bit manner. For example, consider the fourth node from the left in Figure 6.21 at state 11, and assume the received symbol sequence 00 01 01 01 00 have been received at this time. The symbol sequence 00 11 01 has one error into this node from the received sequence, whereas the symbol sequence 11 01 10 has a Hamming distance of 4! Therefore, we can exclude the latter path from consideration since no matter what the subsequent received symbols are, they will affect the distances only over subsequent branches after these two paths have merged. Hence, from this node onward, they are affected in exactly the same way. The same is true for pairs of paths merging at the other three nodes in the same vertical line at state 11. The *survivor* path is the path merging at a given node that has the smaller Hamming distance. At each time period it is necessary to store the minimum Hamming distance symbol sequence from the survivor at each node as well as the value of the Hamming distance.

Then at the next time period the next set of nodes (along a column in Figure 6.21) will have two paths merging along with their stored Hamming distance. Again the survivor is chosen and the path and Hamming distance are stored. The procedure continues in this way storing the Hamming distance and the survivor path. If two paths merge into a node and have the same Hamming distance, either path may be picked by the decoder. This is true since all future sequences would be identical from this same node so that waiting to a later time would not aid the decision process. This last statement is the essence of the algorithm's optimality [34].

The Viterbi decoder is a maximum likelihood decoder only when its decision path memories are infinitely long. However, this implies infinite delay in decoding, which is obviously undesirable. However, convergence may be forced by the inclusion of K or more known bits periodically into the data, as was suggested in the original algorithm description [26]. In fact, however, convergence is rapid enough so that truncation of the survivor memory to four or five constraint lengths will produce only a slight degradation in bit error rate performance. In the original decoding procedure, the output bit at each time step was selected to be the oldest bit associated with the most likely path. Another choice for selection of the data output is the majority function of the oldest bits of each of the 2^{K-1} survivors. A third method is simply to choose a surviving path at random from which to output decoded bits. Layland [33] has simulated all three

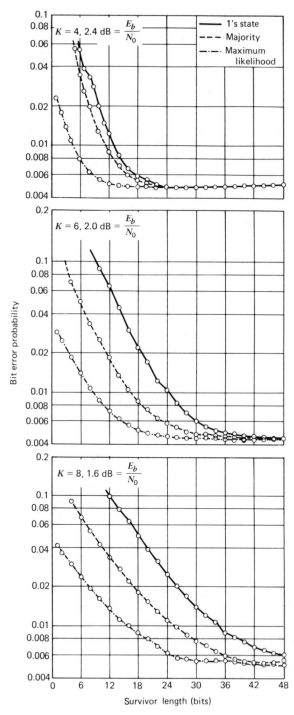

Figure 6.23 Bit error probability versus survivor length (from Layland [33] with permission).

decision schemes for the WGN channel (which we will outline in Problem 11), with the results shown in Figure 6.23. Three constraint lengths were considered: $K = 4$, 6, and 8. The infinite memory bit error rule (BER) was set equal to 5×10^{-3}, which is typical of planetary space missions. For maximum likelihood detection a memory length of 4.5 K incurs virtually no degradation compared to infinite memory decoding!

PROBLEM 11

Show that for the WGN channel that the correct (maximum likelihood) metric for a given bit (assuming bit and symbol synchronization), is given by maximizing

$$\sum_{j=1}^{5K} \sum_{k=1}^{n} y_{jk} x_{jk}$$

where

x_{jk} is the kth code symbol for the jth bit (n symbols per bit).

y_{jk} is the symbol time integrated output (received signal plus noise) of the kth code symbol for the jth bit.

$\sum_{j=1}^{5K}$ is the sum over all bits in a particular path up to a memory of five constraint lengths.

Hint: Consider the log likelihood function for a given bit.

6.7.3   Convolutional Code Properties

Now we consider some basic properties of convolutional codes. First we consider an important distance property of the code, the *free distance*. Since convolutional codes are group codes, there is no loss in generality in computing the distance properties from the all-zero code word to all other code words. Because convolutional codes are group codes, this set of distances is the same as the set of distances from any specific code word to all others. We may redraw Figure 6.21 in terms of their "distances" from the all-zero path, which is shown in Figure 6.24. Suppose that the decoder is located at state 00 at the third time node from the left. Given that the path diverges from the all-zero path, we see that the minimum "distance" to remerge is five symbol errors, and there is just one path that yields this distance, while there are two paths that remerge to the all-zero path at a distance of six symbol errors. There are other paths with increasing numbers of symbol errors to re-emerge to the all-zero paths. In the case of the five symbol errors path, the corresponding number of bit errors is 1. The *minimum free distance* (d) among all paths for this encoder is seen to be 5. It is clear that any code transmission with one or two symbol errors can be correctly decoded, while three or more symbol errors will lead to incorrect decoding.

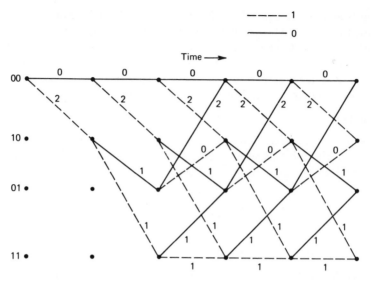

Figure 6.24 Trellis diagram with distances from the all-zero path.

A *systematic convolutional code* is a convolutional code for which one of the symbols (*n* symbols/bit) is set equal to the input bit. Although they have the property that they directly produce the transmitted bit (if no error occurs), there is a penalty in that the BER performance is inferior to nonsystematic codes [35]. If the code is not systematic, it is *nonsystematic*.

The event that a finite number of channel symbol errors can cause an infinite number of data bit errors to be decoded is called *catastrophic error*. Gilhousen *et al.* [31], have shown that catastrophic errors occur if and only if any closed-loop path in the state diagram has zero weight. Let us consider this phenomenon by way of an example illustrated in Figure 6.25, along with its associated state diagram.

PROBLEM 12

(a) Determine the minimum free distance for the systematic convolutional encoder shown below.

(b) Can one do better (improve *d*) if the restriction that this be a systematic code is removed?

Assume that the all-zero bit sequence is the correct bit sequence, that is, the transmitted sequence. Then going through the states in the sequence 00, 10, 11, 11,..., 11, 01, 00 requires six symbol errors. But, 2 plus the number of 11 states entered would equal the number one-bits decoded! Therefore, a finite number of symbol errors *could* lead to an unbounded number of decoded bit errors.

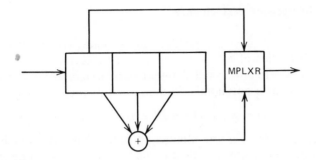

PROBLEM 13

(a) Show that if each adder of the encoder has an even number of connections, then the code will be catastrophic.

(b) Show that a systematic code cannot be catastrophic.

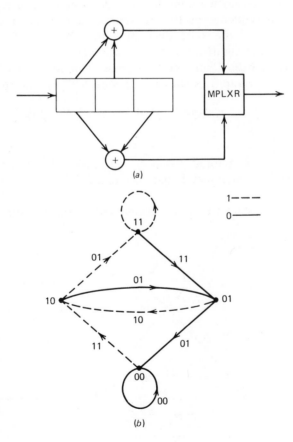

Figure 6.25   An example of a catastrophic code. (a)Encoder. (b) State diagram.

6.7.4   Viterbi Decoder Performance

In this section we discuss a procedure to estimate, via an overbound, the bit error probabilities of a Viterbi decoder. This section follows reference 36 quite closely. First we shall define a *path weight enumerator* or code transfer function as a polynomial in the (dummy) variable D:

$$T(D) = a_0 + a_1 D + a_2 D^2 + a_3 D^3 + a_4 D^4 + \cdots \qquad (6.7\text{-}3)$$

where the coefficient a_i is the number of paths that have weight i (Hamming distance) going from the all-zero state to the all-zero state (point a to point b in Figure 6.26). For example, if $T(D)$ contained a term $2D^2$, that would mean that there are two paths with weight 2 going away from the all-zero state and back again. Consider now the determination of $T(D)$ from the state diagram. As an example, consider the state diagram of Figure 6.22.

First we take the 00 state apart since a closed loop, upon itself, implies correct reception. Again we consider the group code property, and consider the all-zero sequence as the transmitted sequence. In addition, we label each branch of the state diagram with either D^2, D^1, or $D^0 = 1$, according to a Hamming distance of 2, 1, or 0. The redrawn state diagram is shown in Figure 6.26. Using the flow graph techniques of Chapter 9 we can reduce this flowgraph, as shown in Figure 6.27. It can be shown that

$$T(D) = \frac{D^5}{1 - 2D} = D^5 + 2D^6 + 2^2 D^7 + 2^3 D^8 + \cdots + 2^K D^{5+K} + \cdots$$

$$(6.7\text{-}4)$$

and hence there is one path at distance 5 and two paths at distance 6, and so on. It should be mentioned that this method becomes quite tedious beyond $K = 4$.

Viterbi [36] has shown that an "augmented" state diagram is useful to

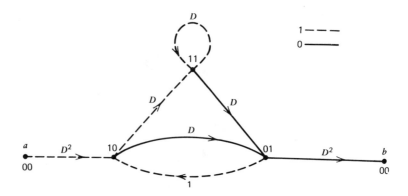

Figure 6.26   State diagram used to compute $T(D)$.

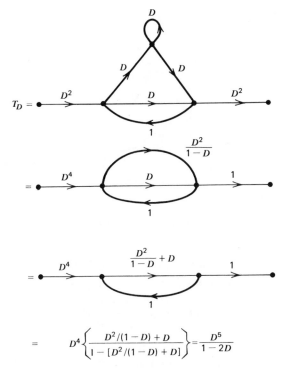

$$= D^4 \left\{ \frac{D^2/(1-D)+D}{1-[D^2/(1-D)+D]} \right\} = \frac{D^5}{1-2D}$$

Figure 6.27 Flowgraph reduction to the transfer function.

determine the number of decoded "one" bits, N, and the number of bits (or branches) that have been inputted, L. The number of bits in the decoder, L, is useful since practical Viterbi decoders only utilize a storage of L equal to about five constraint lengths. Depicting the same example as before, Figure 6.28 illustrates the augmented state diagram.

In Problem 14 it is shown that

$$T(D, L, N) = \frac{D^5 L^3 N}{1 - DL(1+L)N} = D^5 L^3 N + D^6 L^4 (1+L) N^2 + D^7 L^5 (1+L)^2 N^3$$

$$+ \cdots + D^{5+K} L^{3+K} (1+L)^K N^{K+1} + \cdots$$

(6.7-5)

Hence we conclude that of the two paths of distance 6, one is of length 4 and one is length 5 and both differ from the two input bits by 2. Hence, we conclude that if either of these paths is decoded, then two bit errors occur.

PROBLEM 14

Using flowgraph techniques on Figure 6.28, show that (6.7-5) is, in fact, correct.

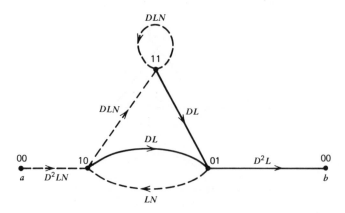

Figure 6.28   Augmented state diagram of Figure 6.19.

Now we are in a position to state BER results that Viterbi [36] obtained for two important channels. For the first case, we consider the BSC. A BSC is a channel in which errors change a channel symbol "zero" to "one" or "one" to "zero" independently, symbol-to-symbol, with probability p. Normally, $p < .5$.

Viterbi [36] has, by means of a union bound, shown, for the case of Q bits input to the encoder per unit time over the BSC, that the BER is bounded by

$$PE_B < \frac{1}{Q} \frac{dT(D, N)}{dN}\Bigg|_{\substack{N=1 \\ D=2\sqrt{p(1-p)}}} \tag{6.7-6}$$

$$T(D, N) = T(D, N, L)|_{L=1} \tag{6.7-7}$$

All of our examples have assumed that $Q = 1$.

An example of this result is our encoder of Figure 6.19. Equation 6.7-5 is the expression for $T(D, N)$, and for $L = 1$ is given by

$$T(D, N) = \frac{D^5 N}{1 - 2ND} \tag{6.7-8}$$

Differentiating, we obtain

$$\frac{\partial T(D, N)}{\partial N}\Bigg|_{N=1} = \frac{D^5}{(1 - 2D)^2} \tag{6.7-9}$$

Hence we conclude

$$PE_B < \frac{2^5 [p(1-p)]^{5/2}}{[1 - 4\sqrt{p(1-p)}]^2} \tag{6.7-10}$$

Due to the singularity at $p = .067$, this bound is only useful when $p \ll .067$. Union bounds are known to be accurate only at low bit error probabilities, so this result is not surprising.

The other channel we shall consider is the additive white Gaussian noise channel (AWG) with biphase modulated data. The upper bound [36] is given by

$$PE_B < Q\left(\sqrt{\frac{2dE_s}{N_0}}\right) \exp\left(\frac{dE_s}{N_0}\right) \frac{dT(D, N)}{dN}\bigg|_{\substack{N=1 \\ D=\exp(-E_s/N_0)}} \tag{6.7-11}$$

where E_s/N_0 is the symbol energy-to-noise spectral density, d is the minimum free distance. Therefore, for our encoder of Figure 6.19, using (6.7-10), we obtain

$$PE_b < \frac{Q(\sqrt{5E_b/N_0})}{[1 - 2\exp(-E_b/2N_0)]^2} \tag{6.7-12}$$

This bound is only useful for $E_b/N_0 \gg 1.42$ dB, since there is a singularity at $E_b/N_0 = 1.42$ dB.

In Figures 6.29 through 6.32 the BER, as determined by simulation, and

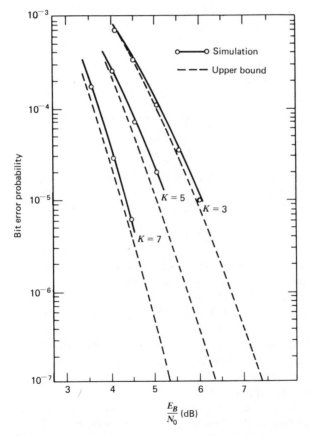

Figure 6.29 Bit error rate versus E_B/N_0 for rate $\frac{1}{2}$ Viterbi decoding. Eight-level quantized simulations with 32-bit paths, and infinitely quantized transfer function bound, $k = 3, 5, 7$ (from Heller and Jacobs [37] with permission).

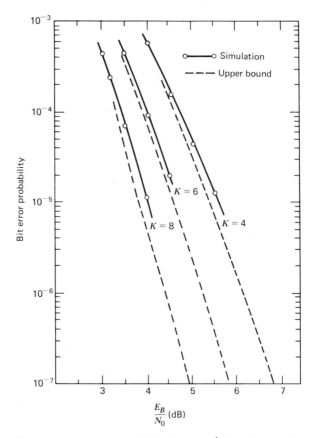

Figure 6.30 Bit error rate versus E_B/N_0 for rate $\frac{1}{2}$ Viterbi decoding. Eight-level quantized simulations with 32-bit paths, and infinitely finely quantized transfer function bound, $k = 4, 6, 8$ (from Heller and Jacobs [37] with permission).

the bounds, are plotted for either eight-level or two-level quantization (BSC) as a function of E_b/N_0 for various optimum codes [31, 38], for BPSK modulation. The probability-of-error bounds are based on the assumption of infinitely quantized, matched filter, outputs, whereas the simulation curves are based on at most three-bit quantization. Therefore, the upper bounds *appear* to be lower bounds. It has been determined that about 0.2–0.25 dB should be subtracted off the (infinite bit) upper bound curves to make them approximately three-bit upper bound curves. When this is done they become true upper bounds.

Some "optimal" convolutional codes [30] are listed in Table 6.2 for rate $\frac{1}{2}$ ($V = 2$) and rate $\frac{1}{3}$ ($V = 3$) codes. A rate R code has $1/R$ symbols per bit. The generator coefficients determine the tap connections. For example, in

Table 6.2 Optimal nonsystematic convolutional codes

K	V	Generator Coefficients (G_1, G_2)
3	2	111, 101
4	2	1111, 1101
5	2	11101, 10011
6	2	111101, 101011
7	2	1111001, 1011011
8	2	11111001, 10100111
3	3	111, 111, 101
4	3	1111, 1101, 1011
5	3	11111, 11011, 10101
6	3	111101, 101011, 10011
7	3	1111001, 1100101, 1011011
8	3	11110111, 11011001, 10010101

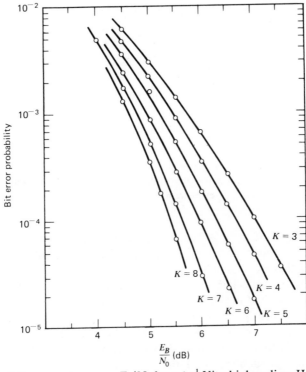

Figure 6.31 Bit error rate versus E_B/N_0 for rate $\frac{1}{2}$ Viterbi decoding. Hard quantized received data with 32-bit paths; $k = 3$ through 8 (from Heller and Jacobs [37] with permission).

265

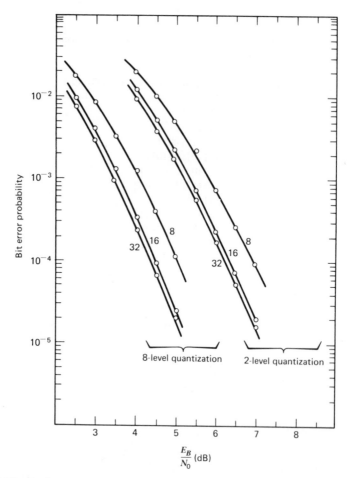

Figure 6.32 Performance comparison of Viterbi decoding using rate $\frac{1}{2}$, $k = 5$ code with 8-, 16-, and 32-bit path lengths and two- and eight-level quantization (from Heller and Jacobs [37] with permission).

the second code ($K = 4$, $V = 2$), the encoder appears as shown in Figure 6.33. There are many other topics that are of concern, but they are beyond the scope of this book. The reason for considering coding in the first place is the fact that more reliable data, that is, lower BERs, can be achieved.

We have assumed in the above discussion that the channels were WGN channels. However, when errors tend to occur in bursts due to pulse type jamming, for example, then *interleaving* [39] is used. Basically, interleaving is accomplished by a device that allows storage and provides scrambling of the code symbols such that no input code symbol is closer than some fixed number of code symbols. At the other end of the channel, a *de-interleaver* must be used to unscramble the channel symbols. This process does

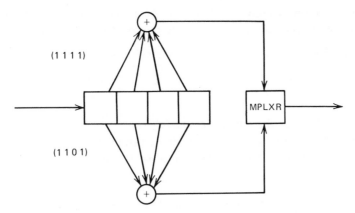

Figure 6.33   The $k = 4$, $v = 2$ convolutional encoder of Table 6.1.

require a delay in decoding. More details on Viterbi decoding can be found in references [31] and [36].

PROBLEM 15

Consider the encoder of Figure 6.33. Determine the state diagram. Obtain an upper bound on the infinitely quantized WGN channel. Obtain a bound on the BSC. What is d for this code?

The Viterbi algorithm has applications (see [40]) other than coding, including demodulation of MSK, text recognition, sequential ranging, intersymbol interference, and pattern recognition.

Sequential decoding [29] can also be used for convolutional codes and is very competitive at very low error probabilities.

6.8   THE EFFECTS OF IMPERFECT CHANNELS

In this section we briefly consider degradation to the channel due to imperfect carrier reference and intersymbol interference. Other imperfections of the channel, such as nonlinear amplitude characteristics, phase distortion, and AM/PM effects, are not considered here.

The imperfect reference problem will be broken down into three cases: (1) uncoded BPSK, (2) Viterbi decoding, and (3) uncoded quadriphase, since these are often used in contemporary communication systems.

6.8.1   Imperfect Carrier Reference for Uncoded and Coded BPSK

The purpose of this section is to model the effects of imperfect carrier synchronization on the BER (or probability of bit error). Let us consider the model shown in Figure 6.34, where the received residual carrier signal is corrupted by AWGN. Expanding the received signal, we have

$$y(t) = \underbrace{\sqrt{2P}\,\cos(\theta)\cos(\omega_0 t + \theta_0)}_{\text{carrier}} - \underbrace{\sqrt{2P}\,\sin(\theta)\,d(t)\,sc(t)\sin(\omega_0 t + \theta_0)}_{\text{modulation}} + \underbrace{n(t)}_{\text{noise}}$$

$$(6.8\text{-}1)$$

where $sc(t)$ denotes the subcarrier waveform.

The carrier loop tracks the carrier component in (6.8-1) to provide a reference for the modulation given by ($\hat{\theta}$ is the carrier loop estimate)

$$r(t) = \sqrt{2}\,\sin(\omega_0 t + \hat{\theta}) \tag{6.8-2}$$

assuming a perfect subcarrier tracking loop (for our discussion) we have, as our demodulated waveform, the term

$$z(t) = \sqrt{P}\,\sin\theta\,d(t)\cos[\phi(t)] + n'(t) \tag{6.8-3}$$

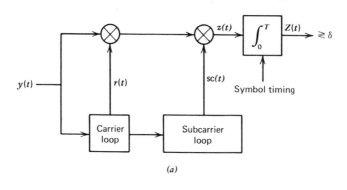

(a)

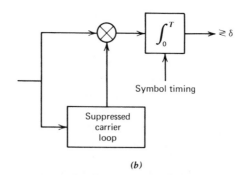

(b)

Figure 6.34 Models to demonstrate an imperfect carrier reference. (a) cw Carrier phase reference model $y(t) = \sqrt{2P}\,\cos[\omega_0 t + \theta_{sc}(t)d(t) + \theta_0] + n(t)$. (b) Suppressed carrier phase reference model.

where the corrupting noise process $n'(t)$ is given by

$$n'(t) = n_s(t) \cos[\phi(t)] - n_c(t) \sin[\phi(t)] \qquad (6.8\text{-}4)$$

where the phase error process is given by

$$\phi(t) = \theta_0 - \hat{\theta}(t) \qquad (6.8\text{-}5)$$

and where the receiver noise $n(t)$ has been represented by

$$n(t) = \sqrt{2}\, n_c(t) \cos(\omega_0 t + \theta_0) + \sqrt{2}\, n_s(t) \sin(\omega_0 t + \theta_0) \qquad (6.8\text{-}6)$$

also, θ is the modulation angle; $d(t)(\pm 1)$ is the modulation; and P is the signal power. Under these conditions the output of the integrator, assuming

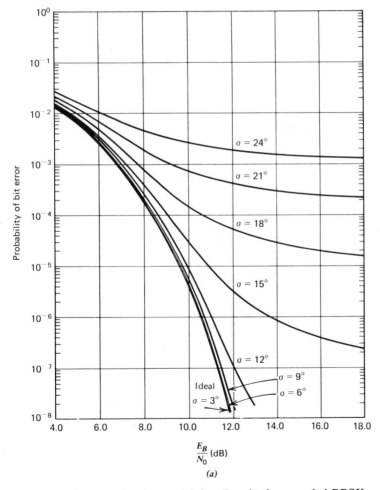

Figure 6.35 Steady state phase error = 0 deg. Results for uncoded BPSK cw loop. Steady state phase error (a) 0°.

perfect bit synchronization, is given by

$$Z(T) = \int_0^T \sqrt{P} \, \sin(\theta) \, d(t) \, \cos[\phi(t)] \, dt + \int_0^T n'(t) \, dt \qquad (6.8\text{-}7)$$

At this point the process $\phi(t)$ is a stochastic process that, in general, will not be constant during the T-sec integration period.

There are two relatively simple cases that we consider here. The first is the case that the phase error process varies rapidly during a bit or symbol period. In the limiting case we have

$$Z(T) \cong \sqrt{P} \, T \, \sin(\theta) \, d(T^-) \, \overline{\cos \phi} + \int_0^T n'(t) \, dt \qquad (6.8\text{-}8)$$

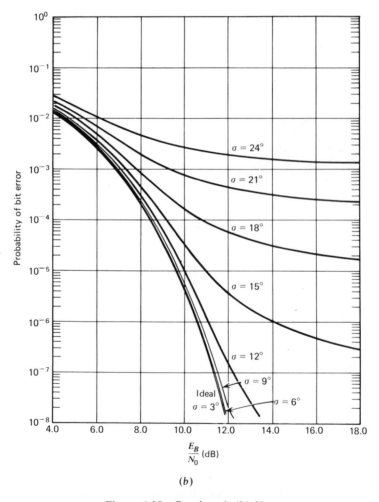

(b)

Figure 6.35 Continued. *(b)* 3°.

[where $d(T^-)$ denotes the value of $d(t)$ prior to the bit change at $t = T.$] Therefore, imperfect synchronization in this case causes an SNR degradation of

$$L = \overline{(\cos \phi)}^2 \qquad (6.8\text{-}9)$$

and the probability of error for BPSK, assuming that $n'(t)$ has the same statistics as $n(t)$, is (see Problem 3)

$$\text{PE}_B \cong \tfrac{1}{2}\,\text{erfc}[\sqrt{R_B}\,\overline{(\cos \phi)}] \qquad R_B = \frac{PT \sin^2 \theta}{N_0} \qquad (6.8\text{-}10)$$

For the Tikhonov distribution (assuming a first-order cw tracking loop) it

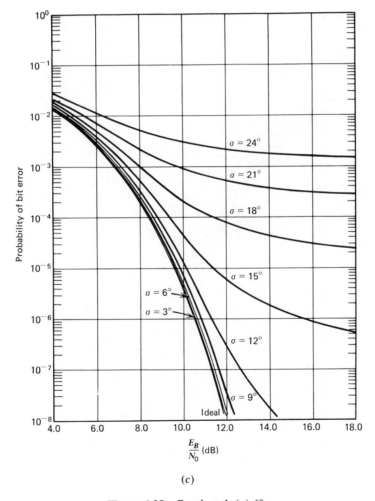

(c)

Figure 6.35 Continued. (c) 6°.

can be shown that [3]

$$\overline{\cos \phi} = \frac{I_1(\rho)}{I_0(\rho)} \tag{6.8-11}$$

where $I_n(\rho)$ is the modified Bessel function of order n and argument ρ (loop SNR). It should be noted that the distribution of ϕ should include both thermal and oscillator phase noise components, with $\rho^{-1} = \sigma_\phi^2 + \sigma_{osc}^2 = \sigma^2$ being a useful approximation to evaluate (6.8-11). (σ_ϕ^2 is the thermal noise component, and σ_{osc}^2 is the oscillator component, of the phase error variance.)

In the other extreme we assume that the phase error process $\phi(t)$ of (6.8-7) is essentially constant during one bit time. Then, again assuming that

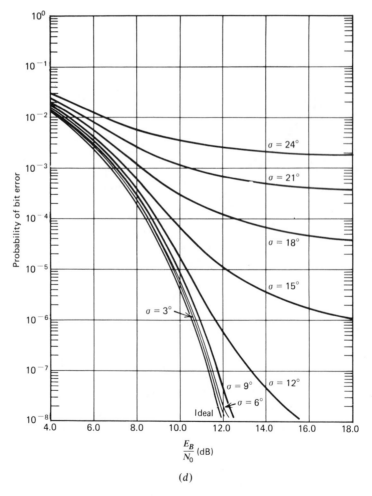

Figure 6.35　Continued. (d) 9°.

$n'(t)$ has the same statistics as $n(t)$, we have, for the BPSK case, that

$$PE_B(\phi) = \tfrac{1}{2}\,\mathrm{erfc}(\sqrt{R_B}\,\cos\phi) \qquad (6.8\text{-}12)$$

Hence the average bit error probability is given by

$$PE_B = \tfrac{1}{2}\int_{-\pi}^{\pi}\mathrm{erfc}[\sqrt{R_B}\,\cos\phi]\,\frac{1}{2\pi I_0(\rho)}\,e^{\rho\cos\phi}\,d\phi \qquad (6.8\text{-}13)$$

This has been evaluated as an infinite series [3]. However, it is just about as easy to do the integration on a digital computer. The results are shown [41] in Figure 6.35a through 6.35g for the BPSK case with static phase errors from 0 to 18° in increments of 3°. These curves are useful for BER

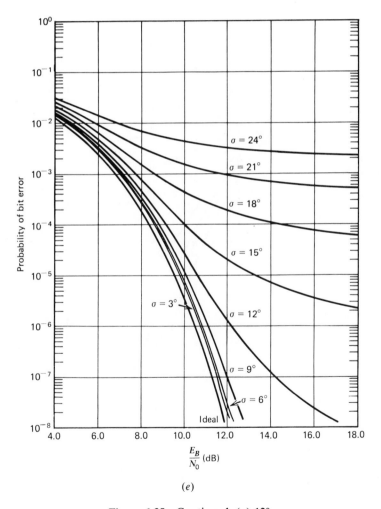

(e)

Figure 6.35 Continued. (e) 12°.

degradation estimates. Notice that as a function of rms jitter, there is an irreducible error!* In these curves $\sigma^2 = 1/\rho$, where ρ is the loop SNR.

For a Costas or a squaring loop for tracking, the phase error expression is approximately Tikhonov (for a first-order loop), so that

$$p(\phi) \cong \frac{1}{\pi I_0(\rho_e)} \exp[\rho_e \cos(2\phi)] \qquad \frac{\pi}{2} \le \phi \le \frac{\pi}{2} \qquad (6.8\text{-}14)$$

where ρ_e is the equivalent loop SNR (accounting for both squaring loss and S-curve compression). In fact for small ϕ we can show, from 6.8-14, that

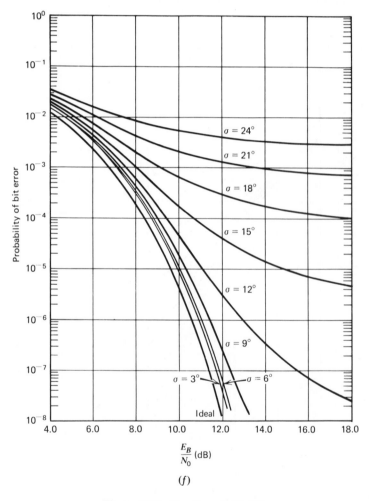

Figure 6.35 Continued. (f) 15°.

*In an actual system, both σ reduces and E_B/N_0 increases together.

the linearized variance σ^2, equals $(4\rho_e)^{-1}$! In Figures 6.36a through 6.36g [41], the suppressed carrier BPSK bit error probabilities are shown with $\rho_e = 1/(4\sigma^2)$ or $\sigma = \sqrt{1/(4\rho_e)}$. In other words $\rho_e^{-1} = \sigma_{\Phi e}^2$ (see 5.2-9) and $(4\rho_e)^{-1} = \sigma^2 = \sigma_\phi^2$.

One approximation that is used to estimate the bit error rate degradation when the phase error process is broadband is to assume that all the phase noise below the symbol rate is essentially constant during a bit time and all the phase error variance due to phase noise spectra above the symbol rate produces a fast variation. Let $\phi(t) = \phi_L + \phi_H(t)$, where ϕ is the total phase noise process, ϕ_L is a random variable representing the phase process below the symbol rate, and $\phi_H(t)$ is a random process representing the phase error process about the symbol rate. Since we are describing a

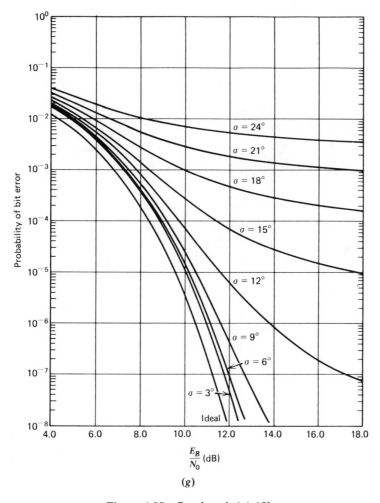

(g)

Figure 6.35 Continued. (g) 18°.

decomposition based on frequency, we define

$$\sigma^2_{\phi_L} = 2 \int_0^{R_S} \mathscr{S}_\phi(f)\, df \qquad (6.8\text{-}15)$$

$$\sigma^2_{\phi_H} = 2 \int_{R_S}^{\infty} \mathscr{S}_\phi(f)\, dt \qquad (6.8\text{-}16)$$

where it is assumed that both variances are bounded, $\mathscr{S}_\phi(f)$ is the spectral density of the phase error process and R_S is the symbol rate. Now note that

$$Z(T) = \sqrt{P}\, d(T^-) \int_0^T \cos[\phi_L + \phi_H(t)]\, dt + \int_0^T n'(t)\, dt \qquad (6.8\text{-}17)$$

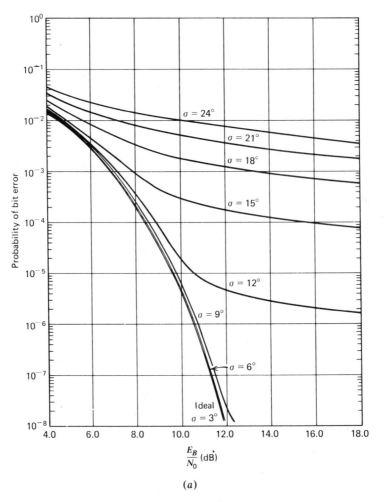

Figure 6.36 Results for uncoded BPSK Costas loop. Steady phase error (a) 0°.

or

$$Z(T) \cong \sqrt{P}\, T\, d(T^-)\{\cos \phi_L \overline{\cos[\phi_H(t)]} - \sin \phi_L \overline{\sin[\phi_H(t)]}\}$$
$$+ \int_0^T n'(t)\, dt \qquad (6.8\text{-}18)$$

In many applications, the sine product terms are small compared to the cosine product terms {if $\phi_H(t)$ is zero mean and has a symmetric density function, then $\overline{\sin[\phi_H(T)]} = 0$}. Assuming that the sine product terms are zero, we have, for BPSK cw demodulation, that

$$PE_B \cong \frac{1}{2} \int_{-\pi}^{\pi} \operatorname{erfc}[\sqrt{R_B}\,(\overline{\cos \phi_H})\cos \phi_L] \frac{1}{2\pi I_0(\rho)} \exp[\rho \cos(\phi_L)]\, d\phi_L$$
$$(6.8\text{-}19)$$

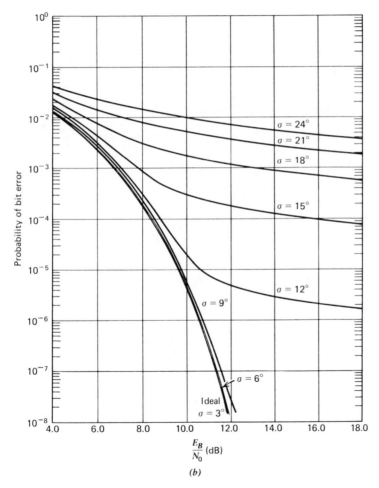

(b)

Figure 6.36 Continued. (b) 3°.

with ρ equal to the inverse total variance, and hence BER degradation can be estimated using the curves of Figures 6.35a through 6.35g and (6.8-11).

The method can be used for suppressed carrier demodulation. Many actual system applications lead to the result that $\cos \phi_H \approx 1$, so that only the low-frequency component of phase noise (phase error process) is important in BER degradation estimation.

Next we consider Viterbi decoding with an imperfect carrier phase reference. As in the BPSK case, if the phase error changes rapidly during a bit time, then the approximation leading to (6.8-11) can be used so that the degradation is

$$L = 20 \log\left[\frac{I_1(\rho)}{I_0(\rho)}\right] \qquad dB \qquad (6.8\text{-}20)$$

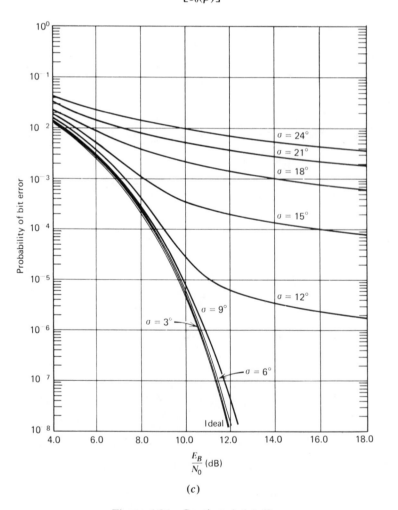

Figure 6.36   Continued. (c) 6°.

In the other extreme, if we assume that the phase error process is constant over the Viterbi decoder memory time, then the average probability of error is given by (assuming cw tracking)

$$PE_B = \int_{-\pi}^{\pi} PE\left(\frac{E_B}{N_0}\cos^2\phi\right)\frac{1}{2\pi I_0(\rho)}\exp(\rho\cos\phi)\,d\phi \qquad (6.8\text{-}21)$$

where $PE[(E_B/N_0)\cos^2\phi]$ can be, for example, a curve fit polynomial that fits the known simulation data without phase error present. This procedure has been used on the constraint length 7, rate $\frac{1}{2}$ code used in Figure 6.29 ($G_1 = 1111001$, $G_2 = 1011011$). However, since it is known that $PE_B \le .5$, a

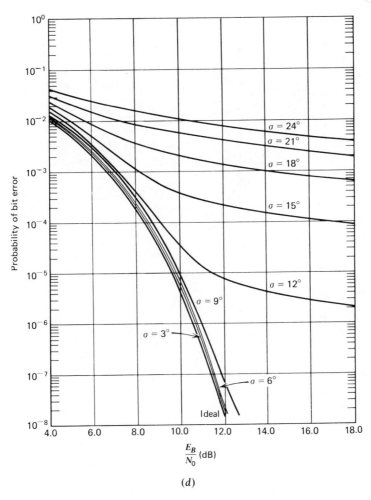

Figure 6.36 Continued. (*d*) 9°.

statement was inserted in the program to set $PE_B = .5$ whenever the curve
fit predicted $PE_B \geq .5$. The results for both cw tracking ($\rho = 1/\sigma^2$) and
Costas loop [$\rho_e = 1/(4\sigma^2)$] (or squaring loop) tracking [42], are shown in
Figures 6.37a through 6.37g and Figures 6.38a through 6.38g, respectively,
again with static phase errors of 0, 3, 6, 9, 12, 15, and 18°. Again we notice
the irreducible error phenomenon. As before, the assumption that the
phase error process is constant over the decoder memory produces the
greatest BER degradation as compared to the case the phase error varies
rapidly over the memory time.

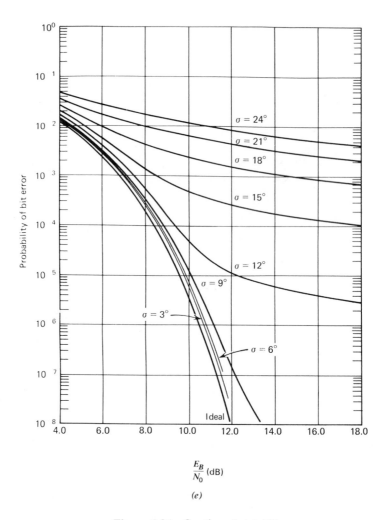

$$\frac{E_B}{N_0} \text{ (dB)}$$

(e)

Figure 6.36 Continued. (e) 12°.

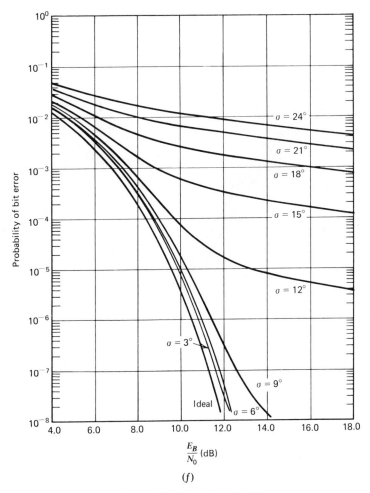

Figure 6.36 Continued. (*f*) 15°.

6.8.2 Imperfect Carrier Reference for Uncoded QPSK Systems

As an example of a QPSK system we consider the case when the data rates on the I channel are n times the data rates on the Q channel. In this quadriphase case, imperfect phase tracking causes degradation from two sources. First consider that the phase error degrades the desired data signal by $\cos[\phi(t)]$ and secondly that it causes interchannel interference so that both the desired data and the opposite channel data appear in the desired data matched filter (where it can add or subtract to the filter output).

In the analysis to follow we assume perfect bit synchronization, un-coded BPSK modulation, random data, and additive WGN interference. The analysis presented here follows that of Osborne [43]. A block diagram

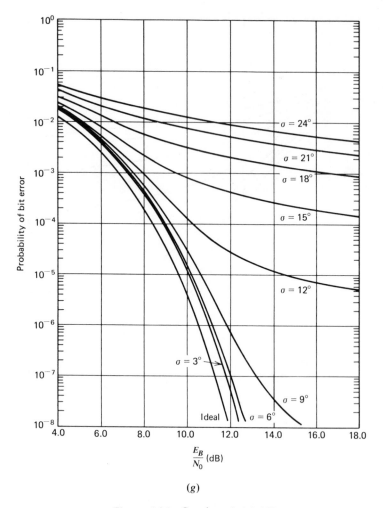

(g)

Figure 6.36 Continued. (g) 18°.

for demodulating unbalanced QPSK (UBPSK) is shown in Figure 6.39. The received signal plus noise is modeled as

$$x(t) = \sqrt{2P_I}\, d_T(t) \sin(\omega_0 t + \theta_0) + \sqrt{2P_Q}\, d_Q(t) \cos(\omega_0 t + \theta_0) \quad (6.8\text{-}22)$$

where P_I and P_Q are the I channel and Q channel power, $d_I(t)$ and $d_Q(t)$ are data sequences with symbol duration of T_I and T_Q sec, respectively. We consider the case that

$$\frac{R_Q}{R_I} = n \qquad n \text{ an integer} \qquad (6.8\text{-}23)$$

where R_Q and R_I are respective data rates of each channel. In Figure 6.40, the relationship between the two channel symbols is shown. The received

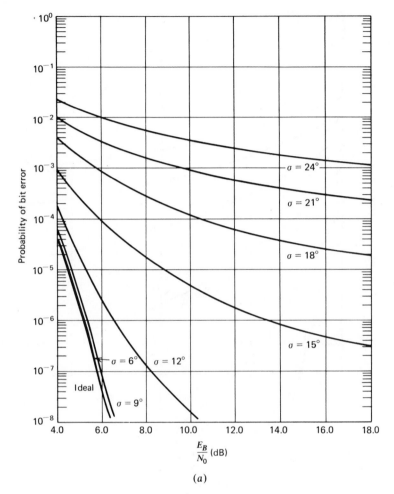

Figure 6.37 Results for cw PLL, Viterbi decoding ($G_1 = 1111001$, $G_2 = 1011011$). Steady state phase error (a) 0°.

signal is modeled as

$$y(t) = x(t) + \sqrt{2}\, n_c(t) \cos(\omega_0 t + \theta_0) + \sqrt{2}\, n_s(t) \sin(\omega_0 t + \theta_0) \quad (6.8\text{-}24)$$

where the input noise and each baseband noise term [$n_c(t)$ and $n_s(t)$] has a two-sided spectral density of $N_0/2$. After coherent demodulation (see Figure 6.39), the two baseband demodulated waveforms are

$$d_I(t) = \sqrt{P_I}\, d_I(t) \cos[\phi(t)] - \sqrt{P_Q}\, d_Q(t) \sin[\phi(t)]$$
$$- n_c(t) \sin[\phi(t)] + n_s(t) \cos[\phi(t)] \quad (6.8\text{-}25)$$
$$d_Q(t) = \sqrt{P_Q}\, d_Q(t) \cos[\phi(t)] + \sqrt{P_I}\, d_I(t) \sin[\phi(t)]$$
$$+ n_c(t) \cos[\phi(t)] + n_s(t) \sin[\phi(t)] \quad (6.8\text{-}26)$$

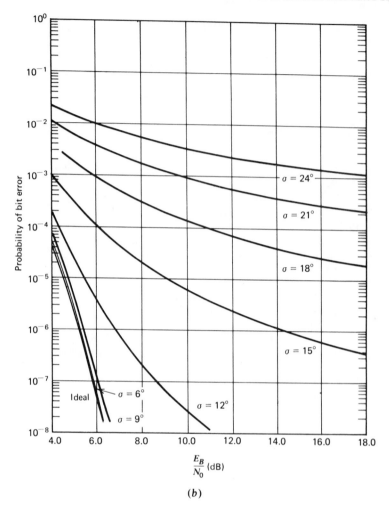

Figure 6.37 Continued. (b) 3°.

where $\phi(t)$ is the phase error process. Since our interest is in carrier demodulation degradation, we shall assume perfect bit synchronization. As can be seen from (6.8-25) and (6.8-26), cross coupling of the data occurs due to imperfect carrier synchronization. Denote the data streams by

$$d_i(t) = \sum_{m=-\infty}^{\infty} d_{i_m} p_i(t - mT_i) \qquad i = \text{I or Q} \qquad (6.8\text{-}27)$$

where d_{i_m} are independent sequences of ± 1's and $p_I(t)$ and $p_Q(t)$ are the symbol pulses that are nonzero only for $t \in (0, T_i)$, and that, we assume here, are either NRZ or Manchester symbols. The outputs of the integrate-and-dump circuits at the kth and lth signaling interval in each channel are

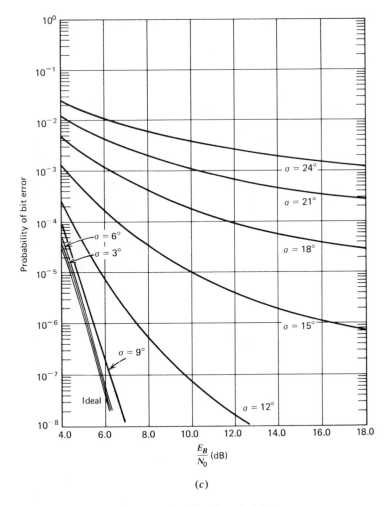

(c)

Figure 6.37 Continued. (c) 6°.

given by [we assume ϕ is constant over $\max(T_I, T_Q)$],

$$v_I = \sqrt{P_I}\, T_I\, d_{I_k} \cos\phi - \sqrt{P_Q} \sin\phi \int_{(k-1)T_I}^{kT_I} d_Q(t)p_I[t-(k-1)T_I]\,dt + N_I(k)$$
(6.8-28)

$$v_Q = \sqrt{P_Q}\, T_Q\, d_{Q_l} \cos\phi + \sqrt{P_I} \sin\phi \int_{(l-1)T_Q}^{lT_Q} d_I(t)p_Q[t-(l-1)T_Q]\,dt + N_Q(k)$$
(6.8-29)

$$N_I(k) = \int_{(k-1)T_I}^{kT_I} [-n_c(t)\sin\phi + n_s(t)\cos\phi]p_I[t-(k-1)T_I]\,dt$$
(6.8-30)

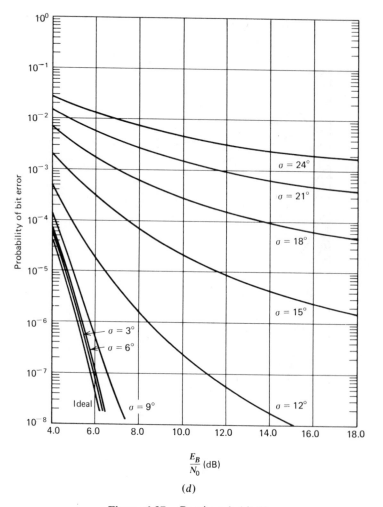

Figure 6.37   Continued. (d) 9°.

$$N_Q(k) = \int_{(l-1)T_Q}^{lT_Q} [n_c(t)\cos\phi + n_s(t)\sin\phi]p_Q[t-(l-1)T_Q]\,dt$$

$$(6.8\text{-}31)$$

The data estimates are simply

$$\hat{d}_{I_k} = \text{sgn}(v_I) \tag{6.8-32}$$

$$d_{Q_l} = \text{sgn}(v_Q) \tag{6.8-33}$$

Since both noise processes in (6.8-30) and (6.8-31) can be shown to be band-limited white noise processes, we have that the mean and variance of

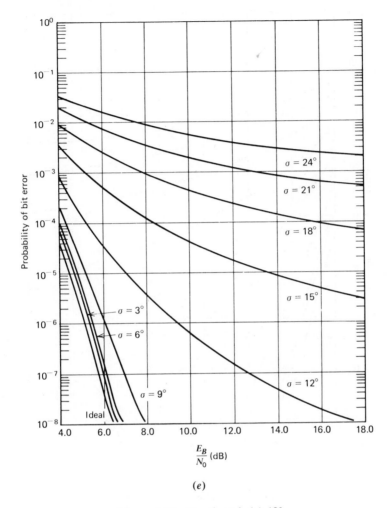

Figure 6.37 Continued. (e) 12°.

the integrator statistics (conditioned on the sequences and ϕ) are given by

$$E(v_1 \mid \phi, d_1, d_Q) = \sqrt{P_1} \, T_1 \, d_{1_k} \cos \phi - \sqrt{P_Q} \sin \phi \int_{(k-1)T_1}^{kT_1} d_Q(t)$$
$$\times p_1[t - (k - 1)T_1] \, dt \tag{6.8-34}$$

$$E(v_Q \mid \phi, d_1, d_Q) = \sqrt{P_Q} \, T_Q \, d_{Q_l} \cos \phi + \sqrt{P_1} \sin \phi \int_{(l-1)T_Q}^{lT_Q} d_1(t)$$
$$\times p_Q[t - (l - 1)T_Q] \, dt \tag{6.8-35}$$

$$\sigma_1^2 = \frac{N_0}{2} T_1 \tag{6.8-36}$$

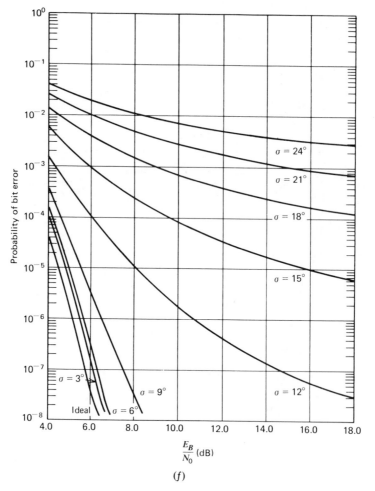

Figure 6.37 Continued. (*f*) 15°.

$$\sigma_Q^2 = \frac{N_0}{2} T_Q \tag{6.8-37}$$

Therefore, the conditional probabilities of error are given by

$$P(E_I \,|\, \phi) = \sum_{d_{Q_m}} P(E_I \,|\, d_{Q_m}, \phi) P(d_{Q_m}) \tag{6.8-38}$$

$$P(E_Q \,|\, \phi) = \sum_{d_{I_m}} P(E_Q \,|\, d_{I_m}, \phi) P(d_{I_m}) \tag{6.8-39}$$

where

$$P(E_I \,|\, d_{Q_m}, \phi) = Q \left\{ \sqrt{\frac{2P_I T_I}{N_0}} \cos \phi - \sqrt{\frac{2P_Q T_I}{N_0}} \sin \phi \, \frac{1}{T_I} \right.$$

$$\left. \times \int_{(k-1)T_I}^{kT_I} d_Q(t) p_I[t - (k-1)T_I] \, dt \right\} \tag{6.8-40}$$

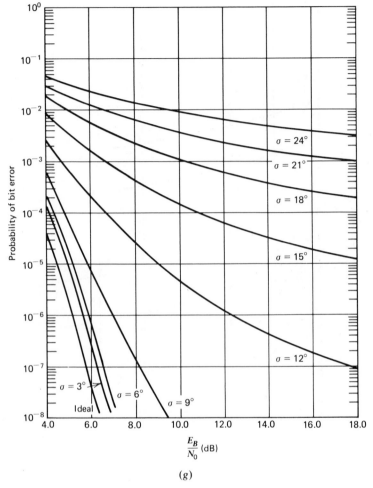

(g)

Figure 6.37 Continued. (g) 18°.

$$P(E_Q \mid d_{I_m}, \phi) = Q\left\{ \sqrt{\frac{2P_Q T_Q}{N_0}} \cos \phi + \sqrt{\frac{2P_I T_Q}{N_0}} \sin \phi \, \frac{1}{T_Q} \right.$$
$$\left. \times \int_{(l-1)T_Q}^{lT_Q} d_I(t) p_Q[t - (l-1)T_Q] \, dt \right\} \qquad (6.8\text{-}41)$$

where

$$Q(x) = \frac{1}{\sqrt{2\pi}} \int_x^\infty e^{-t^2/2} \, dt \qquad (6.8\text{-}42)$$

and the summation in (6.8-38) is over all possible combinations of sequences of $d_I(t)$ over T_Q sec, and the summation in (6.8-39) is over all possible combinations of $d_Q(t)$ over T_I sec. Over the period T_I there are n

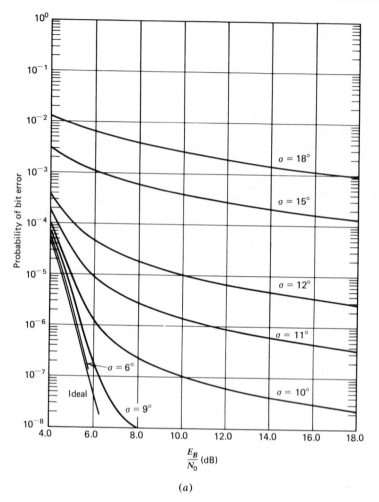

(a)

Figure 6.38 Results for Costas loop, Viterbi decoding. Steady state phase error (a) 0°.

high data rate symbols or 2^n possible combinations, whereas in T_Q sec there are only two possible combinations. We now consider the four possible cases of NRZ and Manchester data on the I and Q channels conditioned on the phase error ϕ. Then we average over ϕ assuming a Costas loop for carrier tracking.

Case II: NRZ_I, NRZ_Q

First consider the output of the I channel. From (6.8-40), noting that there are n symbols of T_Q during T_I sec, we have

$$P(E_I \mid \phi) = \frac{1}{2^n} \sum_{i=0}^{n} \binom{n}{i} Q\left[\sqrt{\frac{2P_I T_I}{N_0}} \cos \phi - \left(\frac{n-2i}{n}\right) \sqrt{\frac{2P_Q T_I}{N_0}} \sin \phi \right] \quad (6.8\text{-}43)$$

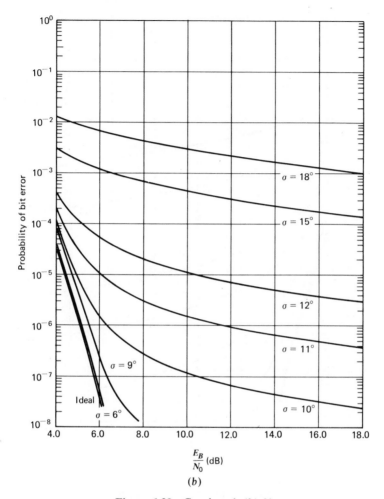

Figure 6.38 Continued. (b) 3°.

since, where there are $n - i$ positive symbols there are i negative symbols and the integral in (6.8-38) is equal to $[(n - 2i)/n]$. Furthermore, this value occurs with a probability $\binom{n}{i}/2^n$.

In the Q channel only two possibilities exist, so that we have

$$P(E_Q \mid \phi) = \tfrac{1}{2}\left[Q\left(\sqrt{\frac{2P_Q T_Q}{N_0}} \cos \phi + \sqrt{\frac{2P_Q T_1}{N_0}} \sin \phi \right) \right.$$
$$\left. + Q\left(\sqrt{\frac{2P_Q T_Q}{N_0}} \cos \phi - \sqrt{\frac{2P_Q T_1}{N_0}} \sin \phi \right) \right] \qquad (6.8\text{-}44)$$

Case II: NRZ$_I$, Manchester$_Q$

When NRZ data is on the low data rate channel and Manchester is on the high, there is no cross channel interference. Therefore, the conditional

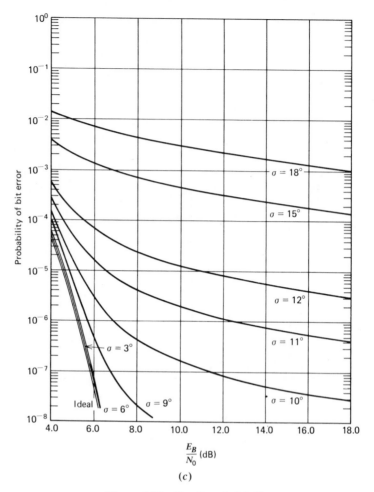

Figure 6.38   Continued. (c) 6°.

probabilities of error are given by

$$P(E_I \mid \phi) = Q\left(\sqrt{\frac{2P_I T_I}{N_0}} \cos \phi\right) \tag{6.8-45}$$

$$P(E_Q \mid \phi) = Q\left(\sqrt{\frac{2P_Q T_Q}{N_0}} \cos \phi\right) \tag{6.8-46}$$

In this case, the curves of Figures 6.36a through 6.36f are applicable. The remaining cases are left as a problem.

PROBLEM 16

Complete the analyses for the two remaining cases for the conditional bit error probability; that is, Manchester$_I$-NRZ$_Q$ and Manchester$_I$-Manchester$_Q$.

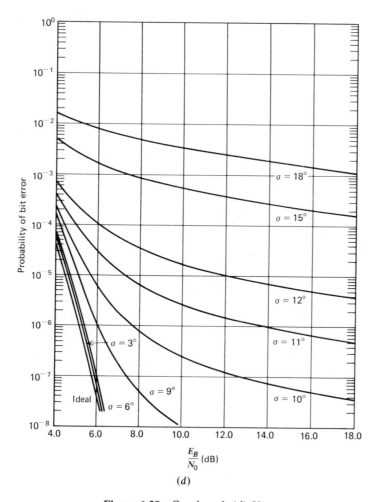

$$\frac{E_B}{N_0} \text{ (dB)}$$

(d)

Figure 6.38 Continued. (d) 9°.

To finally obtain the BER we must average over the phase error distribution under the assumption that the phase error process is essentially constant over max(T_I, T_Q). If this is not true, then the results presented here become lower (worst case) bounds. Assuming that a Costas loop is used for tracking when $n = 4$, we model the phase error probability density by

$$p(\phi) = \frac{\exp(\rho_e \cos 2\phi)}{\pi I_0(\rho_e)}, \quad |\phi| \le \frac{\pi}{2} \qquad (6.8\text{-}47)$$

which is a Tikhonov distribution with a simple change of variable mapping the interval $(-\pi, \pi)$ into the interval $(-\pi/2, \pi/2)$ with $\rho_e = 1/(4\sigma^2)$. The results for the 4:1 data rate cases are shown in Figure 6.41a through 6.41c

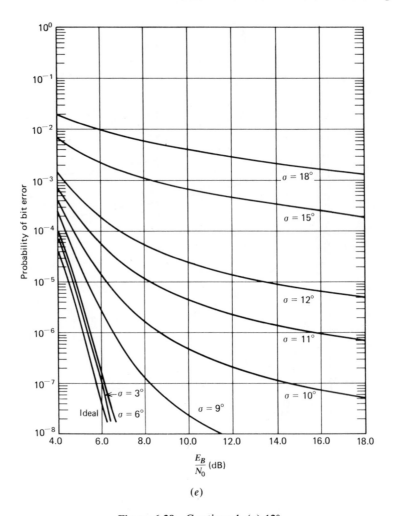

Figure 6.38 Continued. (*e*) 12°.

again with σ^2 the linearized tracking error variance. Notice that there is an increase in the BER degradation in the I channel for low SNR (Figure 6.41*a*).

The complete results for the conditional probabilities are shown in Table 6.3.

6.8.3 Intersymbol Interference Effects on BPSK Signals

So far our discussions have assumed that the channel has no bandwidth limitations, so that all symbol pulses are perfect. However, all real systems limit the signal bandwidth due to transmission bandwidth limitations and

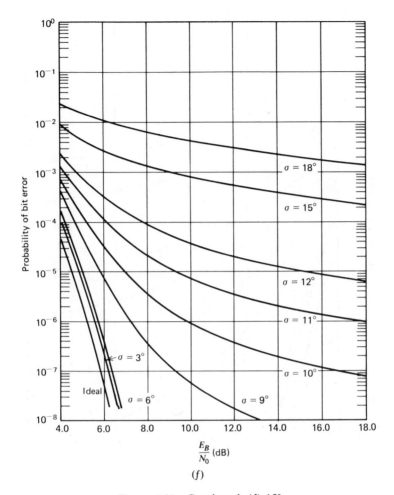

Figure 6.38 Continued. (f) 15°.

receiver bandwidth limitations. The combined effects of these filters, singly and jointly, degrade system performance from the ideal matched filter case due to both intersymbol interference and pulse distortion. Filtering effects that distort the symbol shape cause degradation since the matched filter is designed for "matching" to undistorted symbols (normally), and intersymbol interference causes the earlier symbols to contribute to the present output symbol, causing additional degradation.

Figure 6.42 illustrates the results of band limiting the channel. Besides the delay of the symbols there is distortion and intersymbol interference.

Following Jones [44], we model a BPSK and a QPSK signal by

$$x(t) = \text{Re}[U(t) e^{j2\pi f_0 t}] \tag{6.8-48}$$

with $U(t)$ the complex envelope [45], and f_0 the carrier frequency. The

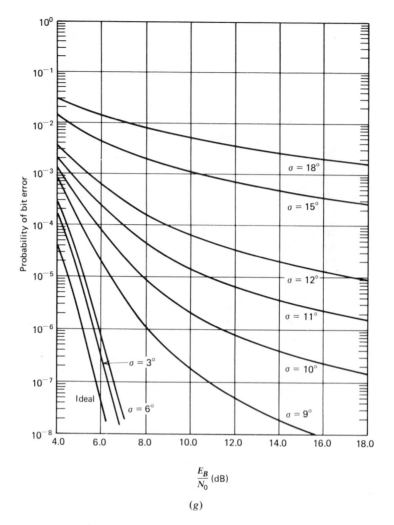

$$\frac{E_B}{N_0} \text{ (dB)}$$

(g)

Figure 6.38 Continued. (g) 18°.

quadriphase signal has a complex envelope given by

$$U(t) = a_i U_r(t) + j b_i U_r(t) = \sqrt{2} \, e^{j(\pi/2)k} \, U_r(t) \qquad (6.8\text{-}49)$$

When $|a_i| = 1 = |b_i|$, for example, and $k = 0, 1, 2, 3$, the signal takes on the four phases ($a_i = \pm 1$, $b_i = \pm 1$). For the BPSK case $b_i = 0$ and $k = 0$ or 2. In passing through the BPF of Figure 6.42a we have the BPSK or QPSK signal output in the form

$$z(t) = p(t) + \sum_{i=1}^{K} \exp\left(\frac{j2\pi N}{M}\right) p(t + iT) e^{j2\pi i f_0 T} \qquad (6.8\text{-}50)$$

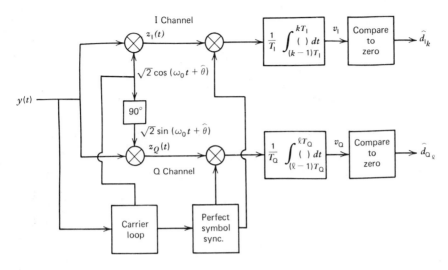

Figure 6.39 Model for UQPSK demodulation.

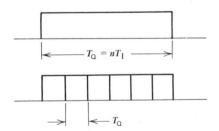

Figure 6.40 The assumed symbol duration relationships shown for NRZ symbols on both channels.

where

$$p(t) = \int_0^T U(\tau)h(t - \tau) \, d\tau \qquad (6.8\text{-}51)$$

with T the symbol duration ($M = 2$ for BPSK and $M = 4$ for QPSK), and $h(t)$ the equivalent lowpass impulse response. Jones then considered both matched filtering detection and two-pole Butterworth detection with sampling at the appropriate time to maximize the output SNR. Letting $K = 1$ (only one symbol memory for interference) Jones obtained BER degradation at $PE_B = 10^{-6}$ for BPSK and QPSK, as shown in Figure 6.42 for transmission filtering. Notice that for $BT < 3$, the two-pole Butterworth filter is more efficient than the integrate-and-dump filter. It should also be observed that these degradations are worse than would be accounted for by power loss through the filter. When a receiver filter is used, that is, when the filtering is done after the noise is added, the degradation is less than shown in Figure 6.42 by about a factor of 0.5 to 0.6 in dBs [44]. More details are given in Jones [44].

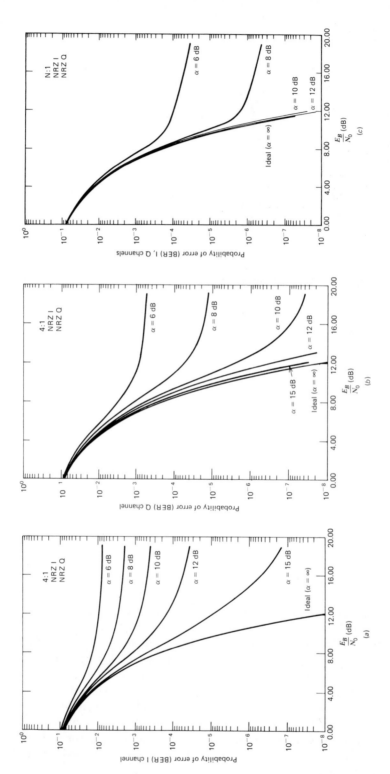

Figure 6.41 Average probability of error for (*a*) I channel versus E_B/N_0 and (*b*) Q channel versus E_B/N_0 for Costas loop, random data, power ratio 4:1 (Q:I), NRZ/NRZ data (*c*) Average probability of error for I, Q channels versus E_B/N_0 for Costas loop, random data, power ratio 4:1 (Q:I), NRZ/biphase data (from Osborne [43] with permission).

Table 6.3 Conditional probabilities of error for I and Q channels when data streams are aligned with I channel weak and Q channel strong[a] (from Osborne [43] with permission).

	NRZ/NRZ (I/Q)	NRZ/Biphase (I/Q)
$PE(d_1 \mid \phi)$ (I channel) $\sigma_1^2 = \dfrac{N_0}{2T_1}$	$\dfrac{1}{2^n} \sum_{i=0}^{n} \binom{n}{i} Q\left(\dfrac{\sqrt{P_I}}{\sigma_1}\cos\phi - \dfrac{(n-2i)}{n}\dfrac{\sqrt{P_Q}}{\sigma_1}\sin\phi\right)$	$Q\left(\dfrac{\sqrt{P_I}}{\sigma_1}\cos\phi\right)$
$PE(d_2 \mid \phi)$ (Q channel) $\sigma_2^2 = \dfrac{N_0}{2T_2}$	$\dfrac{1}{2}\left[Q\left(\dfrac{\sqrt{P_I}}{\sigma_2}\sin\phi + \dfrac{\sqrt{P_Q}}{\sigma_2}\cos\phi\right)\right.$ $\left.+ Q\left(-\dfrac{\sqrt{P_I}}{\sigma_2}\sin\phi + \dfrac{\sqrt{P_Q}}{\sigma_2}\cos\phi\right)\right]$	$Q\left(\dfrac{\sqrt{P_Q}}{\sigma_2}\cos\phi\right)$

[a] From Osborne [43] with permission.

	Biphase/NRZ (I/Q)	
$PE(d_1 \mid \phi)$ (I channel)	$\dfrac{1}{2^{n-1}} \sum_{i=0}^{n-1} \binom{n-1}{i} Q\left(\dfrac{\sqrt{P_I}}{\sigma_1}\cos\phi - \dfrac{(n-1-2i)}{n}\dfrac{\sqrt{P_Q}}{\sigma_1}\sin\phi\right)$	n odd
	$\dfrac{1}{2^n} \sum_{i=0}^{n} \binom{n}{i} Q\left(\dfrac{\sqrt{P_I}}{\sigma_1}\cos\phi - \dfrac{(n-2i)}{n}\dfrac{\sqrt{P_Q}}{\sigma_1}\sin\phi\right)$	n even
$PE(d_2 \mid \phi)$ (Q Channel)	$\dfrac{n-1}{2n}\left[Q\left(\dfrac{\sqrt{P_I}}{\sigma_2}\sin\phi + \dfrac{\sqrt{P_Q}}{\sigma_1}\sin\phi\right)\right.$ $\left.+ Q\left(-\dfrac{\sqrt{P_I}}{\sigma_2}\sin\phi + \dfrac{\sqrt{P_Q}}{\sigma_2}\cos\phi\right)\right]$ $+ \dfrac{1}{n}Q\left(\dfrac{\sqrt{P_Q}}{\sigma_2}\cos\phi\right)$ $\quad n$ odd	
	$\dfrac{1}{2}\left[Q\left(\dfrac{\sqrt{P_I}}{\sigma_2}\sin\phi + \dfrac{\sqrt{P_Q}}{\sigma_2}\cos\phi\right)\right.$ $\left.+ Q\left(-\dfrac{\sqrt{P_I}}{\sigma_2}\sin\phi + \dfrac{\sqrt{P_Q}}{\sigma_2}\cos\phi\right)\right]$ $\quad n$ even	

Table 6.3 (Continued)

	Biphase/Biphase (I/Q)
PE($d_1 \mid \phi$) (I Channel)	$\dfrac{1}{2}\left[Q\left(\dfrac{\sqrt{P_I}}{\sigma_1} \cos\phi - \dfrac{1}{n}\dfrac{\sqrt{P_Q}}{\sigma_1}\sin\phi \right)\right.$ $\left. + Q\left(\dfrac{\sqrt{P_I}}{\sigma_1} \cos\phi + \dfrac{1}{n}\dfrac{\sqrt{P_Q}}{\sigma_1}\sin\phi \right)\right]$ n odd $\quad Q\left(\dfrac{\sqrt{P_I}}{\sigma_1}\cos\phi \right) \qquad n$ even
PE($d_2 \mid \phi$) (Q Channel)	$\dfrac{n-1}{n} Q\left(\dfrac{\sqrt{P_Q}}{\sigma_2}\cos\phi \right) + \dfrac{1}{2n}\left[Q\left(\dfrac{\sqrt{P_I}}{\sigma_2}\sin\phi + \dfrac{\sqrt{P_Q}}{\sigma_2}\cos\phi \right)\right.$ $\left. + Q\left(-\dfrac{\sqrt{P_I}}{\sigma_2}\sin\phi + \dfrac{\sqrt{P_Q}}{\sigma_2}\cos\phi \right)\right]$ n odd $\quad Q\left(\dfrac{\sqrt{P_Q}}{\sigma_2}\cos\phi \right) \qquad n$ even

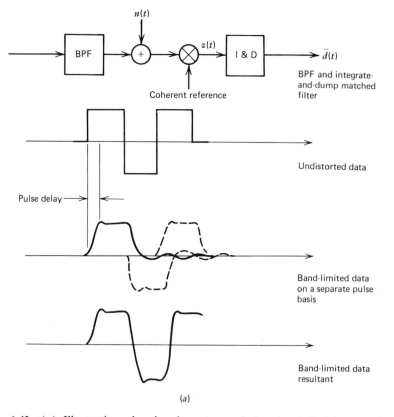

Figure 6.42 (*a*) Illustration showing how transmission band limiting produces intersymbol interference and symbol distortion.

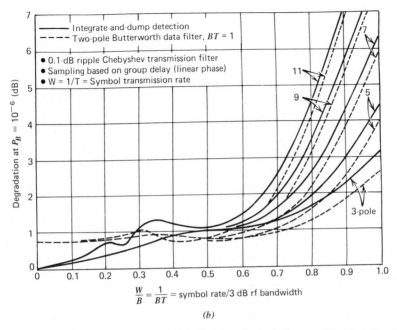

$$\frac{W}{B} = \frac{1}{BT} = \text{symbol rate/3 dB rf bandwidth}$$

(b)

Figure 6.42 Continued. (*b*) Bandwidth limiting degradation of QPSK and BPSK signals (from Jones [44] with permission).

REFERENCES

1 Stiffler, J. J., "Theory of Synchronous Communications," Prentice-Hall, Englewood Cliffs, N.J., 1971.

2 Wozencraft, J. M., and Jacobs, I. M., "Principles of Communication Engineering," Wiley New York, 1965.

3 Lindsey, W. C., and Simon, M. K., "Telecommunication Systems Engineering," Prentice-Hall, Englewood Cliffs, N.J., 1973.

4 Viberbi, A. J., "Advanced Theory of Detection," Lecture notes for Engr. 286A at UCLA, 1965.

5 Helstrom, C. W., "Statistical Theory of Signal Detection," Pergamon, New York, 1960.

6 Weber, C. L., "Elements of Detection and Signal Design," McGraw-Hill, New York, 1968.

7 Viterbi, A. J., "Principles of Coherent Communication," McGraw-Hill, New York, 1966.

8 Golomb, S. W., Ed., "Digital Communications with Space Applications," Prentice-Hall, Englewood Cliffs, N.J., 1964, Chaps. 4 and 7.

9 Birkhoff, G., and MacLane, S., "A Survey of Modern Algebra," MacMillan, New York, p. 223, 1953.

10 Stein, S., and Jones, J. J., "Modern Communication Principles," McGraw-Hill, New York, 1967, Chap. 14.

11 Holmes, J. K., "Equivalence of Bit Error Rates in Word Detection of Quadriphase and Bit Detection of Binary PSK," JPL IOM 331-72-60A, February 8, 1972.

12   Woo, K. T., "Performances of Regenerative and Nonregenerative Satellite Repeaters with MPSK Signaling," ITC 78, Los Angeles, Cal.

13   Smith, E. F., "Attainable Error Probabilities Demodulation of Random Binary PCM/FM Waveform," IRE Trans. Space Electronics and Telemetry, Vol. SET-8, No. 4, Dec. 1962.

14   Kotelnikov, V. A., "The Theory of Optimum Noise Immunity," Transl. by R. A. Silverman, Dover, New York, 1968.

15   Pelchat, M. G., Davis, R. C., and Luntz, M. B., "Coherent Demodulation of Continuous Phase Binary FSK Signals," Proc. Int. Telemetry Conf., Washington, D.C., 1971.

16   Osborne, W. P., and Luntz, M. B., "Coherent and Noncoherent Detection of CPFSK," *IEEE Trans. Communications*, August 1974.

17   DeBuda, R., "Coherent Demodulation of Frequency-Shift Keying with Low Deviation Ratio," *IEEE Trans. Communications*, June 1972.

18   Gronemeyer, S. A., and McBride, A. L., "MSK and Offset QPSK Modulation," *IEEE Trans. Communications*, August 1976.

19   Simon, M. K., "A Generalization of Minimum-Shift-Keying (MSK)-Type Signaling Based Upon Input Data Symbol Pulse Shaping," *IEEE Trans. Communications*, August 1976.

20   Schonhoff, T. A., "Symbol Error Probabilities for M-ary CPFSK: Coherent and Noncoherent Detection," *IEEE Trans. Communications*, June 1976.

21   Forney, G. D., Jr., "The Viterbi Algorithm," *Proc. IEEE*, Vol. 61, pp. 268–278, March 1973.

22   Gronemeyer, S. A., and McBride, A. L., "MSK and Offset QPSK Modulation," *IEEE Trans. Communications*, August 1976.

23   Simon, M. K., "A Generalization of Minimum-Shift-Keying (MSK)-Type Signaling Based Upon Input Data Symbol Pulse Shaping," *IEEE Trans. Communications*, August 1976.

24   Mathwich, H. R., Balcewicz, J. F., and Hecht, M., "The Effect of Tandem Band and Amplitude Limiting on the E_b/N_0 Performance of Minimum (Frequency) Shift Keying (MSK)," *IEEE Trans. Communications*, October 1974.

25   Spilker, J. J., Jr., "Digital Communications by Satellite," Prentice-Hall, Englewood Cliffs, N.J., 1977, Chap. 11.

26   Viterbi, A. J., "Error Bounds for Convolutional Codes and an Asymptotically Optimum Decoding Algirithm," *IEEE Trans. Information Theory*, Vol. IT-13, April 1967.

27   Omura, J. K., "On the Viterbi Decoding Algorithm," *IEEE Trans. Information Theory*, Vol. IT-15, January 1969.

28   Forney, G. D., Jr., "Final Report on a Coding System for Advanced Solar Missions," Contract NASA-3637, Codex Corp., December 20, 1967.

29   McEliece, R. J., "The Theory of Information and Coding," Advanced Book Program, Vol. 3, Addison-Wesley, Reading, Mass., 1977.

30   Batson, B. H., Moorehead, R. W., and Taqui, S. Z. H., "Simulation Results for the Viterbi Decoding Algorithm," NAS S-291, MSC-07027, July 1972.

31   Gilhousen, K. S., Heller, J. A., Jacobs, I. M., and Viterbi, A. J., "Coding Systems Study for High Data Rate Telemetry Links," Linkabit Corp. report, prepared for Contract No. NAS 2-6024, January 1971.

32   Ash, R., "Information Theory," Wiley-Interscience, New York, 1967, Chap. 4.

33   Layland, J. W., "Buffer Parameters and Output Computation in an Optimum Convolutional Decoder," JPL Space Program Summary 37-62, Vol. II, March 31, 1970.

34   Viterbi, A. J., and Odenwalder, J. P., "Further Results on Optimal Decoding of Convolutional Codes," *IEEE Trans. Information Theory*, November 1969.

35 Heller, J. A., and Buchner, E. A., "Error Probability Bounds for Systematic Convolutional Codes," *IEEE Trans. Information Theory*, March 1970.

36 Viterbi, A. J., "Convolutional Codes and Their Performance in Communications Systems," *IEEE Trans. Communication Technology*, Vol. COM-19, No. 5, October 1971.

37 Heller, J. A., and Jacobs, I. M., "Viterbi Decoding for Satellite and Space Communication," *IEEE Communication Technology*, October 1971.

38 Odenwalder, J. D., "Optimum Decoding of Convolutional Codes", Ph.D. Dissertation, System Science Dept., University of California, Los Angeles, Cal. 1970.

39 Ramsey, J. L., "Realization of Optimum Interleavers", *IEEE Trans. Information Theory*, May 1970.

40 Forney, G. D., "The Viterbi Algorithm," *Proc. IEEE*, March 1973.

41 Holmes, J. K., unpublished memo, dated December 12, 1978.

42 Osborne, H. C., "Revised BER Degradation Curves for Viterbi Decoder (BPSK Signal)," TRW IOC 78-7327.01-112, December 15, 1978.

43 Osborne, H. C., "Effect of Noisy Reference on Coherent Detection of Unbalanced Signals," *Proc. NTC 78*, Birmingham, Ala., December 3 to 5, 1978.

44 Jones, J. J., "Filter Distortion and Intersymbol Interference Effects on PSK Signals," *IEEE Trans. Communication Technology*, April 1971.

45 Stein, S., and Jones, J. J., "Modern Communication Principles with Application to Digital Signals," McGraw-Hill, New York, 1967.

46 Booth, R., "Carrier Phase and Bit Sync Regeneration for the Coherent Demodulation of MSK," Session 6.1, NTC 78, Birmingham, Ala., December 3 to 6, 1978.

47 Heller, J., "Short Constraint Length Convolutional Codes," Jet Propulsion Laboratory, SPS 37-50, Vol. III, 1968.

7

AN INTRODUCTION TO LINEAR PSEUDONOISE SEQUENCES

Since a good deal of the following chapters deals with pseudonoise (PN)(maximal length, direct, *m*, etc.) sequences, we develop appropriate theory in this chapter to specify the tap connections of the shift register in order to produce the desired sequences. Most all the theory will be directed toward linear sequences, which will be defined shortly. These sequences have a very well-developed theory associated with them, which is not the case for nonlinear shift register sequences.

Before we delve into the theory of linear sequences we shall mention a few words about their history. Perhaps the first work was done by Gilbert of Bell Labs [1]. However his report had a limited circulation. The work of Golomb [2] and Zierler [3] appeared slightly later. Welch was also active in this area around this time. In addition, Birdsall and Ristenbatt published their work while at the University of Michigan [4]. Golomb [5] credits the original theory of mod *p* addition to Lagrange in the eighteenth century along with a modern mathematical exposition by Hall in 1937. More recently, Gold [6], [7] has developed near optimum shift register sequences for code division multiple access applications.

The following 20 years from the work of Gilbert has shown an amazing growth in the use of linear shift register codes in virtually all types of communication systems, including both military and nonmilitary systems. Recent non military examples include the space shuttle and the tracking data relay satellite system.

Applications of linear shift register sequences include ranging, multiple access, spectrum spreading encipherment, privacy encoding, random bit generators, jamming protection, low detectability radar, and voice and data scrambling. Undoubtedly other applications have been missed, and certainly new ones will arise.

7.1 SHIFT REGISTER SEQUENCES

We now consider in some detail the sequences generated by linear shift registers. The influence of references 2, 4, 5, and 11 is very considerable in this chapter. When referring to a shift register sequence we will mean the sequence of "ones" and "zeros" that will be emitted from the generator, as seen in Figure 7.1*a*. The corresponding time waveform, having values of ± 1, is illustrated in Figure 7.1*b*.

First consider a *simple shift register*, illustrated in Figure 7.2. It is composed of a shift register, for storage and shifting, a mod 2 adder, and a feedback line. Let us define a mod 2 addition. This is most easily done by observing the addition table for the two elements, "zero" and "one," shown in Figure 7.3. Except for $1 \oplus 1 = 0$, the addition follows the rules of ordinary addition. With the mod 2 adder defined to satisfy Figure 7.3 we can commence to discuss how the shift register generator of Figure 7.2 operates. Each storage unit denoted by the numbers 1 through 6 stores a "one" or "zero" until the next clock pulse arrives producing the following changes. The binary digit of cell 1 is transferred to cell 2, the previous contents of cell 2 (not the cell 1 value) is transferred to cell 3, and so on. Finally, the contents of cells 5 and 6 are mod 2 added, and this result is fed back to cell 1. Thus, a shift register is a storage unit that moves its stored contents one position to the right (in this case) for each shift pulse applied to the shift register. The *shift register sequence* is defined to be the output of the last cell.

Assume that the first five cells are loaded with "zeros" and the sixth with a "one." After the first clock pulse, the register will contain 100000, with the "one" of cell 6 being the first digit of the shift register generator (SRG) sequence. The "one" of cell 1 is the result of the mod 2 sum of "zero" and "one." The resulting output sequence will be

$$100000100001100010100111101000111001001011011101100110101011111$$
$$(7.1\text{-}1)$$

We note that there are 63 elements in the sequence, and that it will repeat the same 63 elements over and over again. This is an example of a simple shift register generator (SSRG). In an SSRG, all the feedback signals are returned to a single input. Another type of shift register generator, a multiple-return shift register generator (MRSRG), has adder outputs to two or more input stages. Figure 7.4 illustrates an example of an MRSRG.

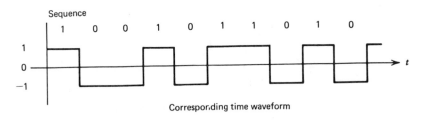

Figure 7.1 Difference between the shift register sequence and waveform.

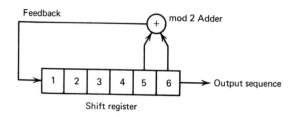

Figure 7.2   A simple shift register generator.

Birdsall and Ristenbatt [4] have shown that every multiple-return generator that has no transients possesses an equivalent SSRG. As a consequence, almost all of the remaining work in this chapter will be based on SSRGs.

If an SRG produces a sequence that goes through a transient series of digits before settling into a periodic sequence, then that generator is said to possess *transients*. Figure 7.5 illustrates two examples of transient generators.

If we denote the contents of each cell in a simple SRG by x_i, with i the cell number, then we say the shift register generator is *linear* if the feedback function can be expressed as a mod 2 sum. Let the Boolean function (a function of n binary input variables and one binary output variable) of the feedback be denoted by $f(x_1, x_2, \ldots, x_n)$, as shown in Figure 7.6.

The shift register is linear since $f(x_1, x_2, x_3, \ldots, x_n)$ is expressible as a mod 2 sum of the contents with $C_i = 0$ or 1 being the feedback connection coefficients. All SRGs in this chapter will be linear unless specifically stated otherwise.

As an example of a nonlinear shift register, consider Figure 7.7. In part

Figure 7.3   Addition table for mod 2 addition.

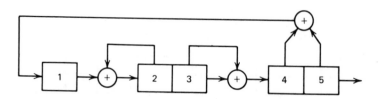

Figure 7.4   An example of a MRSRG.

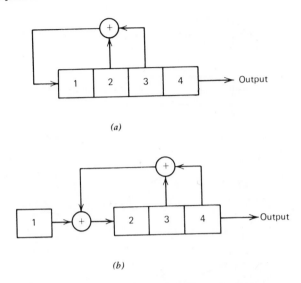

(a)

(b)

Figure 7.5 Two examples of transient generators. (a) Initial condition 1000, output 000101110010111. (b) Initial condition 1000, output 000101110010111.

a of the figure, the SRG is shown, and in part b the state diagram is shown. Starting in state 110 (cell 1 contains a "one" cell 2 a "one" and cell 3 a "zero,") we see that $x_2x_3 = 1 \cdot 0 = 0$, so the new state becomes 011. This process continues until state 000 is reached. From this point on, the SRG will remain in state 000 forever. If, however, the register is loaded with all "ones" so that it is in state 111, then after each clock pulse the register remains in state 111. One more observation is noteworthy, and that is the fact that state 010 can be reached from two distinct predecessors, 101 and 100. Golomb [5] has shown that for linear shift registers every state has exactly one predecessor.

All the material to follow in this chapter deals with linear SRG sequences.

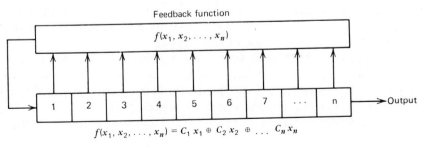

Feedback function

$f(x_1, x_2, \ldots, x_n)$

$$f(x_1, x_2, \ldots, x_n) = C_1 x_1 \oplus C_2 x_2 \oplus \ldots C_n x_n$$

Figure 7.6 A linear SRG.

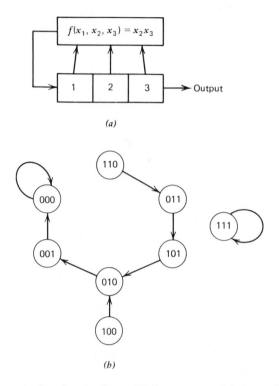

(a)

(b)

Figure 7.7 An example of a nonlinear SRG sequence. (*a*) A nonlinear SRG. (*b*) State diagram of the nonlinear SRG.

7.2 PROPERTIES OF SRG SEQUENCES

A shift register generator produces sequences that depend upon register length, feedback tap connections, and initial conditions. Due to this fact, it is convenient to group the output sequences into two types: *maximal length sequences** and *nonmaximal* length sequences. Maximal sequences have the property that if the SRG has L stages, the sequence length is $2^L - 1$. If the sequence is less than $2^L - 1$, then it is a nonmaximal length sequence. The sequence generated from the SSRG of Figure 7.2 is a maximal length sequence since its length is $2^6 - 1 = 63$. All six-tuples are contained in this sequence except the sequence of six "zeros."

For a given shift register length, the feedback connections determine whether the sequence will be maximal or not. For nonmaximal length sequences, the initial conditions (that is, the initial sequence loaded into the shift register) determine which sequence is generated.

Now consider a theorem important for recognizing maximal sequences.

*Sometimes referred to as maximum length.

Theorem 1 It is not possible to generate a maximal length sequence from an SSRG that has an odd number of taps.

Proof Assume the SRG is loaded with all "ones." Then, clearly, the SRG will remain in the all-one state. Since there are 2^L total possible states and since the all-zero state has period 1 and the all-one state has period 1, the maximum length sequence for an SRG with an odd number of taps satisfies $2^L - 2 < 2^L - 1$. Since this is 1 less than maximal length, we conclude that it cannot be maximal length. In fact, the longest sequence may be much shorter than $2^L - 2$.

If our goal is to find maximal length sequences, we can discard all generators that utilize an odd number of feedback taps, which eliminates about half of the possible SRGs.

Now consider generating the reverse of a specified sequence. In general, if the original sequence is $\ldots, a_k, a_{k+1}, a_{k+2}, \ldots$, then the reverse is $\ldots, a_{k+2}, a_{k+1}, a_k, \ldots$. In some cases the original sequence and the reverse sequence are identical, so the sequence is called *self-reverse*. The procedure for finding the reverse SSRG is easily done. If an n-stage SSRG has feedback taps on stages $n, k, m, \ldots$, the reverse generator will have feedback taps on stages $n, n-k, n-m, \ldots$, etc. $(n > k > m \ldots)$. To illustrate this point, let us consider an example.

In Figure 7.8, the SSRG of Figure 7.2 is reproduced along with the reverse generator. Since the taps are on stages 6 and 5, the reverse generator will have taps on stages 6 and $6-5 = 1$. The sequences from both the original SRG and the reverse SRG are shown in Figure 7.8. As can be seen, the sequences are reverses of one another.

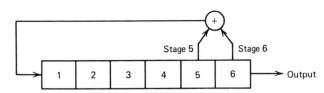

100000100001100010100111101000111001001011011101100110101011111

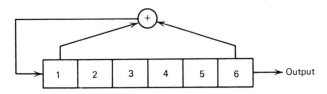

100000111111010101100110111011010010011100010111100101000110000

Figure 7.8 A maximal length SSRG and its reverse SSRG. (*a*) Original SSRG and its sequence. (*b*) Reverse SSRG and its sequence.

If a nonmaximal generator were used in the example, we would find each sequence from the reverse generator to be the reverse of one of the sequences from the other generator. Birdsall and Ristenbatt [4] have shown that for an n-stage SSRG, $2n - 1$ digits observed without error from the SSRG are sufficient to determine uniquely the feedback connections. In communications work, the trouble with this result is that most maximal length sequences have a very low chip energy-to-noise spectral density, so observations without error are unrealistic.

Now we develop the shift-and-add property for linear SSRGs. First denote the sequence $\{a_k\} = \{a_0, a_1, a_2, \ldots\}$ as the output sequence for the SSRG. In Figure 7.9, a linear SSRG is shown with the sequence and feedback weights shown as $\{C_i\}$. We will use $a_k = a(k)$ interchangeably in the remainder of the chapter. The *fundamental linear recursion relation* relating the feedback sequence element a_k and the contents of the register is

$$a(k) = C_1 a(k - 1) + C_2 a(k - 2) + \cdots + C_n a(k - n)$$

$$= \sum_{i=1}^{n} C_i a(k - i) \qquad k \geq n \tag{7.2-1}$$

where C_i is either a "one" if the switch is closed (feedback is connected) or a "zero" if the switch is not closed. The sum is mod 2. The values $a(0)$, $a(1), a(2), \ldots, a(n - 1)$ are termed the *initial conditions*. It follows that the output of a linear SSRG having n stages has the property that any term after the nth term is a fixed linear mod 2 sum of the n preceding terms of the sequence.

We now establish that the mod 2 sum, term by term, of two output sequences from the same shift register is itself another sequence that can be generated from the SRG with appropriate initial conditions.

Theorem 2   If $\{a\}$ and $\{b\}$ are two output sequences of a linear SRG, then so is $\{a\} + \{b\}$ under mod 2 addition.

Proof   We note that $\{a\} + \{b\}$ denotes term by term addition of each sequence. Since $\{a\}$ and $\{b\}$ satisfy the recursion relation, (7.2-1), we have

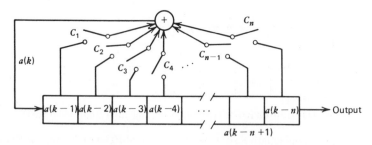

Figure 7.9   Linear SSRG model.

that

$$a(k) = \sum_{i=1}^{n} C_i a(k-i) \qquad (7.2\text{-}2)$$

$$b(k) = \sum_{i=1}^{n} C_i b(k-i) \qquad (7.2\text{-}3)$$

To show that $\{a\}+\{b\}$ is a derived sequence, we must show that $a(k)+b(k)$ satisfies the same recursion relationship. Adding, we have

$$a(k) + b(k) = \sum_{i=1}^{n} C_i[a(k-i)+b(k-i)] \qquad (7.2\text{-}4)$$

Therefore, $\{a\}+\{b\}$ satisfies the same linear recursion. This is clear since $\{a\}+\{b\}=\{d\}$ is an SRG having the initial conditions $a(k-i)+b(k-i)\,(\text{mod } 2)$ for $i=1,\ldots,n$.

We have just proved the closure proprety for linear SRG sequences. We have, in fact, proven the shift-and-add property of maximal SRG sequences:

Theorem 3 If a maximal SRG sequence is added to a proper* phase shift of itself, then the resulting sequence is another shift of the original sequence.

Proof Since a maximal sequence and all its proper $2^n - 2$ phase shifts are the only sequences generated by a maximal SRG, it follows from Theorem 7.2.2 that the sum of the SRG sequence and any proper phase shift must be a phase shift of the same sequence.

One might be tempted to assume that the word maximal· could be removed from the conditions of the theorem. This is not possible since only particular shifts will produce the shift-and-add property.

To illustrate the above points, consider two examples. First, a maximal SRG and its output sequence with a few proper phase shifts are shown in Figure 7.10. The SRG, the state diagram, and the result of the sum of the sequence plus one particular phase are shown. As seen in part c of the figure, the resulting sum sequence is just a shifted version of the original sequence.

In the second example, shown in Figure 7.11, the SRG sequence is not maximal. There are six distinct subsequences that can be obtained by the appropriate initial conditions (Figure 7.11c). Notice one subsequence, when added to the appropriate shift, produces a shifted version of the original sequence (upper portion of Figure 7.11b). However, the lower

---

*A proper phase is all phase shifts less than $2^n - 1$ excluding the zero phase shift.

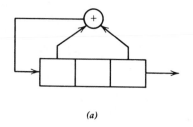

(a)

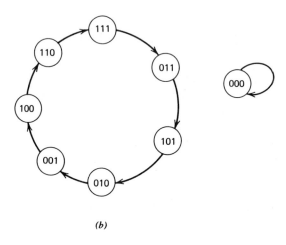

(b)

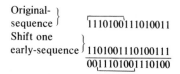

Resulting sequence—two late

Figure 7.10 Demonstration of the shift-and-add property of a maximal SRG sequence. (*a*) A maximal length SRG, $n = 3$. (*b*) The associated state diagram. (*c*) Shift-and-add property of a maximal sequence.

portion of Figure 7.11*b* illustrates a case when the shift-and-add property does not apply. In fact, the sum of subsequence 1 and a shifted version produces subsequence 2, which has a period of 3, although the original sequence has a period of 6.

Before we present the mathematical formulation of SRG sequences, we mention one more property of maximal SRG sequences.

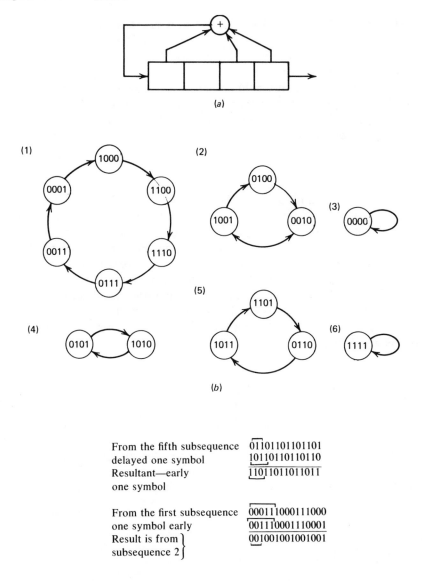

Figure 7.11   An example of a nonmaximal SRG sequence. (a) A nonmaximal SRG. (b) The associated state diagram. (c) Lack of shift-and-add property of a nonmaximal SRG sequence.

The *balance property* of maximal SRG sequences states that there is one more "one" than "zero" in a full period of a maximal sequence. This follows since all states but the all "zero" state are reached in a maximal SRG sequence, hence, there must be 2^{n-1} "ones" and $2^{n-1} - 1$ "zeros"; therefore there is one more "one" than "zero."

7.3 MATHEMATICAL CHARACTERIZATION OF SRGS

In this section the matrix point of view will be exploited to develop the mathematical basis of SRGs [4, 8, 9].

It seems feasible that it would be possible to characterize the action of an SRG by a matrix operation operating on the contents of the register viewed by an n-dimensional vector. Recall that the field of the coefficients will be the integers 0 and 1 (mod 2).

7.3.1 The A Matrix

The A matrix, when multiplied by the n column contents vector, yields the contents of generator after one shift. To form the A matrix, we first number the stages of the register from 1 to n, proceeding in the same direction as 'the contents travels under shifting. For an n-stage SRG, the matrix will be $n \times n$. Identify each row of the A matrix with the input to the corresponding stage in the register. In each row of the matrix, we will enter a "one" for each stage that feeds (including the feedback) the stage corresponding to that row. To clarify the above formulation, reconsider the SRG of Figure 7.11a. We see that we have, from Figure 7.12, that the A matrix is given by

$$A = \begin{bmatrix} 1 & 0 & 1 & 1 \\ 1 & 0 & 0 & 0 \\ 0 & 1 & 0 & 0 \\ 0 & 0 & 1 & 0 \end{bmatrix} \tag{7.3-1}$$

We notice that for a simple SRG that there will always be a diagonal of "ones" just below the main diagonal of the matrix. The method of constructing the A matrix also applies to multiple return generators using

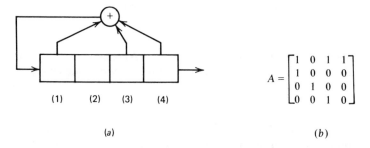

(a)   (b)

Figure 7.12 Example for the construction of the A matrix. (a) Linear SRG used to construct the A matrix. (b) A matrix for the SRG of (a).

the same method. Figure 7.13 illustrates an example of an MRSRG with its
A matrix.

Now denote $x_i(j)$ as the contents of the ith shift register stage after the
jth shift. Then, for an SSRG, we have clearly:

$$
\begin{aligned}
x_1(j+1) &= C_1 x_1(j) + C_2 x_2(j) + C_3 x_3(j) + \cdots + C_n x_n(j) \\
x_2(j+1) &= x_1(j) \\
x_3(j+1) &= \qquad\qquad x_2(j) \\
x_4(j+1) &= \qquad\qquad\qquad\quad x_3(j) \\
&\;\;\vdots \\
x_n(j+1) &= \qquad\qquad\qquad\qquad\qquad\qquad x_{n-1}(j)
\end{aligned}
\tag{7.3-2}
$$

or

$$
x_i(j+1) = \sum_{k=1}^{n} a_{ik} x_k(j) \qquad j = 0, 1, 2, \cdots \tag{7.3-3}
$$

with a_{ik} the elements of the A matrix. In matrix notation

$$
X(j+1) = AX(j) \tag{7.3-4}
$$

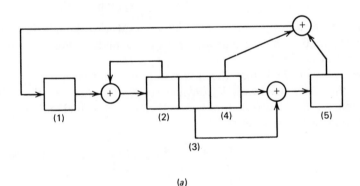

(a)

$$
A = \begin{bmatrix}
0 & 0 & 0 & 1 & 1 \\
1 & 1 & 0 & 0 & 0 \\
0 & 1 & 0 & 0 & 0 \\
0 & 0 & 1 & 0 & 0 \\
0 & 0 & 1 & 1 & 0
\end{bmatrix}
$$

(b)

Figure 7.13 A matrix for an MRSRG. (a) MRSRG. (b) A matrix.

where

$$A = \{a_{jk}\}$$

$$X(j) = \begin{Bmatrix} x_1(j) \\ x_2(j) \\ x_3(j) \\ \vdots \\ x_n(j) \end{Bmatrix} \qquad (7.3\text{-}5)$$

Hence, the state of the kth stage at time j is the kth component in the $X(j)$ column vector.

Since

$$X(j+1) = AX(j) \qquad (7.3\text{-}6)$$

it follows that

$$X(j+m) = [A]^m X(j) \qquad (7.3\text{-}7)$$

It is seen that when $A^m = I$ the vectors $X(j+m) = X(j)$, so that the contents of the register are the same after j shifts as after $j+m$ shifts. Clearly, if the sequence is maximal length, we must have

$$A^L = A^{2n-1} = I \qquad \text{for} \qquad X(j) \neq 0 \qquad \forall j \qquad (7.3\text{-}8)$$

Since the order of a matrix is the lowest power of the matrix that yields the identity matrix, we see that for a maximal length n-stage SRG the order of the matrix is $2^n - 1$.

Now we will show that the inverse of the matrix A is equivalent to running the SRG backwards! Consider an SSRG that has the nth stage connected to the feedback function (otherwise it would not be a proper n-stage register). The general form is

$$A = \begin{bmatrix} C_1 & C_2 & C_3 & \cdots & C_{n-1} & 1 \\ 1 & 0 & 0 & \cdots & & 0 \\ 0 & 1 & 0 & \cdots & & 0 \\ 0 & 0 & 1 & \cdots & & 0 \\ \vdots & \vdots & \vdots & \vdots & & \vdots \\ 0 & 0 & 0 & \cdots & 1 & 0 \end{bmatrix} \quad n \times n \qquad (7.3\text{-}9)$$

We now demonstrate that the inverse A^{-1} is given by

$$A^{-1} = \begin{bmatrix} 0 & 1 & 0 & & \cdots & 0 \\ 0 & 0 & 1 & & \cdots & 0 \\ 0 & 0 & 0 & 1 & \cdots & 0 \\ \vdots & \vdots & \vdots & \vdots & & \vdots \\ 0 & 0 & 0 & 0 & \cdots & 1 \\ 1 & C_1 & C_2 & C_3 & \cdots & C_{n-1} \end{bmatrix} \quad n \times n \qquad (7.3\text{-}10)$$

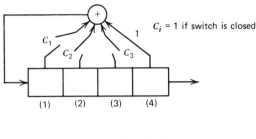

$$A = \begin{bmatrix} C_1 & C_2 & C_3 & 1 \\ 1 & 0 & 0 & 0 \\ 0 & 1 & 0 & 0 \\ 0 & 0 & 1 & 0 \end{bmatrix}$$

(a)

$$A^{-1} = \begin{bmatrix} 0 & 1 & 0 & 0 \\ 0 & 0 & 1 & 0 \\ 0 & 0 & 0 & 1 \\ 1 & C_1 & C_2 & C_3 \end{bmatrix}$$

(b)

Figure 7.14 A four-stage SSRG illustrating A and A^{-1}. (a) Forward operating SSRG and its A matrix. (b) The inverse of A.

Now the product is given by

$$A^{-1}A = \begin{bmatrix} 1 & 0 & 0 & \cdots & 0 \\ 0 & 1 & 0 & \cdots & 0 \\ 0 & 0 & 1 & \cdots & 0 \\ \vdots & \vdots & \vdots & \vdots & \vdots \\ 0 & 0 & 0 & 0 & 0 \\ 2C_1 & 2C_2 & 2C_3 \cdots & 2C_{n-1} & 1 \end{bmatrix}$$

(7.3-11)

But since $2C_i = 0 \pmod 2$, we have

$$A^{-1}A = I \qquad n \times n \qquad (7.3\text{-}12)$$

Hence, A^{-1}, as defined, is the inverse of A. To show that the inverse runs the generator backwards, we consider an example. For simplicity we assume that the SRG has four stages. The shift register is shown in Figure 7.14a, along with its A matrix. At time $j + 1$, the contents of the register is given by

$$X(j+1) = AX(j) = \begin{bmatrix} C_1 & C_2 & C_3 & 1 \\ 1 & 0 & 0 & 0 \\ 0 & 1 & 0 & 0 \\ 0 & 0 & 1 & 0 \end{bmatrix} \begin{bmatrix} x_1(j) \\ x_2(j) \\ x_3(j) \\ x_4(j) \end{bmatrix}$$

(7.3-13)

$$= \begin{cases} C_1 x_1(j) + C_2 x_2(j) + C_3 x_3(j) + x_4(j) \\ x_1(j) \\ x_2(j) \\ x_3(j) \end{cases} \qquad (7.3\text{-}14)$$

Now to back up the contents one stage, we evaluate

$$A^{-1}X(j+1) = \begin{bmatrix} 0 & 1 & 0 & 0 \\ 0 & 0 & 1 & 0 \\ 0 & 0 & 0 & 1 \\ 1 & C_1 & C_2 & C_3 \end{bmatrix} \begin{cases} C_1 x_1(j) + C_2 x_2(j) + C_3 x_3(j) + x_4(j) \\ x_1(j) \\ x_2(j) \\ x_3(j) \end{cases}$$

$$(7.3\text{-}15)$$

or

$$A^{-1}X(j+1) = \begin{cases} x_1(j) \\ x_2(j) \\ x_3(j) \\ C_1 x_1(j) + C_2 x_2(j) + C_3 x_3(j) + x_4(j) \\ + C_1 x_1(j) + C_2 x_2(j) + C_3 x_3(j) \end{cases} = X(j) \qquad (7.3\text{-}16)$$

and therefore we conclude that A^{-1} "runs" the generator backwards.

PROBLEM 1

Consider the register shown below.

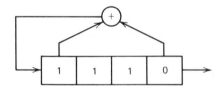

Using the techniques of this section, determine the register loading one clock time earlier and two clock times earlier than the indicated loading.

We now consider the characteristic equation and the characteristic polynomial.

7.3.2 The Characteristic Equation and Polynomial

We will find the characteristic polynomial very important in our following study of sequences. First, we define the characteristic equation. For any $n \times n$ matrix A the *characteristic equation* is formed by setting the determinate of $[A - \lambda I]$ to zero (where I is an $n \times n$ unit matrix, and λ is a parameter), which we denote by

$$|A - \lambda I| = 0 \qquad (7.3\text{-}17)$$

As an example, let

$$A = \begin{bmatrix} C_1 & C_2 & C_3 & 1 \\ 1 & 0 & 0 & 0 \\ 0 & 1 & 0 & 0 \\ 0 & 0 & 1 & 0 \end{bmatrix} \qquad (7.3\text{-}18)$$

Then

$$|A - \lambda I| = \begin{vmatrix} C_1 - \lambda & C_2 & C_3 & 1 \\ 1 & -\lambda & 0 & 0 \\ 0 & 1 & -\lambda & 0 \\ 0 & 0 & 1 & -\lambda \end{vmatrix}$$

$$= \lambda^4 + C_1 \lambda^3 + C_2 \lambda^2 + C_3 \lambda + 1 = 0 \qquad (7.3\text{-}19)$$

since $-1 = +1$, mod 2.

Now consider the general $n \times n$ $[A - \lambda I]$ matrix. The *characteristic equation* is given by [5]

$$F(\lambda) = \begin{vmatrix} C_1 - \lambda & C_2 & C_3 & C_4 & \cdots & 1 \\ 1 & -\lambda & 0 & 0 & \cdots & 0 \\ 0 & 1 & -\lambda & 0 & \cdots & 0 \\ 0 & 0 & 0 & -\lambda & \cdots & \\ \vdots & \vdots & \vdots & \vdots & & \vdots \\ 0 & 0 & & & 1 - \lambda \end{vmatrix} = 0 \qquad (7.3\text{-}20)$$

or

$$F(\lambda) = (C_1 - \lambda)(-\lambda)^{n-1} - C_2(-\lambda)^{n-2} + C_3(-\lambda)^{n-3} + \cdots + (-1)^n \qquad (7.3\text{-}21)$$

after simplifying

$$F(\lambda) = \lambda^n + C_1 \lambda^{n-1} + C_2 \lambda^{n-2} + C_3 \lambda^{n-3} + \cdots + 1 \qquad (7.3\text{-}22)$$

Therefore, we can summarize our result in Figure 7.15b. It is convenient to define $C_0 = 1$ for the summary. As an example, suppose a six-stage SSRG has taps on the first, third, and sixth stages. Therefore, $C_1 = 1$, $C_3 = 1$, and $C_6 = 1$. Using (7.3-22) or Figure 7.15, we find that the characteristic equation is $F(\lambda) = \lambda^6 + \lambda^5 + \lambda^3 + 1 = 0$. Notice that we have assumed that $C_n = 1$. If this were not the case, we would have, at most, an $(n-1)$-stage SSRG.

At this point, it is worth pointing out that by the Caley-Hamilton theorem [11] an $n \times n$ matrix satisfies its own characteristic equation. Therefore

$$F(A) = 0 \qquad (7.3\text{-}23)$$

For example, for the SSRG of Figure 7.8a, we have

$$A^6 + A + I = 0 \qquad (7.3\text{-}24)$$

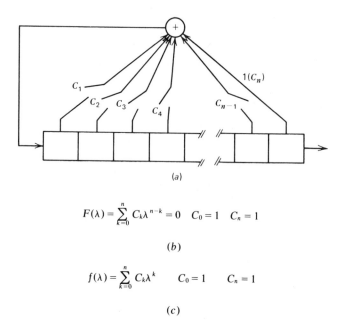

(a)

$$F(\lambda) = \sum_{k=0}^{n} C_k \lambda^{n-k} = 0 \quad C_0 = 1 \quad C_n = 1$$

(b)

$$f(\lambda) = \sum_{k=0}^{n} C_k \lambda^{k} \quad C_0 = 1 \quad C_n = 1$$

(c)

Figure 7.15 The characteristic equation and polynomial of an SSRG. (a) n-Stage SSRG. (b) Corresponding characteristic equation. (c) Corresponding characteristic polynomial.

where I is an $n \times n$ unit matrix. The *characteristic polynomial* is defined* by

$$f(\lambda) = \sum_{k=0}^{n} C_k \lambda^{k} \quad C_0 = 1 \qquad (7.3\text{-}25)$$

This is summarized on Figure 7.15 also. Continuing our example, $C_1 = 1$, $C_3 = 1$, $C_6 = 1$, all the rest of the C_i are zero, which produces the following characteristic polynomial:

$$f(\lambda) = 1 + \lambda + \lambda^3 + \lambda^6 \qquad (7.3\text{-}26)$$

We will find the characteristic polynomial a very important tool for the study of SSRG sequences and their properties.

7.4   THE GENERATING FUNCTION AND THE SEQUENCE LENGTH

As we have seen in the last few sections, the matrix approach was very useful for characterizing the state of the SRG as a function of time. A powerful tool in the analysis and characterization of the output sequence is

*However, reference 4 defines it as $\sum_{k=0}^{n} C_k \lambda^{n-k}$.

the generating function. Again denote the output SSRG sequence by $\{a_m\} = \{a_0, a_1, a_2, \ldots\}$, where the index m denotes time, that is, a_0 occurs initially, then a_1, then a_2, and so on. Then, the *generating function* of the output sequence is given by

$$G(x) = \sum_{k=0}^{\infty} a_k x^k \qquad (7.4\text{-}1)$$

where x is a real variable. The *initial state* of the register is defined as a_{-1}, $a_{-2}, \ldots, a_{-n+1}, a_{-n}$ for an n-stage register. Then, a_0 to a_n are defined by the *recurrence relation* (see Figure 7.16)

$$a_m = \sum_{i=1}^{n} C_i a_{m-i} \qquad (7.4\text{-}2)$$

Using (7.3–1) in (7.3–2), we have

$$\begin{aligned}
G(x) &= \sum_{k=0}^{\infty} \sum_{i=1}^{n} C_i a_{k-i} x^k \\
&= \sum_{i=1}^{n} C_i x_i \sum_{k=0}^{\infty} a_{k-i} x^{k-i} \\
&= \sum_{i=1}^{n} C_i x_i \left(a_{-i} x^{-i} + a_{-i+1} x^{-i+1} + \cdots + a_{-1} x^{-1} + \sum_{k=0}^{\infty} a_k x^k \right) \qquad (7.4\text{-}3)
\end{aligned}$$

The last term on the right of (7.4-3) is $G(x)$, so

$$G(x) = \sum_{i=1}^{n} C_i x^i [a_{-i} x^{-i} + \cdots + a_{-1} x^{-1} + G(x)] \qquad (7.4\text{-}4)$$

Or, solving for the generating function, we obtain

$$G(x) = \frac{\sum_{i=1}^{n} C_i x^i [a_{-i} x^{-i} + a_{-i+1} x^{-i+1} + \cdots + a_{-1} x^{-1}]}{\sum_{i=0}^{n} C_i x^i} \qquad (7.4\text{-}5)$$

We again have used the fact that $+1 = -1 \mod 2$ and conveniently defined

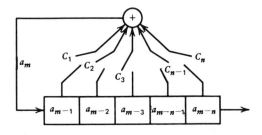

Figure 7.16 SSRG model for the recurrence relation.

$C_0 = 1$ (there is no "zeroth" tap, of course). This form of $G(x)$ has been expressed in terms of the initial state of the register $a_{-1}, a_{-2}, \ldots, a_{-n}$ and the feedback tap coefficients $C_1, C_2, \ldots, C_n$. The denominator is the characteristic polynomial of the sequence $\{a_n\}$ as was defined in Section 7.3.2.

When we perform the long division indicated by the expression $G(x)$, we obtain the series $a_0 + a_1 x + a_2 x^2 + \cdots$ so that the coefficient of x^m is the mth term in the output sequence.

Notice when $a_{-1} = a_2 = \cdots = a_{1-n} = 0$ and $a_{-n} = 1$, $G(x)$ reduces to $(C_n = 1)$

$$G(x) = \frac{1}{f(x)} \tag{7.4-6}$$

Now consider a simple example. Let the SSRG have three stages with feedback tap connections on the first and third stages, as shown in Figure 7.17. Let the initial conditions be given by $a_{-1} = 0$, $a_{-2} = 0$, $a_{-3} = 1$, as shown in the figure. The function $G(x)$ is given by

$$G(x) = \frac{C_1 x(a_{-1}x^{-1}) + C_2 x^2(a_{-2}x^{-2} + a_{-1}x^{-1}) + C_3 x^3(a_{-3}x^{-3} + a_{-2}x^{-2} + a_{-1}x^{-1})}{1 + x + x^3} \tag{7.4-7}$$

Using the initial conditions in (7.4-7) leads to the division

$$
\begin{array}{r}
1 \quad 1 \quad 1 \quad 0 \quad 1 \quad 0 \quad 0 \quad 1 \quad 1 \\
1 + x + x^2 + x^4 + x^7 + x^8 \\
\end{array}
$$

$$
1 + x + x^3 \overline{\bigl)1}
$$

$$
\begin{array}{r}
\underline{1 + x + x^3} \\
x + x^3 \\
\underline{x + x^2 + x^4} \\
x^2 + x^3 + x^4 \\
\underline{x^2 + x^3 + x^5} \\
x^4 + x^5 \\
\underline{x^4 + x^5 + x^7} \\
x^7 \\
\underline{x^7 + x^8 + x^{10}} \\
x^8 + x^{10} \\
\underline{x^8 + x^9 + x^{11}} \\
x^9 + x^{10} + x^{11}
\end{array}
\tag{7.4-8}
$$

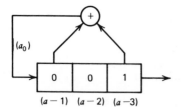

$(a-1) \quad (a-2) \quad (a-3)$

Figure 7.17   SSRG having a maximal length sequence.

Therefore, the output sequence is

$$a_0a_1a_2a_3a_4a_5a_6a_7a_8\cdots$$
$$1\ 1\ 1\ 0\ 1\ 0\ 0\ 1\ 1\ \cdots \qquad (7.4\text{-}9)$$

Notice that the first digit of the sequence is a_0, not a_{-3}! Notice also the variable x^k acts like a k-step delay operator. If we had loaded the register with $a_{-n} = a_{-n+1} = a_{-1} = 0$, the resulting long division would produce $0000000\ldots$, as is easily verified from Figure 7.17.

PROBLEM 2

Determine the mathematical function that allows one to determine the kth symbol in the output sequence. *Hint*: consider the appropriate derivatives.

PROBLEM 3

Using the SSRG shown below,
(a) Find the characteristic equation
(b) Find the characteristic polynomial.
(c) Using (7.3-4), determine the output sequence when the register is initially loaded with $a_{-1} = a_{-2} = a_{-3} = 0$ and $a_{-4} = 1$.
(d) Verify your result by directly computing the states using the recursion relation.

Now we establish a few theorems on the period of the sequence and the polynomial $1 + x^p$. Denote the fact that $f(x)$ *divides* $1 + x^p$ by $f(x)/1 + x^p$.

Theorem 4 If the sequence $\{a_n\}$ has period p, then $f(x)/1 + x^p$.

Proof Denote the numerator of $G(x)$ by $g(x)$, which is, at most, of degree $n - 1$. Pick $a_{-1} = a_{-2} = \cdots = a_{1-n} = 0$ and $a_{-n} = 1$. Then

$$G(x) = \frac{1}{f(x)} = \sum_{n=0}^{\infty} a_n x^n \qquad (7.4\text{-}10)$$

or

$$\frac{1}{f(x)} = a_0 + a_1x + \cdots + a_{p-1}x^{p-1} + x^p(a_0 + a_1x + \cdots + a_{p-1}x^{p-1})$$
$$+ x^{2p}(a_0 + a_1x + \cdots + a_{p-1}x^{p-1}) + \cdots$$
$$= (a_0 + a_1x + a_2x^2 + \cdots + a_{p-1}x^{p-1})(1 + x^p + x^{2p} + \cdots)$$
$$= \frac{a_0 + a_1x + a_2x^2 + \cdots + a_{p-1}x^{p-1}}{1 + x^p} \qquad (7.4\text{-}11)$$

since

$$(1 + x^p + x^{2p} + \cdots) = \frac{1}{1 - x^p} = \frac{1}{1 + x^p} \quad \text{mod } 2 \qquad (7.4\text{-}12)$$

Rearranging

$$\frac{(1+x^p)}{f(x)} = a_0 + a_1x + a_2x^2 + \cdots + a_{p-1}x^{p-1} \tag{7.4-13}$$

It follows that $f(x)/1 + x^p$.

Theorem 5   If $f(x)/1 + x^p$, then $\{a_n\}$ has period p with the initial condition $0, 0, \cdots 0, 1$ if p is the smallest positive integer p such that $f(x)/1 + x^p$.

Proof   We assume $f(x)$ divides $1 + x^p$. Let $q(x)$ be given by

$$q(x) = \frac{1+x^p}{f(x)} \qquad \deg(q) \leq p - 1 \tag{7.4-14}$$

Then

$$G(x) = \frac{1}{f(x)} = \frac{q(x)}{1+x^p} \tag{7.4-15}$$

or

$$G(x) = \frac{1}{f(x)} = \frac{\alpha_0 + \alpha_1x + \cdots + \alpha_{p-1}x^{p-1}}{1+x^p} \tag{7.14-16}$$

where α_i are either 1 or 0. It is clear that

$$G(x) = (\alpha_0 + \alpha_1x + \alpha_2x^2 + \cdots + \alpha_{p-1}x^{p-1})(1 + x^p + x^{2p} + \cdots) \tag{7.4-17}$$

or (7.4-1)

$$\sum_{n=0}^{\infty} a_nx^n = G(x) = (a_0 + a_1x + a_2x^2 + \cdots + a_{p-1}x^{p-1} + a_px^p$$
$$+ a_{p+1}x^{p+1} + a_{2p-1}x^{2p-1} + \cdots) \tag{7.4-18}$$

Equating like powers of x, we see that $\{a_n\} = \{\alpha_n\}$, so that $\{a_n\}$ has period p or same factor of as its period. The period is the smallest positive integer for which $f(x)$ divides $1 + x^p$.

   If the initial conditions are not as stated in Theorem 7.3.2 [so that $G(x) = g(x)/f(x)$, where the numerator $g(x)$ has degree less than the degree of $f(x)$], and $g(x)$ has no common factors with $f(x)$, then the theorem still holds. The smallest p such that the theorem holds when $g(x) \neq 1$ is called the *exponent* of $f(x)$ and is the period of the corresponding sequence.

   As an example of the above results, consider the shift register generator of Figure 7.17. First, let $a_{-1} = a_{-2} = 0$ and $a_{-3} = 1$, so that $g(x) = 1$. Then

$$g(x)(1+x^p) = f(x)(a_0 + a_1x + a_2x^2 + \cdots + a_{p-1}x^{p-1}) \tag{7.4-19}$$

First, by definition, $f(x) = 1 + x + x^3$. Note that this sequence is maximal length, $p = 7$, so that $1 + x^p = 1 + x^7$. Then, with 001 initial register loading, the output sequence is

$$\underset{a_{-3}a_{-2}a_{-1}}{} \overbrace{\underset{1\ 0\ 0\ 1\ 1\ 1\ 0\ 1\ 0\ 0}{}}^{7\ \text{chips}} \tag{7.4-20}$$

so that $a_{-3} = 1$, $a_{-2} = 0$, $a_{-1} = 0$, $a_0 = 1$, $a_1 = 1$, $a_2 = 1$, $a_3 = 0$, $a_4 = 1$, $a_5 = 0$, $a_6 = 0$. Now since the period is 7, we have, in general,

$$G(x) = \frac{g(x)}{f(x)} = \frac{a_0 + a_1 x + a_2 x^2 + \cdots + a_{p-1} x^{p-1}}{1 + x^p} \tag{7.4-21}$$

Since $g(x) = 1$

$$1 + x^p = f(x)(a_0 + a_1 x + \cdots + a_{p-1} x^{p-1}) \tag{7.4-22}$$

or

$$1 + x^7 = (1 + x + x^3)(1 + x + x^2 + 0x^3 + x^4 + 0x^5 + 0x^6)$$
$$= 1 + x^7 \qquad \text{mod } 2 \tag{7.4-23}$$

PROBLEM 4

Using the register of the above example, let the initial conditions be $a_{-1} = 0$, $a_{-2} = 1$, and $a_{-3} = 0$. Show that

(a) $g(x) = x$

(b) $g(x)(1 + x^7) = f(x)(a_0 + a_1 x + \cdots + a_6 x^6)$

PROBLEM 5

Show that the initial state $a_{-1} = 0$, $a_{-2} = 0$, $\cdots a_{-n+1} = 0$, $a_{-n} = 1$ results in the largest period.

Note that if $g(x)$ and $f(x)$ have a common factor (or factors), the result is equivalent to a new shorter characteristic polynomial and a shorter initial condition that will have a shorter period. A polynomial, having coefficients that are 0 or 1, is called *irreducible* if it cannot be factored, that is, divided by another polynomial of degree less than n. For example, $1 + x + x^5$ is reducible since it can be written as

$$1 + x + x^5 = (1 + x^2 + x^3)(1 + x + x^2)$$

that is, it can be factored. However, since $1 + x + x^3$ cannot be factored, it is irreducible.

When $f(x)$ is irreducible, then $g(x)$ and $f(x)$ can have no common factors. *Hence, when $f(x)$ is irreducible, the period of the SSRG sequence does not depend upon the initial condition $[g(x)]$ except for the trivial case when $g(x) = 0$.* This latter case corresponds to a cycle of period 1.

PROBLEM 6

Show that an irreducible polynomial must have an odd number of terms.

Recall that a sequence generated from an n-stage SSRG has *maximal length* if its period is $2^n - 1$ symbols long. We now establish a necessary, but not sufficient, condition for maximal length sequences.

Theorem 6 If the SSRG sequence $\{a_n\}$ has maximal length, its characteristic polynomial $f(x)$ will be irreducible.

Proof   Since the period of the SSRG sequence is maximal, $p = 2^n - 1$. From Theorems 4 and 5 the SSRG sequence has a period p that is the exponent of $f(x)$.

First assume that $f(x)$ has two distinct factors: $f(x) = f_1(x)f_2(x)$. Then

$$\frac{1}{f(x)} = \frac{\beta(x)}{f_1(x)} + \frac{\gamma(x)}{f_2(x)} \tag{7.4-24}$$

by using a partial fraction expansion. Let the degree of $f_1(x)$ be n_1 and the degree of $f_2(x)$ be n_2, so that $n = n_1 + n_2$. Consider $\beta(x)/f_1(x)$ as a separate generating function that produces a sequence having a period of, at most, $2^{n_1} - 1$. Similarly, $\gamma(x)/f_2(x)$ can be viewed as a generating function for a sequence that has a period of, at most, $2^{n_2} - 1$. Then the sum sequence formed from (7.4-24) has a composite period no larger than the least common multiple of the individual periods, which is, at most, the product of the periods

$$p \le (2^{n_1} - 1)(2^{n_2} - 1) \le 2^n - 3 \tag{7.4-25}$$

which contradicts the fact that $p = 2^n - 1 > 2^n - 3$.

Now assume that the factors are not distinct. Then $f(x) = f_1(x)f_1(x)$, so that the resulting period can be shown to be twice that of $f_1(x)$. Hence, $p = 2(2^{n/2} - 1) < 2^n - 1$, contradicting the fact that $p = 2^n - 1$.

It should be stressed that if a polynomial is irreducible, it is not necessarily of maximal length. For example, consider the irreducible characteristic polynomial $f(x) = x^6 + x^3 + 1$. Since $f(x)/1 + x^9$, we see from Theorem 1 that there exists a sequence having period 9, which is not $2^6 - 1$!

We are finally in a position to specify the conditions for maximal length sequences. It has been proved [10] that every irreducible polynomial of degree $n > 1$ divides the polynomial $1 + x^{(2^n - 1)}$, all sums being performed mod 2. From Theorems 1, 2, and 6, we deduce the following.

Theorem 7   Given that the characteristic polynomial is irreducible and of degree n, the period of the SSRG sequence is a factor of $2^n - 1$. As a corollary to this result, we have:

Corollary   If $2^n - 1$ is prime, every irreducible polynomial of degree n corresponds to a shift register sequence of maximal length. However, if we require a maximal length sequence for every n, we must restrict our characteristic polynomials to be primitive. An irreducible polynomial of degree n is called *primitive* if and only if it divides $x^m - 1$ for no m less than $2^n - 1$.

We noticed from Problem 5 that the initial state $000 \cdots 01$ produces the sequence with the largest period p. This was due to the fact $g(x)$, the numerator of $G(x)$, was unity and hence $g(x)$ and $f(x)$ could have no common factors. However, when $g(x)$ and $f(x)$ have common factors, cancellation can occur and a sequence with a period less than p will occur. As an example, consider the SSRG of Figure 7.18a. The characteristic polynomial is given by

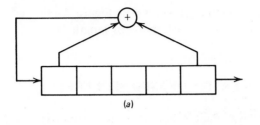

(a)

$$
\left.\begin{array}{l}
0\ 0\ 0\ 0\ 1 \\
1\ 0\ 0\ 0\ 0 \\
1\ 1\ 0\ 0\ 0 \\
1\ 1\ 1\ 0\ 0 \\
1\ 1\ 1\ 1\ 0 \\
1\ 1\ 1\ 1\ 1 \\
0\ 1\ 1\ 1\ 1 \\
1\ 0\ 1\ 1\ 1 \\
0\ 1\ 0\ 1\ 1 \\
1\ 0\ 1\ 0\ 1 \\
0\ 1\ 0\ 1\ 0 \\
0\ 0\ 1\ 0\ 1 \\
1\ 0\ 0\ 1\ 0 \\
1\ 1\ 0\ 0\ 1 \\
0\ 1\ 1\ 0\ 0 \\
0\ 0\ 1\ 1\ 0 \\
0\ 0\ 0\ 1\ 1 \\
1\ 0\ 0\ 0\ 1 \\
0\ 1\ 0\ 0\ 0 \\
0\ 0\ 1\ 0\ 0 \\
0\ 0\ 0\ 1\ 0
\end{array}\right\} p = 21
$$
$$\overline{0\ 0\ 0\ 0\ 1}$$

$$0\ 0\ 0\ 0\ 0\} p = 1$$

$$
\left.\begin{array}{l}
0\ 0\ 1\ 1\ 1 \\
1\ 0\ 0\ 1\ 1 \\
0\ 1\ 0\ 0\ 1 \\
1\ 0\ 1\ 0\ 0 \\
1\ 1\ 0\ 1\ 0 \\
1\ 1\ 1\ 0\ 1 \\
0\ 1\ 1\ 1\ 0
\end{array}\right\} p = 7
$$
$$\overline{0\ 0\ 1\ 1\ 1}$$

$$
\left.\begin{array}{l}
0\ 1\ 1\ 0\ 1 \\
1\ 0\ 1\ 1\ 0 \\
1\ 1\ 0\ 1\ 1
\end{array}\right\} p = 3
$$
$$\overline{0\ 1\ 1\ 0\ 1}$$

(b)

Figure 7.18 (a) SSRG of example having $f(x) = 1 + x + x^5$. (b) Cycles of SSRG having $f(x) = 1 + x + x^5$ (reducible).

$f(x) = 1 + x + x^5$. This polynomial can be factored to

$$1 + x + x^5 = (1 + x + x^2)(1 + x^2 + x^3) \qquad (7.4\text{-}26)$$

It can be shown that the period for reducible polynomials that can be expressed as nonrepeated reducible factors is given by

$$p = \text{LCM}(p_1, p_2) \qquad (7.4\text{-}27)$$

where LCM denotes the least common multiple, p_i is the period of the ith reducible factor, and the initial condition $000 \cdots 01$ is used. We have, from our example

$$p = \text{LCM}(3, 7) = 21 \qquad (7.4\text{-}28)$$

The cycles of the SSRG are shown in Figure 7.18b. As can be seen, the largest cycle contains $00 \cdots 01$ and is of length 21.

7.5   SOME PROPERTIES OF SSRG SEQUENCES

In this section, we shall establish some properties of SSRGs or combinations of them by use of the generating function.

7.5.1   Equivalence of Single and Dual SSRGs

We consider the following theorem.

Theorem 8   Let $f_1(x)$ and $f_2(x)$ be characteristic polynomials. Then, any sequence that is the mod 2 sum of sequences generated by $f_1(x)$ and $f_2(x)$ can be generated by the SSRG having characteristic polynomial $f_1(x)f_2(x)$.

Proof   For simplicity, we drop the explicit dependence on x. Let $a_1 = \{a_{1n}\}$ be the sequence generated by f_1, and $a_2 = \{a_{2n}\}$ be the sequence generated by f_2. Then

$$G_1(x) = \frac{g_1(x)}{f(x)} \qquad \text{and} \qquad G_2(x) = \frac{g_2(x)}{f_2(x)} \qquad (7.5\text{-}1)$$

for some polynomials (with coefficients 1 or 0) with {designating degree of $g(x)$ by $\deg[g(x)]$}

$$\deg[g_i(x)] < \deg[f_i(x)]$$

Then

$$G_1 + G_2 = \frac{g_1}{f_1} + \frac{g_2}{f_2} = \frac{f_1 g_2 + f_2 g_1}{f_1 f_2} \qquad (7.5\text{-}2)$$

Following Gold [11], we see that

$$\deg(f_1 g_2 + f_2 g_1) \leq \max[\deg(f_1 g_2), \deg(f_2 g_1)] \qquad (7.5\text{-}3)$$

This is an upper bound since the highest-order terms could cancel in the sum $f_1 g_2 + f_2 g_1$. Now clearly $\deg(fg) = \deg(f) + \deg(g)$. Therefore

$$\max[\deg(f_1 g_2), \deg(f_2 g_1)] = \max[\deg(f_1) + \deg(g_2), \deg(f_2) + \deg(g_1)]$$
$$< \deg(f_1) + \deg(f_2) = \deg(f_1 f_2) \qquad (7.5\text{-}4)$$

Hence, $G_1 + G_2$ is a proper generator polynomial with its numerator of degree less than the denominator.

Hence, from Theorem 8 we see that any sequence generated from the sum of two SSRGs having, respectively, n_1 and n_2 stages can be generated by a single SSRG having, at most, length $n_1 + n_2$.

PROBLEM 7

Show by direct enumeration of all sequences that the diagram below with two shift registers is equivalent to the single shift register also shown below.

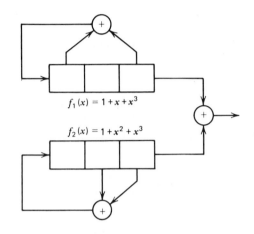

$$f_1(x) = 1 + x + x^3$$

$$f_2(x) = 1 + x^2 + x^3$$

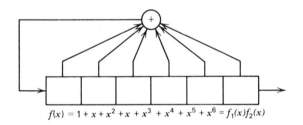

$$f(x) = 1 + x + x^2 + x + x^3 + x^4 + x^5 + x^6 = f_1(x)f_2(x)$$

PROBLEM 8

Prove that if $f_1(x)$ and $f_2(x)$ are relatively prime, then any sequence that can be generated by the shift register corresponding to the polynomial $f_1(x)f_2(x)$ is the sum of sequences generated by shift registers corresponding to $f_1(x)$ and $f_2(x)$. See Gold [11].

7.5.2 Initial Conditions

In this section, we demonstrate how to select initial conditions in order to obtain a sequence from a particular cycle length. We shall demonstrate the method by an example. Consider the SSRG of Figure 7.19. The characteristic polynomial is given by $f(x) = 1 + x^4 + x^5$. This polynomial is not irreducible since we have

$$f(x) = x^5 + x^4 + 1 = (x^2 + x + 1)(x^3 + x + 1) \tag{7.5-5}$$

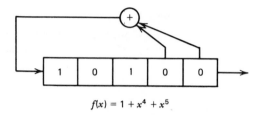

$$f(x) = 1 + x^4 + x^5$$

Figure 7.19 SSRG for initial condition example.

The method depends upon reducing the characteristic polynomial into the product of irreducible polynomials. Now if the term $1 + x + x^2$ can be canceled by the initial conditions polynomial $g(x)$, then $G(x)$ would be given by

$$G(x) = \frac{1}{x^3 + x + 1} \qquad (7.5\text{-}6)$$

Since $(x^3 + x + 1)/(x^7 + 1)$, we see that we will generate a sequence of length 7 (this is a maximal length sequence). Using the definition of $G(x)$ in (7.4-4), we must have

$$g(x) = \sum_{i=1}^{n} C_i x^i (a_{-i} x^{-i} + a_{-i+1} x^{-i+1} + \cdots + a_{-1} x^{-1}) \qquad (7.5\text{-}7)$$

or

$$x^2 + x + 1 = (a_{-4} + a_{-3}x + a_{-2}x^2 + a_{-1}x^3)$$
$$+ (a_{-5} + a_{-4}x + a_{-3}x^2 + a_{-2}x^3 + a_{-1}x^4) \qquad (7.5\text{-}8)$$

Hence

$$\begin{aligned}
a_{-1} &= 0 \\
a_{-1} + a_{-2} &= 0 \Rightarrow a_{-2} = 0 \\
a_{-3} + a_{-2} &= 1 \Rightarrow a_{-3} = 1 \\
a_{-3} + a_{-4} &= 1 \Rightarrow a_{-4} = 0 \\
a_{-4} + a_{-5} &= 1 \Rightarrow a_{-5} = 1
\end{aligned} \qquad (7.5\text{-}9)$$

Therefore, the initial conditions or contents of the register should be 00101, as shown in Figure 7.19, to produce a cycle of length 7.

PROBLEM 9

Verify the results of the above example by showing the state diagram for $f(x) = x^5 + x^4 + 1$. Notice that there are cycles of length 21, 7, 3, and 1.

PROBLEM 10

Following the method of the above example, find the initial state that produces the sequence of length 3.

7.5.3 Phase Shifting the SSRG Sequence

This section illustrates how to phase-shift an SSRG by use of the generating function. Of course one can use direct methods of cycling the register on paper when the register length and the desired shift are small, but that is not practical for large lengths. First we establish a result due to Gold [11].

Theorem 9   Let $G(x) = g(x)/f(x)$, with $f(x)$ the characteristic polynomial of degree n and $g(x)$ a binary polynomial of degree, at most, $n - 1$. Let

$$g(x) = g(0) + g(1)x + g(2)x^2 + \cdots + g(n-1)x^{n-1}$$
$$f(x) = f(0) + f(1)x + f(2)x^2 + \cdots + f(n)x^n$$
$$G(x) = G(0) + G(1)x + G(2)x^2 + \cdots + G(n-1)x^{n-1}$$

Then, given $f(x)$ and $G(x)$, we have

$$[g(0), g(1), \cdots, g(n-1)] = [G(0), G(1), \cdots, G(n-1)]$$

$$\times \begin{bmatrix} f(0), & f(1), & f(2), & \cdots, & f(n-1) \\ 0, & f(0), & f(1), & \cdots, & f(n-2) \\ 0, & 0, & f(0), & \cdots, & f(n-3) \\ \vdots & & & & \\ 0, & 0, & 0, & \cdots & f(0) \end{bmatrix}$$

$$(7.5\text{-}10)$$

Proof   Since

$$G(x) = \frac{g(x)}{f(x)}$$

we have

$$g(x) = G(x)f(x) \qquad (7.5\text{-}11)$$

By multiplying the two polynomials, we obtain

$$\begin{aligned} g(0) &= G(0)f(0) \\ g(1) &= G(1)f(0) + G(0)f(1) \\ g(2) &= G(2)f(0) + G(1)f(1) + G(0)f(2) \\ g(3) &= G(3)f(0) + G(2)f(1) + G(1)f(2) + G(0)f(3) \\ &\vdots \\ g(n-1) &= G(n-1)f(0) + G(n-2)f(1) + \cdots + G(0)f(n-1) \end{aligned} \qquad (7.5\text{-}12)$$

which is the theorem in nonmatrix form. Phase shifting to the right m places is accomplished by multiplying $G(x)$ by x^m and reducing the result mod $f(x)$ if the numerator is of degree n or larger. This method is best explained by taking an example. Let $f(x) = x^4 + x + 1$, which is a primitive polynomial. Let the starting conditions of the output sequence be $G(x) = 1$ and $G(1) = G(2) = G(3) = 0$. Now we determine the starting conditions to shift the sequence to the right 14 places. We do this by solving for $g(x)$, multiplying by x^{14}, and reducing mod $f(x)$:

$$[g(0), g(1), g(2), g(3)] = [1, 0, 0, 0] \begin{bmatrix} 1 & 1 & 0 & 0 \\ 0 & 1 & 1 & 0 \\ 0 & 0 & 1 & 1 \\ 0 & 0 & 0 & 1 \end{bmatrix} \qquad (7.5\text{-}13)$$

$$[g(0), g(1), g(2), g(3)] = [1, 1, 0, 0] \qquad (7.5\text{-}14)$$

Therefore, $g(x)$ is given by $g(x) = 1 + x$. It follows that $G(x)$ is given by

$$G(x) = \frac{1+x}{1+x+x^4} \qquad (7.5\text{-}15)$$

To shift the sequence to the right 14 symbols, we have

$$G'(x) = x^{14}G(x) = \frac{x^{14}+x^{15}}{1+x+x^4} \qquad (7.5\text{-}16)$$

By using

$$\begin{aligned} x^4 &= x+1 \\ x^8 &= x^2+1 \\ x^{12} &= x^3+x^2+x+1 \\ x^{14} &= x^5+x^4+x^3+x^2 \\ x^{15} &= x^6+x^5+x^4+x^3 \end{aligned} \qquad (7.5\text{-}17)$$

we reduce $G'(x)$ to

$$G'(x) = \frac{x^3}{1+x+x^4} \qquad (7.5\text{-}18)$$

To find the resulting output sequence and therefore the starting conditions, we use long division on (7.5-16):

$$\begin{array}{r} x^3+x^4 \\ 1+x+x^4 \overline{\smash{\big)}\, x^3 } \\ \underline{x^3+x^4+x^7} \\ x^4+x^7 \\ \underline{x^4+x^5+x^8} \\ x^8+x^7+x^5 \end{array} \qquad (7.5\text{-}19)$$

Therefore, the output sequence is

$$0 \ 0 \ 0 \ 1 \ 1 \ldots \qquad (7.5\text{-}20)$$

From the output sequence, we obtain the initial conditions $G(0) = G(1) = G(2) = 0$ and $G(3) = 1$. Notice three "zeros" will be generated, then a "one", compatible with the output sequence. To verify our result, Figure 7.20 illustrates the SSRG along with the output sequence directly above the shifted output sequence. As can be seen from the figure, the lower sequence is shifted 14 symbols to the right.

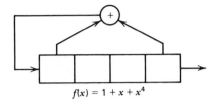

$$f(x) = 1 + x + x^4$$

Output 1 0 0 0 1 1 1 1 0 1 0 1 1 0 0 1 0 0 0 1 1 1 1 0 1 0 1 1 0 0 1 · · ·

Shifted

output 0 0 0 1 1 1 1 0 1 0 1 1 0 0 1 0 0 0 1 1 1 1 0 1 0 · · ·

14 symbols
to the right

Initial condition for sequence 0001
Initial condition for shifted sequence 1000

Figure 7.20   SSRG and assorted output sequences.

7.5.4   Zero Insertion

We will need a sequence in which a "zero" is inserted after every SSRG symbol. Figure 7.21 illustrates an example of a sequence with zeros "stuffed" after every symbol. This turns out to be simply achieved since if we write the original sequence* as

$$a(0) + a(1)x + a(2)x^2 + a(3)x^3 + \cdots \tag{7.5-21}$$

then

$$[a(0) + a(1)x + a(2)x^2 + a(3)x^3 + \cdots]^2 \tag{7.5-22}$$

$$= a(0) + a(1)x^2 + a(2)x^4 + a(3)x^6 + \cdots \tag{7.5-23}$$

since the cross terms are zero mod 2. Therefore, if $G(x) = g(x)/f(x)$ is the representation of the output sequence, then $G^2(x) = g^2(x)/f^2(x)$ is the representation of the "stuffed" sequence. Consider the function $G(x)$ given by

$$G(x) = \frac{1}{1 + x^2 + x^3} \tag{7.5-24}$$

which is illustrated in Figure 7.22a. The associated generating function for

Original sequence   0 1 1 1 1 0 1 0 1 1 0 1 1 · · ·
Stuffed sequence   0 0 1 0 1 0 1 0 1 0 0 0 1 0 0 0 · · ·

Figure 7.21   A Stuffed and an Original Sequence.

*We have used both $G(x)$ and $A(x)$ as our output sequence.

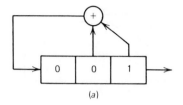

(a)

Output sequence: 1 0 0 1 0 1 1 1 1 0 0 1 0 1 1 $\cdots$

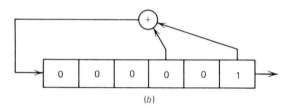

(b)

Output sequence: 1 0 0 0 0 0 1 0 0 0 1 0 $\cdots$

Figure 7.22   (a) SSRG associated with $G(x) = 1 + x^2 + x^3$. (b) SSRG associated with $G'(x) = 1 + x^4 + x^6$.

the stuffed sequence is given by

$$G'(x) = G^2(x) = \frac{1}{1 + x^4 + x^6} \qquad (7.5\text{-}25)$$

The SSRG associated with $G'(x)$ is illustrated in Figure 7.22b.

7.5.5   Interleaving

Next we consider the polynomial representation of two distinct sequences that are interleaved.

Theorem 10   If $G_a(x) = g_a(x)/f_a(x)$ (representing sequence a) and $G_b(x) = g_b(x)/f_b(x)$ (representing sequence b) are generating functions for two sequences, then the interleaved sequence has a generating function given by

$$G_{a,b}(x) = \frac{f_b^2(x)G_a^2(x) + xf_a^2(x)g_b^2(x)}{f_a^2(x)f_b^2(x)}$$

Proof   First note that $g_a^2(x)/f_a^2(x)$ is the generating function that represents sequence a with zeros stuffed (interleaved) between alternate symbols. Secondly, $xg_b^2(x)/f_b^2(x)$ is the generating function that represents sequence b, with zeros stuffed between symbols of sequence b, shifted one symbol to the right. It is clear, then, that the generating function of the interleaved sequence is given by the sum

$$G_{a,b}(x) = \frac{g_a^2(x)}{f_a^2(x)} + \frac{xg_b^2(x)}{f_b^2(x)} \qquad (7.5\text{-}26)$$

or

$$G_{a,b}(x) = \frac{f_b^2(x)g_a^2(x) + xf_a^2(x)g_b^2(x)}{f_a^2(x)f_b^2(x)} \qquad (7.5\text{-}27)$$

Notice that $G_{a,b}(x)$ is a proper generating function since the numerator is at least one degree less than the denominator.

PROBLEM 11

(a) Using the initial SSRG loading of 001, with $f_a(x) = 1 + x + x^3$ and 01 with $f_b(x) = 1 + x + x^2$, use Theorem 10 to compute the interleaved sequence. Check your result with the state diagram.

(b) Repeat the above with $f_a = 1 + x + x^2 + x^3$ and $f_b = 1 + x + x^2$. Notice that sequence a is now not of maximal length.

PROBLEM 12

Show that a sequence with each symbol repeated once has a generating function given by

$$G(x) = \frac{(1 + x)g^2(x)}{f^2(x)}$$

For a discussion of further properties of the generating function references, consult references 5 and 11.

7.6 PSEUDORANDOM SEQUENCES (PN SEQUENCES)

We now come to one of the main reasons for this chapter, that of defining pseudorandom sequences (PN sequences). Random sequences can be generated by making independent samples of a zero-mean noise process possessing a symmetrical density function. PN sequences, on the other hand, are generated by shift register generators.

First, we shall define three randomness properties and then prove that maximal length SRG sequences satisfy all three randomness properties. Based on these results, we will define PN sequences as maximal length sequences.

7.6.1 Postulates of Randomness

We will now list the randomness postulates we desire our PN sequence to satisfy. Since we are concerned with randomlike sequences, we will let the symbols be either $+1$ or -1 rather than 0 or 1. Knowing the sequence of ± 1's implies knowledge of the 0, 1 sequence and *vice versa* since the sequences are equivalent. This section follows Golomb [5] very closely.

P-1 In every sequence period, the number of $+1$'s does not differ from the number of -1's by more than 1.

P-2 For every sequence period, half the runs (of all 1's or all all -1's) have length 1, one-fourth have length 2, one-eight have length 3, etc., as long as the number of runs equals or exceeds 1. Furthermore, for each of the runs there are equally many runs of ± 1's.

To present our third postulate, we must define the autocorrelation function $R(m)$. If the 1, 0 elements are the sequence $\{a_n\}$, then we define the $-1, +1$ sequence $\{b_n\}$ by

$$b_n = 1 - 2a_n \qquad (7.6\text{-}1)$$

The autocorrelation function for the sequence $\{b_n\}$ is given by

$$R(m) = \frac{1}{p} \sum_{n=1}^{p} b_n b_{n+m} \qquad (7.6\text{-}2)$$

P-3 The autocorrelation function $R(m)$ is binary valued*; that is

$$R(m) = \begin{cases} p & m = 0 \\ c & 0 < |m| < p \end{cases} \qquad (7.6\text{-}3)$$

In fact, we will desire $p \gg c$, so that the sequence "looks" white. Basically, in a sequence that very long, random, and independent from symbol to symbol, we would expect that Postulates 1–3 would be essentially satisfied.

Consider as an example the sequence generated from the SSRG of Figure 7.17. Therefore, the output sequence is

$$\{a_n\} = \underbrace{1\ 1\ 1\ 0\ 1\ 0\ 0}_{p\ =\ 7} \cdots \qquad (7.6\text{-}4)$$

or

$$\{b_n\} = \underbrace{-1-1-1\ 1-1\ 1\ 1}_{p\ =\ 7} \cdots \qquad (7.6\text{-}5)$$

We see that there are four -1's and three $+1$'s in the period of (7.6-5), which satisfies Postulate 1. Of the total of four runs, one-half have length 1 and one-fourth have length 2. Notice also there are two runs of $+1$'s and two runs of -1's.

A sequence of ± 1's that satisfies Postulates 1–3 will be called

---

*For a discussion of processes rather than sequences, consult Chapter 8.

a *pseudonoise* (PN), or pseudorandom noise, sequence. We will now show that a maximal length sequence satisfies Postulates 1–3.

Theorem 11   The randomness postulate, 1, is satisfied for all maximal length sequences.

Proof   Since a maximal length SSRG generates all possible sequences (± 1) of length $n(p = 2^n - 1)$ except the all-one (all zero for $\{a_n\}$) sequence, we see that we have exactly (2^{n-1}) -1's and $(2^{n-1} - 1)$ $+1$'s, so that there is one excess -1.

Theorem 12   The run property Postulate 2, holds for all maximal length shift register sequences.

Proof   Consider the runs of -1's first. In the SSRG sequence of length n(SR length), there occurs the all -1 sequence only once. Further, this run must be followed and preceded by $+1$'s, or else there would be runs of length greater than n and the sequence could not be of maximal length.

A $+1$ followed by $(n - 1)$ -1's occurs exactly once. But this has already been accounted for by the run of n -1's. Further, $(n - 1)$ -1's followed by a $+1$ occurs once, and is accounted for by the fact that the run of n -1's must be followed by a $+1$. Therefore, there is no true run of $(n - 1)$ -1's.

Now consider the runs of -1's of length k, where $0 < k < n - 1$. Consider n consecutive terms beginning with 1, then k -1's, then a $+1$, and the remaining $n - k - 2$ terms chosen arbitrarily. This occurs in 2^{n-k-2} ways.

Similar reasoning holds for the number of runs of 1's of length k for $0 < k < n - 1$. There is no run of n consecutive -1's since this would trap the register in the all -1 state. However, -1, followed by $(n - 1)$ $+1$'s must occur for there to be a run of $(n - 1)$ $+1$'s.

We can summarize this result as follows: Denoting $p = 2^n - 1$ as the sequence period, there are $(p + 1)/2$ runs, half of them -1's and half of them $+1$'s. Of the runs of -1's, half of them are of length 1, one-fourth have length 2, one-eight have length 3, etc., and similarly for the $+1$'s. This relationship continued until there is one run of -1's and one run of $+1$'s of length $n - 2$. Finally, then there will be a run of $+1$'s of length $n - 1$ and a block of length n. Now we consider the final property, Postulate 3.

Theorem 13   The two-level autocorrelation property holds for all maximal length sequences.

Proof   By Theorem 3, we see that term-by-term mod 2 adding of two distinct phase shifts of a maximal length code produces a new phase shift of the code. Therefore, since there is one more 0 (or -1), the autocorrelation equation (7.6-2) of any nonzero phase shift is $-1/p$. When the phase shift is zero ($m = 0$), then the autocorrelation is unity.

We see, then, that a maximal length sequence satisfies all three postulates of randomness and therefore is a pseudonoise sequence. Consequently, all our comments about PN sequences are equally applicable to maximal length sequences.

7.6.2 The Number of Irreducible and Maximal Length Polynomials

In this section we shall present two results that allow us to determine the number of irreducible polynomials and the number of maximal length sequences for a given SRG length. To do this, we will have to discuss briefly two number-theoretic functions.

From the unique factorization theorem of arithmetic, every positive integer n greater than 1 can be expressed as the product of powers of primes so that

$$n = \prod_{i=1}^{k} p_i^{\alpha_i} \tag{7.6-6}$$

where p_i is the ith prime and α_i is a positive integer. As an example, we can express 56 by

$$56 = 7 \times 2^3 \qquad p_1 = 2, p_2 = 7, \alpha_1 = 3, \alpha_2 = 1 \tag{7.6-7}$$

Now the Euler ϕ-function is defined by

$$\phi(n) = \begin{cases} 1 & n = 1 \\ \prod_{i=1}^{k} p_i^{\alpha_i - 1}(p_i - 1) & n > 1 \\ p - 1 & \text{if } n = p \text{ a prime} \end{cases} \tag{7.6-8}$$

The last condition, when p is a prime, follows directly from the second condition, so that often the first two conditions are taken as the definition.

PROBLEM 13

Let p and q denote distinct primes. Show that

(a) $$\phi(pq) = (p - 1)(q - 1)$$
(b) $$\phi(p^2) = p(p - 1)$$
(c) $$\phi\left(\prod_{i=1}^{k} p_i\right) = \prod_{i=1}^{k} (p_i - 1) \qquad p_i \text{ distinct primes}$$

Golomb [5] has noted that $\phi(n)$ can be interpreted as the number of positive irreducible fractions not greater than 1, with denominator n.

The Möbius μ-function is defined by

$$\mu(n) = \begin{cases} 1 & n = 1 \\ 0 & \prod_{i=1}^{k} \alpha_i > 1 \\ (-1)^k & \begin{array}{l} n \text{ is the product of} \\ k \text{ distinct primes} \end{array} \end{cases} \tag{7.6-9}$$

with α_i from (7.6-6).

PROBLEM 14

Show that if p and q are distinct primes that

(a) $$\mu(p) = -1$$
(b) $$\mu(pq) = 1$$
(c) $$\mu(p^2) = 0$$

Zierler [3] has shown that the number of irreducible polynomials mod 2 of degree n is given by

$$N_I = \frac{1}{n} \sum_{d|n} 2^d \mu\left(\frac{n}{d}\right) \tag{7.6-10}$$

where the sum is over all positive divisors of n, including 1. As an example, consider $n = 6$. Then, $d = 1, 2, 3, 6$. Hence

$$N_I = \tfrac{1}{6}[2^1\mu(6) + 2^2\mu(3) + 2^3\mu(2) + 2^6\mu(1)]$$
$$= \tfrac{1}{6}(2 - 4 - 8 + 64) = 9 \tag{7.6-11}$$

Therefore, there are nine irreducible polynomials of degree 6. Of course, not all irreducible polynomials produce sequences of maximal length. Zieler [3] has shown that the number of maximal length sequences is given by

$$N_M = \frac{\phi(2^n - 1)}{n} \tag{7.6-12}$$

Using the above example, we obtain

$$N_M = \frac{\phi(64 - 1)}{6} = \frac{\phi(7 \times 3^2)}{6} = \frac{36}{6}$$
$$= 6 \tag{7.6-13}$$

Hence in this case two-thirds of the irreducible polynomials are of maximum exponent. Based on Golomb [5] and Ristenbatt [12] Table 7.1 lists the code length ($L = 2^n - 1$), the number of irreducible polynomials N_I, and the number of maximum exponent polynomials N_M.

7.6.3 A List of PN Code Generators

We shall give a list of primitive polynomials from which we can directly build a PN generator to generate a maximal length code. This table is based in part on the work of Peterson [14]. As can be seen in Table 7.1, it is impossible to list all the maximal length sequences, so we shall only list a few for each shift register length. Reference 14 should be consulted for the complete set. In the use of those tables it should be noted that the reciprocal of a primitive polynomial is primitive, and further, the reciprocal of an irreducible polynomial is irreducible. A *reciprocal polynomial* of

Table 7.1 Number of maximal length (PN)
(primitive) sequences and the number of irreducible polynomials

n	$2^n - 1$	N_m	N_I
1	1^a	1	2
2	3^a	1	1
3	7^a	2	2
4	15	2	3
5	31^a	6	6
6	63	6	9
7	127^a	18	18
8	255	16	30
9	511	48	56
10	1,023	60	99
11	2,047	176	186
12	4,095	144	335
13	$8,191^a$	630	630
14	16,383	756	1,161
15	32,768	1,800	2,182
16	65,535	2,048	4,080
17	$131,071^a$	7,710	7,710
18	262,143	8,064	14,532
19	$524,287^a$	27,594	27,594
20	1,048,575	24,000	52,377
21	2,097,151	84,672	99,858
22	4,194,303	120,032	190,557
23	8,388,607	356,960	364,722
24	16,777,215	276,480	698,870

^{a} Denotes Mersenne prime.

degree n denoted by $f^R(x)$ is defined by

$$f^R(x) = x^n f\left(\frac{1}{x}\right) \tag{7.6-14}$$

For example, $f(x) = x^4 + x + 1$ is primitive. The reciprocal polynomial to
$f(x)$ is given by

$$f^R(x) = x^4(x^{-4} + x^{-1} + 1) = x^4 + x^3 + 1 \tag{7.6-15}$$

No reciprocal polynomials will be listed in Table 7.2 since it is simple to
obtain them from (7.6-14). The taps for all PN generators up to 34 are
given in Table 7.2 based on reference 16. As an example consider a PN
code SRG for $n = 10$. Taking the first entry [3, 10], we have first that the

Table 7.2 Table of PN generators (primitive polynomials)

Register Length n	
2^a	[1, 2]
3^a	[1, 3]
4^a	[1, 4]
5^a	[2, 5] [2, 3, 4, 5] [1, 2, 4, 5]
6^a	[1, 6] (1, 2, 5, 6] [2, 3, 5, 6]
7^a	[3, 7] [1, 2, 3, 7] [1, 2, 4, 5, 6, 7] [2, 3, 4, 7]
	[1, 2, 3, 4, 5, 7] [2, 4, 6, 7] [1, 7] [1, 3, 6, 7]
	[2, 5, 6, 7]
8^a	[2, 3, 4, 8] [3, 5, 6, 8] [1, 2, 5, 6, 7, 8]
	[1, 3, 5, 8] [2, 5, 6, 8] [1, 5, 6, 8]
	[1, 2, 3, 4, 6, 8] [1, 6, 7, 8]
9	[4, 9] [3, 4, 6, 9] [4, 5, 8, 9]
	[1, 4, 8, 9] [2, 3, 5, 9] [1, 2, 4, 5, 6, 9]
	[5, 6, 8, 9] [1, 3, 4, 6, 7, 9] [2, 7, 8, 9]
10	[3, 10] [2, 3, 8, 10] [3, 4, 5, 6, 7, 8, 9, 10]
	[1, 2, 3, 5, 6, 10] [2, 3, 6, 8, 9, 10] [1, 3, 4, 5, 6, 7, 8, 10]
11	[2, 11] [2, 5, 8, 11] [2, 3, 7, 11]
	[2, 3, 5, 11] [2, 3, 10, 11] [1, 3, 8, 9, 10, 11]
12	[1, 4, 6, 12] [1, 2, 5, 7, 8, 9, 11, 12]
	[1, 3, 4, 6, 8, 10, 11, 12] [1, 2, 5, 10, 11, 12]
	[2, 3, 9, 12] [1, 2, 4, 6, 11, 12]
13	[1, 3, 4, 13] [4, 5, 7, 9, 10, 13] [1, 4, 7, 8, 11, 13]
	[1, 2, 3, 6, 8, 9, 10, 13] [5, 6, 7, 8, 12, 13] [1, 5, 7, 8, 9, 13]
14	[1, 6, 10, 14] [1, 3, 4, 6, 7, 9, 10, 14]
	[4, 5, 6, 7, 8, 9, 12, 14] [1, 6, 8, 14]
	[5, 6, 9, 10, 11, 12, 13, 14] [1, 2, 3, 4, 5, 7, 8, 10, 13, 14]
15	[1, 15] [1, 5, 10, 15] [1, 3, 12, 15]
	[1, 2, 4, 5, 10, 15] [1, 2, 6, 7, 11, 15] [1, 2, 3, 6, 7, 15]
16	[1, 3, 12, 16] [1, 3, 6, 7, 11, 12, 13, 16]
	[1, 2, 4, 6, 8, 9, 10, 11, 15, 16] [1, 2, 3, 5, 6, 7, 10, 15, 16]
	[2, 3, 4, 6, 7, 8, 9, 16] [7, 10, 12, 13, 14, 16]
17	[3, 17] [1, 2, 3, 17] [3, 4, 8, 17]
18	[7, 18] [5, 7, 10, 18] [7, 8, 9, 10, 15, 16, 17, 18]
19	[1, 2, 5, 19] [3, 4, 5, 8, 13, 19] [3, 7, 9, 10, 12, 19]
20	[3, 20] [3, 5, 9, 20] [2, 3, 6, 8, 11, 20]
21	[2, 21] [2, 7, 14, 21] [2, 5, 13, 21]
22	[1, 22] [1, 5, 9, 22] [1, 4, 7, 10, 13, 16, 19, 22]
23	[5, 23] [5, 11, 17, 23]
24	[1, 2, 7, 24] [4, 5, 7, 8, 9, 11, 14, 16, 18, 20, 22, 24]
	[1, 4, 5, 9, 10, 13, 14, 15, 16, 17, 18, 19, 21, 24]
25	[3, 25] [1, 2, 3, 25] [3, 4, 12, 25]
26	[1, 2, 6, 26] [1, 3, 4, 5, 8, 10, 11, 12, 16, 21, 22, 26]
	[2, 3, 5, 6, 7, 8, 9, 11, 13, 14, 15, 16, 19, 26]

Table 7.2 (Continued)

27	[1, 2, 5, 27] [3, 4, 5, 9, 10, 11, 18, 27]
	[3, 4, 5, 6, 11, 13, 22, 27]
28	[3, 28] [3, 4, 8, 12, 16, 20, 24, 28]
	[3, 5, 9, 11, 13, 28]
29	[2, 29] [2, 11, 20, 29] [2, 5, 21, 29]
30	[1, 2, 23, 30] [1, 3, 4, 6, 7, 9, 12, 13, 15, 18, 20, 21, 24, 30]
31	[3, 31] [1, 2, 3, 31] [3, 8, 13, 31]
32	[1, 2, 22, 32] [2, 4, 5, 7, 8, 10, 11, 12, 16, 22, 23, 26]
	[1, 5, 6, 9, 10, 11, 14, 16, 18, 19, 28, 32]
33	[13, 33] [11, 13, 22, 33] [13, 17, 29, 33]
34	[1, 2, 27, 34]
	[1, 2, 6, 11, 13, 19, 21, 22, 23, 26, 27, 34]
	[1, 2, 4, 6, 8, 9, 10, 11, 27, 28, 29, 30, 31, 32, 33, 34]
61	[1, 2, 5, 61]
89	[3, 5, 6, 89]

^{a} Includes all primitive polynomials except reciprocals.

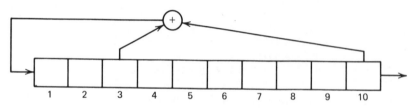

Figure 7.23   PN code shift register generator of length 10 associated with [3, 10].

characteristic polynomial is given by $f(x) = 1 + x^3 + x^{10}$, and further, that the tap connections are in positions 3 and 10. Hence the result is shown in Figure 7.23.

Reference 15 was used to obtain the taps for lengths 61 and 89. Note that from Problem 6 all Table 7.2 entries must have an even number of entries.

REFERENCES

1 Gilbert, E. N., "Quasi-Random Binary Sequences," Bell Telephone Laboratories Memorandum MM-53-1400-42, November 27, 1953.

2 Golomb, S. W., "Sequences with Randomness Properties," Terminal Progress Report under Contract Req. No. 639498, Martin Co., Baltimore, Md., June 1955.

3 Zierler, N., "Several Binary-Sequence Generators," Tech. Report No. 95, Lincoln Labs., MIT, September 12, 1955.

4 Birdsall, T. G., and Ristenbatt, M. P., "Introduction to Linear Shift-Register Generalized Sequences," University of Michigan Research Institute, Dept. of Electrical Engineering Tech. Report No. 90, October 1958.

5 Golomb,, S. W., *Shift Register Sequences*, Holden-Day, 1967.

6 Gold, R., "Optimal Binary Sequences for Spread Spectrum Multiplexing," *IEEE Trans. Information Theory*, October 1967.

7 Gold, R., "Maximal Recursive Sequences with 3-Valued Recursive Cross-Correlation Functions," *IEEE Trans. Information Theory*, Vol. IT-14, January 1968.

8 Zierler, N., "Linear Recurring Sequences," J. Soc. *Industrial and Applied Math.* Vol. 7, pp. 31–48, 1959.

9 Perlman, M., "Theory and Application of Feedback Shift Registers," UCLA Engineering Extension Course No. X463.14, 1975.

10 Van der Waerden, B. L., *Modern Algebra*, Frederick Ungar, New York, 1949, Vol. I and 1950. Vol. 11.

11 Gold, R., "Properties of Linear Binary Encoding Sequences," Lecture notes prepared by Robert Gold Associates, copyright pending 1973, July 1975.

12 Franklin, J. N., *Matrix Theory*, Prentice Hall, Englewood Cliffs, N.J., 1968.

13 Restenbatt, M., "Pseudo-Random Binary Coded Waveforms," Berkowitz, P., Ed., *Modern Radar*, Wiley, New York, 1965, Chap. 4.

14 Peterson, W. W., *Error Correcting Codes*, MIT Press, Cambridge, Mass, 1961, Appendix C.

15 Dixon, R. C., *Spread Spectrum Systems*, Wiley-Interscience, New York, 1976, Chap. 3.

16 Peterson, W. W., and Weldon, E. J., Jr., *Error Correcting Codes*, 2nd, ed., MIT Press, Cambridge, Mass., 1972, Appendix C.

8

PN CODE SPREAD
SPECTRUM SYSTEMS

Spread spectrum modulation is a modulation scheme characterized by the fact that the transmitted bandwidth is normally much larger than the information bearing (modulated) bandwidth. This spread spectrum signal is due to a special modulation technique. The basic types of spread spectrum modulation include frequency hopping (FH), time hopping (TH), and pseudonoise (PN) schemes (also called PRN for pseudorandom noise, PNG for pseudonoise generator, PRG for pseudorandom generator), and direct sequence systems.

A basic transmitter is shown in Figure 8.1. The data are modulated onto the carrier which, in turn, is modulated by the spread spectrum modulator that causes the composite signal to achieve a bandwidth large compared to the modulated data bandwidth.

A coherent spread spectrum receiver has five basic units. First the acquisition circuit has the function of acquiring (synchronizing to) the spread spectrum modulation waveform and then tracking it (maintaining synchronization). Second, the despreader "removes" the PN code and collapses the spectrum back to the data modulated bandwidth. Next the carrier tracking loop is swept (by a sweep circuit or FLL) or preset to the carrier frequency and then locked so that the baseband data modulation is available to the bit synchronizer. The bit synchronizer locks to the data signal, which provides transition timing for the data demodulator. The function of the data demodulator is to retrieve the data from the noise-corrupted data obtained from the carrier tracking loop. (Also frame synchronization must follow the data demodulator).

The primary advantage of spread spectrum receivers is their ability to reject unintentional and intentional interference. This ability to communicate while being "jammed" is termed the antijam (AJ) capability and is very important in military communication systems operating in a hostile environment. This tolerance to spectrally congruent signals leads to multiple access by utilizing (roughly) orthogonal signals in a common frequency band.

Spread spectrum modulation produces a wideband, low power spectral density signal. Consequently the transmitted signal is not readily detected

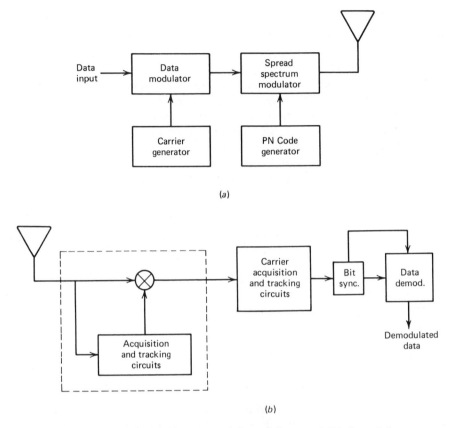

Figure 8.1 Basic spread spectrum (a) modulator and (b) demodulator.

or recognized by alien listeners. Therefore the signals are not easily detected.

Another desirable aspect of spread spectrum signals, especially PN signals, is their inherent ranging capability. High rate PN codes can be used to obtain very accurate range estimates by transmission through a cooperative transponder-equipped satellite.

Our main concern here will be with direct sequence spread spectrum systems, which we discuss in the next section.

8.1 PN CODE SPREAD SPECTRUM SYSTEMS

For the sake of completeness we briefly discuss the other two basic spread spectrum systems: frequency hopping and time hopping.

Frequency hopping is effected by selecting the carrier frequency from a frequency synthesizer, which, by virtue of a PN code, pseudorandomly picks a transmit frequency. These available frequencies are picked at a predetermined rate, the hopping rate, and are arranged to cover a band of frequencies. The receiver for this system must synchronize with the hopping rate and phase and dehop the received signal.

Time hopping obtains spectral spreading by channelizing the carrier via time slots that use a burst transmission within the slot. The slot in which the transmitted pulse occurs is selected under the control of a PN code generator [a form of time division multiple access (TDMA)].

All of these systems utilize a PN code generator to obtain their spread spectrum. However, the actual implementation differs with each one.

A typical model of a direct sequence spread spectrum system transmitter is shown in Figure 8.2.

The modulated biphase waveform can be expressed as

$$x(t) = \sqrt{2P} \, d(t) \, PN(t) \cos(\omega_0 t + \theta_0) \qquad (8.1\text{-}1)$$

where

P is signal power.

$d(t)$ is the data sequence (± 1).

$PN(t)$ is the PN code signal (± 1).

ω_0 is the carrier frequency (rad/sec).

θ_0 is the carrier phase (rad).

Normally the duration of a PN code symbol (chip) is small compared to the duration of a data bit, which produces spectral spreading. Another commonly used modulation scheme for PN code type systems is the quadriphase scheme in which two distinct PN codes or time shifts of

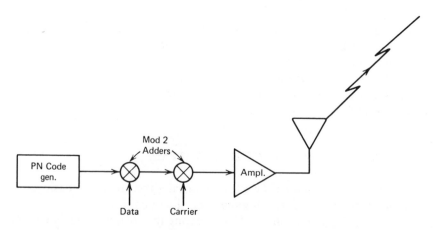

Figure 8.2 Direct sequence transmitter conceptual block diagram.

the same PN code can be used. This modulation scheme is illustrated in Figure 8.3.

The resulting waveform is given by

$$x(t) = \sqrt{2P_I}\, d_I(t) PN_I(t) \cos(\omega_0 t + \theta_0) + \sqrt{2P_Q}\, d_Q(t) PN_Q(t) \sin(\omega_0 t + \theta_0)$$
$$(8.1\text{-}2)$$

where the parameters are the following.

$PN_I(t)$, $PN_Q(t)$ are I channel and Q channel PN sequences.

$d_I(t)$, $d_Q(t)$ are I and Q channel data signals.

P_I, P_Q are I and Q channel power levels.

ω_0 is the carrier frequency (rad/sec).

θ_0 is the carrier phase (rad).

Normally the PN code rates are equal and the epoch times occur simultaneously. In some applications $d_1(t) = d_2(t)$, so each "phase" of the quadriphase signal has the same data stream on it. *Staggered quadriphase PN* refers to a quadriphase PN modulation scheme in which the end of one I-PN code chip occurs at the midpoint of the Q-PN code chip so that there is a difference of one-half chip in their zero crossing times. This implementation has the advantage that when this signal structure is filtered and then limited prior to transmission the sidebands remain suppressed. However for the nonstaggered version this limiting process "restores" the spectrum back to its original shape. The restoration of the spectrum can be

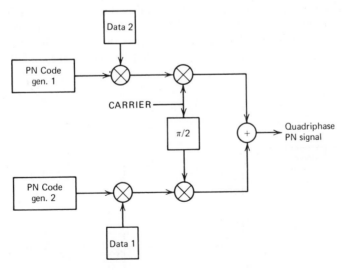

Figure 8.3   Quadriphase direct sequence transmitter.

a source of adjacent channel interference. This will be discussed later in this chapter.

A typical biphase receiver is shown in Figure 8.4. If the I and Q PN sequences are delayed versions of each other, then one despreader would be sufficient to provide acquisition and tracking if this meets performance goals. Otherwise predetection or postdetection combining can be employed to utilize both the I and Q channel energy simultaneously.

First the PN code is "acquired" (PN code time synchronized) and then tracked via the code tracking loop. Next the carrier loop is swept or by some means acquired, conceivably with the aid of the code loop if code and carrier are derived from the same master oscillator. Next the bit synchronizer locks to the zero crossings of the data or symbol stream so that the demodulator and the decoder (if present) can function properly.

8.2 PROCESSING GAIN

The term "processing gain" is often used among engineers in spread spectrum systems to denote a reduction in the effect of an interfering signal. We shall define processing gain for a particular case and caution the reader that this "gain" depends upon the system as well as upon the interfering signal structure. Consider the simplified model of a coherent receiver shown in Figure 8.5 for the case of a noncoherent jammer. Let the input signal plus tone interferer (at the same frequency) be modeled as

$$y(t) = \sqrt{2P}\, d(t)\, \text{PN}(t) \sin \omega_0 t + \sqrt{2P_I} \sin[\omega_0 t + \theta(t)] \qquad (8.2\text{-}1)$$

which includes the desired signal with power P plus a tone interferer with power P_I. The data sequence $d(t)$, has symbol duration T sec. The process $\theta(t)$ is assumed to be of the form $\Omega t + \theta_0$, where θ_0 is a uniform random variable. For our discussion we shall assume that our code tracking loop provides a perfect estimate of the received PN sequence and the carrier

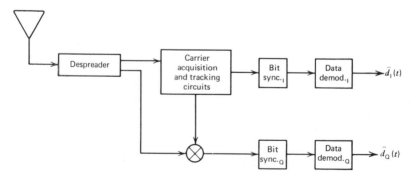

Figure 8.4 Representative quadriphase receiver.

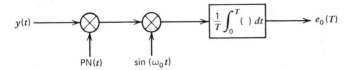

Figure 8.5   Coherent receiver model.

tracking loop provides a perfect estimate of the carrier signal so that the input to the data demodulator, due to the signal component, is baseband data. For convenience we shall assume that the data symbols are NRZ, so that the optimum data demodulator is an integrate-and-dump matched filter. Hence the signal into the matched filter is given by

$$e_i(t) = \sqrt{P}\, d(t) + \sqrt{P_I}\, PN(t) \cos[\theta(t)] \qquad (8.2\text{-}2)$$

with the assumption that the $2\omega_0$ term is effectively removed by the matched filter. When $\theta(t) = \Omega t + \theta_0$ is such that $0 < \Omega/2\pi \ll 1/T_c = R_c$, with T_c the PN code chip duration, the output signal voltage and mean square interference power are given by [assuming the $PN(t)$ is a very long code]

$$\bar{e}_0 = \sqrt{P}\, d(T^-) \qquad (8.2\text{-}3)$$

$$\overline{(e_I - \bar{e}_I)^2} = P_I T_c \left(\frac{1}{T}\right) \qquad (8.2\text{-}4)$$

This result follows since the two-sided spectra of a long PN code (PN code period much greater than T) is $P_I T_c$ near $f = 0$, and the two-sided noise bandwidth of the averager is $1/T$. The two spectra centered at $\pm\Omega$ add in the matched filter bandwidth. The output SNR is given by

$$SNR_0 = \frac{(\bar{e}_0)^2}{\overline{(e_0 - \bar{e}_0)^2}} = \frac{P}{P_I}\left(\frac{R_c}{R}\right) \qquad (8.2\text{-}5)$$

where $R = 1/T$.

The effective reduction of the interference is thus R/R_c. Consequently R_c/R is called the *processing gain*. Hence higher PN code rates reduce the interference to a greater degree. However, realization of the gain depends upon very careful circuit design to combat mixer leakage. More generally we define the processing gain for PSK systems (with coherent demodulation) and a particular interference to be $SNR_0/(P/P_I)$ where SNR_0 is the matched filter output SNR. By this definition short PN codes may offer less processing gain than very long PN codes since the spectral components are, in general, $\Delta f = 1/NT_c$ Hz apart where T_c is the chip time and N is the number of chips in one period. Long PN codes have spectral lines very close together and short PN codes have spectral lines quite far apart. The result is that a long PN code can be modeled to have a continuous power spectral density but in a short code the discrete lines must be taken into account, which limits the processing gain.

It should be noted that other spurious signals or leakage terms generated in the receiver can degrade the processing gain.

PROBLEM 1

Consider a quadriphase signal of the form

$$y(t) = \sqrt{P}\, PN_1(t)\, d(t)\cos(\omega_0 t) + \sqrt{P}\, PN_2(t) d(t)\sin\omega_0 t$$
$$+ \sqrt{P_1}\sin(\omega_0 t + \phi)$$

Consider the demodulator shown below.

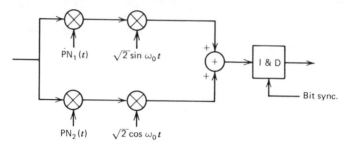

Show that the demodulated SNR is given by

$$SNR_0 = \frac{2P}{P_1}\left(\frac{R_c}{R}\right)$$

independent of ϕ. In this case the processing gain is 3 dB better than the biphase case of $\theta(t) = 0$ in 8.2-1.

8.3 CORRELATION LOSS FROM RF FILTERED PN SIGNALS

The purpose of this section is to develop an expression for the filter distortion loss due to the presence of a BPF preceding the correlation operation. The result allows estimates of the effective autocorrelation function after correlating. The approach follows that of Cahn [1] with some modifications.

Consider Figure 8.6 illustrating the model under consideration.

The received PN modulated carrier $x(t)$ is modeled as

$$x(t) = \sqrt{2P}\, Re[PN(t)\, e^{j\omega_c t}] = \sqrt{2P}\, PN(t)\cos\omega_c t \qquad (8.3\text{-}1)$$

where $Re(Z)$ denotes the real part of the complex number Z. It is convenient to use complex notation in the following calculations. Initially we truncate the PN sequence to length $2T$ sec. Assuming that PN code is very long, we may consider each chip random and statistically independent from each other chip. We may write the input signal as

$$x(t) = \sqrt{2P}\, Re\left[e^{j\omega_c t}\frac{1}{2\pi}\int_{-\infty}^{\infty} A_T(j\omega)\, e^{j\omega t}\, d\omega\right] \qquad (8.3\text{-}2)$$

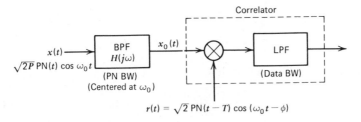

Figure 8.6 Filter model for filtered correlation curve [$H_{BB}(j\omega)$ is the baseband equivalent of $H(j\omega)$].

where $A_T(j\omega)$ denotes the Fourier transform of a $2T$-sec segment of $PN(t)$, ($|t| \leq T$). After filtering by the PN filter (assumed to be centered at ω_c) and denoting $H_{BB}(j\omega)$ as the baseband equivalent of $H(j\omega)$

$$H(j\omega) = H_{BB}[j(\omega - \omega_c)] \qquad \omega > 0$$

we obtain

$$x_{0_T}(t) = \sqrt{2P}\, \mathrm{Re}\left[e^{j\omega_c t} \frac{1}{2\pi} \int_{-\infty}^{\infty} A_T(j\omega) H_{BB}(j\omega) e^{j\omega t}\, d\omega \right] \qquad (8.3\text{-}3)$$

In general the reference signal $r(t)$ will be shifted in time and rf phase from $x_0(t)$ so that is modeled as

$$r(t) = \sqrt{2}\, \mathrm{Re}[PN(t - \tau) e^{j(\omega_c t - \phi)}], \qquad (8.3\text{-}4)$$

With τ the code time shift and ϕ the rf carrier phase difference, alternately we write (8.3-4) as

$$r_T(t) = \sqrt{2}\, \mathrm{Re}\left[e^{j\omega_c t - j\phi} \frac{1}{2\pi} \int_{-\infty}^{\infty} A_T(j\omega) e^{j\omega(t - \tau)}\, d\omega \right] \qquad (8.3\text{-}5)$$

Define the cross correlation function between the two random functions $x_{0_T}(t)$ and $r(t)$ by

$$R(\tau, \phi) = \lim_{T \to \infty} \frac{1}{2T} \int_{-T}^{T} \overline{x_{0_T}(t) r_T(t)}\, dt \qquad (8.3\text{-}6)$$

where the overbar denotes the ensemble average. Using the fact that

$$\mathrm{Re}\{z_1\}\, \mathrm{Re}\{z_2\} = \tfrac{1}{2}\, \mathrm{Re}\{z_1 z_2\} + \tfrac{1}{2}\, \mathrm{Re}\{z_1 z_2^*\} \qquad (8.3\text{-}7)$$

we obtain for the autocorrelation function

$$R(\tau, \phi) = \sqrt{P} \lim_{T \to \infty} \frac{1}{2T}\, \mathrm{Re}$$

$$\times \int_{-T}^{T} \left\{ e^{j\phi} \overline{\left[\frac{1}{2\pi} \int_{-\infty}^{\infty} A_T(j\omega) H(j\omega) e^{j\omega t}\, d\omega \right]\left[\frac{1}{2\pi} \int_{-\infty}^{\infty} A_T^*(j\omega') e^{-j\omega'(t - \tau)}\, d\omega' \right]} \right\} dt$$

$$(8.3\text{-}8)$$

But

$$\frac{1}{2\pi} \int_{-\infty}^{\infty} e^{j(\omega - \omega')t} dt = \delta(\omega - \omega') \tag{8.3-9}$$

From Chapter 2 we recall that

$$\lim_{T \to \infty} \left[\frac{\overline{A_T(j\omega)A^*(j\omega)}}{2T} \right] = \mathcal{S}_{PN}(\omega) \tag{8.3-10}$$

which is the spectral density of the PN signal under the assumption of random, independent NRZ symbols with transition density of $\frac{1}{2}$. Hence we obtain from (8.3-8), (8.3-9), and (8.3-10) our final result:

$$R(\tau, \phi) = \sqrt{P} \, \text{Re}\left[\frac{1}{2\pi} \int_{-\infty}^{\infty} \mathcal{S}_{PN}(\omega) H_{BB}(j\omega) \, e^{j(\omega\tau + \phi)} \, d\omega \right] \tag{8.3-11}$$

Note that we have assumed in effect that the postdetection BPF is of zero bandwidth. This is a good approximation in the case when the PN bandwidth is large compared to the data bandwidth.

Now we show that when $H(j\omega)$ is complex conjugate symmetric [$h(t)$ is real] $R(\tau, \phi)$ is maximized when $\phi = 0$. Since

$$\text{Re}(z) = \frac{z + z^*}{2} \tag{8.3-12}$$

we write

$$R(\tau, \phi) = \frac{\sqrt{P}}{2\pi} \int_{-\infty}^{\infty} \mathcal{S}_{PN}(\omega) \left(\frac{H_{BB}(j\omega) \, e^{j(\omega\tau + \phi)} + H_{BB}^*(j\omega) \, e^{-j(\omega\tau + \phi)}}{2} \right) d\omega \tag{8.3-13}$$

Since $H(j\omega)$ is complex conjugate symmetric, we have

$$2R(\tau, \phi) = \frac{\sqrt{P}}{2\pi} \int_{-\infty}^{\infty} \mathcal{S}_{PN}(\omega) H_{BB}(j\omega) \, e^{j(\omega\tau + \phi)} \, d\omega$$

$$+ \frac{\sqrt{P}}{2\pi} \int_{-\infty}^{\infty} \mathcal{S}_{PN}(\omega) H_{BB}(-j\omega) e^{-j(\omega\tau + \phi)} d\omega \tag{8.3-14}$$

Changing the variables ($f = -f'$) in the last integral produces

$$R(\tau, \phi) = \frac{\sqrt{P}}{2\pi} \int_{-\infty}^{\infty} \mathcal{S}_{PN}(\omega) H_{BB}(j\omega) \, e^{j\omega\tau} \cos \phi \, d\omega \tag{8.3-15}$$

which is clearly maximized by setting $\phi = 0$ or, in fact more generally, $\phi = 2\pi n$, where $n = 0, \pm 1, \pm 2$, and so on. In other words when the bandpass filter preceding the correlator is symmetric about the carrier frequency the correlated output signal is given by (8.3-15) and is maximized when the reference carrier is in phase with the input carrier. We also notice that the cross correlation voltage depends on $\cos \phi$, with ϕ the rf phase error between the received and reference carriers. In general, if the filter is not complex conjugate symmetric, the maximum magnitude will not necessarily occur at $\phi = 0$. Further, this asymmetry can cause a code loop tracking error bias (see Chapter 10).

PROBLEM 2

Show that for a symmetric filter, that is $H_{BB}^*(j\omega) = H_{BB}(-j\omega)$, starting with (8.3-11) that

$$R(\tau, 0) = \text{Re}\left[\int_{-\infty}^{\infty} \mathscr{S}_{PN}(\omega) H_{BB}(jf)\, e^{j\omega\tau} \frac{d\omega}{2\pi}\right] \sqrt{P}$$

$$= \left[\int_{-\infty}^{\infty} \mathscr{S}_{PN}(\omega) H_{BB}(jf)\, e^{j\omega\tau} \frac{d\omega}{2\pi}\right] \sqrt{P} \quad \text{is a real number.}$$

In order to obtain an estimate of the correlation loss due to prefiltering the PN sequence consider the value of $R(0, 0)$ for a symmetric multipole filter modeled as an ideal filter with linear phase across the band $- B/2$ to $B/2$:

$$H_{BB}(j\omega) = e^{-j\omega\alpha} \quad |\omega| < \pi B$$
$$= 0 \quad \text{otherwise} \tag{8.3-16}$$

When $\tau = \alpha$, the time delay introduced by the filter, we have from (8.3-11)

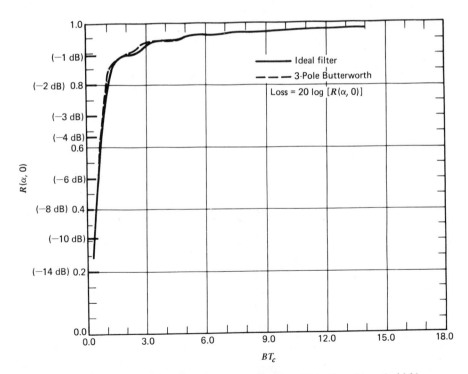

Figure 8.7 Correlation loss due to prefiltering (B is the rf bandwidth).

that

$$R(\alpha, 0) = \frac{\sqrt{P}}{2\pi} \int_{-\pi B}^{\pi B} \mathscr{S}_{PN}(\omega) \, d\omega \tag{8.3-17}$$

For very long PN codes using NRZ symbols yields

$$R(\alpha, 0) = \frac{\sqrt{P}}{2\pi} \int_{-\pi B}^{\pi B} T_c \left[\frac{\sin(\omega T_c/2}{(\omega T_c/2)} \right]^2 d\omega \tag{8.3-18}$$

In Figure 8.7 $R(\alpha, 0)$ is plotted for various values of BT_c, the channel bandwidth to chip rate ratio. For the commonly used value of $BT_c = 2$ the loss is 0.92 dB. Also plotted in Figure 8.7 is the maximum correlation value (maximized over τ) for a three-pole Butterworth filter with B denoting the 3-dB bandwidth. Notice that the two filters have almost identical peak correlation values. In Figure 8.8 the correlation curve with prefiltering of $BT_c = 2$ (three-pole Butterworth) is shown. Note both that the peak value is diminished by filtering and the curve is somewhat broadened, and further that the correlation curve is delayed by about 0.35 chips. This delay will be tracked out by a code tracking loop. Greater filtering ($BT_c < 2$) will produce even greater delays.

It is to be noted that the noise is spread as is shown in Section 8.8: the precorrelation filter decreases the correlation signal but also decreases the interfering noise power spectral density. The result for symmetric PN filters is given by

$$\left(\frac{C}{N_0}\right)_{\text{filtered}} = \left(\frac{C}{N_0}\right)_{\text{unfiltered}} \cos^2 \phi \frac{\left| T \int_{-\infty}^{\infty} \frac{\sin^2(\omega T/2)}{(\omega T/2)^2} H_{BB}(\omega) \, e^{j\omega\tau} \frac{d\omega}{2\pi} \right|^2}{\left[T \int_{-\infty}^{\infty} \frac{\sin^2(\omega T/2)}{(\omega T/2)^2} |H_{BB}(\omega)|^2 \frac{d\omega}{2\pi} \right]} \tag{8.3-19}$$

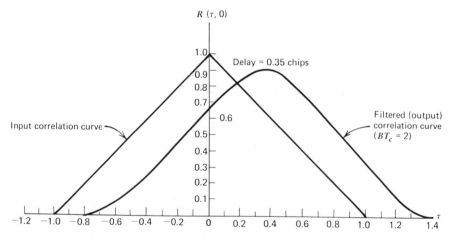

Figure 8.8 Unfiltered and filtered ($BT = 2$) voltage correlation curves for three-pole Butterworth filter.

8.3.1   Correlation Loss with Nonlinear Distortion

We now consider the case when the filter distorts the PN signal by virtue of its nonlinear phase but has an ideal filter amplitude response. Using our baseband equivalent filter, we can write

$$\theta(\omega) = \omega \left. \frac{d\theta}{d\omega}\right|_{\omega=0} + \frac{1}{6}\omega^3 \left.\frac{d^3\theta}{d\omega^3}\right|_{\omega=0} + \cdots \qquad (8.3\text{-}20)$$

since a real impulse response implies that θ is an odd function of ω. We assume here that $\theta(\omega)$ is real. Since the first term corresponds to a fixed time delay and does not distort the signal, we neglect it in our calculation. The second term, however, does cause distortion (parabolic time delay distortion). Our resulting value of the correlation peak loss, using (8.3-11), is given by

$$R(0,0) = \frac{\sqrt{P}\,T_c}{2\pi}\int_{-\pi B}^{\pi B}\left[\frac{\sin(\omega T_c/2)}{(\omega T_c/2)}\right]\cos(\alpha\omega^3)\,d\omega \qquad (8.3\text{-}21)$$

This result is plotted in Figure 8.9 for various values of phase distortion in radians at the first null [$f = 1/T_c$ (one sided phase shift)] as a function of BT_c, the rf bandwidth to chip rate ratio.

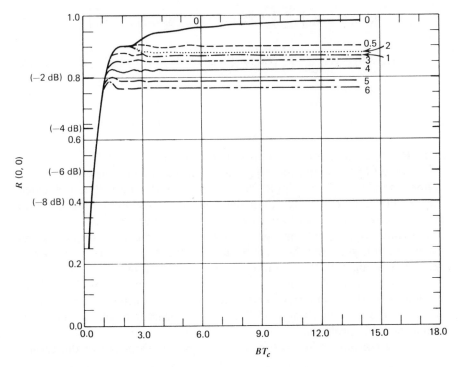

Figure 8.9   Correlation loss due to parabolic time delay distortion.

As can be seen from the figure a phase shift from band center (with parabolic delay distortion) to the first null of about $\frac{1}{2}$ rad (NRZ code symbols) produces about 1 dB degradation from $BT_c = 1.6$ to the infinite bandwidth case. Further, phase shifts of 5 rad produce over 2-dB degradation. Notice that over some regions increasing the bandwidth slightly can decrease the correlation value.

8.3.2 Correlation Loss due to Amplitude and Phase Distortion

We have considered nonlinear phase distortion and seen that it causes a degradation in the peak correlation value. Now we consider two additional particular types of distortion, one of which is amplitude distortion and one of which is a phase distortion.

First consider a first-order sinusoidal amplitude distortion in the filter transfer function of the form (see [9])

$$H_{BB}(\omega) = 1 + a \cos \alpha\omega \tag{8.3-22}$$

which is an even function of frequency as the amplitude response must be. Recall $H_{BB}(\omega)$ is the baseband equivalent of the precorrelation filter as illustrated in Figure 8.6. Using (8.3-11), we obtain (letting $P = 1$ for convenience)

$$R_a(\tau, 0) = \text{Re}\left[\frac{1}{2\pi} \int_{-\infty}^{\infty} \mathcal{S}_{PN}(\omega)(1 + a \cos \alpha\omega) e^{j\omega\tau} d\omega\right] \tag{8.3-23}$$

or

$$R_a(\tau, 0) = \text{Re}\left[\frac{1}{2\pi} \int_{-\infty}^{\infty} \mathcal{S}_{PN}(\omega)\left(1 + \frac{a}{2} e^{j\alpha\omega} + \frac{a}{2} e^{-j\alpha\omega}\right) e^{j\omega\tau} d\omega\right] \tag{8.3-24}$$

or

$$R_a(\tau, 0) = R(\tau, 0) + \frac{a}{2} R(\tau + \alpha, 0) + \frac{a}{2} R(\tau - \alpha, 0) \tag{8.3-25}$$

where $R(\tau, 0)$ is the undistorted correlation function.

The result of the amplitude ripple in the transfer function is to produce "echos," replicas of the ripple-free correlation function $R(\tau, 0)$ located α sec earlier and α sec later than the nonecho correlation function with amplitude one-half the transfer function ripple amplitude. If, for example, the amplitude ripple goes through one cycle between center frequency and $1/T_c$ Hz, then $\alpha = T_c$ and the "echos" are located T_c sec apart, as shown in Figure 8.10 along with the resultant correlation function.

The main effect of amplitude ripple (for small ripple amplitude) is to broaden the correlation curve. This in turn does not introduce a bias in the S-curve of the code tracking loop. False lock, for large ripple amplitude, is another possible problem.

Now we consider a first-order sinusoidal phase distortion of the form

$$H(\omega) = e^{j\theta \sin \omega\alpha} \tag{8.3-26}$$

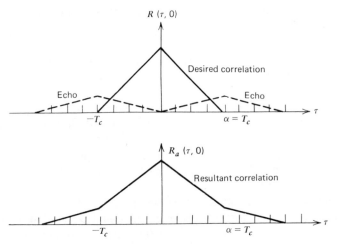

Figure 8.10 Echo correlation functions and resultant correlation function due to amplitude distortion.

which has a phase function $\theta \sin(\omega\alpha)$, which is an odd function of frequency, as it must be for a real impulse response. Again, using (8.3–11), we obtain

$$R_\theta(\tau, 0) = \mathrm{Re}\left[\frac{1}{2\pi}\int_{-\infty}^{\infty} \mathcal{S}_{PN}(\omega)\, e^{j\theta \sin \omega\alpha}\, e^{j\omega\tau}\, d\omega\right] \qquad (8.3\text{-}27)$$

Using the fact that

$$e^{j\theta \sin \omega\alpha} = \sum_{k=-\infty}^{\infty} J_k(\theta)\, e^{jk\omega\alpha} \qquad (8.3\text{-}28)$$

we obtain

$$R_\theta(\tau, 0) = \mathrm{Re}\left[\frac{1}{2\pi}\int_{-\infty}^{\infty} \mathcal{S}_{PN}(\omega) \sum_{k=-\infty}^{\infty} J_k(\theta)\, e^{j(k\omega\alpha + \omega\tau)} d\omega\right] \qquad (8.3\text{-}29)$$

or

$$R_\theta(\tau, 0) = \sum_{k=-\infty}^{\infty} R(\tau + k\alpha) J_k(\theta) \qquad (8.3\text{-}30)$$

Noting that

$$J_{-k}(\theta) = (-1)^k J_k(\theta) \qquad (8.3\text{-}31)$$

we obtain

$$\begin{aligned}
R_\theta(\tau, 0) = \; & J_0(\theta)R(\tau, 0) + J_1(\theta)R(\tau + \alpha, 0) - J_1(\theta)R(\tau - \alpha, 0) \\
& + J_2(\theta)R(\tau + 2\alpha, 0) + J_2(\theta)R(\tau - 2\alpha, 0) \\
& + J_3(\theta)R(\tau + 3\alpha, 0) - J_3(\theta)R(\tau - 3\alpha, 0) \\
& + J_4(\theta)R(\tau + 4\alpha, 0) + J_4(\theta)R(\tau - 4\alpha, 0) \\
& + J_5(\theta)R(\tau + 5\alpha, 0) - J_5(\theta)R(\tau - 5\alpha, 0) + \cdots
\end{aligned} \qquad (8.3\text{-}32)$$

When θ is small (which is the normal case) we have approximately

$$R_\theta(\tau, 0) \cong J_0(\theta)R(\tau, 0) + J_1(\theta)R(\tau + \alpha, 0) - J_1(\theta)R(\tau - \alpha, 0) \quad (8.3\text{-}33)$$

and therefore "echos" occur displaced α sec apart from the main correlation curve but are antisymmetric, whereas the amplitude ripple case caused symmetric echos.

The resultant correlation function becomes a distorted one. As an example the case $\alpha = T_c$ and $\theta = 0.4$ rad is shown in Figure 8.11 where second-order and higher terms have been neglected.

As can be seen from the figure (for the case considered) the resultant correlation function is broadened and distorted in a nonsymmetrical manner so that a code loop will track with a timing error bias.

In conclusion we see that sinusoidal phase distortion is worse than sinusoidal amplitude distortion. Further sinusoidal phase distortion at low frequency has more deleterious effects than at high frequency. False lock is a potential problem with either amplitude or phase distortion.

PROBLEM 3

Using the same techniques as in the above, show that for both amplitude and phase distortion, that is,

$$H(\omega) = (1 + a \cos \alpha\omega) \, e^{j\theta \sin \gamma\omega}$$

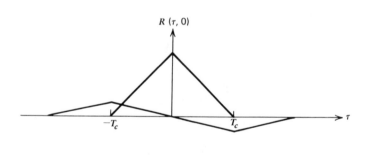

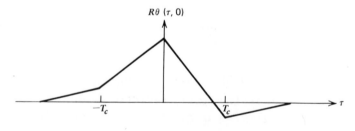

Figure 8.11 Echo and resultant correlation functions for the case $\theta = 0.4$ rad with phase distortion.

with small θ and a, the output correlation function is given by

$$R_{a\theta}(\tau, 0) = J_0(\theta)R(\tau, 0) + J_1(\theta)R(\tau + \gamma, 0) - J_1(\theta)R(\tau - \gamma, 0)$$

$$+ \frac{a}{2} J_0(\theta)R(\tau + \alpha, 0) + \frac{a}{2} J_0(\theta)R(\tau - \alpha, 0)$$

$$+ \frac{a}{2} J_1(\theta)R(\tau + \alpha + \gamma, 0) - \frac{a}{2} J_1(\theta)R(\tau + \alpha - \gamma, 0)$$

$$+ \frac{a}{2} J_1(\theta)R(\tau - \alpha + \gamma, 0) - \frac{a}{2} J_1(\theta)R(\tau - \gamma - \alpha, 0)$$

8.4 PN REFERENCE SIGNAL SHOULD BE FILTERED

In this section we establish the fact that if the PN modulated signal is filtered, then the best reference PN signal would be one filtered in the same manner. To show this consider Figure 8.12, where two candidate systems are shown in baseband form (which is sufficient for our discussion).

In the two figures $d(t)$ is the filtered PN signal, $d(t)$ is the unfiltered PN signal (± 1), and $n(t)$ is a sample function of a WGN process. The signal to noise ratio (SNR) is given by the quantity (mean signal)^2/(variance at time T). For case a we have that

$$\widetilde{SNR} = \frac{\left\{ \int_0^T [\tilde{d}(t)]^2 \, dt \right\}^2}{\frac{N_0}{2} \int_0^T [\tilde{d}(t)]^2 \, dt} = \frac{2}{N_0} \int_0^T [\tilde{d}(t)]^2 \, dt \qquad (8.4\text{-}1)$$

For case b we have that

$$SNR = \frac{\left\{ \int_0^T [\tilde{d}(t)d(t)] \, dt \right\}^2}{(N_0/2)T} \qquad (8.4\text{-}2)$$

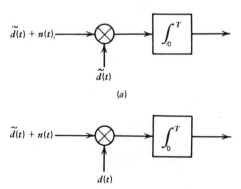

(a)

(b)

Figure 8.12 Two candidate correlators. (a) Matched case. (b) Unmatched case.

Now using the Schwartz inequality we deduce that

$$\text{SNR} \le \frac{\int_0^T [\tilde{d}(t)]^2 \, dt}{N_0/2} = \widetilde{\text{SNR}} \tag{8.4-3}$$

with equality only being achieved when $\tilde{d}(t) = Kd(t)$, that is, when there is no filtering of the input PN signal.

We conclude that better performance can be achieved in any application for which maximizing the SNR is the appropriate measure, by prefiltering the correlating PN signal by a filter equivalent to that which filters the input PN signal.

8.5   EFFECTS OF THE AGC CIRCUITS

When designing spread spectrum systems the range of signal amplitude must be considered so that thresholds for acquisition and lock detection (see Chapter 9) can be properly set. First we consider the power level that the signal can change given that the AGC detector bandwidth is SNR. The function of the AGC is to keep the total power in the AGC bandwidth constant, to avoid saturation as well as other design problems. Assume an ideal AGC that maintains the total output power at level P_T. If we define the power level of the input signal at threshold as P_{S_1} and the input noise power as P_N, then

$$g_1(P_{S_1} + P_N) = P_T \tag{8.5-1}$$

where g_i is the gain of the AGC for the ith signal. At a stronger input signal level we have

$$g_2(P_{S_2} + P_N) = P_T \tag{8.5-2}$$

Now the output signal ratio is given by

$$\frac{P_{S_2}^0}{P_{S_1}^0} = \frac{g_2 P_{S_2}}{g_1 P_{S_1}} = \frac{1 + (P_N/P_{S_1})}{1 + (P_N/P_{S_2})} \tag{8.5-3}$$

By definition

$$\text{SNR}_i = \frac{P_{S_i}}{P_N} \tag{8.5-4}$$

Therefore it follows that

$$\frac{P_{S_2}^0}{P_{S_1}^0} = \frac{1 + (\text{SNR}_1)^{-1}}{1 + (\text{SNR}_2)^{-1}} \tag{8.5-5}$$

which is our general result. Note that when SNR_1 is the threshold SNR and SNR_2 is infinite that the output signal power has a range given by

$$\frac{P_{S_2}^0}{P_{S_1}^0} = 1 + (\text{SNR}_1)^{-1} \tag{8.5-6}$$

For example, if the SNR is $-20\,\text{dB}$ in the AGC bandwidth at threshold, then the output signal power at infinite SNR is 101 times larger than the threshold signal output power. Further, the output signal voltage ratio is a little over 10 to 1.

Equation 8.5-5 is useful in determining loop bandwidth variations for carrier and code tracking loops.

8.6 A SELF-NORMALIZING REFERENCE CIRCUIT FOR ACQUISITION

If the signal and noise power did not vary in level, it would be quite feasible to design a detection circuit for synchronizing the local PN code to the incoming PN signal by using the system modeled in Figure 8.13. We assume in this discussion that both the data and the PN code are phase shift keyed (PSK) onto the carrier. When the local PN code, $PN(t-T)$, is more than a chip in error from the received code, there is no correlation (virtually), so the only output is due to thermal noise. Thus the mean output after a "dwell" (or integration) of τ_D sec is given by

$$\bar{e}_0 = N_0 B \tau_D - \delta \qquad (8.6\text{-}1)$$

where N_0 is the one-sided noise spectral density, δ is a reference voltage, and τ_D is the integration time. The bias δ can be picked so that $\delta > N_0 B \tau_D$ and a suitable false alarm probability could be obtained. Suppose now that due to thermal effects N_0 increases and the filter bandwidth widens so that $N_0 B \tau_D < \delta$. The result is that almost every integration (or dwell) would exceed the threshold, and the acquisition time would increase dramatically, to make acquisition virtually impossible due to the preponderance of false alarms.

A method to diminish the effect of changing noise power relative to the threshold is discussed below [2]. By using a reference channel to produce an estimate of $N_0 B$ (properly scaled), it is possible to make detection decisions on the normalized detection variable given by

$$e_0 = \text{sgn}\left(\frac{V_S - V_R}{V_R} - \text{THD}\right) \qquad (8.6\text{-}2)$$

where V_S is the signal plus noise channel output component, V_R is the

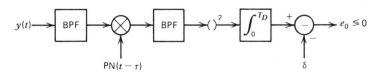

Figure 8.13 Model of a PN code fixed dwell time acquisition system.

reference channel output component (noise only), and THD is the threshold value. Both V_S and V_R refer to the output of the T_D-sec integrator. When the signal is not present (code timing is in error) V_S is proportional to V_R, so that even if the noise spectral density (N_0) increases or decreases the threshold setting is still correct since the change effects the numerator and denominator of the test statistic equally.

Consider now how to implement this statistic $\zeta = (V_S - V_R)/V_R$ First write the test statistic as

$$e_0 = \text{sgn}\left\{\frac{1}{V_R}[V_S - V_R - (\text{THD})\, V_R]\right\} \qquad (8.6\text{-}3)$$

Since $V_R > 0$, it being the square of a noise process,* we have

$$e_0 = \text{sgn}(V_S - K V_R) \qquad (8.6\text{-}4)$$

where $K = 1 + \text{THD}$. Now consider the block diagram of the circuit that provides the self-normalizing feature shown in Figure 8.14.

The direct channel or despread channel has a correlator (PN signal multiplier and BPF combination) and squarer (detector) with the bandpass filter (BPF) having a bandwidth equal to approximately the sum of the modulation bandwidth and the maximum doppler frequency.

The reference channel's function is to estimate the value of $N_0 B$ and hence the presence of the lowpass filter (LPF) following the detector. This can be achieved by using a frequency-offset filter to pass (sample) noise only. If we denote V_S as the direct channel component out of the integrator

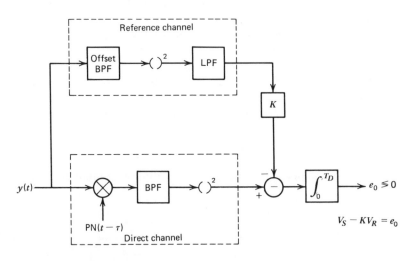

Figure 8.14 Self-normalizing threshold circuit.

---

*Actually $V_R = 0$ with probability zero.

and V_R as the reference voltage component out of the integrator, we have

$$e_0 = \text{sgn}(V_S - K V_R) \tag{8.6-5}$$

or

$$e_0 = \text{sgn}\left(\frac{V_S - V_R - \text{THD } V_R}{V_R}\right) \tag{8.6-6}$$

or

$$e_0 = \text{sgn}\left(\frac{V_S - V_R}{V_R} - \text{THD}\right) \tag{8.6-7}$$

and hence the circuit provides self-normalizing features.

8.7 AN ALTERNATE REFERENCE CHANNEL

A slightly different approach to obtaining the reference bias term $N_0 B$ can be obtained [3] by the diagram shown in Figure 8.15.

When there are no interfering signals this alternate scheme shown in Figure 8.15 is essentially the same as the one discussed in Section 8.6. However if, for example, a tone jammer were present at the input, then the system of Figure 8.14 would not function properly since the tone would be spread in the direct channel. It would not be spread in the reference channel and consequently an incorrect estimate of $N_0 B$ would be obtained that would lower the probability of detection. Consequently, the acquisition time would increase.

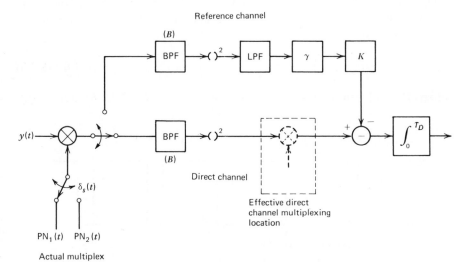

Figure 8.15 Alternate reference channel approach.

The scheme of Figure 8.15 avoids this problem by multiplying two distinct PN codes, one the desired PN code and another one that has low cross correlation to the first. Normally the desired code, $PN_1(t)$, is multiplexed on a large percentage of the time, and the "orthogonal" code is on the remaining time. The multiplexing function $s(t)$ routes the desired code to the direct channel and "orthogonal" code to the reference channel. In Figure 8.16 the reference signal is illustrated.

In order to appraise the effect of multiplexing on the detection system we first assume that multiplexing on the input is equivalent to multiplexing on the output. This will be essentially true when the multiplexing period is large compared to $1/B$, where B is the bandwidth of the BPFs. Using Davenport and Root [4], we can show for an unmodulated carrier that the power spectral density of the noise, after square law detection, has the shape shown in Figure 8.17.

Since the noise process and the "effective" multiplexing process are independent, the resulting multiplexed output spectrum is the following convolution:

$$\mathscr{S}_0(f) = \mathscr{S}_\delta(f) * \mathscr{S}_N(f) \tag{8.7-1}$$

By using reference 5, the output spectra can be written as

$$\mathscr{S}_0(f) = \int_{-\infty}^{\infty} \mathscr{S}_\delta(u)\mathscr{S}_N(f - u)\, du$$
$$= \int_{-\infty}^{\infty} \left(\frac{T-\tau}{T}\right)^2 \sum_{n=-\infty}^{\infty} \text{sinc}^2\left[\frac{\pi n(T-\tau)}{T}\right] \delta\left(u - \frac{n}{T}\right)\mathscr{S}_N(f - u)\, du \tag{8.7-2}$$

where $\mathscr{S}_N(f)$ is the noise spectral density shown in Figure 8.17. Under the assumption that $BT \gg 1$ we obtain, to a good approximation, the spectral density at $f = 0$:

$$\mathscr{S}_0(0) \cong \left(\frac{T-\tau}{T}\right)^2 \sum_{-\infty}^{\infty} \text{sinc}^2\left[\frac{\pi n(T-\tau)}{T}\right] \mathscr{S}_N\left(\frac{n}{T}\right) \cong \frac{T-\tau}{T}\mathscr{S}_N(0) \tag{8.7-3}$$

when $B \gg 1/T$. In (8.7-3) we have used series No. 571 of Jolley [6].

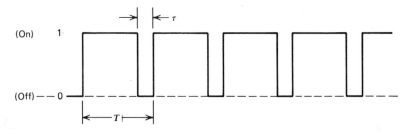

Figure 8.16 Multiplexing function waveform to gate $PN_1(t)$ on and off.

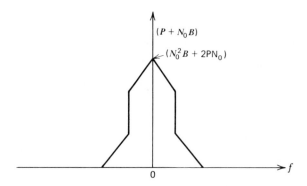

Figure 8.17 Power spectral density out of the direct channel squarer before equivalent multiplexing.

Furthermore the mean signal output to the integrator is reduced by $(T - \tau)/T$ so that the effect of the multiplexing operation is to diminish the mean signal to noise spectral density by $(T - \tau)/T$. Consequently a reference channel utilizing a 10% on-time rate degrades the output SNR out of the direct channel integrator (an LPF) by about 0.5 dB under the assumption that $BT \gg 1$. For this degradation the system designer obtains considerable immunity to interfering signals. Note that the "gain" of the reference channel must be multiplied by $\gamma = (T - \tau)/\tau$ to allow the $N_0 B$ estimate to be scaled correctly since it is on only τ out of T sec.

8.8   NOISE SPREADING BY CORRELATION

In appraising the acquisition or tracking performance of a PN spread spectrum system it is necessary to take into account the fact that the noise process (modeled as band-limited white noise) when multiplied by the local PN code is spectrally spread, with a resultant decrease in the spectral density at the IF frequency.

Consider our model shown in Figure 8.18. The noise-only input is modeled as

$$n(t) = \sqrt{2}\, n_c(t) \cos \omega_0 t + \sqrt{2}\, n_s(t) \sin \omega_0 t \qquad (8.8\text{-}1)$$

where $n_c(t)$ and $n_s(t)$, baseband sample functions of band-limited WGN, each has a two-sided spectral density of $N_0/2$ W/Hz. The multiplier output

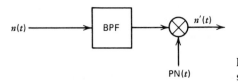

Figure 8.18   Model for estimating noise spreading.

noise $N'(t)$ is expressed by

$$n'(t) = n(t) PN(t) \tag{8.8-2}$$

The autocorrelation function of the output is given by*

$$R_{n'}(\tau) = E[n'(t)n'(t + \tau)] \tag{8.8-3}$$

or

$$R_{n'}(\tau) = R_{PN}(\tau)R_{n_c}(\tau) \cos \omega_0\tau + R_{PN}(\tau)R_{n_s}(\tau) \cos \omega_0\tau \tag{8.8-4}$$

Consequently the power spectral density at ω is given by

$$\mathscr{S}_e(\omega) = \int_{-\infty}^{\infty} R_{PN}(\tau)R_{n_c}(\tau) \cos \omega_0\tau\, e^{-j\omega\tau}\, d\tau + \int_{-\infty}^{\infty} R_{PN}(\tau)R_{n_s}(\tau) \cos \omega_0\tau\, e^{-j\omega\tau}\, d\tau \tag{8.8-5}$$

Since $R_{n_c}(\tau) = R_{n_s}(\tau)$, we have

$$\mathscr{S}_n(\omega) = \int_{-\infty}^{\infty} R_{PN}(\tau)R_{n_c}(\tau)(e^{j\omega_0\tau} + e^{-j\omega_0\tau})\, e^{-j\omega\tau}\, d\tau \tag{8.8-6}$$

Assuming that the spectral density at ω_0 due to the spectra at $-\omega_0$ is negligible, we have

$$\mathscr{S}(\omega_0) = \int_{-\infty}^{\infty} R_{PN}(\tau)R_{n_c}(\tau)\, d\tau \tag{8.8-7}$$

Since the integration of the product of two autocorrelation functions is equal to integration of the product of their respective power spectral densities, we have

$$\mathscr{S}(\omega_0) = \frac{1}{2\pi} \int_{-\infty}^{\infty} \mathscr{S}_{PN}(\omega)\, \mathscr{S}_{n_c}(\omega)\, d\omega \tag{8.8-8}$$

Since the power spectral density of a very long PN code is given by

$$\mathscr{S}_{PN}(f) = T_c \left(\frac{\sin \pi f T_c}{\pi f T_c} \right)^2 \tag{8.8-9}$$

we have

$$\mathscr{S}(\omega_0) = \frac{N_0}{2} \int_{-\infty}^{\infty} T_c \left(\frac{\sin \pi f T_c}{\pi f T_c} \right)^2 |H_{BB}(f)|^2\, df \tag{8.8-10}$$

where $H_{BB}(f)$ is the baseband equivalent of the filter preceding the multiplier. If we assume that the filter is a sharp cutoff BPF (an ideal BPF) having bandwidth B, we have

$$\mathscr{S}(\omega_0) = \frac{N_0}{2} \int_{-BT_c/2}^{BT_c/2} \left(\frac{\sin \pi x}{\pi x} \right)^2\, dx \tag{8.8-11}$$

---

*Again we assume that the epoch times are randomized to make the process stationary.

Table 8.1 Reduction in effective noise spectral density

PN Bandwidth (BT_c)	Reduction in $N_0/2$
6	0.22 dB
4	0.27 dB
3	0.39 dB
2	0.51 dB
1	1.02 dB
0.5	3.28 dB

For example with $BT_c = 2$ we obtain

$$\mathcal{S}(\omega_0) = 0.89 \left(\frac{N_0}{2}\right)$$ (8.8-12)

Therefore in this case the noise density at ω_0 is reduced by about 0.51 dB. Table 8.1 lists noise spectral density reduction as a function of the rf bandwidth (for an ideal BPF) to code rate ratio BT_c.

Although it is true that prefiltering the noise and then correlating with the PN signal reduces the noise spectral density in the data bandwidth, it also reduces the signal (see Section 8.3).

8.9 SPECTRAL EFFECTS OF HARD LIMITING QUADRIPHASE PN SIGNALS

As was discussed in Section 8.1 quadriphase PN signaling is a four-phase signal that can be described by

$$x(t) = \sqrt{2P_\mathrm{I}}\, d_\mathrm{I}(t) PN_\mathrm{I}(t) \cos(\omega_0 t + \theta_0) + \sqrt{2P_\mathrm{Q}}\, d_\mathrm{Q}(t) PN_\mathrm{Q}(t) \sin(\omega_0 t + \theta_0)$$
(8.9-1)

where $d_\mathrm{I}(t)$ and $d_\mathrm{Q}(t)$ are the I and Q data streams, $PN_\mathrm{I}(t)$ and $PN_\mathrm{Q}(t)$ are the I and Q PN sequences, with the I signals on a carrier shifted 90° from the Q carrier. If the epoch (zero crossing) times are common to both the I and Q signals, we have a conventional quadriphase signal (QPSK). However, if the epoch times are separated by one-half a PN code symbol, the result is called staggered quadriphase PN.

Repeater terminals are often used in satellite communication systems which bandpass-filters and hard-limit the signal before transmission. When conventional QPSK PN is employed in these satellite channels the hard limiting squares up the modulation, which tends to spread the modulation back to the prefiltered spectral shape and thereby cause cross channel interference. Staggered quadriphase PN to a certain extent minimizes this

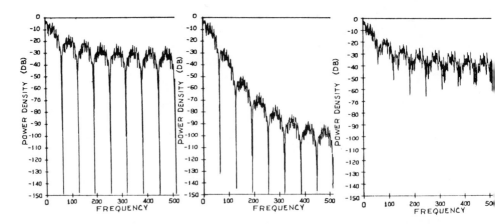

Figure 8.19 Power spectra obtained from simulation of conventional QPSK signals (from Rhodes [7] with permission). (*a*) Unfiltered. (*b*) Filtered, $BT = 1$. (*c*) Filtered and limited.

spectral spreading effect [7] at the expense of possibly producing false lock on the code acquisition process (see Chapter 11).

Figures 8.19 and 8.20 illustrate the effect prefiltering a QPSK PN signal with a BT_c product of 2, with B the rf (positive frequency) bandwidth [7]. In Figure 8.19 conventional QPSK was prefiltered and hard limited. It is clear that the resulting spectra are essentially those of the unfiltered NRZ signal. In Figure 8.20 the case when staggered quadriphase is used is shown. Notice in Figure 8.20*c* the filtered and limited staggered quadriphase PN is still essentially filtered after the hard limiting. Rhodes

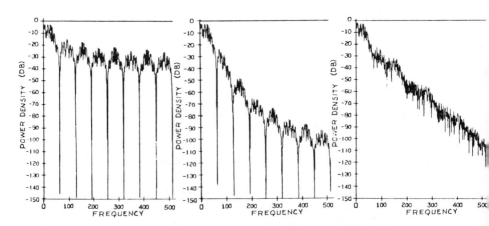

Figure 8.20 Power spectra obtained from simulation of offset QPSK signals (from Rhodes [7] with permission). (*a*) Unfiltered. (*b*) Filtered, $BT = 1$. (*c*) Filtered and limited.

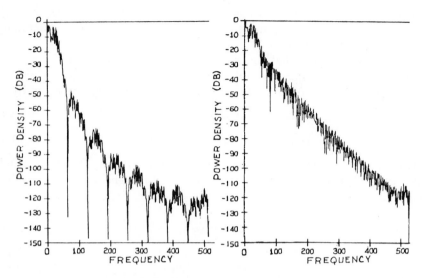

Figure 8.21 Spectra for offset QPSK with reduced filter bandwidth. (from Rhodes [7] with permission). (a) Filtered, $BT = 0.5$. (b) Filtered and limited.

[7] also observed that with other BT_c products the same effects were observed; see Figure 8.21 for the case $BT_c = 1$. Hence filtered staggered quadriphase PN leaves the sidelobes virtually unaltered after limiting. The hard limiting is an approximate model of a traveling wave tube (TWT) driven into saturation.

8.10 SNR IMPROVEMENT BY SIGNAL COMBINING

Many communication systems utilize a quadriphase signal structure. For these systems it is possible to utilize the power in both phases by properly combining each phase. In fact it will be shown that even if one phase has a much higher power level than the other one, it is possible, when the data are coherent, to improve the SNR by properly weighting each component. The general concept of pre- or postdetection combining can be applied to acquisition circuits or code tracking circuits equally well.

8.10.1 Predetection Combining

First consider the (coherent) case when the data are identical and in time synchronization on both phases. The quadriphase signal for the coherent data can be modeled as

$$x(t) = \sqrt{2P_\mathrm{I}}d(t)\,\mathrm{PN_I}(t)\cos(\omega_0 t + \theta) + \sqrt{2P_\mathrm{Q}}d(t)\mathrm{PN_Q}(t)\sin(\omega_0 t + \theta_0) + n(t)$$

$$(8.10\text{-}1)$$

Again $PN_Q(t)$ and $PN_I(t)$ are assumed to be essentially orthogonal, with P_I and P_Q the respective component signal powers. A predetection combining system is shown in Figure 8.22.

It will be assumed that the correlating codes $PN_I(t)$ and $PN_Q(t)$ are in-phase with their respective codes of the received quadriphase signal.

The correlated outputs, assuming perfect code tracking, are given by

$$y_I(t) = \sqrt{2P_I}\, d_I(t) \cos \omega_0 t + n_I(t) \tag{8.10-2}$$

$$y_Q(t) = \sqrt{2P_Q}\, d(t) \sin \omega_0(t) + n_Q(t) \tag{8.10-3}$$

where $n_I(t)$ and $n_Q(t)$ are independent processes (when $BT_c \ll 1$), which have two-sided spectral densities equal to $N_0/2$. The respective SNRs (neglecting uncorrelated noise) in each channel are given by

$$SNR_I = \frac{\alpha P_I}{N_0 B} \tag{8.10-4}$$

$$SNR_Q = \frac{\alpha P_Q}{N_0 B} \tag{8.10-5}$$

where B is the noise bandwidth of either BPF and α is the relative power loss of each signal due to the BPF filtering. Since there is no loss in generality, let $W_Q = 1$, and $W_I = \beta$. After adding coherently by shifting the carrier 90°, we obtain the total combined output SNR, given by

$$SNR_T = \frac{(\sqrt{\alpha P_Q} + \beta \sqrt{\alpha P_I})^2}{N_0 B (1 + \beta^2)} \tag{8.10-6}$$

Let

$$P_I = \gamma P_Q \tag{8.10-7}$$

Then we have

$$SNR_T = \frac{\alpha P_Q}{N_0 B} \left(\frac{1 + 2\beta\sqrt{\gamma} + \beta^2 \gamma}{1 + \beta^2} \right) \tag{8.10-8}$$

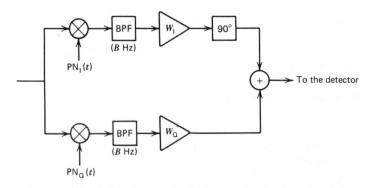

Figure 8.22   Predetection combining model.

Differentiating with respect to β to find the optimum weight β_0, we obtain

$$\frac{d\,\text{SNR}_T}{d\beta_0} = \frac{\alpha P_Q}{N_0 B}\left[\frac{2\sqrt{\gamma}+2\beta_0\gamma+2\sqrt{\gamma}\beta_0^2+2\gamma\beta_0^3-2\beta_0-4\beta_0^2\sqrt{\gamma}-2\beta_0^3\gamma}{(1+\beta_0^2)^2}\right]$$
(8.10-9)

Therefore β_0 is a root of

$$2\sqrt{\gamma}+2\beta_0\gamma-2\beta_0^2\sqrt{\gamma}-2\beta_0=0 \qquad (8.10\text{-}10)$$

One root is clearly $\beta_0=\sqrt{\gamma}$. From Descartes' rule of signs [8] it is clear that there is only one positive root, which is therefore $\beta_0=\sqrt{\gamma}$.

In Problem 4 it is established that $\beta_0=\sqrt{\gamma}$ yields a unique maximum.

PROBLEM 4

Show that $\beta_0=\sqrt{\gamma}$ is indeed the maximum by showing that

$$\text{SNR}_T(\sqrt{\gamma})-\text{SNR}_T(\beta)=\frac{\alpha P_Q}{N_0 B}\left[\frac{(\beta-\sqrt{\gamma})^2}{1+\beta^2}\right]\geq 0 \qquad (8.10\text{-}11)$$

Therefore since $\gamma=P_I/P_Q$, the optimum weighting is

$$\frac{W_I}{W_Q}=\sqrt{\gamma} \qquad (8.10\text{-}12)$$

that is the weights should be proportional to the square root of the relative power levels!

Notice that for the optimum weighting

$$\text{SNR}_T^0=\frac{\alpha P_Q}{N_0 B}\left[\frac{(1+\sqrt{\gamma}\sqrt{\gamma})^2}{1+\gamma}\right] \qquad (8.10\text{-}13)$$

or

$$\text{SNR}_T^0=\frac{\alpha P_Q}{N_0 B}(1+\gamma)=\text{SNR}_Q+\text{SNR}_I \qquad (8.10\text{-}14)$$

where SNR^0 denotes the optimally combined SNR. Therefore optimum combining results in the combined SNR being just the sum of the individual I and Q SNRs!

The SNR for the case of equal weighting is given by

$$\text{SNR}_T^e=\frac{\alpha P_Q}{N_0 B}\frac{(1+\sqrt{\gamma})^2}{2} \qquad (8.10\text{-}15)$$

Therefore by comparison to the optimum SNR, we have

$$\text{SNR}_T^e=\left[\frac{1+2\sqrt{\gamma}+\gamma}{2(1+\gamma)}\right]\text{SNR}_T^0 \qquad (8.10\text{-}16)$$

The bracketed term is the "loss" from the optimum weighting and is tabulated in Table 8.2.

Table 8.2 Equal weight
combining loss (from
optimum weighting-
coherent combining)

$\dfrac{P_Q}{P_I}$	$L(dB)$
1	0.0
1.5	0.04
2	0.13
3	0.30
4	0.46
5	0.59
10	1.04
20	1.47
∞	3.0

8.10.2 Postdetection Combining

In this section we consider quadriphase PN modulation in which different data sequences are modulated on the two quadriphased carriers. Since the data streams are assumed independent it is not possible to combine before detection and recapture some of the power that would be lost if only one phase were used. Postdetection combining as well as predetection combining can be used on code acquisition and code tracking as well as on code tracking lock indication.

Consider the postdetection combining detector model shown in Figure 8.23.

Again we assume that the cross correlation between the I and Q channels is negligible. Without loss of generality we let $W_I = 1$ and $W_Q = \beta$ and also $P_I = \gamma P_Q$.* The correlated inputs to the square law detectors, assuming perfect code tracking, are given by

$$y_I(t) = \sqrt{2P_I}\, d_I(t) \cos \omega_0 t + n_I(t) \qquad (8.10\text{-}17)$$

$$y_Q(t) = \sqrt{2P_Q}\, d_Q(t) \sin \omega_0 t + n_Q(t) \qquad (8.10\text{-}18)$$

The signal to noise spectral density ratio at the output is given by

$$\left(\frac{C}{N_0}\right)_{\text{eff}} = \frac{\alpha^2 (\gamma P_Q + P_Q \beta)^2}{[N'_{0_I} + \beta^2 N'_{0_Q}]} = \frac{(\text{mean signal})^2}{\text{spectral density}} \qquad (8.10\text{-}19)$$

where $P_I = \gamma P_Q$ and where N'_{0_I} and N'_{0_Q} are the respective one-sided effective noise spectral densities and α is the filtering loss in each BPF (assuming equal data rates). Based on reference 4 we have

$$N'_{0_I} = N_0^2 B + 2\alpha \gamma P_Q N_0 \qquad (8.10\text{-}20)$$

---

*Note that this is the opposite definition of β compared to the coherent combining case notation.

$$N'_{0_Q} = N_0^2 B + 2\alpha P_Q N_0 \qquad (8.10\text{-}21)$$

so that using (8.10-19) through (8.10-21) yields

$$\left(\frac{C}{N_0}\right)_{\text{eff}} = \frac{\alpha P_Q}{N_0}\left[\frac{\gamma^2 + 2\gamma\beta + \beta^2}{\frac{1}{\rho} + 2\gamma + \beta^2\left(\frac{1}{\rho} + 2\right)}\right] \qquad (8.10\text{-}22)$$

where ρ is the input SNR in the Q channel defined by

$$\rho = \frac{\alpha P_Q}{N_0 B} \qquad (8.10\text{-}23)$$

In order to determine the optimum value of β we differentiate with respect to β to obtain, after some algebra,

$$\left(4\gamma + \frac{2\gamma}{\rho}\right)\beta^2 + \left(4\gamma^2 + \frac{2\gamma^2}{\rho} - 4\gamma - \frac{2}{\rho}\right)\beta - \left(\frac{2\gamma}{\rho} + 4\gamma^2\right) = 0 \quad (8.10\text{-}24)$$

Although this can be solved, it is somewhat messy. Let us consider the solution at both high-input SNR $(\rho \to \infty)$ and low-input SNR $(\rho \to 0)$.

At high SNR we obtain $(\rho \to \infty)$

$$\beta^2 + (\gamma - 1)\beta - \gamma = 0 \qquad (8.10\text{-}25)$$

or

$$\beta = \frac{1 - \gamma \pm \gamma + 1}{2} = 1, -\gamma \qquad (8.10\text{-}26)$$

The only admissible root is 1; hence at high SNR in a postdetection combining system equal weights are optimum!

Now consider the low SNR case; we obtain

$$\beta^2 - \left(\frac{1}{\gamma} - \gamma\right)\beta - 1 = 0 \qquad (8.10\text{-}27)$$

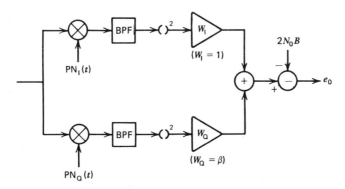

Figure 8.23 Postdetection combining model.

or

$$\beta = \frac{\left(\frac{1}{\gamma} - \gamma\right) \pm \left(\gamma + \frac{1}{\gamma}\right)}{2} = \frac{1}{\gamma}, -\gamma \tag{8.10-28}$$

The only admissible root is $= 1/\gamma$, so that in a low noise environment each channel should be weighted in proportion to its carrier power! This observation follows from the fact that if both weights are multiplied by γ, we find that the Q channel is weighted by unity and the I channel is weighted by γ.

Now consider the high SNR, optimal weighting postcorrelation detector shown in Figure 8.24.

The signal to noise spectral density ratio is given by

$$\left(\frac{C}{N_0}\right)_{\text{eff}} = \frac{\alpha^2 (P_I + P_Q)^2}{N_0^2 B + 2P_I N_0 + N_0^2 B + 2P_Q N_0} \tag{8.10-29}$$

or

$$\left(\frac{C}{N_0}\right)_{\text{eff}} = \frac{\alpha^2 P_T^2}{2N_0^2 B + 2N_0 P_T} \tag{8.10-30}$$

and therefore equal-weight postdetection combining produces a carrier-to-noise density ratio independent of the ratio of P_I to P_Q! Consequently, where SNR is important, for example, in code tracking, this weighting yields an SNR independent of the I to Q power ratio ($P_T = P_I + P_Q$).

In order to assess the performance of the postdetection combining as compared to single-channel processing, note that we can write $(C/N_0)_{\text{eff}}$ as

$$\left(\frac{C}{N_0}\right)_{\text{eff}} = \frac{\alpha P_T}{N_0} \left(\frac{1}{2/\rho + 2}\right) \tag{8.10-31}$$

where ρ, the total input SNR, is defined by

$$\rho = \frac{\alpha P_T}{N_0 B} \tag{8.10-32}$$

Now if all the power were in one phase of the quadriphase signal, we

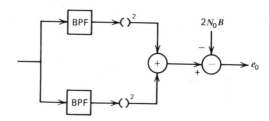

Figure 8.24 Equal weight postdetection combiner.

would obtain a carrier-to-noise spectral density of

$$\frac{C}{N_0} = \frac{\alpha P_T}{N_0} \left[\frac{1}{(1/\rho) + 2} \right] \qquad (8.10\text{-}33)$$

Hence we see that the ratio of C/N_0 for quadriphase processing to C/N_0 for single-channel processing is given by

$$\frac{(C/N_0)_{\text{eff}}}{C/N_0} = \left[\frac{(1/\rho) + 2}{(2/\rho) + 2} \right] = L \qquad (8.10\text{-}34)$$

where L denotes the loss relative to the case that all the power is in one channel.

Therefore we conclude that at low SNR ($\rho \ll 1$)

$$L = -3 \, \text{dB}$$

and at high SNR ($\rho \gg 1$)

$$L = 0 \, \text{dB}$$

and at unity SNR

$$L = -1.25 \, \text{dB}$$

Therefore at high-input SNRs postdetection combining with quadriphase signal processing is as efficient as the case for biphase signaling with the same power, at least as far as the carrier-to-effective-noise spectral density ratio is concerned.

8.11   PRE- AND POSTCORRELATOR SPURIOUS INTERFERENCE

When designing PN code tracking systems, in fact, virtually any type of communication system, it is necessary to assess the effect of spurious interferences. If these interferences arise prior to the despreader in PN type spread spectrum systems, then their effect is diminished by virtue of the fact that the correlator spreads them out. If a spur (spurious response) occurs after the despreader, then the effect on performance depends upon the data detector.

8.11.1   Precorrelator Spurious Interference

First we consider the case in which the interference occurs prior to the correlator. The received signal, interferer, and noise are modeled as

$$y(t) = \sqrt{2P} \, \text{PN}(t) \, d(t) \sin(\omega_0 t + \theta_0) + \sqrt{2P_1} \cos(\omega_1 t + \theta_1) + n(t) \qquad (8.11\text{-}1)$$

where P is the signal power, P_1 the interference, $\text{PN}(t)$ the PN sequence, $d(t)$ the data sequence, and $n(t)$ is modeled as WGN. We also assume that $\omega_0 \neq \omega_1$ and that θ_1 is a uniform random variable in the interval $(0, 2\pi)$.

The despreader, assuming perfect tracking, can be modeled as shown in Figure 8.25.

After ideal despreading, the interfering signal is spread and the resulting signal is given by

$$e_0(t) = \sqrt{2P}\ d(t)\sin(\omega_0 t + \theta_0) + \sqrt{2P_1}\ PN(t)\cos(\omega_1 t + \theta_1) + n'(t)$$
$$(8.11\text{-}2)$$

where $n'(t)$ is given by

$$n'(t) = PN(t)n(t) \tag{8.11-3}$$

Now consider the autocorrelation function of the interference:

$$R_1(\tau) = 2P_1\ \overline{PN(t)\,PN(t+\tau)\cos(\omega_1 t + \theta_1)\cos[\omega(t+\tau)+\theta_1]}$$
$$(8.11\text{-}4)$$

where the overbar denotes ensemble averaging, and we model the PN sequence as a unit-amplitude random stationary NRZ waveform.

Evaluating, we obtain

$$R_1(\tau) = P_1 R_{PN}(\tau)\cos\omega_1\tau \tag{8.11-5}$$

The spectral density at $f_1 = \omega_1/2\pi$ is obtained from

$$\mathscr{S}_1(f_1) = P_1 \int_{-\infty}^{\infty} R_{PN}(\tau) \left(\frac{e^{i\omega_1\tau} + e^{-i\omega_1\tau}}{2} \right) e^{i\omega_1\tau}\, d\tau \tag{8.11-6}$$

$$\mathscr{S}_1(f_1) = \frac{P_1}{2}[\mathscr{S}_{PN}(2\omega_1) + \mathscr{S}_{PN}(0)] \cong \frac{P_1}{2}\mathscr{S}_{PN}(0) \tag{8.11-7}$$

In the usual case that the PN code chip rate is much greater than the data rate, we have $\mathscr{S}_{PN}(0) = T_c$ so that

$$\mathscr{S}_1(f_1) \cong \frac{P_1 T_c}{2} \tag{8.11-8}$$

If it is required that this interfering signal produce negligible degradation, then the noise spectral density $N_0/2$ must be much larger than $\mathscr{S}_1(f_1)$ so that

$$\frac{N_0'}{2} \gg \frac{P_1 T_c}{2} \tag{8.11-9}$$

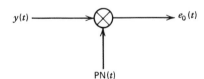

$y(t) \longrightarrow \bigotimes \longrightarrow e_0(t)$

$PN(t)$

Figure 8.25 Despreader interference model.

where N_0' is the one-sided spectral density of $PN(t)n(t)$. Assuming that this is essentially N_0, we have the requirement

$$N_0 B_{PN} \gg P_I \qquad (8.11\text{-}10)$$

where $B_{PN} = 1/T_c$ is the noise bandwidth of a random (unit amplitude) sequence NRZ waveform (two-sided). We have assumed that $|f_0 - f_1| T_c \ll 1$ and that $f_0 \neq f_1$ in the above analysis.

8.11.2 Postcorrelator Spurious Interference

In order to access the effect of postcorrelator spurious interference it is necessary to specify the symbol waveform shape (NRZ, Manchester, etc.) so that the filtering effect of the symbol detector can be accounted for. To provide a concrete example consider a Manchester data detector.

The received signal is modeled as

$$\begin{aligned}
y(t) = \sqrt{2P}\, d(t) \cos(\omega_0 t) + \sqrt{2P_I} \cos(\omega_1 t + \theta) \\
+ \sqrt{2}\, n_c(t) \cos(\omega_0 t) + \sqrt{2}\, n_s(t) \sin(\omega_0 t)
\end{aligned} \qquad (8.11\text{-}11)$$

where P is the signal power, P_I is the interference power, and $n_c(t)$ and $n_s(t)$ are independent, white, band-limited Gaussian random processes.

After demodulation by a carrier tracking loop, the baseband signal is given by

$$y_0(t) = \sqrt{P}\, d(t) \cos \phi + \sqrt{P_I} \cos(\phi + \Delta\omega t + \theta) + n'(t) \qquad (8.11\text{-}12)$$

with $n'(t)$ being an essentially band-limited WGN process, and $\Delta\omega = \omega_1 - \omega_0$. Out of the Manchester matched filter at the end of a symbol time (T_s sec), assuming $\phi = 0$, we have

$$\begin{aligned}
e_0(T) = \sqrt{P}\, T_s d(T_s) + \sqrt{P_I} \left[\int_0^{T_s/2} \cos(\Delta\omega t + \theta) dt - \int_{T_s/2}^{T_s} \cos(\Delta\omega t + \theta)\, dt \right] \\
+ e_n(T)
\end{aligned} \qquad (8.11\text{-}13)$$

where $e_n(t)$ is a Gaussian random variable* having mean zero and variance equal to $(N_0/2)T_s$.
Alternately we can write

$$e_0(T) = \sqrt{P}\, T_s d(T_s) + \frac{\sqrt{P_I}}{\Delta\omega} \mathrm{Re}\left\{ e^{j\theta} \left(\int_0^{\Delta\omega T_s/2} e^{jx} dx - \int_{\Delta\omega T_s/2}^{\Delta\omega T_s} e^{jx} dx \right) \right\} + e_n(T) \qquad (8.11\text{-}14)$$

After some algebra we obtain

$$e_0(T) = \sqrt{P}\, T_s d(T_s) + \frac{\sqrt{P_I}}{\Delta\omega} \mathrm{Im}\left[e^{j\theta} e^{j\Delta\omega T_s/2} \left(2 - 2\cos \frac{\Delta\omega T_s}{2} \right) \right] + e_n(T) \qquad (8.11\text{-}15)$$

---

*Note that $d(T_s)$ is defined to be the value of $d(t)$ at $t = T_s^-$, that is, just before the transition time.

which can finally be written as

$$e_0(T) = \sqrt{P}\, T_s d(T_s) + \sqrt{P_1}T_s\, \frac{\sin^2(\Delta\omega T_s/4)\, \sin[(\Delta\omega T_s/2) + \theta]}{(\Delta\omega T_s/4)} + e_n(T_s)$$

(8.11-16)

Hence under worst case conditions when $\theta = 0$ the interference voltage is only suppressed to $0.66\sqrt{P_1}\, T_s$ at $fT_s \cong 0.55$, or about -3.6 dB. We see that interferers occurring after the correlator can, in general, be much more troublesome from the point of view of a bit error rate degradation than interferers before the correlator.

PROBLEM 5

(a) For NRZ data show that postcorrelator spurious (line spectra) interference produces, with coherent demodulation, a detection variable of the form ($\Delta\omega = \omega_0 - \omega_1$)

$$e_0(T) = \sqrt{P}\, T_s d(T_s) + \sqrt{P_1}\, T_s\, \frac{\sin(\Delta\omega T_s/2)}{(\Delta\omega T_s/2)} \cos(\Delta\omega T_s/2 + \theta) + e_n(T_s)$$

where $e_n(T_s)$ is a Gaussian random variable having zero mean and a variance of $(N_0/2)T_s$.

Assume that the signal plus interference is modeled as

$$y(t) = \sqrt{2P}\, d(t)\cos\omega_0 t + \sqrt{2P_1}\cos(\omega_1 t + \theta) + n(t)$$

8.12 SPECTRA OF THE DESPREAD PN WAVEFORM

In this section we consider the spectra of the received despread signal at the point just after the multiplier denoted by $C(t, \epsilon)$, as illustrated in Figure 8.26.

In the model $PN(t - T)$ is the received PN waveform with delay T sec. In addition $\hat{T}$ is the local estimate of T, with $\epsilon = T - \hat{T}$ being the timing difference or error between codes. As long as the modulation is noncoherent with the PN waveform, the PSK modulated spectra are just translated versions of the baseband spectra. Most of the results in this section are due to the work of Gill [11] and Gill and Spilker [10]. An alternative

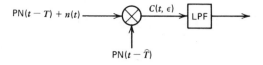

Figure 8.26 Baseband model of despread signal, $C(t, \epsilon)$, $\epsilon = T - \hat{T}$.

viewpoint of the despread spectra, for the random data case, can be found in Cartier [12].

Before we compute the spectra of the despread signal, $C(t, \epsilon)$, we determine the spectra of the original PN waveform, $PN(t)$.

8.12.1 Power Spectral Density of a PN Waveform

It is convenient to compute first the autocorrelation function of $PN(t)$ and then by means of the Fourier transform obtain the power spectral density. Note that for a PN sequence the autocorrelation is either 1 or $-1/N$ (N is the sequence length), depending on whether the time displacement between the sequence and a shifted version is zero mod N or nonzero mod N. The autocorrelation function is given by

$$R(\tau) = \frac{1}{NT_c} \int_0^{NT_c} PN(t)\, PN(t + \tau)\, dt \qquad (8.12\text{-}1)$$

where T_c is the chip time. For a fixed value of τ such that $-T_c < \tau < T_c$ it is clear that the value of the autocorrelation is a linear combination of the autocorrelation for $\tau = 0$ and $\tau = T_c$. For example, if the time displacement is one-half of a chip (see Figure 8.27), one-half of the product will have a correlation value of 1 (shaded areas) and one-half of the product will have a correlation value of $-1/N$. Further, when $|\tau| \geq T_c$ and $|\tau| < NT_c$, one-half of the correlation value will be $-1/N$ and the other half will also have a correlation value of $-1/N$. The resulting sum is just $-1/N$. Therefore we conclude that the autocorrelation function is as shown in Figure 8.28.

To obtain the power spectral density we require the Fourier transform. First define the "triangle function" $\underset{T_c}{\Lambda}(\tau)$ as

$$\underset{T_c}{\Lambda}(\tau) = \begin{cases} 1 - \dfrac{|\tau|}{T_c} & |\tau| \leq T_c \\ 0 & \text{elsewhere} \end{cases} \qquad (8.12\text{-}2)$$

Then the Fourier transform of $\underset{T_c}{\Lambda}(\tau)$ is given by [13]

$$\mathscr{F}\left[\underset{T_c}{\Lambda}(\tau)\right] = T_c \, \text{sinc}^2(\pi f T_c) \qquad (8.12\text{-}3)$$

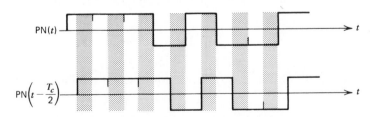

Figure 8.27 PN waveforms used to compute $R_{PN}(\tau)$.

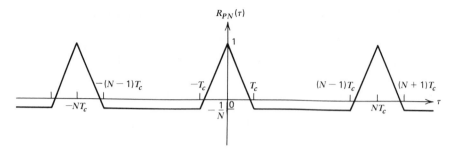

Figure 8.28   PN waveform autocorrelation function.

where

$$\text{sinc}(x) = \frac{\sin x}{x} \tag{8.12-4}$$

Since $R_{PN}(\tau)$ is periodic, it can be expressed as the sum of a dc term and an infinite series of triangle functions separated by the period of NT_c sec. Thus

$$R_{PN}(\tau) = -\frac{1}{N} + \frac{N+1}{N} \Lambda_{T_c}(\tau) * \sum_{m=-\infty}^{\infty} \delta(\tau + NmT_c) \tag{8.12-5}$$

where the $*$ denotes the convolution operation. Using the result [13]

$$\mathscr{F}\left[\sum_{m=-\infty}^{\infty} \delta(\tau + mNT_c)\right] = \frac{1}{NT_c} \sum_{m=-\infty}^{\infty} \delta\left(f + \frac{m}{NT_c}\right) \tag{8.12-6}$$

we obtain

$$\mathscr{S}_{PN}(f) = -\frac{1}{N} \delta(f) + \frac{N+1}{N^2} \text{sinc}^2(\pi f T_c) \sum_{m=-\infty}^{\infty} \delta\left(f + \frac{m}{NT_c}\right) \tag{8.12-7}$$

Combining terms at dc yields our result:

$$\mathscr{S}_{PN}(f) = \frac{1}{N^2} \delta(f) + \frac{N+1}{N^2} \sum_{\substack{m=-\infty \\ m\neq 0}}^{\infty} \text{sinc}^2\left(\frac{m\pi}{N}\right) \delta\left(f + \frac{m}{NT_c}\right) \tag{8.12-8}$$

A sketch of this spectral density is shown in Figure 8.29.

If the PN waveform is phase-shift-keyed onto a carrier and if the carrier frequency and the code are not coherent, the resulting spectral

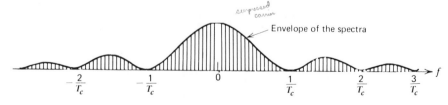

Figure 8.29   Power spectral density of a PN waveform.

density is given by

$$\mathscr{S}_c(f) = \frac{P_c}{2}\,\mathscr{S}_{PN}(f + f_0) + \frac{P_c}{2}\,\mathscr{S}_{PN}(f - f_0) \qquad (8.12\text{-}9)$$

where P_c is the unmodulated carrier power; that is, the baseband spectra are shifted up to f_0 and down to $-f_0$.

8.12.2 Power Spectral Density of a Decorrelated PN Waveform

Now we consider the power spectral density of $C(t, \epsilon)$ as depicted in Figure 8.26. In order to proceed it is convenient to utilize a decomposition introduced by Gill and Spilker [10]. Consider the decomposition

$$C(t, \epsilon) = p(t, \epsilon) + q(t, \epsilon) \qquad (8.12\text{-}10)$$

where $p(t, \epsilon)$ is a binary-valued periodic function and $q(t, \epsilon)$ is a ternary-level pseudorandom function. An example of this decomposition is shown in Figure 8.30.

The first row is a PN sequence of "ones" and "zeros" of length 15. The next two rows are the waveforms $PN(t)$ and $PN(t + T_c/4)$. In the fourth row the product is depicted for the case when the timing error is $\epsilon = T_c/4$. The fifth row is a plot of a periodic binary-valued waveform denoted by $p(t, T_c/4)$. The sixth row is the ternary-valued waveform $q(t, T_c/4)$ for

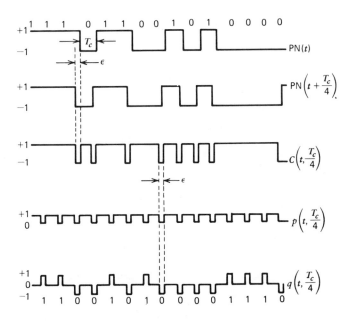

Figure 8.30 Decomposition of the product of PN sequences. $C(t) = p(t) + q(t)$, for $\epsilon = T_c/4$.

which its nonzero values are a shifted version of the original PN sequence. This clever decomposition of $C(t, \epsilon)$ makes the evaluation of the autocorrelation function a relatively straightforward task, as will be seen below. The autocorrelation function $R_c(\tau, \epsilon)$ is given by

$$R_c(\tau, \epsilon) = R_p(\tau, \epsilon) + R_q(\tau, \epsilon) + R_{pq}(\tau, \epsilon) + R_{pq}(-\tau, \epsilon) \qquad (8.12\text{-}11)$$

To obtain the power spectral density we shall evaluate each term separately and then take the Fourier transform.

Consider the case when $0 \le |\epsilon| \le T_c/2$. First we determine the autocorrelation function of the periodic component, $R_p(\tau, \epsilon)$. In Figure 8.31 $R_p(\tau, \epsilon)$ is plotted, again, for the case $\epsilon = T_c/4$, along with $R_q(\tau, \epsilon)$ and $R_{pq}(\tau, \epsilon)$.

The maximum correlation is $(1 - |\epsilon|/T_c)$ and it drops off to $(1 - 2|\epsilon|/T_c)$ until it increases to $(1 - |\epsilon|/T_c)$ at $|\tau| = T_c$, and so on. Therefore the autocorrelation function is given by

$$R_p(\tau, \epsilon) = \left(1 - \frac{2|\epsilon|}{T_c}\right) + \frac{|\epsilon|}{T_c} \Lambda_{|\epsilon|}(\tau) * \sum_{m=-\infty}^{\infty} \delta(\tau + mT_c) \qquad (8.12\text{-}12)$$

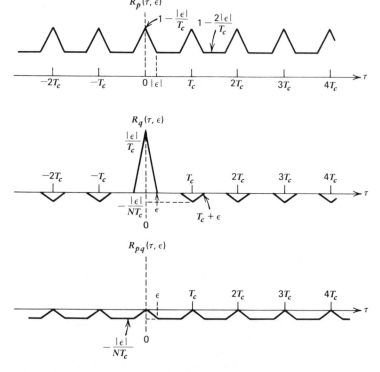

Figure 8.31 Component autocorrelation functions for $0 \le |\epsilon| \le T_c/2$.

where $\Lambda(\tau)$ is defined in (8.12-2). Now we consider $R_q(\tau, \epsilon)$. Since the nonzero values form a shifted version of the PN sequence with a maximum correlation value of $|\epsilon|$ and since the correlation is zero where the timing difference is such that the "pulses" do not overlap, the resulting autocorrelation function is shown in the second figure of Figure 8.31. Therefore

$$R_q(\tau, \epsilon) = \frac{|\epsilon|}{T_c} \Lambda_{|\epsilon|}(\tau) * \sum_{m=-\infty}^{\infty} \delta(\tau + mNT_c) - \frac{|\epsilon|}{NT_c} \Lambda_{|\epsilon|}(\tau) * \sum_{\substack{m=-\infty \\ m \neq 0 \pm N, \pm 2N, \dots}}^{\infty} \delta(\tau + mT_c)$$

$$(8.12\text{-}13)$$

adding and subtracting triangle functions at $0, \pm N, \pm 2N, \dots$, we obtain

$$R_q(\tau, \epsilon) = \frac{N+1}{N} \frac{|\epsilon|}{T_c} \Lambda_{|\epsilon|}(\tau) * \sum_{m=-\infty}^{\infty} \delta(\tau + mNT_c) - \frac{|\epsilon|}{NT_c} \Lambda_{|\epsilon|}(\tau) * \sum_{m=-\infty}^{\infty} (\tau + mT_c)$$

$$(8.12\text{-}14)$$

Now consider the cross correlation between $p(t, \epsilon)$ and $q(t, \epsilon)$. Since $R_{pq}(\tau, \epsilon) = R_{pq}(-\tau, \epsilon)$, it is necessary only to consider $R_{pq}(\tau, \epsilon)$. From Figure 8.30 it is clear that the cross correlation is given by the last figure of Figure 8.31. At $\tau = 0$, $R_{pq}(0, \epsilon) = 0$, but as τ increases the value of $R_{pq}(\tau, \epsilon)$ decreases until it reaches $-|\epsilon|/NT_c$. From this point it remains constant until τ reaches $T_c - |\epsilon|$. Again it decreases linearly toward zero as τ approaches T_c, and so on. Therefore we have

$$2R_{pq}(\tau, \epsilon) = -\frac{2|\epsilon|}{NT_c} + \frac{2|\epsilon|}{NT_c} \Lambda_{|\epsilon|}(\tau) * \sum_{m=-\infty}^{\infty} \delta(\tau + mT_c) \qquad (8.12\text{-}15)$$

Combining these four component correlation functions produces

$$R_c(\tau, \epsilon) = \left(1 - \frac{2|\epsilon|}{T_c} - \frac{2|\epsilon|}{NT_c}\right) + \left(\frac{|\epsilon|}{T_c} + \frac{|\epsilon|}{NT_c}\right) \Lambda_{|\epsilon|}(\tau) * \sum_{m=-\infty}^{\infty} \delta(\tau + mT_c)$$

$$+ \frac{|\epsilon|}{T_c}\left(\frac{N+1}{N}\right) \Lambda_{|\epsilon|}(\tau) * \sum_{m=-\infty}^{\infty} \delta(\tau + mNT_c) \qquad (8.12\text{-}16)$$

Taking Fourier transforms produces

$$\mathcal{S}_c(f) = \left(1 - \frac{2|\epsilon|}{T_c} - \frac{2|\epsilon|}{NT_c}\right) \delta(f) + \left(\frac{|\epsilon|}{T_c}\right) \frac{N+1}{N} \frac{|\epsilon|}{T_c} \text{sinc}^2(\pi f \epsilon) \sum_{m=-\infty}^{\infty} \delta\left(f + \frac{m}{T_c}\right)$$

$$+ \frac{|\epsilon|}{T_c}\left(\frac{N+1}{N}\right) \frac{|\epsilon|}{NT_c} \text{sinc}^2 \pi f \epsilon \sum_{m=-\infty}^{\infty} \delta\left(f + \frac{m}{NT_c}\right) \qquad (8.12\text{-}17)$$

Combining the terms at dc, we obtain, finally,

$$\mathcal{S}_c(f, \epsilon) = \left(1 - \frac{|\epsilon|}{T_c} \frac{N+1}{N}\right)^2 \delta(f) + \frac{N+1}{N}\left(\frac{\epsilon}{T_c}\right)^2 \sum_{\substack{m=-\infty \\ m \neq 0}}^{\infty} \text{sinc}^2\left(\frac{\pi m \epsilon}{T_c}\right) \delta\left(f + \frac{m}{T_c}\right)$$

$$+ \frac{N+1}{N^2}\left(\frac{\epsilon}{T_c}\right)^2 \sum_{\substack{m=-\infty \\ m \neq 0}}^{\infty} \text{sinc}^2\left(\frac{\pi m \epsilon}{NT_c}\right) \delta\left(f + \frac{m}{NT_c}\right) \qquad 0 \leq |\epsilon| \leq T_c/2$$

$$(8.12\text{-}18)$$

For $T_c/2 \le |\epsilon| \le T_c$ the autocorrelation functions are plotted in Figure 8.32.

Using the same techniques, we obtain

$$R_p(\tau, \epsilon) = \left(1 - \frac{|\epsilon|}{T_c}\right) \underset{T_c - |\epsilon|}{\Lambda} (\tau) * \sum_{m=-\infty}^{\infty} \delta(\tau + mT_c) \qquad (8.12\text{-}19)$$

$$R_q(\tau, \epsilon) = \left(\frac{|\epsilon|}{T_c} + \frac{|\epsilon|}{NT_c}\right) \underset{|\epsilon|}{\Lambda} (\tau) * \sum_{m=-\infty}^{\infty} \delta(\tau + mNT_c)$$

$$- \frac{|\epsilon|}{NT_c} \underset{|\epsilon|}{\Lambda} (\tau) * \sum_{m=-\infty}^{\infty} \delta(\tau + mT_c) \qquad (8.12\text{-}20)$$

$$2R_{pq}(\tau, \epsilon) = -2\frac{1}{N}\left(1 - \frac{|\epsilon|}{T_c}\right) + \frac{2}{N}\left(1 - \frac{|\epsilon|}{T_c}\right) \underset{T_c - |\epsilon|}{\Lambda} (\tau) * \sum_{m=-\infty}^{\infty} \delta(\tau + mT_c)$$

$$(8.12\text{-}21)$$

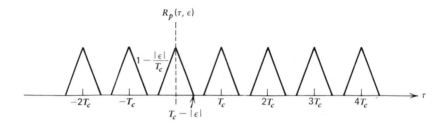

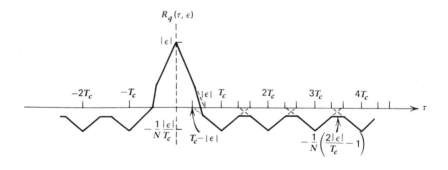

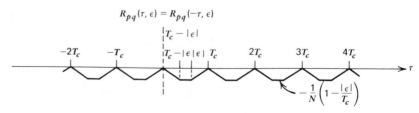

Figure 8.32 Correlation functions for $T_c/2 \le |\epsilon| \le T_c$.

Combining and taking Fourier transforms, we obtain

$$\mathscr{S}_c(f, \epsilon) = -\frac{2}{N}\left(1 - \frac{|\epsilon|}{T_c}\right)\delta(f) + \frac{N+2}{N}\left(1 - \frac{|\epsilon|}{T_c}\right)^2 \text{sinc}^2 \pi f(T_c - |\epsilon|) \sum_{m=-\infty}^{\infty} \delta\left(f + \frac{m}{T_c}\right)$$

$$-\frac{1}{N}\left(\frac{\epsilon}{T_c}\right)^2 \text{sinc}^2(\pi f\epsilon) \sum_{m=-\infty}^{\infty} \delta\left(f + \frac{m}{T_c}\right)$$

$$+\frac{N+1}{N^2}\left(\frac{\epsilon}{T_c}\right)^2 \text{sinc}^2 \pi f\epsilon \sum_{m=-\infty}^{\infty} \delta\left(f + \frac{m}{NT_c}\right) \tag{8.12-22}$$

Note that

$$\text{sinc}^2[\pi f(T_c - |\epsilon|)] \sum_{m=-\infty}^{\infty} \delta\left(f + \frac{m}{T_c}\right) = \sum_{m=-\infty}^{\infty} \frac{\sin^2 m\pi(1 - |\epsilon|/T_c)}{m\pi(1 - |\epsilon|/T_c)^2}\delta\left(f + \frac{m}{T_c}\right)$$

$$= \delta(f) + \frac{(\epsilon/T_c)^2}{(1 - |\epsilon|/T_c)^2} \sum_{\substack{m=-\infty \\ m \neq 0}}^{\infty} \text{sinc}^2\left(\frac{m\pi\epsilon}{T_c}\right)\delta\left(f + \frac{m}{T_c}\right) \tag{8.12-23}$$

Combining the terms at $f = 0$, we finally obtain the same result as obtained in (8.12-18).

$$\mathscr{S}_c(f, \epsilon) = \left(1 - \frac{|\epsilon|}{T_c}\frac{N+1}{N}\right)^2 \delta(f) + \frac{N+1}{N}\left(\frac{\epsilon}{T_c}\right)^2 \sum_{\substack{m=-\infty \\ m \neq 0}}^{\infty} \text{sinc}^2\left(\frac{\pi m\epsilon}{T_c}\right)\delta\left(f + \frac{m}{T_c}\right)$$

$$+\frac{N+1}{N^2}\left(\frac{\epsilon}{T_c}\right)^2 \sum_{\substack{m=-\infty \\ m \neq 0}}^{\infty} \text{sinc}^2\left(\frac{\pi m\epsilon}{NT_c}\right)\delta\left(f + \frac{m}{NT_c}\right) \qquad T_c/2 \leq |\epsilon| \leq T_c \tag{8.12-24}$$

We conclude that the above result holds for all ϵ such that $|\epsilon| \leq T_c$. This result was first obtained by Gill [11]. The resulting power spectral densities for the case that $|\epsilon| = 0$, $T_c/10$, $T_c/2$, and T_c are shown in Figure 8.33. Notice that when $|\epsilon| = T_c$ the result degenerates to the power spectral density of a PN waveform, as it should, and when $\epsilon = 0$ all the power resides at dc.

Hence we see that small errors reduce the correlation value and produce a spread signal wider than the original spectrum. As the timing error increases towards $\epsilon = T_c$ the spectrum approaches the spectrum of an unadulterated PN waveform.

The above calculations are much easier to make in the case the NRZ waveform is random rather than pseudorandom. For example, consider the case $|\epsilon| \leq T_c$. The same decomposition for random sequences can be made with $q(t, \epsilon)$ now having random (with probability $= 1/2$) nonzero amplitudes (± 1). The autocorrelation function is given by

$$R_c(\tau, \epsilon) = E\langle[p(t, \epsilon) + q(t, \epsilon)][p(t + \tau, \epsilon) + q(t + \tau, \epsilon)]\rangle \tag{8.12-25}$$

where E is the ensemble average and $\langle\cdot\rangle$ indicates a time average. We obtain

$$R_c(\tau, \epsilon) = R_p(\tau, \epsilon) + R_q(\tau, \epsilon) \tag{8.12-26}$$

since the cross terms have the value zero.

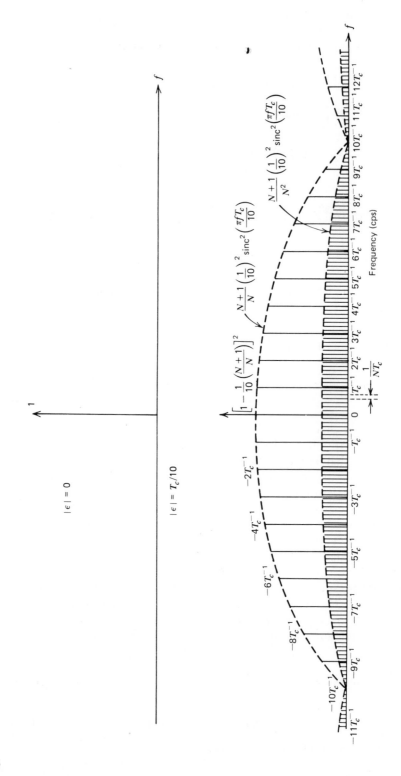

$|\epsilon| = 0$

$|\epsilon| = T_c/10$

$\left[1 - \frac{1}{10}\left(\frac{N+1}{N}\right)^2\right]$

$\frac{N+1}{N}\left(\frac{1}{10}\right)^2 \operatorname{sinc}^2\left(\frac{\pi f T_c}{10}\right)$

$\frac{N+1}{N^2}\left(\frac{1}{10}\right)^2 \operatorname{sinc}^2\left(\frac{\pi f T_c}{10}\right)$

Frequency (cps)

$\frac{1}{NT_c}$

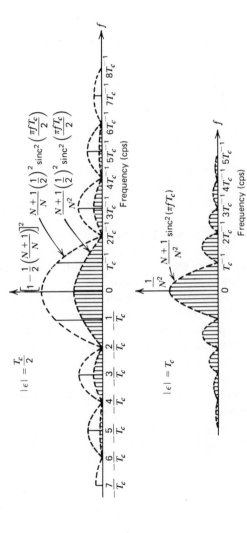

Figure 8.33 Decorrelated spectra for various timing errors.

Consider $T_c/2 \le |\epsilon| \le T_c$. Since $p(t, \epsilon)$ is periodic we have, as before

$$R_p(\tau, \epsilon) = \left(1 - \frac{|\epsilon|}{T_c}\right) \underset{T_c-|\epsilon|}{\Lambda}(\tau) * \sum_{m=-\infty}^{\infty} \delta(\tau + mT_c) \qquad (8.12\text{-}27)$$

Since all digits have been assumed to be statistically independent, we have

$$R_q(\tau, \epsilon) = \frac{|\epsilon|}{T_c} \underset{|\epsilon|}{\Lambda}(\tau) \qquad (8.12\text{-}28)$$

The resulting power spectral density is given by, after converting to sinc terms of $|\epsilon|$,

$$\mathcal{S}_c(f, \epsilon) = \left(1 - \frac{|\epsilon|}{T_c}\right)^2 \delta(f) + \left(\frac{\epsilon}{T_c}\right)^2 \sum_{\substack{m=-\infty \\ m \ne 0}}^{\infty} \text{sinc}^2\!\left(\frac{m\pi\epsilon}{T_c}\right) \delta\!\left(f + \frac{m}{T_c}\right)$$

$$+ \frac{\epsilon^2}{T_c} \text{sinc}^2(\pi f \epsilon) \qquad \frac{T_c}{2} \le |\epsilon| \le T_c, \qquad \forall f \qquad (8.12\text{-}29)$$

Since this result can be shown to hold for $0 \le |\epsilon| \le T_c/2$ also, it applies to the case where the timing error is less than one chip. Notice that there are both a discrete line spectrum and a continuous spectrum in this random symbol model.

Also notice that for large N the results are essentially identical to the PN case except that the last term for the PN spectrum, with discrete lines spaced apart by $1/T_c$ Hz, is replaced with a continuous spectrum having virtually the same average spectral density.

We conclude that when the code length is large the random waveform model is quite adequate, as intuition would suggest. It should be cautioned that for very narrowband systems the discrete line spectra would have to be accounted for; a continuous spectrum model should not be assumed.

Now we consider the case of random NRZ code symbols rather than pseudorandom symbols for the case timing errors are greater than one chip. Then the resulting product $C(t, \epsilon)$ can be written as the infinite sum of two random waveforms:

$$C(t, \epsilon) = \sum_{m=-\infty}^{\infty} a_m u(t - mT_c) + \sum_{n=-\infty}^{\infty} b_n v(t - nT_c) \qquad (8.12\text{-}30)$$

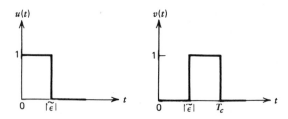

Figure 8.34 $u(t)$ and $v(t)$ waveforms.

where $u(t)$ and $v(t)$ are shown in Figure 8.34, and a_m and b_n are independent random sequences taking on the values of $+1$ and -1. $|\bar{\epsilon}|$ is the magnitude of the timing reduced mod T_c. Since the sequences are independent, the power spectral density is just the sum of the power spectral density of each series. Therefore, from Chapter 2

$$\mathcal{S}_c(f, \epsilon) = \frac{1}{T_c} |S_u(f)|^2 + \frac{1}{T_c} |S_v(f)|^2 \qquad (8.12\text{-}31)$$

where $S_u(t)$ is the Fourier transform of $u(t)$ and $S_v(t)$ is the Fourier transform of $v(t)$. We find that

$$S_u(f) = |\bar{\epsilon}| \operatorname{sinc}(\pi f |\bar{\epsilon}|)\, e^{-j\pi f |\bar{\epsilon}|} \qquad (8.12\text{-}32)$$

$$S_v(f) = (T_c - |\bar{\epsilon}|) \operatorname{sinc}[\pi f(T_c - |\bar{\epsilon}|)] \exp[-j\pi f(|\bar{\epsilon}| + T_c)] \qquad (8.12\text{-}33)$$

so that

$$\mathcal{S}_c(f, \epsilon) = \frac{|\bar{\epsilon}|^2}{T_c} \operatorname{sinc}^2(\pi f \bar{\epsilon}) + \frac{(T_c - |\bar{\epsilon}|)^2}{T_c} \operatorname{sinc}^2[\pi f(T_c - |\bar{\epsilon}|)] \qquad |\epsilon| \geq T_c$$
$$(8.12\text{-}34)$$

Hence we see that, for example, when $|\bar{\epsilon}| = T_c/2$ the spectral width is doubled over the original random NRZ sequence. We notice that the spectra is continuous with no discrete components.

PROBLEM 6

Let $|\epsilon| < T_c$ and use the fact that for random data the output $C(t, \epsilon)$ can be

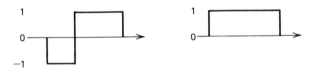

viewed as having one of two symbols ($\epsilon > 0$). Then using the result for purely random data (Chapter 2) obtain the result obtained for $|\epsilon| \leq T_c$, above. Note that for $\epsilon < 0$ two different but related symbols arise, but the same spectrum results in either case (see [12]).

8.13 SURFACE ACOUSTIC WAVE DEVICES

Surface acoustic wave (SAW) devices have certain general characteristics that make them ideal choices for use in spread spectrum communication systems, which must process high data rate signals.

SAW technology, relatively new, is based on an analog device utilizing Rayleigh wave propagation on lithium niobate ($LiNbO_3$) or quartz crystals. Convolvers having a bandwidth of 50 MHz with time bandwidths on the order of 1000 have been built [14, 15]. Furthermore, SAWs have a dynamic range approaching 100 dB [16].

A SAW is an elastic wave that travels along the surface of a solid and is confined to the vicinity of that surface. Bulk waves, on the other hand, occupy the entire cross section of the medium. Because SAWs propagate with a velocity four to five orders of magnitude lower than that of electromagnetic waves, SAW technology dramatically reduces the size and weight of matched filters.

A transducer, commonly used to convert electrical energy to mechanical energy and *vice versa*, is a piezoelectric material which, when compressed or elongated, produces an electric field inside the material. Further, the application of an electric field to the material results in compression or elongation of the material. Therefore by connecting metal contacts to a piece of piezoelectric material in the form of a plate at both ends we form an interdigital transducer (IDT) on the piezoelectric material, commonly referred to as a substrate. A typical IDT is shown in Figure 8.35. Metal strips are fastened to the substrate as shown in Figure 8.35, forming fingers. The usual width of a finger is $\lambda/4$, and the spacing between windows is also $\lambda/4$, where λ is the wavelength of the substrate. When an rf voltage is applied to the input terminals an electric field is set up between the adjacent fingers simultaneously. Consequently the piezoelectric material has a wave traveling toward the output fingers and toward the output terminals. The more fingers placed on the substrate, the more efficient the electrical to acoustical conversion. The detection process, that is, the conversion from acoustical to electrical energy, requires the same configuration at the output end of the substrate. It can be shown that although a greater number of finger pairs produces greater efficiency it also causes a reduction in bandwidth.

Propagation losses range from 0.2 dB/μsec delay to 0.4 dB μsec delay at VHF for materials like quartz and lithium noibate [17] and go up to 1 dB/μsec delay for lithium niobate at 1 GHz.

Another device that can be easily implemented is the transversal filter. A transversal filter is a linear combination of the signal weighted and

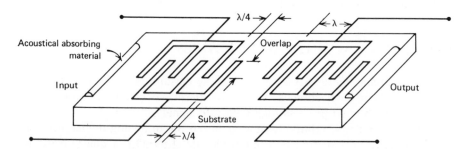

Figure 8.35 An interdigital transducer with the nominal finger spacing.

phased so that the transfer function is given by

$$H(\omega) = \sum_{k=0}^{N-1} a_k e^{-j\omega kT} \qquad (8.13-1)$$

where the a_k are real coefficients. By adjusting the amount of overlap of the finger pairs the various magnitudes can be obtained. Hence by using a single IDT with variable overlaps the composite voltage produced will be the sum of the contributions due to each finger pair. The transversal filter is a cheap and simple way to implement a bandpass filter. The main limitation of SAW filters arises from the finite size of the SAW device or from the attenuation of the wave. Both have a broadbanding effect that makes it imposible to achieve a very narrowband response or a shape factor very close to unity with a passive filter [16]. Although lithium niobate has a higher maximum Q than quartz, its temperature coefficient excludes its use in most high Q requirements [16]. At greater than 55 dB out-of-band rejections have been reported with amplitude ripples of ± 1 dB [18].

SAW filter are presently used in TDMA communication links of the AWACS program, as well as in the channelized receivers of the Westinghouse jamming pod for the ALQ-131 update [18]. Further, SAW bandpass filters have been used in deep-space transponders and spread spectrum receivers [20]. Finally in the commercial sector, SAW devices are being used in TV and CATV as exciters, modulators, and processors, and for minimizing BER and interpulse interference in digital communications equipment [18].

Other applications of SAW devices include oscillators. SAW oscillators typically have higher phase noise levels than crystal oscillators under a vibration-free environment; however under reasonable levels of vibration (8 G rms) SAW oscillators have an unaffected phase noise level, whereas crystal oscillators, except with sophisticated suspense configurations, increase their phase noise level by 30 to 40 dB. Oscillator frequencies have been made in the 20 MHz to 1.56 GHz range [16].

Other applications include matched filtering, pulse compression, programmable correlators, and Fourier transformations.

Now we consider some different implementations of programmable correlators. The programmable matched filter that has been most extensively used for processing biphase coded waveforms is simply a tapped-delay line with taps that transfer portions of the signal to a summer with a phase shift controlled by electronic circuitry. The impulse response is set to be the time reverse of the waveform to be detected. Bandwidth limitations are on the order of 50 MHz [16].

Another type of programmable device that will process any waveform for a given bandwidth and time duration is the SAW convolver. This device exploits the acoustoelectric interaction of a SAW with the free carriers on a semiconductor surface. A layer of semiconductor (e.g., silicon) on a piezoelectric substrate (e.g., lithium niobate) is used to form a space-charge coupled SAW device. The SAW induces charge separation, and hence it carries an electric field with it that exists both inside and outside of the substrate. The electric field outside disturbs the carriers in the neighboring semiconductor and thereby produces "space" charge. Nonlinear interaction between oppositely directed surface waves produces an output that is proportional to the convolution of two input signals (waves). More precisely, the output, for time-limited signals, is of the form [19].

$$e_0(t) = \int_{t-t_0}^{t+t_0} f(u)g(2t - u)\, du \qquad (8.13\text{-}2)$$

Thus we see there is an output time compression of 2. These real time analog signal processors act as programmable filters or correlators in the sense that one convolver can be used for many types of signals. The ease of programmability is an advantage over the transversal filter.

8.14 CHARGE COUPLED DEVICES

Since its invention in 1969 the charge coupled device (CCD) has been capturing an increasing share of the electronics market. It has successfully been used for signal processing as well as optical and infrared imaging and digital memory applications [21].

As delay lines and transversal filters CCDs perform many of the same functions as SAW devices except in a different range of time delay and bandwidth. Most practical CCDs are limited to a bandwidth of 10 MHz whereas SAWs can be much wider. Practical SAW devices are limited to about 100-μsec delays, whereas CCDs can achieve up to 1 sec of delay if they are temperature controlled.

Now let us consider how a CCD delay line works. The CCD consists of an array of closely spaced capacitors. When a clock voltage is applied to the metal (aluminum) side of one of these capacitors, a potential well is created at the semiconductor (silicon) surface. The CCD signal is just the number of electrons placed in the potential well. The range of the electron count can go from zero to several million electrons, so the information is analog. Typical dynamic range is around 80 dB [21].

In Figure 8.36 the surface potential (voltage) is shown as a function of position across the CCD. The well below the CCD device illustrates the case where Φ_2 is "on" and the other two clocks (Φ_1 and Φ_3) are off. The electrons residing within the cross hatched area are indicated by the cross hatched area. In the second well diagram Φ_3 is turned on so that the electrons fill up the expanded well to a lower level. In the last well diagram Φ_2 is turned off, pushing the remaining electrons into the Φ_3 potential well. Hence by sequentially clocking the electrodes in this manner the signal (charge) is passed from one clock phase to the next. A two-phase device requires only two voltage changes rather than three [21].

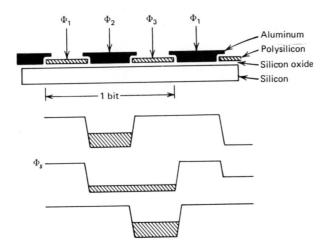

Figure 8.36 Cross section of a polysilicon/aluminum three-phase CCD showing charge transfer to the right.

Two imperfections arise that limit the CCDs operational usefulness. First where charge is transferred, only 99.99% of the charge is transferred. Secondly, thermal leakage continually fills the wells with a leakage current, thus increasing the signal (charge). Thermal leakage goes exponentially with temperature so that if operation at 85°C is required, the maximum time delay is on the order of 10–100 msec [21].

The CCD device easily adapts to implement transversal filters, with the CCD providing the delay of the sampled signal.

REFERENCES

1 Cahn, C. R., "Spread Spectrum Applications and State-of-the-Art Equipments," Paper 5, AGARD-NATO Lecture Series, No. 58, May 28, 1973.

2 Pergal, F., and Huang, M., private communication.

3 Pergal, F., private communication.

4 Davenport, W. B., and Root, W. L., *An Introduction to the Theory of Random Signals and Noise*, McGraw-Hill, New York, 1958.

5 Panter, P., *Modulation, Noise, and Spectral Analysis*, McGraw-Hill, New York, 1965.

6 Jolley, L. B. W., *Summation of Series*, 2nd ed., Dover, 1961.

7 Rhodes, S. A., *Effects of Hardlimiting on Bandlimited Transmissions with Conventional and Offset QPSK Modulation*, NTC, pp. 20F-1 to 20F-7, December 1972.

8 Rosenbach and Whitman, *College Algebra*, Education Manual MC425, Gunn, 308, 1949.

9 Goldman, S., *Frequency Analysis, Modulation, and Noise*, McGraw-Hill, New York, 1948, Section 4.12.

10 Gill, W. J., and Spilker, J. J., Jr., "An Interesting Decomposition Property for the Self-Products of Random or Pseudorandom Binary Sequences," *IEEE Trans. Communication Systems*, June, 1963.

11 Gill, W. J., "Effect of Synchronization Error in Pseudorandom Carrier Communications," First Annunal IEEE Comm. Conv. Conference Record, Denver, pp. 187–191, June 7–9, 1965.

12 Cartier, D. E., "A Frequency Domain Approach to Shuttle PN Code Acquisition," National Telemetry Conference, pp. 713–717, 1974.

13 Carlson, A. B., *Communication Systems*, McGraw-Hill, New York, pp. 43, 52, 1968.

14 Das, P., "Microwave Acoustics in Layered Structure," RADC-TR76-104, Final Technical Report, 1976.

15 Cafarella, J. H., Brown, W. M. Jr., Stern, E., and Alusow, J. A., "Acoustoelectric Convolvers for Programmable Matched Filtering in Spread-Spectrum System," *Proceedings IEEE*, Vol. 64, p. 756, 1976.

16 Maines, J. D., and Paige, E. G. S., "Surface-Acoustic-Wave Devices for Signal Processing Applications," Proceedings of the IEEE, Vol. 64, No. 5, May 1976.

17 O'Clock, G. D., "Matched Filters Boost Receiver Gain, Electronic Design," Vol. 11, May 24, 1979.

18 Collins, Jeff, and Owens, John, "Saw Devices Meet High-Rel Systems Needs," *Microwave Systems Design*, Vol. 7, No. 10, October 1977.

19 Milstein, L. G., and Das, P. K., "Surface Acoustic Wave Devices," *IEEE Communications Mag.* September, 1979.

20 Milstein, L. G., and Das, P. K., "Spread Spectrum Receiver Using Surface Acoustic Wave Technology," *IEEE Trans. Communications*, No. 8, August, 1977.

21 Buss, D. D., Hewes, C. R., de Wit, M., and Brodersen, R. W., "CCDs: Versatility with Integration-A Winning Combination," MSN, October, 1977.

9

PN CODE ACQUISITION
AND LOCK DETECTION
PERFORMANCE

The first subject of this chapter is the PN code acquisition process and the associated acquisition time calculations and mathematical techniques required. The second subject is lock detector performance, which uses Markov chain techniques for an analysis similar to those used in the acquisition theory section.

Two distinct types of code searching will be discussed; (1) fixed integration time systems including both single and double dwell times, and (2) a variable integration time system commonly known as sequential detection.

9.1 THE ACQUISITION PROBLEM

Although this topic has been covered somewhat in the previous chapter, a short discussion will be given for continuity.

A typical PN modulator and noncoherent demodulator are shown in Figure 9.1.

The modulator modulates both data and the PN signal onto the transmitted carrier, normally by means of phase modulation. In order to demodulate data the receiver PN coder must search through all possible PN code phases until the receiver PN coder and the received PN code waveform are (essentially) aligned. After the two codes are aligned the PN code is (essentially) removed from the modulated signal. Once a PN correlation is found and verified, the code tracking loop (see Chapter 10) is employed to maintain PN code lock. The remaining portion of the demodulator is conventional: a bit synchronizer provides bit synchronization, and the matched filter demodulates the data. If the data is coded, the matched filter output (soft quantization) is fed into a decoder.

9.2   ACQUISITION TIME VIA THE GENERATING FUNCTION

The first problem we now consider is how long it takes to acquire the code (align the receiver code generator to within a fraction ($\frac{1}{2}$ or $\frac{1}{4}$ typically) of a chip. (A chip is a PN code symbol.) Since we are assuming a white Gaussian noise (WGN) channel, the observed time is a random variable and therefore statistical techniques are needed to model and analyze the statistics of the acquisition time. Before delving into the theory let us determine the acquisition time in a very simple case. Assume the code has N PN symbols, the probability of detection is unity, the probability of false alarm is zero, and the dwell time (integration time) is T_D sec (Figure 9.1b). Then assuming no doppler shifts and no oscillator instabilities, the time to search all N chips in half chip increments ($2N$ cells) is

$$T_{acq} = 2NT_D \qquad \text{sec} \qquad\qquad (9.2\text{-}1)$$

and the mean acquisition time is just

$$\bar{T}_{acq} = NT_D \qquad \text{sec} \qquad\qquad (9.2\text{-}2)$$

When the detection probability P_D is not unity and the false alarm probability, P_{FA} is not zero, the calculation is no longer quite so simple.

PROBLEM 1

Show that if $P_D \neq 1$ and $P_{FA} = 0$, then

$$\bar{T}_{acq} = \frac{2 - P_D}{P_D} NT_D$$

by summing the appropriate series. Notice that when $P_D = 1$ the result agrees with (9.2-2).

Before we embark on a discussion of the first technique — using a flow graph to obtain acquisition times—we present a brief discussion of flow graph theory.

9.2.1   Signal Flow Graphs

A signal flow graph is a topological representation of the simultaneous equations describing a system function. Signal flow graphs were developed by Mason [2, 3] based on some preliminary work of Tustin [4]. Once a system is represented by a flow graph it becomes a relatively simple matter to determine the closed-loop transfer function. One natural application of flow graphs is to control theory. However, many other applications exist also as long as a transfer function is defined in the application. We follow references 2–5 in what follows.

Denote variables by *nodes* (small dots) and consider the following equation:

$$x_i = T_{ji}x_j \qquad\qquad (9.2\text{-}3)$$

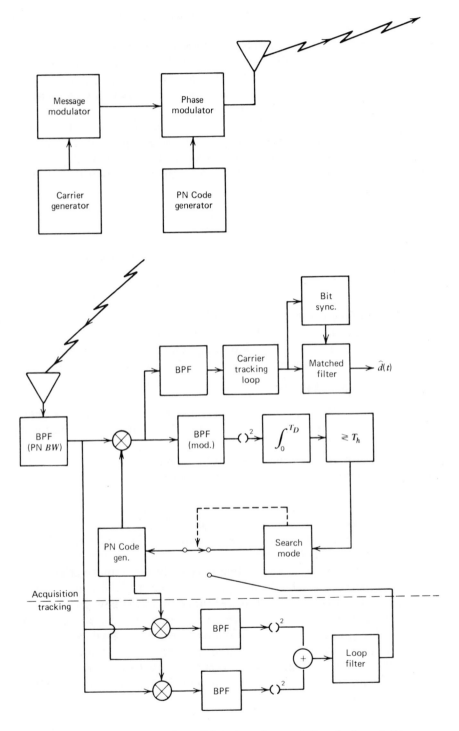

Figure 9.1 (a) PN modulator. (b) Noncoherent PN code demodulator.

where the variables x_i and x_j can be functions of a variable such as time or can be states, for example, in a state diagram. The nodes are connected by directed *branches* that represent the operator T_{ji} mapping x_j into x_i. These operators are called *transmission functions* (or branch transmittances). Figure 9.2 illustrates the above equation.

Branches are always unidirectional, with the direction of flow denoted by the arrowhead. Every variable in a flow graph is represented by a node and every transmission function by a branch. The value of the variable represented by a node is equal to the sum of all signals entering the node. Let

$$x_i = \sum_{j=1}^{n} T_{ji}x_j \qquad (9.2\text{-}4)$$

Then the representation is given in Figure 9.3. As an example consider the two equations

$$y = 5x \qquad x = 2w \qquad (9.2\text{-}5)$$

The associated flow graph is shown in Figure 9.4.

A series connection of $n - 1$ branches and n nodes can be replaced by a single branch with a new transmission function equal to the product of the old original transmission functions. The proof is obvious.

Signal Flow Graph Definitions

We have already defined nodes and branches and some elementary operations. Now we consider some definitions useful in signal flow graph theory and signal flow graph reduction.

An *open path* is a continuous, unidirectional succession of branches along which no node is passed more than once. In Figure 9.5 x_2 to x_3 and back to x_2 is an example of an open path. Also x_1 to x_2 to x_4 to x_5 is another example; so is x_1 to x_2 to x_3 to x_4 to x_5. An *input node* (or source node) is a node with only outgoing branches. In Figure 9.5 x_1 is the only input node. An *output node* (or sink node) is a node with only incoming branches. In Figure 9.5 x_5 is the only output node.

Sometimes a variable in the system is actually the output variable. For example, in the classical feedback system the output is also the variable that is fed back to the feedback transfer function. In the flow graph for this example the output node would have an outgoing branch contrary to the definition of an output node. This dilemma can be solved by adding a branch with a unit transmission function entering a dummy node.

A *feedback loop* (or feedback path) is a path that starts and terminates on the very same node. In Figure 9.5 the path from x_2 to x_3 and back to x_2 is a feedback path.

x_j　　　　　　　　　　　T_{ji}　　　　x_i　　　　**Figure 9.2**　A simple signal flow graph
with one branch.

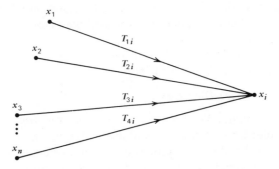

Figure 9.3   Pictorial representation of addition.

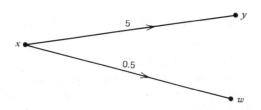

Figure 9.4   Signal flow graph for $y = 5x$ and $x = 2w$.

A *forward path* is an open path from the input node to the *output node*. In our example, x_1 to x_2 to x_3 to x_4 to x_5 and x_1 to x_2 to x_4 to x_5 are the two forward paths of Figure 9.5.

The *gain of a branch* is the transmission function of that branch when the transmission function is a multiplicative operator.

The *loop gain* is the product of the branch gains of the loop. Again referring to Figure 9.5, we mention that the loop gain of the feedback from x_2 to x_3 and back to x_2 is $T_{23}T_{32}$.

Nontouching loops are a set of loops that have no nodes in common. An example will be given later to illustrate this definition.

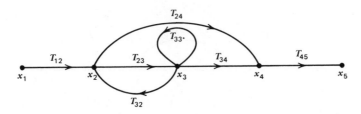

Figure 9.5   A signal flow graph.

The *path gain* is the product of the branch gains found in traversing a path. In Figure 9.5 the path gain of the forward path from x_1 to x_2 to x_3 to x_4 to x_5 is $T_{12}T_{23}T_{34}T_{45}$.

A *self-loop* is a feedback loop composed of a single branch. For example, in Figure 9.5, T_{33} is a self-loop.

Flow Graphs from Block Diagrams

Given a block diagram, a flow graph can easily be constructed by observing the following rule: Construct every variable of the block diagram as a node and each block in the block diagram as a branch. For example, consider the basic control loop diagram shown in Figure 9.6a. Notice that since the output variable C' was constructed using the branch gain of 1, C' is a proper output node. More general block diagrams or systems of equations can be handled in a similar manner.

Signal Flow Graph Reduction and Mason's Gain Formula

In the study of flow graphs it is desirable to reduce the original signal flow graph to two nodes and one branch, in particular, to the input and output nodes and the overall gain. There are two ways to achieve this goal. First, one can use reduction methods; secondly, one can use Mason's gain formula. Although reduction techniques will not be described in detail here, Table 9.1 illustrates some equivalents that are useful in flow graph reduction.

In Table 9.1, Part *a*, for example we have

$$x_2 = T_{12}x_1$$
$$x_3 = T_{23}x_2 = T_{12}T_{23}x_1$$

(9.2-6)

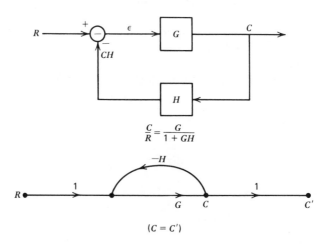

$$\frac{C}{R} = \frac{G}{1 + GH}$$

$$(C = C')$$

Figure 9.6 Block diagram and the associated signal flow diagram.

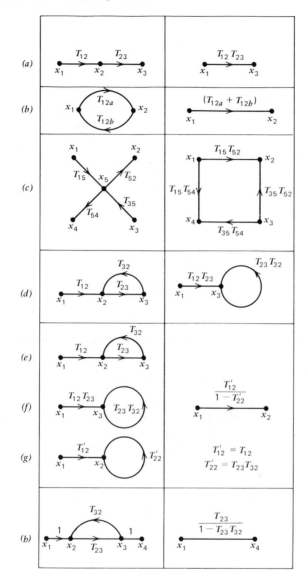

Table 9.1   Flow graph equivalents.

For the parallel paths of Table 9.1, part b we have

$$x_2 = (T_{12a} + T_{12b})x_1 \qquad (9.2\text{-}7)$$

In Part c a node is eliminated by the following equations:

$$x_2 = T_{52}x_5$$
$$x_4 = T_{54}x_5 \qquad (9.2\text{-}8)$$
$$x_5 = T_{15}x_1 + T_{35}x_3$$

now eliminating the x_5 variable:

$$x_2 = T_{52}T_{15}x_1 + T_{35}T_{52}x_3$$
$$x_4 = T_{15}T_{54}x_1 + T_{35}T_{54}x_3$$

(9.2-9)

which is shown as the equivalent in the table, Part c. This is an example of node elimination. The equivalence of Table 9.1d is left as an exercise. The equivalence in f follows directly from e using d. To obtain equivalence g, write from e

$$x_2 = T_{12}x_1 + T_{32}x_3$$
$$x_3 = T_{23}x_2$$

(9.2-10)

eliminate x_3,

$$x_2 = T_{12}x_1 + T_{32}T_{23}x_2$$

(9.2-11)

or

$$x_2 = \frac{T_{12}}{1 - T_{32}T_{23}}x_1$$

PROBLEM 2

Show the equivalence in the two signal flow graphs in Parts d and h of Table 9.1.

The second method of reduction of a signal flow graph is the use of Mason's gain formula. In many cases it is possible and much less time consuming to write down the input-output relationship by a few simple rules applied directly to the flow graph.

If we denote the transfer function from input to output by TF, then *Mason's gain formula* states that

$$\text{TF} = \frac{\sum\limits_i T_i \Delta_i}{\Delta_g}$$

(9.2-12)

where

T_i is the ith forward path gain.

Δ_g is the signal flow graph determinate, given by

$$\Delta_g = 1 - \sum_i T_i^1 + \sum_i T_i^2 - \sum_i T_i^3 + \cdots$$
$$+ \cdots$$

(9.2-13)

$\sum_i T_i^1$ is the sum of all feedback loop gains.

$\sum_i T_i^2$ is the sum of all gain products of two nontouching feedback loops.

and in general

$\Sigma_i T_i^n$ is the sum of all gain products of n nontouching feedback loops.

Δ_i is Δ_g evaluated with all feedback loops touching the ith forward path eliminated.

Although the formula appears formidable, it is actually quite simple to use. Consider the example shown in Figure 9.7.

There is only one signal path from input to output. Hence

$$T_1 = T_{12}T_{23}T_{34}T_{45} \tag{9.2-14}$$

There are three feedback loops:

$$x_3 \to x_2 \to x_3 \qquad T_1' = T_{23}T_{32} \tag{9.2-15}$$

$$x_4 \to x_3 \to x_4 \qquad T_2' = T_{34}T_{43} \tag{9.2-16}$$

$$x_4 \to x_5 \to x_4 \qquad T_3' = T_{45}T_{54} \tag{9.2-17}$$

There are two nontouching feedback loops

$$x_3 \to x_2 \to x_3 \quad \text{and} \quad x_4 \to x_5 \to x_4 \tag{9.2-18}$$

since all three feedback loops touch the path from output to input. Now the signal flow graph determinate is given by

$$\Delta_g = 1 - (T_{23}T_{32} + T_{34}T_{43} + T_{45}T_{54})$$
$$+ T_{23}T_{32}T_{45}T_{54} \tag{9.2-19}$$

Therefore

$$\text{TF} = \frac{T_{12}T_{23}T_{34}T_{45}}{1 - T_{23}T_{32} - T_{34}T_{43} + T_{23}T_{32}T_{45}T_{54}} \tag{9.2-20}$$

PROBLEM 3

Using Mason's gain formula show that the transfer function for the flow graph shown below is given by

$$\text{TF} = \frac{G_1(G_2 + G_3)}{1 - G_1H_1 + G_1G_2H_2 + G_1G_3H_2}$$

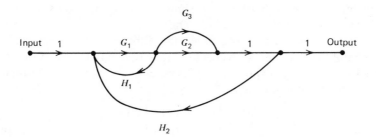

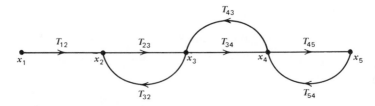

Figure 9.7   Signal flow graph example.

9.2.2   Discrete Time Invariant Markov Processes and Flow Graphs

Although discrete Markov processes have been characterized by many techniques such as difference equations or matrix equations, it will be convenient in this section to describe them by state transition diagrams. From these diagrams, what we will call the generating function flow graphs will be established, which will lead to the generating function of the process [6–10] under consideration. This section is based primarily on references 6 and 7.

Markov processes with a finite (or countable) number of states and time invariant transition probabilities can be described with the aid of a state transition diagram. The state transition diagram is characterized by states indicated by dots that are connected by directed lines that indicate the probability of going from the originating dot to the terminating dot.

Consider as an example a two-state time invariant discrete Markov process which is characterized by two states:

$$P(S2 \mid S1) = \tfrac{3}{8}$$

state 1:   (9.2-21)

$$P(S1 \mid S1) = \tfrac{5}{8}$$

$$P(S1 \mid S2) = \tfrac{1}{2}$$

state 2:   (9.2-22)

$$P(S2 \mid S2) = \tfrac{1}{2}$$

where $S2$ denotes state 2 and $S1$ denotes state 1. The state transition diagram is shown in Figure 9.8. Suppose that at time $t = n$ the process is in state 1; then at time $t = n + 1$ the process will be in state 1 with probability $\tfrac{5}{8}$ and in state 2 with probability $\tfrac{3}{8}$. A similar relationship holds if we start at state 2, except the probabilities are $\tfrac{1}{2}$ and $\tfrac{1}{2}$.

We note that we may write down the difference equations that relate the probabilities of being in each state. Let $p_1(n)$ and $p_2(n)$ be the probabilities of being in states 1 and 2, respectively, at $t = n$. Then one time unit later the probabilities are

$$p_1(n + 1) = \tfrac{5}{8} p_1(n) + \tfrac{1}{2} p_2(n)$$
$$p_2(n + 1) = \tfrac{3}{8} p_1(n) + \tfrac{1}{2} p_2(n) \qquad (9.2\text{-}23)$$

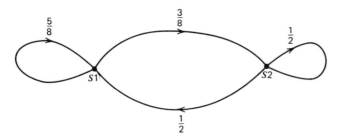

Figure 9.8 State transition diagram for the example.

Since the process is determined by a set of linear, constant coefficient, difference equations, one wonders whether a linear system technique such as flow graphs could be used to obtain the desired probabilities. The answer is yes, as we shall see shortly, and they are used to great advantage.

Let us view our probabilities $p_1(n)$ and $p_2(n)$ as electrical signals and note that the difference equation shows that these signals are devised as linear combinations of themselves after a unit time delay. Hence we can associate with Figure 9.8 an associated linear system flow graph shown in Figure 9.9.

The signal at each node is the probability $p_1(n)$ and $p_2(n)$, respectively. The presence of the z variable on the branches along with the transition probability gains denote the delay of one unit time in going from one node to the next. The reason this diagram leads to the generating function will be proven shortly. Define $p_{ij}(n)$ as the (time invariant) probability of going from state i to state j in n units of time; this is the n-step transition probability. Usually $p_{ij}(1)$ is written as p_{ij}.

As is usual with linear systems, operations in the transform domain are most convenient. We introduce the z transform, also called the *generating function* or the *geometric transform*, of the discrete probability function $p_{ij}(n)$:

$$P_{ij}(z) = \sum_{n=0}^{\infty} z^n p_{ij}(n) \tag{9.2-24}$$

This series converges for all z inside the unit circle.

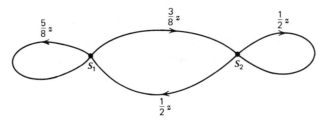

Figure 9.9 Associated linear system (generating function) flow graph of Figure 9.8.

Notice that

$$\frac{dP_{ij}(0)}{dz} = \sum_{n=0}^{\infty} n p_{ij}(n) = \bar{T}_{ij} \tag{9.2-25}$$

where $\bar{T}_{ij}$ is the mean time to go from state i to state j. It is left to the reader to show that

$$\text{var}(T_{ij}) = \frac{d^2 P_{ij}(1)}{dz^2} + \frac{dP_{ij}(1)}{dz} - \left[\frac{dP_{ij}(1)}{dz}\right]^2 \tag{9.2-26}$$

PROBLEM 4

Show that

$$\text{var}(T_{ij}) = \frac{d^2 P_{ij}(1)}{dz^2} + \frac{dP_{ij}(1)}{dz} - \left[\frac{dP_{ij}(1)}{dz}\right]^2$$

Hence we see that $P_{ij}(z)$ is the generating function of the transition process. In addition to the moments it is possible to find the probability $p_{ij}(n)$ by the following formula:

$$p_{ij}(n) = \frac{1}{n!} \frac{d^n P_{ij}(0)}{dz^n} \tag{9.2-27}$$

Consider now using the generating function flow graph of Figure 9.9 in an example. Suppose that we are given that the system is in state 1 at $t = 0$ and we desire the probability of starting in state 1 at $t = 0$ and arriving in state 1 at $t = n$. The approach to the problem is to view it as a "system," with the input composed of a unit sample at $t = 0$ to node 1 and the output a tap off of node 1. We accomplish our end by determining the transfer function $P_{11}(z)$, which is the z transform of the required probability signal.

We note that in general that with arbitrary initial values we apply to each node, i, of the system at $t = 0$, samples equal to the prescribed initial values $P_i(0)$. Using the superposition property of linear systems, such a problem can be solved by adding linear sums of simple node to node responses. Hence if $P_{ij}(z)$ is the transfer function from node i to node j, then the signal at node j in the general case is given by the linear sum

$$P_j(z) = \sum_i p_i(0) P_{ij}(z) \tag{9.2-28}$$

In our example $p_1(0) = 1$, $p_2(0) = 0$, and $P_1(z) = P_{11}(z)$. Many problems have the above characteristic, that, say, $p_1(0) = 1$ and $p_i(0) = 0$, $i \neq 1$. Continuing with our example, we obtain the generating function flow graph of Figure 9.10.

Using Mason's gain formula, we find that

$$T_1 = 1, \qquad \Delta_1 = \tfrac{z}{2}$$
$$\Delta g = 1 - \tfrac{5}{8} z - \tfrac{1}{2} z - \tfrac{3}{16} z^2 + \tfrac{5}{16} z^2 \tag{9.2-29}$$

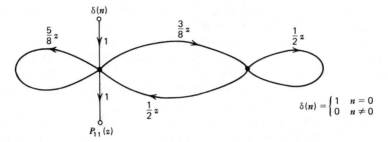

Figure 9.10 Generating function flow graph for the example following problem 4.

so that

$$P_{11}(z) = \frac{1-\frac{z}{2}}{1-\frac{9}{8}z+\frac{1}{8}z^2} \qquad (9.2\text{-}30)$$

Notice that $P_{11}(0)=1$, as it should be by the initial conditions of our problem. It is to be noted that if the Markov process has a finite number of states, then the resulting transfer function will be a rational function.

To obtain $p_{ij}(n)$ we may differentiate $P_{11}(z)$, use long division to obtain the series, or use partial fractions to obtain separate series. We shall utilize the latter technique. We write

$$P_{11}(z) = \frac{1-\dfrac{z}{2}}{(1-z)(1-z/8)} \qquad (9.2\text{-}31)$$

or expanding in partial fractions

$$P_{11}(z) = \frac{4/7}{1-z} + \frac{3/7}{1-z/8} \qquad (9.2\text{-}32)$$

so that

$$P_{11}(z) = \tfrac{4}{7}(1+z+z^2+z^3+\cdots) + \tfrac{3}{7}\left(1+\frac{z}{8}+\frac{z^2}{8^2}+\frac{z^3}{8^3}+\cdots\right) \qquad (9.2\text{-}33)$$

Hence we see directly that

$$p_{11}(n) = \tfrac{4}{7} + \tfrac{3}{7}(\tfrac{1}{8})^n \qquad n \ge 0 \qquad (9.2\text{-}34)$$

PROBLEM 5

Find $p_{11}(n)$ for the model described by the state transition diagram below

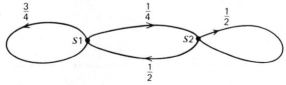

using the techniques of this section.
Show that

$$P_{11}(n) = \tfrac{2}{3} + \tfrac{1}{3}(\tfrac{1}{4})^n \qquad n \ge 0$$

As another example illustrating the techniques, consider a simple random walk problem with absorbing boundaries. Suppose a particle has probability 1/2 of moving to the right or the left. The process ends when the particle reaches either boundary located two steps away. One problem is to find the probability that the process ends in n steps, and the other is to find the mean time to end the process. We assume that we start in state 0. The generating function flow graph is shown in Figure 9.11. Since we start at state (node) zero, we apply a unit sample at $t = 0$ to our system. Node 3 corresponds to the event the process terminates and hence is our required output point to determine $P_{03}(z)$, the generating function for this absorbing random walk problem.

Using Mason's gain formula we find that

$$T_1 = \tfrac{1}{4} z^2$$
$$\Delta_1 = 1$$
$$T_2 = \tfrac{1}{4} z^2 \qquad (9.2\text{-}35)$$
$$\Delta_2 = 1$$
$$\Delta_g = 1 - \tfrac{1}{4}z^2 - \tfrac{1}{4}z^2$$

so that

$$P_{03}(z) = \frac{\tfrac{1}{2} z^2}{1 - \tfrac{1}{2} z^2} \qquad (9.2\text{-}36a)$$

Writing the denominator as a power series, we have

$$P_{03}(z) = \tfrac{1}{2} z^2 - (\tfrac{1}{2} z^2)^2 + (\tfrac{1}{2} z^2)^3 - \cdots \qquad (9.2\text{-}36b)$$

so that the probability of ending on the nth toss is

$$p_{03}(n) = \begin{cases} (\tfrac{1}{2})^{n/2} & n = 2, 4, 6, 8, \ldots \\ 0 & \text{otherwise} \end{cases} \qquad (9.2\text{-}37)$$

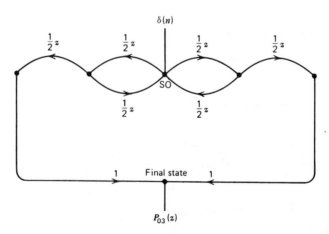

Figure 9.11 Generating function flow graph for particle absorption.

Note that $p_{03}(n)$ is the probability of being in the final state, starting at state 0, at time n. Now the mean time time for the process to terminate is given by

$$\bar{T}_{03} = \frac{dP_{03}(z)}{dz}\bigg|_{z=1} = \frac{1}{(1-\frac{1}{2}z^2)^2}\bigg|_{z=1} = 4 \qquad (9.2\text{-}38)$$

Hence the process terminates, on the average, in four steps. The standard deviation of the time to terminate the process is, from (9.2-26),

$$\sigma_{T_{03}} = \sqrt{\frac{d_2 P_{03}(1)}{dz^2} + \bar{T}_{03} - (\bar{T}_{03})^2} = 2 \qquad (9.2\text{-}39)$$

PROBLEM 6

Consider the Markov process to define the state transition diagram shown below.

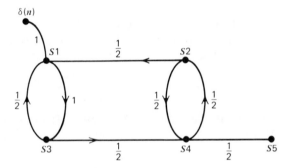

State 5 is an absorption state. Show that the mean time to be absorbed is nine units of time.

PROBLEM 7

Consider a fair coin having a head (H) and a tail (T), each having probability $\frac{1}{2}$ of occurrence.

(a) Show that the generating function flow graph for the event, "obtaining a head (H) for the first time on the nth toss," is given by

so that the desired probability is $(\frac{1}{2})^n$.

(b) Show that the generating function flow graph for the event "obtaining the third head on the nth toss of a coin," is given by

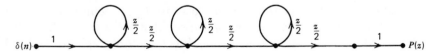

so that

$$P(z) = \left(\frac{z/2}{1 - z/2}\right)^3$$

and the desired probability is given by $(n - 1)(n - 2)(\frac{1}{2})^{n+1}$, $(n \geq 3)$.

For additional examples and applications of the flow graph technique, references 6 and 7 should be consulted.

Mathematical Basis for Generating Function Flow Graph Techniques

We have seen many examples of the generating function flow graph technique for solving, time invariant, finite state, Markov chain problems. Here we present a proof establishing the transition from a state transition diagram to a generating function flow graph. This section is based on reference 9.

Recall that we defined $p_{ij}(n)$ as the probability of going from state i to state j in n steps. It is shown in any text [10] on Markov chains that the following recursion formula

$$p_{ij}(n + 1) = \sum_{k=1}^{N} p_{ik}(n)\, p_{kj} \qquad 1 \leq i, j \leq N, \qquad n = 0, 1, \ldots \qquad (9.2\text{-}40)$$

is satisfied where N is the number of states. Denote $p_{kj}(1) = p_{kj}$ for convenience. Multiplying this equation by z^n and summing over n produces

$$\frac{1}{z}[P_{ij}(z) - p_{ij}(0)] = \sum_{k=1}^{N} p_{kj} P_{ik}(z) \qquad (9.2\text{-}41)$$

where $P_{ij}(z)$ is the generating function and is defined by (9.2-24) since $p_{ij}(0) = \delta_{ij}$ (it allows the recursion relation hold at $n = 0$), where δ_{ij} is the Kronecker δ and is defined by

$$\delta_{ij} = \begin{cases} 1 & i = j \\ 0 & i \neq j \end{cases} \qquad (9.2\text{-}42)$$

Solving for $P_{ij}(z)$, we obtain

$$P_{ij}(z) = \sum_{k=1}^{N} P_{ik}(z)\, p_{kj} z + \delta_{ij} \qquad 1 \leq i, j \leq N \qquad (9.2\text{-}43)$$

This relationship is the set of equations that describes the N-node flow graph of Figure 9.12. This flow graph is just the state transition diagram of the Markov chain with each of its branches labeled with $p_{ij}z$ instead of p_{ij}. The generating function $P_{ij}(z)$ (z transform) is the transformed "signal" at node j when a unit input is applied only to node i; hence the δ_{ij} term applied at time $n = 0$. Since the desired transform $P_{ij}(z)$ requires an input at node i and the output taken at node j, we see that the flow graph for the N-state Markov chain and our generating function flow graph are one and

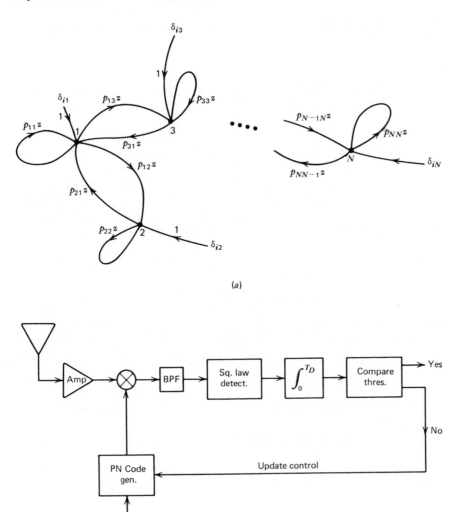

(a)

(b)

Figure 9.12 (a) Generating function flow graph of N-state Markov chain. (b) PN code acquisition system.

the same. Therefore the generating function flow graph technique represents precisely the same Markov chain as the state transition diagram.

9.2.3 Acquisition Time for the Single Dwell Time Search

Armed with our background of moment generating function flow graph applications we are ready to tackle an important problem, the computation of the acquisition time for a single dwell time, pseudonoise, spread spectrum, system.

Acquisition Model

Consider the simplified filter, square, and integrate detector acquisition circuit shown in Figure 9.12. We follow the method of references 11 and 13.

Our model is as follows. Assume that there are q cells to be searched. Now q may be equal to the length of the PN code to be searched or some multiple of it (for example, if the update size is one-half chip, q will be twice the code length to be searched). Further assume that if a "hit" (output is above threshold) is detected by the threshold detector, the system goes into a verification mode that may include both an extended duration dwell time and an entry into a code loop tracking mode. In any event, we model the "penalty" of obtaining a false alarm as $K\tau_D$ sec, and the dwell time itself as τ_D sec. If a true hit is observed, the system has acquired the signal, and the search is completed. Assume the false alarm probability P_{FA} and probability of detection P_D are given. See Section 9.3 for the Gaussian approximation to the probability of detection and false alarm. Clearly the time to acquire, that is, to obtain a true hit (not a false alarm) is a random variable. Although the problem of determining the distribution function of the time to acquire can be obtained in principle, it is very difficult to obtain in practice, at least in closed form. Consequently we shall be content with the mean and the variance of the acquisition time.

Initially we will assume that doppler is not present; later we will relax this assumption. Also we will assume that the correlation error will be fixed at one-half the update size, which in practice is often taken as one-half a chip (PN code symbol).

This assumption implies that P_D is constant (time invariant) and as a consequence, the analysis of the model follows our previous theory.

Analysis

Let each cell be numbered from left to right so that the kth cell has an *a priori* probability of having the signal present, given that it was not present in cells 1 through $k-1$, of

$$P_k = \frac{1}{q+1-k} \tag{9.2-44}$$

The generating function flow diagram is given in Figure 9.13 using the rule that at each node the sum of the probability emanating from the node equals unity.

Consider node 1. The probability *a priori* of having the signal present is $P_1 = 1/q$, and the probability of it not being present in the cell is $1-P_1$. Suppose the signal were not present. Then we advance to the next node (node 1*a*); since it corresponds to a probabilistic decision and not a unit

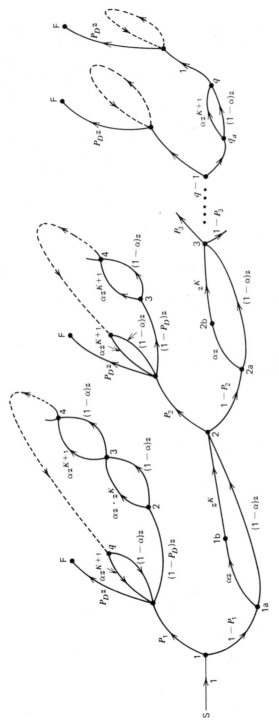

K = penalty due to false alarm ($K\tau_D$ sec)
P_D = probability of detection
P_{FA} = probability of false alarm = α

Figure 9.13 Generating function flow graph for acquisition time.

time delay, no z multiplies the branch going to it. At node $1a$ either a false alarm occurs, with probability $P_{FA} = \alpha$, which requires one unit of time to determine (τ_D sec) and then K units of time ($K\tau_D$ sec) are needed to determine that there is no false alarm or there is no false alarm with probability $(1 - \alpha)$, which takes one dwell time to determine, which requires the $(1 - \alpha)z$ branch going to node 2.

Now consider the situation at node 1 when the signal does occur there. If a hit occurs, then acquisition, as we have defined it, occurs and the process is terminated, hence the node F denoting "finish." If there was no hit at node 1 (the integrator output was below the threshold), which occurs with probability $1 - P_D$, one unit of time would be consumed. This is represented by the branch $(1 - P_D)z$, leading to node 2.

At node 2 in the upper left part of the diagram either a false alarm occurs with probability α and delay $(K + 1)$ or a false alarm does not occur with a delay of 1 unit. The remaining portion of the generating function flow graph is a repetition of the portion just discussed with the appropriate node changes.

In order to obtain the transfer function of the flow graph we shall reduce it by combining sections at a time. Let

$$H(z) = \alpha z^{K+1} + (1 - \alpha) z \qquad (9.2\text{-}45)$$

Then the flow graph can be drawn as shown in Figure 9.14.

Letting $\beta = 1 - P_D$ and

$$Q(z) = \frac{z}{1 - \beta (H)^{q-1}} \qquad (9.2\text{-}46)$$

we can reduce the flow graph to that of Figure 9.15; since $Q(z)$ takes into account the feedback loops. In Figure 9.16 the generating function flow graph is redrawn in a slightly simpler form.

In Figure 9.16 we define

$$B_l(z) = \frac{P_l(1 - \beta)z}{1 - \beta z H^{q-1}} \qquad (9.2\text{-}47)$$

$$A_l(z) = (1 - P_l) H(z) \qquad (9.2\text{-}48)$$

By inspection the flow graph is given by

$$U(z) = B_1 + A_1 B_2 + A_1 A_2 B_3 + A_1 A_2 A_3 B_4 + \cdots + A_1 \cdots A_{q-1} B_q$$

$$= \frac{(1 - \beta)z}{1 - \beta z H^{q-1}} [P_1 + (1 - P_1) H P_2 + (1 - P_1)(1 - P_2) P_3 H^2$$

$$+ \cdots + (1 - P_1)(1 - P_2) \cdots (1 - P_{q-1}) H^{q-1}] \qquad (9.2\text{-}49)$$

Writing P_i and $1 - P_i$ in terms of q, we have

$$U(z) = \frac{(1 - \beta)z}{1 - z\beta H^{q-1}} \left(\frac{1}{q} + \frac{q-1}{q} \frac{H}{q-1} + \frac{q-1}{q} \frac{q-2}{q-1} \frac{1}{q-2} H^2 + \cdots + \frac{H^{q-1}}{q} \right)$$

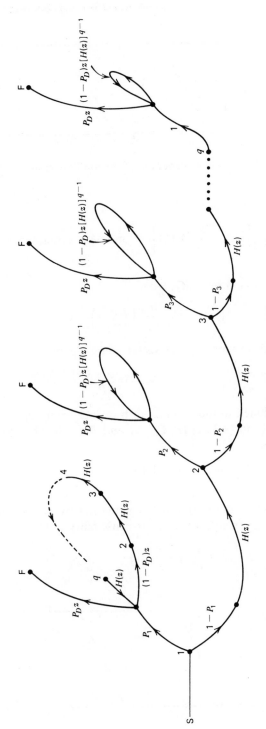

Figure 9.14 Reduced flow graph diagram.

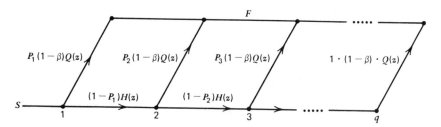

Figure 9.15   Reduction of generating function flow graph continued.

so

$$U(z) = \frac{(1-\beta)}{1-\beta z H^{q-1}} \frac{1}{q} \left[\sum_{l=0}^{q-1} H^l(z)\right] \qquad \text{moment generating function}$$

(9.2-50)

As a check $U(1)$ should be unity.

$$U(1) = \frac{(1-\beta)/q}{1-\beta} \left(\sum_{l=0}^{q-1} 1\right) = 1$$

(9.2-51)

The <u>mean acquisition time</u> is given (after some algebra) by

$$\bar{T} = \frac{\partial \ln U}{\partial z}\bigg|_{z=1} (\tau_D) = \frac{2 + (2 - P_D)(q-1)(1 + KP_{FA})}{2P_D} \tau_d$$

(9.2-52)

with τ_D being included in the formula to translate from our unit time scale. As a partial check on (9.2-52) let $P_D = 1$ and $P_{FA} = 0$. Then we have from (9.2-52)

$$\bar{T} = \frac{1+q}{2} \tau_D$$

(9.2-53)

This result can be obtained by direct calculation by noting that the mean time to acquire is given by (the *a priori* probability is $1/q$)

$$\bar{T} = \frac{1}{q} \tau_D + \frac{2}{q} \tau_D + \cdots + \frac{q}{q} \tau_D$$

(9.2-54)

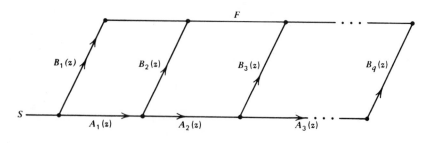

Figure 9.16   Final flow graph.

Summing, we obtain

$$\bar{T} = \frac{\tau_D}{q} \frac{q(q+1)}{2} = \frac{(q+1)\tau_D}{2} \tag{9.2-55}$$

For the usual case when $q \gg 1$, $\bar{T}$ is given by

$$\bar{T} \cong \frac{(2 - P_D)(1 + KP_{FA})}{2P_D} (q\tau_D) \qquad \text{mean acquisition time} \tag{9.2-56}$$

The variance of the acquisition time is given by

$$\sigma^2 = \left[\frac{d^2 U}{dz^2} + \frac{dU}{dz} - \left(\frac{dU}{dz} \right)^2 \right]_{z=1} \tag{9.2-57}$$

or alternatively by the equivalent, but usually simpler formula [when $U(1) = 1$]

$$\sigma^2 = \left(\frac{d^2 \ln U}{dz^2} + \frac{d \ln U}{dz} \right)_{z=1} \tag{9.2-58}$$

Using the latter expression, it can be shown that the expression for σ^2 is $(q \gg 1)$

$$\sigma^2 = \tau_D^2 \left\{ (1 + KP_{FA})^2 \, q^2 \left(\frac{1}{12} - \frac{1}{P_D} + \frac{1}{P_D^2} \right) + 6q \left[K(K+1)P_{FA}(2P_D - P_D^2) \right. \right.$$
$$\left. \left. + (1 + P_{FA}K)(4 - 2P_D - P_D^2) \right] + \frac{1 - P_D}{P_D^2} \right\} \tag{9.2-59}$$

In addition when $K(1 + KP_{FA}) \ll q$, then

$$\sigma^2 = \tau_D^2 \, (1 + KP_{FA})^2 \, q^2 \left(\frac{1}{12} - \frac{1}{P_D} + \frac{1}{P_D^2} \right) \tag{9.2-60}$$

As a partial check on the variance result, let $P_{FA} \to 0$ and $P_D \to 1$. Then we have

$$\sigma^2 = \frac{(q\tau_D)^2}{12} \tag{9.2-61}$$

which is the variance of a uniformly distributed random variable, as one would expect for the limiting case.

The above results provide a useful estimate of acquisition time for an idealized PN type system. Two basic modifications should be made to make the estimates reflect actual hardware or software systems.

First, doppler effects should be taken into account. The result of code doppler is to smear the relative code phase during the acquisition dwell time, which increases or reduces the probability of detection depending on the code phase and the algebraic sign of the code doppler rate. The doppler also affects the effective code sweep rate, which in the extreme case can reduce it to zero to cause the search time to increase greatly. This topic will be discussed in the next section.

The second refinement to the model concerns the handover process between acquisition and tracking. Typically after a "hit" the code tracking loop is turned on to attempt to pull the code into tight lock. Further, often in low-SNR systems where both acquisition (pull-in) bandwidth and tracking bandwidth are used, multiple code loop bandwidths will be employed in order to soften the transition between acquisition and track modes. Consequently, the probability of going from the acquisition mode to the final code loop bandwidth in the tracking mode occurs with some probability less than 1.

The estimation of this probability is at best a very difficult problem (although, some approximate results have been developed). At high SNRs this probability quickly approaches 1, so it is not a problem. At low SNRs the above formula for acquisition time should replace P_D with P_D', where

$$P_D' = P_D P_{HO} \tag{9.2-62}$$

with P_{HO} being the probability of handover. In the S-band shuttle system at TRW Systems it was found that at threshold $(C/N_0 = 51 \, dB - Hz)$ P_{HO} varied from .06 to .5 depending upon the code doppler. Without code doppler P_{HO} was .25, which, if not taken into account in the acquisition time equation, would predict the mean acquisition time to be about four times too fast!

Single Dwell Acquisition Time Formula with Doppler

It is possible to account for doppler as far as effective sweep rate as follows. Define

N = number of chips to be searched

$\Delta T_c/T_c$ = step size of search in fractions of a chip (typically $\frac{1}{2}$)

Δf_c = code doppler in chips

$\Delta f_c \tau_D$ = PN code phase timing shift due to code doppler during dwell time

$\Delta f_c K \tau_D$ = code phase shift during hit verification

P_{FA} = probability of a false alarm

P_D = probability of detection

Since the mean (search phase) update timing change is

$$\mu = \frac{\Delta T_c}{T_c} + \tau_D \Delta f_c + K \tau_D \Delta f_c P_{FA} \tag{9.2-63}$$

we have

$$\bar{T} \cong \frac{(2 - P_D)(1 + K P_{FA})(N \tau_D)}{2 P_D \left| \dfrac{\Delta T_c}{T_c} + \tau_D \Delta f_c + K \tau_D \Delta f_c P_{FA} \right|} \tag{9.2-64}$$

$$\sigma^2 \cong \frac{N^2(1+kP_{FA})^2 N^2(1/12 - 1/P_D + 1/P_D^2)}{(\Delta T_c/T_c + \tau_D \Delta f_c + k\tau_D \Delta f_c P_{FA})^2} \tag{9.2-65}$$

In the above equations the algebraic sign of the code doppler must be assigned so that when the search is speeded up by the code doppler, it is positive and *vice versa*. Note when $\Delta f_c = 0$ then $q = N/(\Delta T_c/T_c)$, so that the result collapses to the previous result. Actually the two doppler effects tend to counteract each other to a certain extent, since when the search is speeded up (due to doppler) the correlation values tend to be reduced. It is to be cautioned that (9.2-65) is only an approximation. It breaks down as the denominator approaches zero, which implies $\bar{T} \to \infty$. This is not correct. However, the acquisition time does increase significantly when the denominator approaches zero. The magnitude sign in the mean acquisition time expression is used to take into account the sweep direction symmetry.

9.2.4 Acquisition Time for the Double Dwell Time Search

A simple generalization of the single dwell time system produces the double dwell time system. It has two integration periods, one to search the code phases quickly (τ_{D_1}, small) and the second to provide a "better" estimate of whether the "in-sync" code phase has been found. The basic idea is to apportion some of the false alarm protection in the first integration and place the remaining (usually greater) protection in the second integration. Generally speaking this double dwell time procedure reduces acquisition time. The amount of improvement depends upon the parameters involved. The system model is shown in Figure 9.17. The system can perhaps best be understood by looking at an acquisition track algorithm shown in Figure 9.18. Typically the search starts by advancing the reference code phase to the extremity of the range ambiguity region; then a trial integration of τ_{D_1}, sec is made on the received signal plus noise.

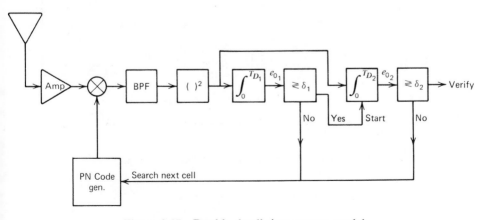

Figure 9.17 Double dwell time system model.

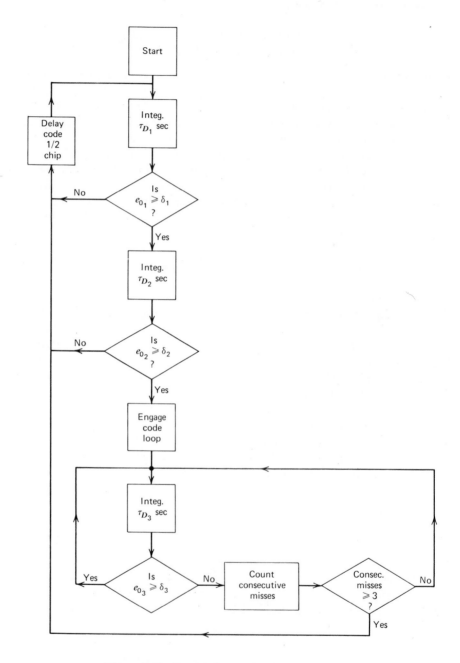

Figure 9.18 Double integration time flow graph.

If the threshold is not exceeded, then the reference (system) code is delayed, for example, one-half chip, and again the dwell (integration) time is τ_{D_1} sec. The process continues in this manner until a hit (threshold is exceeded) occurs. Then, without changing the code phase, the integration time is increased to τ_{D_2} sec. This dwell provides both a higher probability of detection and a lower probability of false alarm. If the second threshold is exceeded, typically the code loop is activated and a third integration of the input signal plus noise is performed (in practice it could be the same duration as τ_{D_2}). If the threshold were not exceeded at this point, the search would continue. This final integration period forms the basis of a lock detector, of which more will be said in Section 9.4. Briefly, the idea is to detect (say) three consecutive times in which the threshold does not exceed the threshold. When this event occurs the code loop is declared to be out of lock and the search is started over.

A slight modification of DiCarlo's [12] results produces the following results for this double dwell time scheme:

$$\bar{T} = \frac{2 - P_D}{2P_D} [\tau_{D_1} + \tau_{D_2} P_{FA_1}(1 + KP_{FA_2})] q \qquad q \gg 1 \qquad (9.2\text{-}66)$$

$$\sigma^2 = \tau_{D_1}^2 + \tau_{D_2}^2 P_{FA_1}^2 (1 + KP_{FA_2})^2 q^2 \left(\frac{1}{12} - \frac{1}{P_D} + \frac{1}{P_D^2}\right) \qquad q \gg P_{FA_2} K(K+1)$$
$$(9.2\text{-}67)$$

where

$\tau_{D_1} =$ first dwell time

$\tau_{D_2} =$ second dwell time

$P_D = P_{D_1} P_{D_2}$, product of detection probabilities of dwell one and two

$P_{FA_1} =$ false alarm probability of the first dwell

$P_{FA_2} =$ false alarm probability of the second dwell

$q =$ number of cells to be searched

$k =$ penalty for a false alarm at the second detector (number of τ_{D_2} units of time)

In the above equations the conditions $q \gg 1$ and $q \gg P_{FA_2} K(K+1)$ are normally met. The more exact results, however, if needed, can be found in DiCarlo [12]. Notice, as a check, if $\tau_{D_1} = 0$ and $P_{FA_1} = 1$, then this system collapses to the single dwell time system and the analytic results agree with those of the single dwell system.

PROBLEM 8

Derive the expression for the mean and the variance of the acquisition time for the double dwell system as described above.

In the same manner as the single dwell acquisition system the double dwell system can be modified to include doppler, with the mean code search phase update being

$$\mu = \frac{\Delta T_c}{T_c} + \tau_{D_1}\Delta f_c + P_{FA_1}\tau_{D_2}\Delta f_c + P_{FA_1}P_{FA_2}K\tau_{D_2}\Delta f_c \qquad (9.2\text{-}68)$$

Therefore the mean and variance of the acquisition time with doppler, letting N be the number of chips to be searched, become

$$\bar{T} = \frac{2 - P_D}{2P_D} \frac{[\tau_{D_1} + \tau_{D_2}P_{FA_1}(1 + KP_{FA_2})]N}{|(\Delta T_c/T_c) + \tau_{D_1}\Delta f_c + P_{FA_1}\tau_{D_2}\Delta f_c + K\tau_{D_2}\Delta f_c P_{FA_1}P_{FA_2}|} \qquad (9.2\text{-}69)$$

$$\sigma^2 = \frac{N^2[\tau_{D_1} + \tau_{D_2}P_{FA_1}(1 + kP_{FA_2})]^2(\frac{1}{12} - 1/P_D + 1/P_D^2)}{[(\Delta T_c/T_c) + \tau_{D_1}\Delta f_c + \tau_{D_2}\Delta f_c P_{FA_1} + k\tau_{D_2}\Delta f_c P_{FA_1}P_{FA_2}]^2} \qquad (9.2\text{-}70)$$

with the algebraic sign of the code doppler assigned such that when the search is speeded up by the code doppler Δf_c is positive and *vice versa*.

PROBLEM 9

Consider the code correlation curve shown below.

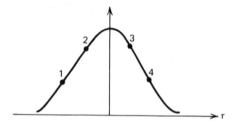

Denoting $P_D^{(i)}$ as the probability of detection at the ith point on the correlation curve ($i = 1, 2, 3, 4$), show that the total probability of detection, including all four points, is given by

$$P_D^T = P_D^{(1)} + P_D^{(2)}(1 - P_D^{(1)}) + P_D^{(3)}((1 - P_D^{(1)})(1 - P_D^{(2)}))$$
$$+ P_D^{(4)}(1 - P_D^{(1)})(1 - P_D^{(2)})(1 - P_D^{(3)})$$

9.3 PROBABILITY OF DETECTION AND FALSE ALARM APPROXIMATIONS

The purpose of this section is to develop approximate expressions for both the probability of detection P_D and the probability of false alarm P_{FA} for a fixed dwell integration period following a square law detector whose input is either band-limited Gaussian noise or band-limited Gaussian noise plus signal. We consider both the integrator and a single-pole RC lowpass postdetection filter.

9.3.1 Integrator Postdetection Filter

The basic model is shown in Figure 9.19. The statistics desired are those of $z(t)$, the output of the τ_D sec integrator. Our input signal is modeled as follows:

$$y(t) = \sqrt{2P} \, \sin \omega_0 t + n(t) \tag{9.3-1}$$

where the signal is a sine wave having power P and the noise process is a sample function of a WGN process having two-sided noise spectral $N_0/2$. Davenport and Root [14] have shown that the mean output of the square law detector to be given by (that is, just the sum of the noise power and the signal power)

$$\mu = N_0 B(1 + \text{SNR}) \tag{9.3-2}$$

where B is the noise bandwidth of the bandpass filter (BPF), $\text{SNR} = P/N_0 B$, and N_0 is the one-sided noise spectral density. Therefore out of the integrator, after τ sec, the mean output is given

$$\mu_0 = N_0 B(1 + \text{SNR}) \, \tau \tag{9.3-3}$$

Davenport and Root [14] also have shown that the two-sided spectral density at the origin out of the detector is given by

$$\mathcal{S}(0) = N_0^2 B + 2PN_0 \tag{9.3-4}$$

Since the transfer function of a τ-sec integrator is given by

$$H(f) = \tau \frac{\sin \pi f \tau}{\pi f \tau} e^{-j\omega\tau/2} \tag{9.3-5}$$

we find that the variance out of the integrator is given by

$$\sigma_0^2 \cong \int_{-\infty}^{\infty} \mathcal{S}(0) \, \tau^2 \left(\frac{\sin \pi f \tau}{\pi f \tau} \right)^2 df \tag{9.3-6}$$

with the assumption that $1/\tau \ll B$, so that only the spectrum near $f = 0$ is

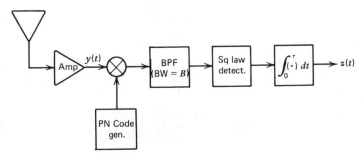

Figure 9.19 Square law detector model.

important in the evaluation of the integral. Evaluating, we obtain

$$\sigma_0^2 = N_0^2 B(1 + 2\text{SNR})\,\tau \tag{9.3-7}$$

Now if the BPF has sharp cutoff skirts, then samples placed B^{-1} sec apart are essentially independent. Further if $B\tau \gg 1$, the output variable, $y(t)$ being a large sum of independent variates, tends toward a normal random variable by the central limit theorem. With the approximation made, the probability of detection (probability of being above threshold given the signal is present) is given by

$$P_D = \int_\delta^\infty \frac{1}{\sqrt{2\pi}\,\sigma_0} \exp\left[\frac{-(y - \mu_0)^2}{2\sigma_0^2}\right] dy \tag{9.3-8}$$

or

$$P_D = Q\left(\frac{\delta - \mu_0}{\sigma_0}\right) \tag{9.3-9}$$

with

$$Q(t) = \int_t^\infty \frac{1}{\sqrt{2\pi}} \exp\left(\frac{-u^2}{2}\right) du \tag{9.3-10}$$

In order to simplify this expression, consider the result for the false alarm. Again assuming Gaussian statistics, we have for the false alarm probability [the probability that $y(\tau) > \delta$, given the signal was not present]

$$P_{FA} = \int_\delta^\infty \frac{1}{\sqrt{2\pi}\,\sigma} \exp\left[\frac{(y - \mu)^2}{2\sigma^2}\right] dy \tag{9.3-11}$$

where

$$\left.\begin{array}{l} \mu = N_0 B\tau \\ \sigma^2 = N_0^2 B\tau \end{array}\right\} P = 0 \tag{9.3-12}$$

Hence we obtain

$$P_{FA} = Q\left(\frac{\delta - N_0 B\tau}{N_0\sqrt{B\tau}}\right) \tag{9.3-13}$$

Define

$$\beta = \frac{\delta - N_0 B\tau}{N_0\sqrt{B\tau}} \tag{9.3-14}$$

Then

$$P_D = \text{erfc}\left[\frac{\delta - N_0 B\tau}{N_0\sqrt{B\tau}\,(1 + 2\text{SNR})^{1/2}} - \frac{BT\,\text{SNR}}{\sqrt{B\tau}\,(1 + 2\text{SNR})^{1/2}}\right] \tag{9.3-15}$$

or finally

$$P_D = Q\left(\frac{\beta - \sqrt{.B\tau}\,\text{SNR}}{\sqrt{1 + 2\,\text{SNR}}}\right) \tag{9.3-16}$$

and

$BK = N$

$$P_{FA} = Q(\beta) \tag{9.3-17}$$

Consequently once the false alarm is specified, so is β; and knowing $B\tau$ and SNR, one can find P_D. Based on the author's experience on a NASA program this approximation is quite accurate. The advantage of this approximation is that it is very simple to use when estimating acquisition times based on using the results of Sections 9.2.3 and 9.2.4. This result still holds approximately when the signal is modulated as long as the loss due to the bandpass filter is taken into account. In other words, the SNR would be defined as

$$SNR = \frac{\alpha P}{N_0 B} \tag{9.3-18}$$

where α denotes the fraction of the modulated signal that is passed through the BPF.

An important point to note is the following, which has also been verified experimentally. As the data rate increases, B increases, usually in proportion. Now if the power also increases by the same proportion, the SNR remains constant. However, for the same false alarm probability, P_D turns out to be greater because of the larger $B\tau$ product! Equation 9.3-16 verifies this observation.

9.3.2 RC Single-Pole Filter

In Problem 10 the results using the RC single-pole lowpass filter are developed. That no dumping circuits are required simplifies the hardware. The results from Problem 10 are

$$P_D = \mathrm{erfc}\left(\frac{\beta}{\sqrt{1+2\,SNR}} - \frac{\sqrt{2B\tau}\,SNR(1-\bar{e}^{\,T/\tau})}{\sqrt{1+2\,SNR}}\right) \tag{9.3-19}$$

$$P_D = \mathrm{erfc}(\beta) \tag{9.3-20}$$

where

$$\beta = \frac{\delta - N_0 B}{\sqrt{N_0^2 B/(2\tau)}} \tag{9.3-21}$$

with δ the threshold, N_0 the one-sided spectral density, and $\tau = RC$ the RC time constant. From Problem 10 the optimized value of τ in relation to T (the dwell time) is $\tau = 0.8T$. The optimized time constant produces the following probability of detection and false alarm probabilities

$$P_D = Q\left(\frac{\beta - .894\,\sqrt{BT}\,SNR}{\sqrt{1+2\,SNR}}\right) \tag{9.3-22}$$

$$P_{FA} = Q(\beta)$$

We conclude that at low SNR this system is about 0.5 dB [10 log (.894)], inferior to the integrate-and-dump postdetection filter. This may well be worth the cost in a system with adequate margin.

PROBLEM 10

Using the normal approximations that were employed for the integrate-and-dump circuit show that

(a)   The mean output out of the RC lowpass filter is

$$\mu = N_0 B + N_0 B(\text{SNR})(1 - e^{-T/\tau})$$

where T is the dwell time, $\tau = RC$, $N_0 B$ is the mean noise power out of the bandpass filter, and SNR is the signal to noise ratio out of the low pass filter, $\text{SNR} = P/N_0 B$.

(b)   The variance in steady state is given by

$$\sigma^2 = \frac{N_0^2 B + 2 P N_0}{2\tau}$$

(c)   That

$$P_D = Q\left[\frac{\beta - \text{SNR}\sqrt{2B\tau}(1 - e^{-T/\tau})}{\sqrt{1 + 2\,\text{SNR}}}\right]$$

$$P_{FA} = Q[\beta]$$

$$\beta = \frac{\delta - N_0 B}{\sqrt{N_0^2 B/(2\tau)}}$$

$$Q(x) = \int_x^\infty \frac{1}{\sqrt{2\pi}} e^{-t^2/2}\, dt$$

PROBLEM 11

Using the result of Problem 9, demonstrate that summing two consecutive correlation integrator values and comparing to a modified threshold (to keep P_{FA} constant) improves the total probability of acquisition and therefore reduces acquisition time. Assume $\text{SNR} = 0\,\text{dB}$, $BT = 100$, and $P_{FA} = 10^{-4}$. Note that the equation for P_{FA} and P_D must be modified for this new case. Assume that each correlation value out of the integrator is essentially independent and has a Gaussian distribution.

9.4   LOCK DETECTOR THEORY AND ABSORBING MARKOV CHAINS

In this section we will introduce a matrix approach to absorbing Markov chains by which, by the use of some simple matrix manipulations, we can obtain the mean and variance of the time it takes to indicate out-of-lock. Code loop detectors are a very important part of any acquisition system since any carrier acquisition scheme must wait until the lock detector verifies that the code is in synchronization. Consequently, the verification time of the code lock detector is one component of the total time it takes to acquire the code.

The method taken here is to follow the Markov chain approach of Kemeny and Snell [15] quite closely and apply it to the lock detector problem in a manner similar to Hopkins [16]. First we develop some basic results of absorbing Markov chains.

9.4.1 Absorbing Markov Chains

The states of a Markov chain can be divided into *transient states* and *persistent states* (also the terms "recurrent" and "ergodic" are used in place of persistent). Transient states, once left, can never be entered again, while persistent states, once entered, can never be left. A set of persistent states with only one element is called an *absorbing state*, which is characterized by a transition probability of $p_{ii} = 1$. In this case all the other entries in the ith row of the transition matrix must be zero. A Markov chain consisting of a single persistent set is called a *persistent Markov* chain. A Markov chain all of whose nontransient states are absorbing is called an *absorbing Markov chain*. Every Markov chain must have a persistent set of states, but it does not have to have a set of transient states. If a Markov chain contains only one state, then it must be an absorbing chain (this is not a very interesting chain however!). As an example to clarify the above definitions, consider a Markov chain characterized by the state transition matrix of the form

$$
P_1 = \begin{array}{c} \\ 1 \\ 2 \\ 3 \\ 4 \\ 5 \end{array}
\begin{array}{ccccc}
1 & 2 & 3 & 4 & 5 \\
\left[\begin{array}{ccccc}
1 & 0 & 0 & 0 & 0 \\
q & 0 & p & 0 & 0 \\
0 & q & 0 & p & 0 \\
0 & 0 & q & 0 & p \\
0 & 0 & 0 & 0 & 1
\end{array}\right]
\end{array}
\tag{9.4-1}
$$

where p is the probability of the process going to a higher state and q that of going to a lower state. Notice once the process reaches state 1 or 5 it remains there forever; hence states 1 and 5 are absorbing states. States 1, 2, and 3 are transient states, since eventually the process leaves them and remains in the persistent states (1 and 5). Note that this is an absorbing Markov chain.

If we modify our example to have "reflecting boundaries" at states 1 and 5, the transition matrix becomes

$$
P_2 = \begin{array}{c} \\ 1 \\ 2 \\ 3 \\ 4 \\ 5 \end{array}
\begin{array}{ccccc}
1 & 2 & 3 & 4 & 5 \\
\left[\begin{array}{ccccc}
0 & 1 & 0 & 0 & 0 \\
q & 0 & p & 0 & 0 \\
0 & q & 0 & p & 0 \\
0 & 0 & q & 0 & p \\
0 & 0 & 0 & 1 & 0
\end{array}\right]
\end{array}
\tag{9.4-2}
$$

Now there are no transient states; they are all persistent. Since they all communicate (any state can be reached by another state), there is one persistent set. Hence we have a persistent Markov chain. The rest of this section will concern absorbing Markov chains.

Theorem 1   In any Markov chain the probability that the process is in a persistent state tends to 1 as the number of steps tends to infinity, no matter where the starting point.

Proof   Once a persistent set is reached it can never be left (by definition of a persistent set of states). Suppose then that the process starts in a transient state. Suppose that from any transient state it is possible to reach an ergodic state in not more than n steps. There is a positive probability p of entering a persistent state from any transient state in n steps. Therefore the probability of not reaching a persistent state in n steps is $1 - p < 1$. Consequently, the probability of not reaching a persistent state in mn steps is less than or equal to $(1 - p)^m$, which approaches zero as $m \to \infty$. From this theorem we see that if the chain is absorbing, the process will eventually end up in an absorbing state with probability 1. Most of the lock detector problems that we will be concerned with will be modeled as absorbing chains with a single absorbing state.

In order to utilize our matrix results it is convenient to utilize a *canonical form* of the transition matrix P. First group all the persistent states together and then all the transient states. Assume there are n total states and t transient states and therefore $n - t$ persistent states. The rearranged matrix becomes

$$P* = \begin{array}{c} \\ n-t\{ \\ t\{ \end{array} \overset{\displaystyle n-t \quad t}{\left[\begin{array}{cc} S & O \\ R & Q \end{array}\right]} \qquad (9.4\text{-}3)$$

where S deals with the process after it has reached the persistent states, O is an $(n - t) \times t$ all-zero submatrix; the $t \times (n - t)$ submatrix R deals with the transition from transient to persistent states, and the submatrix Q deals with the process as long as it stays in the transient states.

The above theorem allows us to conclude that the powers of Q (representing A as time evolves) tends to a $t \times t$ zero matrix, and consequently all the elements of the last t columns of $P^n \to 0$ as $n \to \infty$.

For an absorbing chain it is obvious that the canonical form is

$$P* = \begin{array}{c} \\ n-t\{ \\ t\{ \end{array} \overset{\displaystyle n-t \quad t}{\left[\begin{array}{cc} I & O \\ R & Q \end{array}\right]} \qquad (9.4\text{-}4)$$

since, by definition, if we have $n - t$ absorbing states, then I represents the states as an $(n - t) \times (n - t)$ unit submatrix.

Let us consider the first example of this section, which has the transition matrix

$$
P_1 = \begin{array}{c} \\ 1 \\ 2 \\ 3 \\ 4 \\ 5 \end{array}
\begin{array}{c}
\begin{array}{ccccc} 1 & 2 & 3 & 4 & 5 \end{array} \\
\left[\begin{array}{ccccc}
1 & 0 & 0 & 0 & 0 \\
q & 0 & p & 0 & 0 \\
0 & q & 0 & p & 0 \\
0 & 0 & q & 0 & p \\
0 & 0 & 0 & 0 & 1
\end{array}\right]
\end{array}
\tag{9.4-5}
$$

Since states 1 and 5 are absorbing, the canonical form is given by

$$
P_1^* = \begin{array}{c} \\ 1 \\ 5 \\ 2 \\ 3 \\ 4 \end{array}
\begin{array}{c}
\begin{array}{ccccc} 1 & 5 & 2 & 3 & 4 \end{array} \\
\left[\begin{array}{cc:ccc}
1 & 0 & 0 & 0 & 0 \\
0 & 1 & 0 & 0 & 0 \\
\hdashline
q & 0 & 0 & p & 0 \\
0 & 0 & q & 0 & p \\
0 & p & 0 & q & 0
\end{array}\right]
\end{array}
\tag{9.4-6}
$$

Hence

$$
I = \begin{bmatrix} 1 & 0 \\ 0 & 1 \end{bmatrix} \qquad
O = \begin{bmatrix} 0 & 0 & 0 \\ 0 & 0 & 0 \end{bmatrix}
$$

$$
R = \begin{bmatrix} q & 0 \\ 0 & 0 \\ 0 & p \end{bmatrix} \qquad
Q = \begin{bmatrix} 0 & p & 0 \\ q & 0 & p \\ 0 & q & 0 \end{bmatrix}
\tag{9.4-7}
$$

If we do not care at which state the absorbing set of states is entered we may combine the two absorbing states together to obtain

$$
\hat{P}_1^* = \begin{bmatrix}
1 & 0 & 0 & 0 \\
q & 0 & p & 0 \\
0 & q & 0 & p \\
p & 0 & q & 0
\end{bmatrix}
\tag{9.4-8}
$$

If we are interested in what happens as we go from state j to state k we may make state k an absorbing state.

We will find that absorbing chains answer the type of questions we have regarding the time to drop out of lock as well as the time to enter lock.

9.4.2 The Fundamental Matrix

A very important matrix will be used in the following development. First we prove a theorem required to prove that the matrix exists.

Theorem 2 If A^n tends to 0 (zero matrix) as n tends to infinity, then $[I - A]$ has an inverse and further

$$
[I - A]^{-1} = \sum_{k=0}^{\infty} A^k \qquad A^0 = I
\tag{9.4-9}
$$

Proof Consider the following equality:

$$(I - A)(I + A + A^2 + \cdots + A^{n-1}) = I - A^n \qquad (9.4\text{-}10)$$

We wish to show that $I - A$ has a nonzero determinate since this is a sufficient condition for the matrix to have an inverse. By hypothesis the right side of the above equation converges to the unit matrix I. Hence the determinate of the right side is nonzero for sufficiently large n. Now the determinate of the product of two matrices is the product of the two determinates. Therefore $I - A$ cannot have a zero determinate, which is what we set out to show; that is, $I - A$ has an inverse. Now premultiply both sides of the equation by $[I - A]^{-1}$:

$$[I + A + A^2 + A^3 + \cdots + A^{n-1}] = [I - A]^{-1}[I - A^n] \qquad (9.4\text{-}11)$$

Hence as n becomes unbounded

$$[I - A]^{-1} = \sum_{k=0}^{\infty} A^k \qquad (9.4\text{-}12)$$

From Theorem 1 we see that $Q^n \to 0$. Hence we have the following result.

Corollary 1

$$[I - Q]^{-1} = \sum_{k=0}^{\infty} Q^k \qquad Q^0 = I \qquad (9.4\text{-}13)$$

Definition We define the *fundamental matrix* for an absorbing Markov chain to be the following quantity:

$$N = [I - Q]^{-1} \qquad (t \times t) \qquad (9.4\text{-}14)$$

Definition Define n_j to be the total number of times that the process is in state j (denoted by S_j), where S_j is a transient state. Also define U_j^k as a function that is equal to 1 if the process is in state S_j after k steps, and is zero otherwise.

Theorem 3 Letting T denote the set of transient states, we have

$$\{E_i[n_j]\} = N \qquad S_i, S_j \in T \qquad (9.4\text{-}15)$$

where $\{E_i[n_j]\}$ denotes the matrix composed of elements formed by the mean number of total times the process is in state j starting in state i. Hence this result shows that the average number of times that the process is in a particular transient state j starting at state i is the ijth element of the matrix N. Furthermore it establishes the fact that the mean number of times the process is in a particular state is finite!

Proof First we note that

$$n_j = \sum_{k=0}^{\infty} u_j^k \qquad (9.4\text{-}16)$$

Therefore

$$\{E_i[n_j]\} = \left\{ E_i \left[\sum_{k=0}^{\infty} u_j^k \right] \right\} = \left\{ \sum_{k=0}^{\infty} E_i u_j^k \right\} \tag{9.4-17}$$

Now since $p_{ij}^{(k)}$ is defined as the probability of going from state i to state j in k steps, we have

$$\{E_i U_j^k\} = \left\{ \sum_{k=0}^{\infty} (1 - p_{ij}^{(k)}) \cdot 0 + p_{ij}^{(k)} \cdot 1 \right\} = \left\{ \sum_{k=0}^{\infty} p_{ij}^{(k)} \right\} \tag{9.4-18}$$

But $\{p_{ij}^{(k)}\} = [P*]^k$ for all states i, j. Recall that

$$[P*] = \begin{bmatrix} S & O \\ R & Q \end{bmatrix} \tag{9.4-19}$$

But since it is clear by matrix multiplication that

$$[P*]^n = \begin{bmatrix} S^n & O \\ R' & Q^n \end{bmatrix} \tag{9.4-20}$$

where R' is in general different from R but is of no concern here and S^n and Q^n are just the nth powers of their respective matrices. But since the states i and j were restricted to transient states, we have

$$\left\{ \sum_{k=0}^{\infty} p_{ij}^{(k)} \right\} = \sum_{k=0}^{\infty} Q^k \qquad S_i, S_j \text{ transient states} \tag{9.4-21}$$

By our Corollary 1 we have

$$\left\{ \sum_{k=0}^{\infty} p_{ij}^{(k)} \right\} = [I - Q]^{-1} = N \qquad \text{QED} \tag{9.4-22}$$

To illustrate the application of the matrix $[N]$ consider the canonical form of the transition matrix

$$P* = \begin{matrix} \\ 1 \\ 2 \\ 3 \end{matrix} \begin{matrix} 1 & 2 & 3 \\ \begin{bmatrix} 1 & 0 & 0 \\ \frac{1}{2} & 0 & \frac{1}{2} \\ \frac{1}{4} & \frac{1}{4} & \frac{1}{2} \end{bmatrix} \end{matrix} \tag{9.4-23}$$

Therefore

$$Q = \begin{matrix} 2 \\ 3 \end{matrix} \begin{matrix} 2 & 3 \\ \begin{bmatrix} 0 & \frac{1}{2} \\ \frac{1}{4} & \frac{1}{2} \end{bmatrix} \end{matrix} \tag{9.4-24}$$

$$N = [I - Q]^{-1} = \begin{matrix} 2 \\ 3 \end{matrix} \begin{matrix} 2 & 3 \\ \begin{bmatrix} \frac{4}{3} & \frac{4}{3} \\ \frac{2}{3} & \frac{8}{3} \end{bmatrix} \end{matrix} \tag{9.4-25}$$

Hence the mean number of times the process is in state 2 starting in state 3 before entering state i is $\frac{2}{3}$, whereas the mean time in state 3 starting at state 3 before entering state 1 is $\frac{8}{3}$. Notice that the mean number of times the process is in a nonabsorbing state, starting in state 3, is $\frac{10}{3}$, and the mean number of times the process is in a nonabsorbing state starting in state 2 is $\frac{8}{3}$. Now we are at a point where we can now prove our main results. The results are a slight generalization of those of Kemeny and Snell [15].

Definition Let n be the function giving the time including the original position, in which the process is in a transient state.

Theorem 4 The mean value of n, given that the process starts in state i is given by

$$\{E_i[n]\} = [N]\,\mathbf{1} \qquad S_i \in T \tag{9.4-26}$$

where $[N]$ is the matrix $[I-Q]^{-1}$ and $\mathbf{1}$ is a t component unit column vector (recall Q is $t \times t$), each component of which represents the time (unity) in that state.

Proof From our definition of n_j we have

$$n = \sum_{S_j \in T} n_j \tag{9.4-27}$$

with T denoting the set of transient states, so that

$$\{E_i[n]\} = \left\{ \sum_{s_j \in T} E_i[n_j] \right\} \tag{9.4-28}$$

But in Theorem 3 we have shown that

$$\{E_i[n_j]\} = N \tag{9.4-29}$$

Hence

$$E_i[n] = N\,\mathbf{1} \tag{9.4-30}$$

since this expression gives the row sums of N.

Corollary It follows from Theorem 4 and Problem 12 that the mean time the process is in a transient state, starting in state i, is given by

$$E_i[\xi] = N\tau \qquad \tau = \left\{ \begin{array}{c} T_1 \\ T_2 \\ \vdots \\ T_t \end{array} \right\} \tag{9.4-31}$$

where τ is a t-component column vector whose components are the respective "dwell" times and ξ denotes time. This result generalizes equation 9.4-26 to the case the dwell times are not equal.

Theorem 5

$$\mathrm{var}_i[n] = (2N - I)N\,\mathbf{1} - (N\,\mathbf{1})_{\mathrm{sq}} \qquad S_i \in T \tag{9.4-32}$$

where $(N\,\mathbf{1})_{\mathrm{sq}} =$ square of each component of $N\,\mathbf{1}$. Hence the above equation gives the variance of the number of times (n) that the process is in transient states before being absorbed, given that the process started in state i. This result holds for unit dwell times.

Proof First we note that

$$\{\text{var}_i[n]\} = \{E_i[n^2]\} - \{E_i[n]\}^2 \tag{9.4-33}$$

Hence it remains to evaluate $E_i[n^2]$. Denote $\tilde{T}$ as the set of nontransient states (persistent states). Then

$$\{E_i[n^2]\} = \left\{ \sum_{S_k \in \tilde{T}} p_{ik} \cdot 1 + \sum_{S_k \in T} p_{ik} E_k[(n+1)^2] \right\} \tag{9.4-34}$$

so

$$\{E_i[n^2]\} = \left\{ \sum_{S_k \in \tilde{T} \cup T} p_{ik} \cdot 1 + \sum_{S_k \in T} p_{ik} E_k[n^2] + 2 \sum_{S_k \in T} p_{ik} E_k[n] \right\} \tag{9.4-35}$$

Since the probability of going from state i to any of the possible states is 1, we have

$$\{E_i[n^2]\} = \left\{ 1 + \sum_{S_k \in T} p_{ik} E_k[n^2] + 2Q(N\,1) \right\} \tag{9.4-36}$$

where the last term was obtained by noting $\{E_k[n]\} = N\tau$ and $\sum_{S_k \in T} p_{ik} = Q$, as was shown before. Finally we have

$$\{E_i[n^2]\} = 1 + 2QN\,1 + QE_i[n^2]$$

or

$$\{E_i[n^2]\} = (I - Q)^{-1}(2QN\,1 + 1) \tag{9.4-37}$$

or

$$\{E_i[n^2]\} = 2NQN\,1 + N\,1 \tag{9.4-38}$$

It is not difficult to show that

$$QN = NQ = N - I \tag{9.4-39}$$

so that

$$\{E_i[n^2]\} = 2[N - I]N\,1 + N\,1 = (2N - I)\,N\,1 \tag{9.4-40}$$

Finally we have

$$\{\text{var}_i[n]\} = (2N - I)N\,1 - \{E_i[n]\}^2$$
$$= (2N - I)\,N\,1 - (N\,1)_{sq} \qquad \text{QED} \tag{9.4-41a}$$

Corollary 5 It follows from Theorem 5 and Problem 12 that the variance of the time ξ, that the process is in a transient state, is given by

$$\text{var}_i(\xi) = 2[I - Q]^{-1}[T]Q[I - Q]^{-1}\tau + [I - Q]^{-1}(\tau_{sq}) - ([I - Q]^{-1}\tau)_{sq} \tag{9.4-41b}$$

$$[T] = \begin{bmatrix} T_1 & 0 & 0 & 0 & 0\ldots 0 \\ 0 & T_2 & 0 & 0 & 0\ldots 0 \\ 0 & 0 & T_3 & 0 & 0\ldots 0 \\ \vdots & \vdots & & & \vdots \\ 0 & 0 & \cdots & 0 & 0 \quad T_t \end{bmatrix}$$

when the dwell times are not equal, that is having dwell time τ. The corollaries to Theorems 4 and 5 are the main results of the section and generalize Kemeny and Snells [15] results to unequal dwell times. Using the transition matrix (canonical form) of the last example let us consider computing the mean and variance of the time that the process is in a transient state given that it started in state i. Assume that the "dwell" in each state is 1 sec. Recall that we obtained

$$Q = \begin{array}{c} \\ 2 \\ 3 \end{array}\!\!\begin{array}{c} 2 \quad 3 \\ \begin{bmatrix} 0 & \frac{1}{2} \\ \frac{1}{4} & \frac{1}{2} \end{bmatrix} \end{array}$$

$$N = \begin{array}{c} \\ 2 \\ 3 \end{array}\!\!\begin{array}{c} 2 \quad 3 \\ \begin{bmatrix} \frac{4}{3} & \frac{4}{3} \\ \frac{2}{3} & \frac{8}{3} \end{bmatrix} \end{array} \qquad (9.4\text{-}42)$$

so

$$\{E_i[\xi]\} = \begin{bmatrix} \frac{4}{3} & \frac{4}{3} \\ \frac{2}{3} & \frac{8}{3} \end{bmatrix}\begin{Bmatrix} 1 \\ 1 \end{Bmatrix} = \frac{2}{3}\begin{Bmatrix} \frac{8}{3} \\ \frac{10}{3} \end{Bmatrix}$$

Therefore the mean time in a transient state, starting in state 2, before being absorbed in $\frac{8}{3}$ sec and starting in state 3 is $\frac{10}{3}$ sec. This agrees with our interpretation of the N matrix in our last example. To compute the variance we must compute (using equation 9.4-32)

$$\{var_i[\xi]\} = (2N - I)\,N\mathbf{1} - (N\mathbf{1})_{sq}$$

$$= \left\{ \begin{bmatrix} \frac{8}{3} & \frac{8}{3} \\ \frac{4}{3} & \frac{16}{3} \end{bmatrix} - \begin{bmatrix} 1 & 0 \\ 0 & 1 \end{bmatrix} \right\}\begin{Bmatrix} \frac{8}{3} \\ \frac{10}{3} \end{Bmatrix} - \begin{Bmatrix} \frac{64}{9} \\ \frac{100}{9} \end{Bmatrix} \qquad (9.4\text{-}43)$$

or

$$\{var_i[\xi]\} = \frac{2}{3}\begin{Bmatrix} \frac{56}{9} \\ \frac{62}{9} \end{Bmatrix}$$

Consequently the variance of the number of times the process is in transient states, starting in state 3, is larger than starting in state 2. The standard deviation is just the square root of the respective variances.

PROBLEM 12

Establish (9.4-31) and (9.4-41b), which generalize Kemeny and Snell's equal dwell time results to unequal dwell times.

The above theory allows us to readily compute the mean and variance of the number of steps (including the original position), and time it takes to be absorbed (reach an absorbing state).

In general to study the mean time (or variance) it takes to reach a particular state, it is convenient simply to modify the transition matrix or state transition diagram so that the state under study is an absorbing one! The resulting process so described will be a Markov process with one absorbing state. The behavior of the process before absorption is exactly the same as the original process.

Another interesting result due to Kemeny and Snell [15] concerns the *probability* of starting in a transient state and ending up in a persistent state. This result yields no time information. Furthermore, with only one absorbing state the probabilities are all 1! Therefore the result is primarily useful when there are two or more absorbing states.

Theorem 6 The probability that the Markov process starting in a transient state i ends up in an absorbing state j is b_{ij}, where

$$\{b_{ij}\} = B = NR \qquad (t \times n - t) \text{ matrix} \qquad (9.4\text{-}44)$$

R is the matrix defined by

$$P^* = \begin{bmatrix} S & O \\ R & Q \end{bmatrix} \qquad (9.4\text{-}45)$$

and deals with the transition from transient to absorbing states.

Proof Starting in state i, the process may be absorbed in state j in one or more steps. For a single step the probability of absorption is p_{ij}. If the process is not absorbed, the process can move to either another absorbing state (and cannot reach state j) or to a transient state k. In this latter case the probability of being absorbed in state j is b_{jk} for any transient state k. Therefore we can write

$$b_{ij} = p_{ij} + \sum_{S_k \in T} p_{ik} b_{kj} \qquad (9.4\text{-}46)$$

which in matrix form is

$$B = R + QB \qquad (9.4\text{-}47)$$

since the submatrix R contains the transition probabilities of the transient states (the set T) to the absorbing states (the set $\tilde{T}$), and the submatrix Q contains the transition probabilities of transient states to transient states. Solving

$$B = [I - Q]^{-1}R = NR \qquad \text{QED}$$

PROBLEM 13

Show that for the process defined by the matrix shown below that

(a)
$$\{E_i[n]\} = \begin{Bmatrix} \frac{17}{5} \\ \frac{18}{5} \\ \frac{11}{5} \end{Bmatrix} \qquad \{var_i[n]\} = \begin{Bmatrix} 168/25 \\ 144/25 \\ 120/25 \end{Bmatrix}$$

(b)

$$R = \begin{array}{c} \\ 2 \\ 3 \\ 4 \end{array}\overset{\displaystyle 1 \qquad 5}{\begin{bmatrix} \frac{1}{3} & 0 \\ 0 & 0 \\ 0 & \frac{2}{3} \end{bmatrix}}$$

$$B = \begin{array}{c} \\ 2 \\ 3 \\ 4 \end{array}\overset{\displaystyle 1 \qquad\quad 5}{\begin{bmatrix} 7/15 & 8/15 \\ 1/5 & 4/5 \\ 1/15 & 14/15 \end{bmatrix}}$$

where

$$P = \begin{array}{c} \\ 1 \\ 2 \\ 3 \\ 4 \\ 5 \end{array}\overset{\displaystyle 1 \quad 2 \quad 3 \quad 4 \quad 5}{\begin{bmatrix} 1 & 0 & 0 & 0 & 0 \\ \frac{1}{3} & 0 & \frac{2}{3} & 0 & 0 \\ 0 & \frac{1}{3} & 0 & \frac{2}{3} & 0 \\ 0 & 0 & \frac{1}{3} & 0 & \frac{2}{3} \\ 0 & 0 & 0 & 0 & 1 \end{bmatrix}}$$

Notice as a byproduct of the solution, the probability of eventually being absorbed, starting in any state is 1!

PROBLEM 14

Let

$$P = \begin{array}{c} \\ 1 \\ 2 \\ 3 \\ 4 \\ 5 \end{array}\overset{\displaystyle 1 \quad 2 \quad 3 \quad 4 \quad 5}{\begin{bmatrix} 1 & 0 & 0 & 0 & 0 \\ \frac{1}{2} & 0 & \frac{1}{2} & 0 & 0 \\ 0 & \frac{1}{2} & 0 & \frac{1}{2} & 0 \\ 0 & 0 & \frac{1}{2} & 0 & \frac{1}{2} \\ 0 & 0 & 0 & 0 & 1 \end{bmatrix}}$$

and show that

$$N = \begin{bmatrix} \frac{3}{2} & 1 & \frac{1}{2} \\ 1 & 2 & 1 \\ \frac{1}{2} & 1 & \frac{3}{2} \end{bmatrix}$$

$$\{E_i[n]\} = \begin{Bmatrix} 3 \\ 4 \\ 3 \end{Bmatrix}$$

$$\{\text{var}_i[n]\} = \begin{Bmatrix} 8 \\ 8 \\ 8 \end{Bmatrix}$$

$$B = \begin{bmatrix} \frac{3}{4} & \frac{1}{4} \\ \frac{1}{2} & \frac{1}{2} \\ \frac{1}{4} & \frac{3}{4} \end{bmatrix}$$

PROBLEM 15

Obtain the same quantities as in Problem 14 for the following matrix:

$$
P = \begin{array}{c} \\ 1 \\ 2 \\ 3 \\ 4 \\ 5 \end{array}
\begin{array}{c}
\begin{array}{ccccc} 1 & 2 & 3 & 4 & 5 \end{array} \\
\left[\begin{array}{ccccc}
1 & 0 & 0 & 0 & 0 \\
q & 0 & p & 0 & 0 \\
0 & q & 0 & p & 0 \\
0 & 0 & q & 0 & p \\
0 & 0 & 0 & 0 & 1
\end{array} \right]
\end{array}
$$

PROBLEM 16

Consider the following state transition diagram:

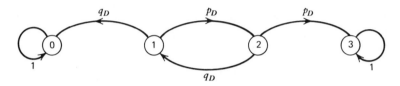

The diagram represents a simplified model to determine the probability of lock [16], where $p_D = P(\text{hit} \mid \text{search mode})$ and $q_D = P(\text{no hit} \mid \text{search mode})$.
(a) Show that the probability of reaching state 3 starting in state 1 is given by

$$
P_L = \frac{p_D^2}{1 - p_D q_D}
$$

(b) Show that the probability of reaching state 3 or state 0 from state 1 is unity!

As can be seen from the above problem, this theory can be applied to acquisition schemes as well as to lock detection.

9.4.3   Code Lock Detector Performance

Lock detectors are devices that indicate when a PN code acquisition system has succeeded in obtaining lock. The results developed here are applicable to Lock detectors that can be characterized as Markov chains, with the transition probabilities being time independent. Lock detection is typically accomplished after lock is tentatively obtained and the code tracking loop is turned on.

Typically either a code lock detector utilizes a long-duration integration or a series of integrations based on a counter that declares the system to be in lock until some number of consecutive below-threshold indications occur, at which time the system is declared out of lock. One possible problem with simply extending the dwell time to ascertain the in- or the out-of-lock state is due to the fact that if a drop-out occurs because of antenna switching, deep fades, etc., a single integration might become unreliable (due to high AGC gain induced noise effects) indicating that the system is in lock when it is out or *vice versa*. Counter type lock detectors will be the subject of this section since a single drop-out does not seriously affect their performance. Many systems employ counters with shorter integration times. The actual implementation can be in hardware or software or a combination of both.

We shall utilize the results of this section to solve a trivial type of lock detector problem. Consider a lock detector based on an integrator with dwell time T, that integrate the square-law-detected signal as shown in Figure 9.17. We shall assume that when the signal is present the probability of correct detection is p and the miss detection probability is q. One miss detection causes the detector to declare that the system is out of lock. The state transition diagram is shown below in Figure 9.20.

The process remains in state 1 T sec, then the system is tested again and either remains in state 1 for T more seconds or is absorbed in state 2.

The transition matrix is given by

$$P = \begin{array}{c} 1 \\ 2 \end{array}\begin{array}{cc} 1 & 2 \\ \begin{bmatrix} p & q \\ 0 & 1 \end{bmatrix} \end{array} \tag{9.4-48}$$

In canonical form it becomes

$$P^* = \begin{array}{c} 2 \\ 1 \end{array}\begin{array}{cc} 2 & 1 \\ \begin{bmatrix} 1 & 0 \\ q & p \end{bmatrix} \end{array} \tag{9.4-49}$$

The mean time to absorption is (using the corollary to Theorem 9.4.5)

$$E_1[\xi] = N\tau = (1 - p)^{-1}T \qquad \text{one-dimensional vector}$$

or

$$E_1[\xi] = \frac{T}{1 - p} \tag{9.4-50}$$

If, for example, $p = .99$, then $E_1[\xi] = 100T$, which is the mean time to declare the system is out of lock when in fact it is in lock. Note the results also apply to the case when the signal is absent. In this case p would

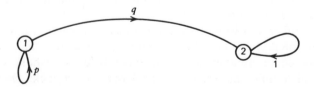

Figure 9.20 Simple single-count lock detector state transition diagram.

normally be near zero and q near 1, so that $\bar{T} = E_1[\xi] \cong T$! This simple example can also be solved by a straightforward calculation. Let P_k be the probability of being absorbed on the kth trial. Then $(p = 1 - q)$

$$\bar{T} = \sum_{k=1}^{\infty} kTP_k = \sum_{k=1}^{\infty} k(1-q)^{k-1}qT \tag{9.4-51}$$

Since

$$\sum_{1}^{\infty} k(1-q)^{k-1} = \frac{1}{q^2} \tag{9.4-52}$$

the result is

$$\bar{T} = \frac{T}{q} = \frac{T}{1-p} \tag{9.4-53}$$

as was obtained above. However, as the number of states increases this direct enumeration of the kth absorption probability becomes much more difficult. Although not computed, the variance can also be easily obtained for the example.

PROBLEM 17

Show that the variance of the above described lock detector is given by

$$\sigma^2 = \frac{p}{q^2} T^2$$

Use both the matrix method and the direct enumeration method.

As a more complex example of a lock detector consider the following. Require that the lock detector have three consecutive missed detections before the system is declared out of lock. It is clear that the state transition diagram associated with this detector is given in Figure 9.21. The state transition matrix is given by

$$
P = \begin{array}{c}
\\
1 \\
2 \\
3 \\
4
\end{array}
\begin{array}{c}
\begin{array}{cccc} 1 & 2 & 3 & 4 \end{array} \\
\left[\begin{array}{cccc}
p & q & 0 & 0 \\
p & 0 & q & 0 \\
p & 0 & 0 & q \\
0 & 0 & 0 & 1
\end{array}\right]
\end{array} \tag{9.4-54}
$$

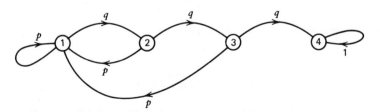

Figure 9.21 "Three consecutive counts" state transition diagram.

In canonical form we have

$$P^* = \begin{array}{c} \\ 4 \\ 1 \\ 2 \\ 3 \end{array}\begin{array}{c} \begin{array}{cccc} 4 & 1 & 2 & 3 \end{array} \\ \begin{bmatrix} 1 & 0 & 0 & 0 \\ 0 & p & q & 0 \\ 0 & p & 0 & q \\ q & p & 0 & 0 \end{bmatrix} \end{array} \tag{9.4-55}$$

since the fourth state is the absorbing one. The mean value vector is given by

$$\{E_i[\xi]\} = \left\{ \begin{bmatrix} 1 & 0 & 0 \\ 0 & 1 & 0 \\ 0 & 0 & 1 \end{bmatrix} - \begin{bmatrix} p & q & 0 \\ p & 0 & q \\ p & 0 & 0 \end{bmatrix} \right\}^{-1} \begin{bmatrix} 1 \\ 1 \\ 1 \end{bmatrix} T \tag{9.4-56}$$

where T is the duration of each dwell time. Evaluating, we have

$$\bar{T} = \frac{1 + q + q^2}{(1-p)^3} T \tag{9.4-57}$$

which is the mean time for the lock detector to declare out of lock when the signal is present. Making p sufficiently close to 1 causes $\bar{T}$ to be much greater than T.

It has been proven by Holmes [17] that the "n consecutive count" lock detector has a mean false dismal time of

$$\bar{T} = \frac{1 + q + \cdots + q^{n-1}}{q^n} T \qquad \begin{array}{l} n+1 \text{ nodes on the state} \\ \text{transition diagram} \end{array} \tag{9.4-58}$$

Another type of lock detector is the up-down counter type that starts in state 1 and remains there unless a miss detection occurs, at which time the counter drops to state 2. A hit returns the counter to state 1 whereas a miss detection sends the counter to state 3 from state 2. Finally when the nth state is reached the lock detector declares the system is out of lock.

Consider the state transition diagram for a three-state counter lock detector shown in Figure 9.21.

The canonical form of the matrix is

$$P^* = \begin{bmatrix} 1 & 0 & 0 & 0 \\ 0 & p & q & 0 \\ 0 & p & 0 & q \\ q & 0 & p & 0 \end{bmatrix} \tag{9.4-59}$$

It can be shown that (Problem 18) the mean time to declare a false dismissal is given by

$$\bar{T} = \frac{1 + 2q^2}{q^3} T \tag{9.4-60}$$

which does not differ significantly from the other three-state (four-node) detector. Notice as a check if $q = 1$ and $p = 0$, $\bar{T} = 3T$, as is obvious from

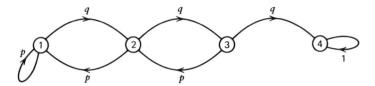

Figure 9.22 "Three-state up-down counter" lock detector state transition diagram.

Figure 9.22. Note that, as before, (9.4-60) holds when the signal is not present also.

In general it appears that there is not much to choose between the two types of lock detectors discussed here.

PROBLEM 18

Referring to Figure 9.22, which illustrates the three-state up-down lock detector counter state transition diagram, show that the mean time to indicate the out of lock state when the system is in lock is given by

$$\bar{T} = \frac{1+2q^2}{q^3} T$$

where q is the probability of the T-sec integrator being below the threshold and p is the probability of it being above the threshold.

Note that this result holds both when the signal is present and when it is absent. Of course a different p and q apply to each case.

PROBLEM 19

Compute the mean times to declare out of lock for the lock detector type that requires five consecutive below threshold observations before declaring out of lock.

9.5 SEQUENTIAL DETECTION

In this chapter we have been considering acquisition schemes that utilize fixed integration times. It can be shown that for low SNR the noncoherent square law detector scheme that we have considered is essentially optimum when a fixed integration time scheme is used. However, if one is willing to accommodate a variable integration time it is possible to make the average integration time, to achieve a given probability of detection and false alarm, less than the corresponding fixed time scheme, and in fact may be appreciably less.

Applications involving sequential detection (the dwell or integration time is a random variable) are now being used in military and nonmilitary systems to achieve faster acquisition times. The genesis of the theory is due to Wald [18]. Further work has been contributed by Bussgang and

Middleton [19], Wald and Wolfowitz [20], and others [21–23]. Numerous references on sequential detection and analysis are given at the end of reference 19. DiFranco and Rubin [24] and Hancock and Wintz [25] have texts with chapters on sequential detection.

9.5.1 Sequential Probability Ratio Test

Consider a process $x(t)$ for which successive samples are labeled x_1, x_2, $x_3, \ldots, x_m$. The condition probability density function of the m samples is denoted by $p(\mathbf{x}_m \mid \boldsymbol{\theta})$, where $\boldsymbol{\theta}$ is a k-dimensional vector having components θ_1, θ_2, $\theta_3, \ldots, \theta_k$ and $\mathbf{x}_m$ is an m-dimensional vector having components x_1, $x_2, \ldots, x_m$. We define H_0 to be the simple hypothesis $\boldsymbol{\theta} = \boldsymbol{\theta}_0$ and H_1 to be the alternate simple hypothesis $\boldsymbol{\theta} = \boldsymbol{\theta}_1$. The classical radar case is then defined: H_1 denoting signal present, $\boldsymbol{\theta} = \boldsymbol{\theta}_1$; and H_0 denoting the absence of signal, $\boldsymbol{\theta} = \boldsymbol{\theta}_0$. The likelihood ratio

$$\Lambda(\mathbf{x}_m) = \frac{p(\mathbf{x}_m \mid \theta_1)}{p(\mathbf{x}_m \mid \theta_0)} \tag{9.5-1}$$

which is a function of the sample size, m.

The sequential test of hypothesis H_1 against H_0 is as follows. Denote A and B as two positive constants such that $A > B$. Then at the mth sample accept H_0 if

$$\Lambda(\mathbf{x}_m) \leq B \tag{9.5-2}$$

Accept H_1 if

$$\Lambda(\mathbf{x}_m) \geq A \tag{9.5-3}$$

and continue observing samples if

$$B < \Lambda(\mathbf{x}_m) < A \tag{9.5-4}$$

Based on this algorithm we can visualize one realization of a sequential text, as shown in Figure 9.23. Denote the nth sample as the sample for which the following is satisfied:

$$\Lambda(\mathbf{x}_n) \geq A \tag{9.5-5}$$

Then

$$p(\mathbf{x}_n \mid \theta_1) \geq A p(\mathbf{x}_n \mid \boldsymbol{\theta}_0) \tag{9.5-6}$$

Let Γ_1 denote the set of samples $\mathbf{x}_n$ that lead to the acceptance of H_1. Integrating, we have

$$\int_{\Gamma_1} p(\mathbf{x}_n \mid \boldsymbol{\theta}_1) d\mathbf{x}_n \geq A \int_{\Gamma_1} p(\mathbf{x}_n \mid \boldsymbol{\theta}_0) \, d\mathbf{x}_n \tag{9.5-7}$$

Now the integral on the left is P_D, the probability of detection, and the integral on the right side is the probability of false alarm, so

$$P_D \geq A P_{FA} \tag{9.5-8}$$

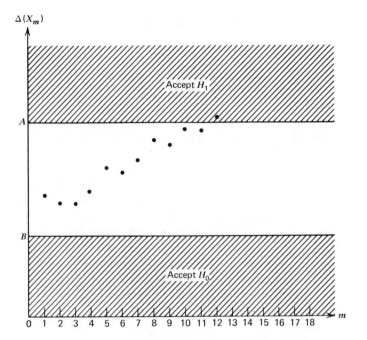

Figure 9.23 An example of a sequential test result.

Proceeding in an analogous manner we conclude that the termination of the sequential ratio test with the acceptance of hypothesis H_0 yields

$$1 - P_D \leq B(1 - P_{FA}) \tag{9.5-9}$$

Since discrete samples are used in the test it is quite possible that exact equality may never occur. This phenomenon is called the excess over the boundaries problem and is discussed in detail by Wald [18].

Let $(\delta, \beta) > 0$ be the excess over the boundaries such that

$$P_D = AP_{FA} + \delta \tag{9.5-10}$$
$$1 - P_D = B(1 - P_{FA}) - \beta$$

Solving, we obtain

$$P_{FA} = \frac{1 - B + \delta - \beta}{(A - B)}$$
$$P_D = \frac{A(B - 1) + \delta B - \beta A}{(B - A)} \tag{9.5-11}$$

Therefore when the excess over the boundaries is negligible, the thresholds should be set at

$$A \cong \frac{P_D}{P_{FA}}$$
$$B \cong \frac{1 - P_D}{1 - P_{FA}} \tag{9.5-12}$$

And further, the error probabilities are given by

$$P_{FA} \cong \frac{1-B}{A-B}$$

$$P_D \cong \frac{A(B-1)}{B-A}$$

(9.5-13)

Note if $P_D > \frac{1}{2}$ and $P_{FA} < \frac{1}{2}$, then $A > 1$ and $B < 1$. When the value of n, the value of time, at which the sample exceeds the threshold, is large it is normally assumed that $\delta = \beta = 0$.

Let us now consider an example. Consider using a sequential test to determine the mean of a sequence of independent observations of a Gaussian random process with known variance σ^2. We wish to test the null hypothesis $H_0:\bar{\mathbf{x}} = 0$ against the hypothesis $H_1: \bar{\mathbf{x}} = \mu$.

Again let A be the upper boundary and B the lower boundary. The sequential ratio procedure is to compare $\Lambda(\mathbf{x}_m)$ to A and B. We could just as well compare any monotone function of these variables. Since $\Lambda(\mathbf{x}_m)$, A, and B are all nonnegative quantities, an especially convenient monotonic function is the natural logarithm. Consequently, the stochastically equivalent test becomes accept H_0 if

$$\ln \Lambda(\mathbf{x}_m) \le \ln B$$

accept H_1 if

$$\ln \Lambda(\mathbf{x}_m) \ge \ln A$$

and continue sampling otherwise. For the case in hand after the mth sample we have

$$\ln \Lambda(\mathbf{x}_m) = \ln \frac{\dfrac{1}{(2\pi)^{m/2}\sigma^m} \exp\left(-\dfrac{1}{2\sigma^2} \sum_{i=1}^{m} (x_i - \mu)^2\right)}{\dfrac{1}{(2\pi)^{m/2}\sigma^m} \exp\left(-\dfrac{1}{2\sigma^2} \sum_{i=1}^{m} x_i^2\right)}$$

or

$$\ln \Lambda(\mathbf{x}_m) = \frac{\mu}{\sigma^2} \sum_{i=1}^{m} \left(x_i - \frac{\mu}{2}\right)$$

Therefore our test can be written

$$\sum_{i=1}^{m} x_i \ge \frac{\sigma^2}{\mu} \ln A + \frac{m}{2}\mu \qquad \text{accept } H_1$$

$$\sum_{i=1}^{m} x_i \le \frac{\sigma^2}{\mu} \ln B + \frac{m}{2}\mu \qquad \text{accept } H_0$$

(9.5-14)

$$\frac{\sigma^2}{\mu} \ln B + \frac{m}{2}\mu < \sum_{i=1}^{m} x_i < \frac{\sigma^2}{\mu} \ln A + \frac{m}{2}\mu \qquad \text{take another sample}$$

Note this last inequality can also be written as

$$\frac{\sigma^2}{m\mu} \ln B + \frac{\mu}{2} < \bar{x}_m < \frac{\sigma^2}{m\mu} \ln A + \frac{\mu}{2}$$

(9.5-15)

where

$$\bar{x}_m = \frac{1}{m} \sum_{i=1}^{m} x_i$$

Wald and Wolfowitz [20] have shown that the sequential probability ratio test is optimum in the sense that no other test, on the average, can achieve the same error probabilities with fewer samples.

9.5.2 The Operating Characteristic Function

Wald [18] has shown that a sequential test will terminate with probability 1 if the samples x_i are statistically independent. Additionally he has shown that a sequential test will terminate with unity probability for a large class of distributions when the samples are not statistically independent. In this section on sequential detection only tests that terminate are considered.

The operating characteristic function (OCF), $L(\theta)$, is defined as the probability that the sequential test terminates with the acceptance of hypothesis H when θ is the true state of the unknown parameter. It follows from this definition that

$$L(\theta_0) = 1 - P_{FA}$$
$$L(\theta_1) = 1 - P_D(\theta) \qquad (9.5\text{-}16)$$

A typical operating function of one variable is shown in Figure 9.24. An optimal OCF when the one-dimensional parameter θ is signal to noise is a dirac delta function of unit weight located at $\theta = \theta_0$ and zero elsewhere. Then any signal $(\theta > \theta_0)$ would always be detected.

In general a sequential ratio test is designed for a specific set of design point parameters denoted by θ_D. However, the entire operating characteristic $L(\theta)$ is needed in order to provide the average sample number (see Section 9.5.3 for all values of θ. The method used to determine it is approximate since the excess over the boundaries is neglected. This method is due to Wald [18]. We will henceforth consider the one-dimensional case of $\theta_1 = \theta_D$ and $\theta_0 = 0$, which is the standard detection of signal in noise case.

Consider the expression

$$[\Lambda(\mathbf{x}_m)]^h = \left[\frac{p(\mathbf{x}_m \mid \theta_D)}{p(\mathbf{x}_m \mid 0)} \right]^h \qquad (9.5\text{-}17)$$

where $h = h(\theta, \theta_D)$ is a real valued function such that the expected value of (9.5-17) is unity, that is,

$$h(\theta, \theta_D) \neq 1 \qquad (9.5\text{-}18a)$$

$$\overline{[\Lambda(\mathbf{x}_m)]^h} = \int_{-\infty}^{\infty} \cdots \int_{-\infty}^{\infty} \left[\frac{p(\mathbf{x}_m \mid \theta_D)}{p(\mathbf{x}_m \mid 0)} \right]^h p(\mathbf{x}_m \mid \theta) d\mathbf{x}_m = 1 \qquad (9.5\text{-}18b)$$

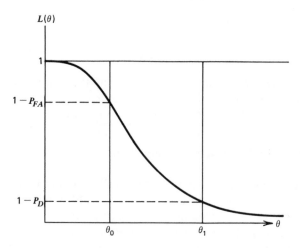

Figure 9.24 A typical OCF for a one-dimensional signal parameter.

which is an m-fold integration. The overbar denotes the ensemble average. Wald [18] has shown a unique solution for $h(\theta, \theta_D)$ exists. We see that the integrand of the above integral $g(\mathbf{x}_m \mid \theta)$, where we have

$$g(\mathbf{x}_m \mid \theta) = \left[\frac{p(\mathbf{x}_m \mid \theta_D)}{p(\mathbf{x}_m \mid 0)} \right]^h P(\mathbf{x}_m \mid \theta) \qquad (9.5\text{-}19)$$

is formally a proper density function in the sense that it is nonnegative and it integrates to unity. First consider the case that $h > 0$. We define H_g to be the hypothesis that the true distribution of $\mathbf{x}_m$ is $g(\mathbf{x}_m \mid \theta)$ and H_p the hypothesis that the true distribution of $\mathbf{x}_m$ is $p(\mathbf{x}_m \mid \theta)$. Now consider a sequential probability ratio test with the following rules. Continue taking observations when

$$B^h < \frac{g(\mathbf{x}_m \mid \theta)}{p(\mathbf{x}_m \mid \theta)} < A^h$$

Accept H_p when

$$\frac{g(\mathbf{x}_m \mid \theta)}{p(\mathbf{x}_m \mid \theta)} \le B^h$$

Accept H_g when

$$\frac{g(\mathbf{x}_m \mid \theta)}{p(\mathbf{x}_m \mid \theta)} \ge A^h \qquad (9.5\text{-}20)$$

Now note that

$$\frac{g(\mathbf{x}_m \mid \theta)}{p(\mathbf{x}_m \mid \theta)} = \left[\frac{p(\mathbf{x}_m \mid \theta_D)}{p(\mathbf{x}_m \mid 0)} \right]^h \qquad (9.5\text{-}21)$$

Therefore our test is equivalent to the following:

$$B < \frac{p(\mathbf{x}_m \mid \theta_D)}{p(\mathbf{x}_m \mid 0)} \quad \text{accept } H_p$$

$$\frac{p(\mathbf{x}_m \mid \theta_D)}{p(\mathbf{x}_m \mid 0)} > A \quad \text{accept } H_g \qquad (9.5\text{-}22)$$

$$B < \frac{p(\mathbf{x}_m \mid \theta_D)}{p(\mathbf{x}_m \mid 0)} < A \quad \text{take another sample}$$

Hence we see that if the test for H_p against H_g leads to the acceptance of H_p, then the test for H_0 against H_1 leads to the acceptance of H_0. Alternately, if the test for H_p against H_g leads to acceptance of H_g, then the test for H_0 against H_1 leads to the acceptance of H_1. It therefore follows that the probability of accepting H_0 when θ is true, that is, $L(\theta)$, is the same as the probability of accepting H_p when $p(\mathbf{x}_m \mid \theta)$ is the true distribution of $\mathbf{x}_m$. In the test of H_p against H_g the probability of accepting H_p when $p(\mathbf{x}_m \mid \theta)$ is true $1 - P_{FA}$, where the false alarm probability is related to the thresholds A^h and B^h via

$$A^h = \frac{P_D'}{P_{FA}'}$$

$$B^h = \frac{1 - P_D'}{1 - P_{FA}'} \qquad (9.5\text{-}23)$$

where P_D' and P_{FA}' are the probabilities associated with the test of H_p against H_g. Solving now for $L(\theta) = 1 - P_{FA}'$, we find that

$$L(\theta) = \frac{A^h - 1}{A^h - B^h} \qquad A = P_D/P_{FA}$$

$$B = (1 - P_D)/(1 - P_{FA}) \qquad (9.5\text{-}24)$$

where $h = h(\theta, \theta_D)$. The case for $h < 0$ can be solved in a similar manner with the same result. By setting $\theta = 0$ in (9.5-18) we find that $h(0, \theta_D) = 1$; and by setting $\theta = \theta_D$, we find that $h(\theta_D, \theta_D) = -1$.

PROBLEM 20

Show that when $h(\theta', \theta_D) = 0$ the value of $L(\theta')$ is given by

$$L(\theta') = \frac{\ln A}{\ln(A/B)} \qquad A = P_D/P_{FA}$$

$$B = (1 - P_D)/(1 - P_{FA})$$

The usefulness of this technique will be clarified when we consider the noncoherent detection case.

9.5.3 The Average Sample Number (ASN)

Let $\bar{n}(\theta)$ denote the average number of samples required to terminate the text, that is, to reach or exceed either boundary. Recall that our sequential ratio test procedure is to accept H_0 when $\Lambda(\mathbf{x}_m) \le B$, accept H_1 when $\Lambda(\mathbf{x}_m) \ge A$, and continue taking samples otherwise.

The average value of $\Lambda(\mathbf{x}_m)$ at test termination is approximately given by (neglecting the excess over the boundaries) log B times the probability of accepting H_0 plus log A times the probability of accepting H_1. Therefore for a terminated test of length n, given θ, the mean value is

$$E[\Lambda(\mathbf{x}_n)] \cong L(\theta) \ln B + [1 - L(\theta)] \ln A \qquad (9.5\text{-}25)$$

when the samples are independent

$$\Lambda(\mathbf{x}_n) = \sum_{i=1}^{n} \Lambda_i \qquad (9.5\text{-}26)$$

where Λ_i is the logarithm of the likelihood ratio of the ith sample:

$$\Lambda_i = \ln\left[\frac{p(x_i \mid \theta_D)}{p(x_i \mid 0)}\right] \qquad (9.5\text{-}27)$$

Based on stationary, independent, observations is the following:

$$E[\Lambda(\mathbf{x}_m \mid \theta)] = E\left[\sum_{i=1}^{n} \Lambda_i(\theta)\right] = \bar{n}(\theta)\, E[\Lambda_i(\theta)] \qquad (9.5\text{-}28)$$

Therefore from (9.5-25) and (9.5-28) we obtain our result for the ASN:

$$\bar{n}(\theta) \cong \frac{L(\theta) \ln B + [1 - L(\theta)] \ln A}{E[\Lambda_i(\theta)]} \qquad (9.5\text{-}29)$$

In reference 19 it was shown that when the observations are not statistically independent it is possible to obtain a relationship for $\bar{n}(\theta)$ provided that $\Lambda(\mathbf{x}_n)$ is a linear function of $\bar{n}(\theta)$. When θ is set to zero, $\bar{n}(0)$ is the mean number of (samples) steps required to terminate the test when the signal is not present. Further, when θ is set equal to θ_D, the design point $\bar{n}(\theta_D)$ is the mean number of samples needed when the signal is at the design point level.

In the next section we will utilize $\bar{n}(\theta_D)$ and $\bar{n}(0)$ to estimate the time it takes to obtain acquisition based on a noncoherent detector.

9.5.4 Average Sample Number for Noncoherent Detection

Consider the noncoherent detection of a sine wave signal of amplitude A (see Figure 9.25) corrupted by WGN of two-sided spectral density of $N_0/2$. The probability density function of the envelope out of a noncoherent

Figure 9.25 Noncoherent envelope detector.

envelope detector [19] is given by

$$p(r \mid \text{SNR}_i) = r \exp\left(-\frac{r^2 + 2\text{SNR}_i}{2}\right) I_0(\sqrt{2\,\text{SNR}_i}\, r) \qquad r \geq 0 \tag{9.5-30}$$

where SNR_i is the input SNR to the envelope detector and is given by

$$\text{SNR}_i = \frac{P}{N_0 B} \tag{9.5-31}$$

where $P = A^2/2$ is the input sine wave power and B is the bandpass filter noise bandwidth.

The sequential ratio test variable is Λ_i, where

$$\Lambda_i = \ln\left[\frac{p(r_i \mid \text{SNR}_D)}{p(r_i \mid 0)}\right] = -\text{SNR}_D + \ln[I_0(\sqrt{2\,\text{SNR}_D}\, r_i)] \tag{9.5-32}$$

with SNR_D being the design SNR out of the BPF. The optimum sequential detector then compares

$$\sum_{i=1}^{m} \Lambda_i$$

against the threshold A and B as indicated previously. When the input SNR is very small we have, by [19] expanding $\ln I_0(x)$,

$$\Lambda_i = \text{SNR}_D + \frac{\text{SNR}_D r_i^2}{2} - \frac{\text{SNR}_D^2\, r_i^4}{16} + O(\text{SNR}_D^3 r_i^6) \tag{9.5-33}$$

Blasbalg [26] and Bussgang [27] have shown that the contribution of the fourth-order term is significant in computing the mean of Λ_i. They have shown that the mean of the third term (with the signal set to zero) must be included in the statistic. DiFranco and Rubin [24] have shown that

$$\overline{r_i^2} = 2(1 + \text{SNR}_i)$$
$$\overline{r_i^4} = 2(4 + 2\text{SNR}_i + 2\text{SNR}_i^2) \tag{9.5-34}$$

Therefore

$$E(\Lambda_i \mid \text{SNR}_i) = \text{SNR}_D \cdot \text{SNR}_i - \frac{\text{SNR}_D^2}{2} + O(\text{SNR}_D^2\,\text{SNR}_i) \tag{9.5-35}$$

Bussgang and Middleton [19] then recommend that the low SNR approximation for Λ_i be

$$\Lambda_i \cong -\text{SNR}_D\left(1 + \frac{\text{SNR}_D}{2}\right) + \frac{\text{SNR}_D r_i^2}{2} \tag{9.5-36}$$

so that

$$E(\Lambda_i \mid 0) = -\frac{\text{SNR}_D^2}{2}$$

$$E(\Lambda_i \mid \text{SNR}_i) = -\frac{\text{SNR}_D^2}{2} + \text{SNR}_D\,\text{SNR}_i \tag{9.5-37}$$

$$E(\Lambda_i \mid \text{SNR}_D) = \frac{\text{SNR}_D^2}{2}$$

All the above approximations only hold when $\text{SNR}_D \ll 1$ and $\text{SNR} \ll 1$. Notice that the test statistic is symmetrically biased about the signal-present and the signal-absent cases. Since we assumed noncoherent detection we have considered the envelope statistic.

PROBLEM 21

Determine the asymptotic form of the optimum detector assuming high, rather than low, SNR.

In order to obtain the ASN it is necessary to solve

$$L(\text{SNR}_i) = \frac{A^h - 1}{A^h - B^h} \tag{9.5-38}$$

where h is determined from

$$\int_{-\infty}^{\infty} \left[\frac{p(r_i \mid \text{SNR}_D)}{p(r_i \mid 0)} \right]^h p(r_i \mid \text{SNR}_i) \, dr_i = 1 \tag{9.5-39}$$

where we have set $m = 1$ in (9.5-18) since we are assuming independent samples. Evaluating (9.5-39), we desire the solution of h from

$$\int_0^{\infty} r_i I_0(\sqrt{2\text{SNR}_D}\, r_i)[I_0(2\text{SNR}_D r_i)]^h \, e^{-r_i^2} \, dr_i = \exp(\text{SNR}_i + h\,\text{SNR}_D) \tag{9.5-40}$$

Bussgang and Middleton [19] have shown that h is given approximately by

$$h \cong 1 - 2\frac{\text{SNR}_i}{\text{SNR}_D} \tag{9.5-41}$$

and

$$L(\text{SNR}_i) = \frac{\left(\dfrac{P_D}{P_{FA}}\right)^h - 1}{\left(\dfrac{P_D}{P_{FA}}\right)^h - \left(\dfrac{1-P_D}{1-P_{FA}}\right)^h} \tag{9.5-42}$$

for $P_{FA} \ll 1$ and $\text{SNR}_d \ll 1$. The average sample number is given by, assuming that $\text{SNR}_D \ll 1$,

$$\bar{n}_{\text{SNR}_i} \cong \frac{L(\text{SNR}_i)\log\left[\dfrac{1-P_D}{1-P_{FA}}\right] + [1 - L(\text{SNR}_i)]\log\left[\dfrac{P_D}{P_{FA}}\right]}{\text{SNR}_D(\text{SNR}_i - \text{SNR}_D/2)} \tag{9.5-43}$$

except when $\text{SNR}_i = \text{SNR}_D/2$, where L'Hospital's rule must be used ($h = 0$)!

Assuming that an appropriate bandwidth of the envelope process can be defined so that each sample is essentially independent, the ASN would produce the average sample time AST defined by

$$\bar{T}_{\text{SNR}_i} = \bar{n}_{\text{SNR}_i}\Delta T \tag{9.5-44}$$

where ΔT is the time between independent samples. The actual acquisition time would then be given by an appropriate linear combination of

$$\bar{n}\bigg|_{SNR_i=0} + \bar{n}\bigg|_{SNR_i=SNR_d}$$

Unfortunately calculations based on the above theory do not agree well with either simulation or actual hardware systems.

One of the problems in the present theory is that it does not account for a finite truncation time. All practical implementations, of course, utilize this feature. Bussgang and Middleton [19] have obtained tighter bounds than Wald [18] for the truncated case.

In light of the above problems we now consider one particular practical realization of a sequential detector in the next section.

9.5.5 A Sequential Detector Implementation

The purpose of this section is to describe a practical sequential detector system for a PSK modulated signal that utilizes the essence of Wald's sequential ratio detector. Consider Figure 9.26. The center channel (with the two summers) is the basic "direct" channel. Initially an arbitrary local reference (PN GEN) timing relationship is established. The signal, after being multiplied by this reference baseband PN signal is filtered to the data modulation bandwidth (B) and square law detected. The reference channel provides an estimate of $N_0B + P_{SP}$ (offset sufficiently in frequency from the data modulation,) where N_0B is the mean noise power out of the BPF and P_{SP} is the spread signal power. There is always some spread signal power except when the local PN code and the received PN code are exactly in phase.

Now the reference channel output, after heavy lowpass filtering and a scale change, is subtracted as a bias (b) from the direct channel output $x(t)$ to yield an integrator input of $x(t) - b$. The output of the integrator has another value, the threshold, added to it. This final signal, $y(t)$, is compared to zero up to a truncation time T_{TR}. If the signal $y(t)$ has not yet dropped to zero after T_{TR} sec, the signal is declared present. If, during this time, the output $y(t)$ drops to zero, the signal is declared not present.

Figure 9.27 illustrates three sample functions, one with no signal and two with different levels of signal power. All three sample functions start from the value "threshold" with "nominal" slopes $\pm P/2$, where P is the power level of that particular signal. This is based on a nominal optimum bias $= N_0B + P_0/2$, with P_0 being the threshold power.* The third sample function is shown with the assumption that the signal is at a higher power level. The noise crosses the zero value line at T_D, the dismissal time.

---

*Notice the essential optimum bias $= N_0B(1 + SNR_D/2)$ where SNR_D is the design (threshold) SNR out of the bandpass filter of bandwidth B.

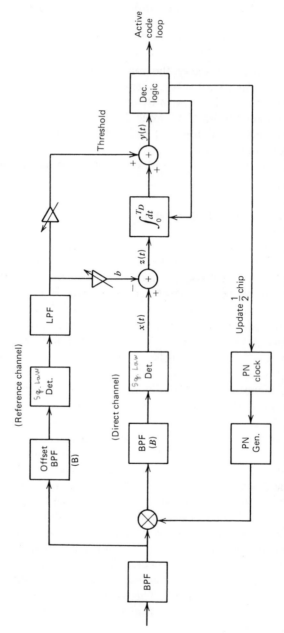

Figure 9.26 Sequential detection acquisition circuit. (Biphase signal).

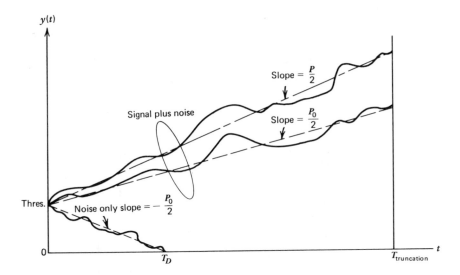

Figure 9.27 Integrator plus threshold output (sequential detector).

In Figure 9.28 the probability of false alarm and probability of detection are plotted versus truncation time for three different bias settings, based on a simulation due to Ricker *et al.* [28] of TRW. As can be seen, P_D saturates with truncation time but P_{FA} decreases as truncation time increases. Both false alarm (P_{FA}) probability and the probability of detection (P_D) are strongly dependent upon the bias setting, however. Ricker's program has verified that the optimum (minimum acquisition time) bias is about $N_0B + P_0/2$. Three biases were considered in the simulation, b_{opt}, the optimum value, and b_1, which was 94% of b_{opt}, and b_2, which was 105% of b_{opt}.

Acquisition time is not nearly as sensitive to threshold as it is to bias, as can be seen in Figure 9.29. The actual sensitivities depend upon the input signal-to-noise ratio of course.

9.5.6 Acquisition Time of a Sequential Detector

This section describes how the computer simulation results [27] of false alarm probability and probability of detection can be related to the acquisition time for a given probability of acquisition assuming a uniform *a priori* distribution.

Denote the time to acquire with probability P as T_P. Then we can write

$$T_P = T_{P\bar{S}} + T_{PS} \qquad (9.5\text{-}45)$$

where $T_{P\bar{S}}$ is the time required to search that part of the uncertainty region where the signal is not present and T_{PS} is the time required to search the cell where the signal is located. Let $\bar{T}_{DIS}$ denote the mean dismissal time

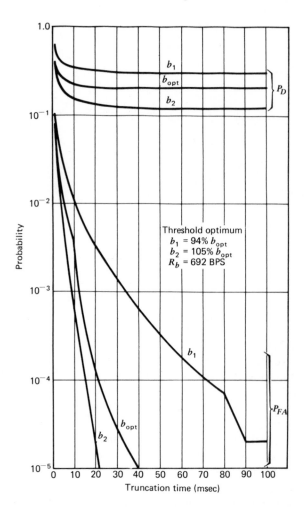

Figure 9.28 P_{FA} and P_D versus truncation time.

per cell required for the sequential detector to dismiss the hypothesis that the signal is not present. Further denote by T_{VR} the time required to verify that a false alarm occurred and T_{TR} the time it takes to reach truncation, and therefore declare that the signal is present (it could be a false alarm of course). We have

$$T_{P\bar{S}} \cong [\bar{T}_{DIS} + P_{FA}(T_{TR} + T_{VR})]N(q-1) \qquad (9.5\text{-}46)$$

where N is the number of passes through the uncertainty region and q is the total number of cells to be searched in the uncertainty region. Hence $q-1$ is the number of cells to be searched where the signal is not present.

Let the probability of one cell (or the total probability of the four

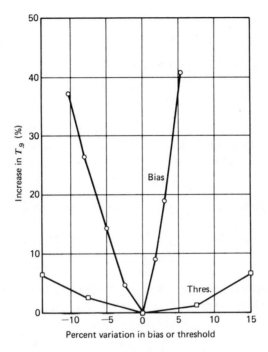

Figure 9.29 Acquisition time sensitivity.

correlation values) be denoted by P_D. Then it is shown in Problem 22 that

$$N = \left\{ (K-1) + \frac{P - \hat{P}(K-1)}{\hat{P}(K) - \hat{P}(K-1)} \right\}$$

$$K = \min[m \ni \hat{P}(m) > P, \qquad m = 1, 2, 3, \ldots]$$

(9.5-47)

and to a good approximation

$$N \cong \frac{\ln(1-P)}{\ln(1-P_D)} \qquad \text{when} \qquad N \geq 2$$

PROBLEM 22

Determine the number of sweeps through the uncertainty region by noting that for n sweeps through the complete uncertainty region the acquisition probability is (n an integer)

$$P_{\text{acq}}(n) = P_D + (1-P_D)P_D + (1-P_D)^2 P_D + \cdots + (1-P_D)^{n-1} P_D$$

Further note that when n is not an integer the probability of acquisition increases linearly (between integers) so that

$$N = \left\{ (K-1) + \frac{P - \hat{P}(K-1)}{\hat{P}(K) - \hat{P}(K-1)} \right\}$$

with

$$K = \min(k \ni \hat{P}(k) > P \qquad k = 1, 2, 3, 4, \dots)$$

Also show that for $n \geq 2$ we have approximately

$$N = \frac{\ln(1 - P)}{\ln(1 - P_D)}$$

When the signal is present it is not always detected in the first pass because $P_D < 1$ in general. Denote $\bar{T}_{\text{miss}}$ as the average time required to reject the desired signal. Then

$$T_{PS} \cong (N - 1)\bar{T}_{\text{miss}} + T_{TR} \qquad (9.5\text{-}48)$$

We have finally that T_p is given by, using (9.5-46), (9.5-47), and (9.5-48)

$$T_P \cong N\bar{T}_{\text{dis}}\left[1 + \frac{P_{FA}(T_{TR} + T_{VR})}{\bar{T}_{\text{dis}}}\right](q - 1) + (N - 1)\bar{T}_{\text{miss}} + T_{TR} \qquad (9.5\text{-}49)$$

Under the normal assumptions that $q \gg 1$, that $\bar{T}_{\text{dis}}$ is not more than a few orders of magnitude smaller than $\bar{T}_{\text{miss}}$ or T_{TR}, and that $N \geq 2$, we have

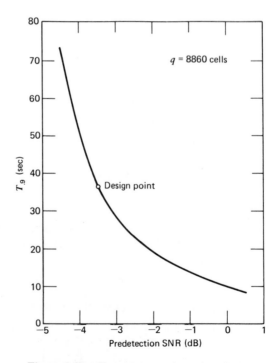

Figure 9.30 $T_{.9}$ versus predetection SNR.

approximately

$$T_P \cong \bar{T}_{\text{dis}}\left[1 + \frac{P_{FA}(T_{TR} + T_{VR})}{\bar{T}_{\text{dis}}}\right] \cdot q \cdot \frac{\ln(1-P)}{\ln(1-P_D)} \qquad (9.5\text{-}50)$$

For this approximation then the design parameters P (probability of acquisition desired), T_{TR}, T_{VR}, and the simulation values T_{dis}, P_{FA}, and P_D are needed to estimate the acquisition time, with probability P. For the more exact expressions $\bar{T}_{\text{miss}}$ and T_{TR} are also needed explicitly. Note also that if W chips are to be searched with the despreader updating one-half chip at a time, $2W$ cells must be searched.

In Figure 9.30 the .9 probability of acquisition is displayed with $q = 8860$ cells, $T_{TR} = 50$ msec, $T_{VR} = 50$ msec, a BPF bandwidth of 8 kHz, and a data rate of 1400 NRZ symbols per second. The design point SNR in 8 kHz, where bias and threshold are optimized, is -3.5 dB. This supports a 10^{-5} bit error rate using a rate $\frac{1}{3}$, constraint length 7, Viterbi decoder at an $E_B/N_0 = 4.6$ dB. Acquisition times for different cell uncertainties can be scaled accordingly.

Generally speaking, a properly designed sequential detector should be anywhere from two to ten times faster than a properly designed single dwell despreader.

9.6 PASSIVE MATCHED FILTERS

Considerable attention has been paid to matched filter development [29] in the past 20 years. However, until recently, many of the proposed analog techniques were impractical because of the size and weight of transversal filters (based on tapped coaxial or stripline transmission lines). Digital implementation of matched filters has been attempted but some signal degradation occurs when the received rf or IF signal is demodulated. In addition, as bit rates and clock speeds are increased, conventional digital logic becomes costly and requires a great deal of power. The main advantage of matched filtering is that it has the potential of very fast acquisition times.

With the advent of charge coupled devices and surface acoustic wave devices, certain classes of matched filters become much more flexible [30].

All the methods of acquisition up until this section are referred to as active correlation techniques. In this section we briefly outline the mathematical basis for a matched filter matched to a PN sequence, which is a passive correlation technique. A matched filter is one which maximizes the filter output SNR when the signal is immersed in WGN. The impulse response of a matched filter is well-known and is of the form

$$h(t) = s(T - t) \qquad 0 \le t \le T \qquad (9.6\text{-}1)$$

where $s(t)$ is the signal to be detected.

Hence, we see that the matched filter impulse response is a time reversed form of the signal. The Fourier transform of the matched filter is given by

$$H(\omega) = S^*(\omega)\, e^{-j\omega T} \qquad (9.6\text{-}2)$$

where $S(\omega)$ is the Fourier transform of $s(t)$.

Now we consider a matched filter implementation for a PN type encoded PSK signal. We model the N chip PSK waveform by

$$s(t) = \sum_{n=1}^{N} a_n p[t-(n-1)T] \qquad (9.6\text{-}3)$$

where a_n is ± 1 according to the particular sequence involved and $p(t)$ denotes a unit amplitude NRZ pulse of duration T sec. The Fourier transform of (9.6-3) produces

$$S(\omega) = P(\omega) \sum_{n=1}^{N} a_n\, e^{-i\omega(n-1)T} \qquad (9.6\text{-}4)$$

where $P(\omega)$ is the Fourier transform of the the basic pulse $p(t)$. The signal $s(t)$ could be generated by a tapped delay line generator [30] as shown in Figure 9.31.

Using (9.6-2) and (9.6-4), we conclude that the matched filter transfer function is given by

$$H(\omega) = P^*(\omega) \sum_{n=1}^{N} a_n\, e^{-i\omega(N-n+1)} \qquad (9.6\text{-}5)$$

Now $P(\omega)$ can be expressed as

$$P(\omega) = \int_{-T/2}^{T/2} \sqrt{2}\, A \cos \omega_0 t\, e^{-j\omega t}\, dt \qquad (9.6\text{-}6)$$

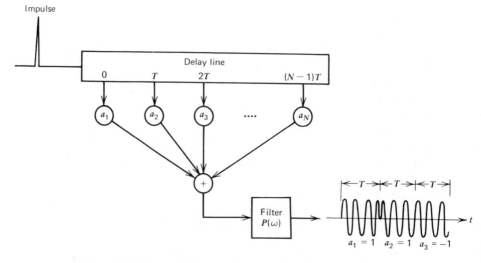

Figure 9.31   Tapped delay implementation of a signal generator.

which, when evaluated, becomes

$$P(\omega) = \frac{AT}{2}\left\{\frac{\sin[(\omega - \omega_0)T/2]}{[(\omega - \omega_0)T/2]} + \frac{\sin[(\omega + \omega_0)T/2]}{(\omega + \omega_0)T/2}\right\} \qquad (9.6\text{-}7)$$

and clearly is real, so

$$P^*(\omega) = P(\omega) \qquad (9.6\text{-}8)$$

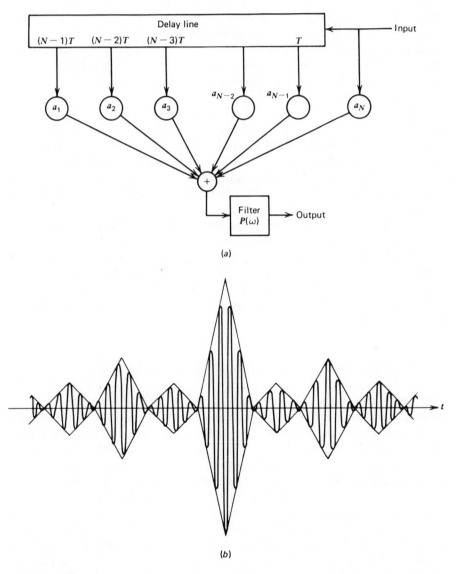

(a)

(b)

Figure 9.32 (a) Matched filter for coded PSK sequence. (b) Matched filter output of a PN type code.

We therefore conclude that the matched filter transfer function $H(\omega)$ can be written as

$$H(\omega) = P(\omega) \sum_{n=1}^{N} a_n e^{-i\omega(N-n+1)} \qquad (9.6\text{-}9)$$

Figure 9.32 illustrates the matched filter for our coded PSK signal, with $P(\omega)$ defined in (9.6-7). Notice from Figure 9.31 the signal runs "backwards" through the delay line. In Figure 9.32, the output of a matched filter under consideration is sketched showing the peak along with residual responses. Realizable filters are approximations to (9.6-7). An envelope detector can be used to detect the maximum output of the matched filter. This matched filter "timing pulse" can be used to start a PN generator running at the PN code rate, which can be used to drive a PN code loop and thereby continue the tracking process.

The basic approach indicated above can be used when the PN code length is not large (typically 512 chips) and the SNRs are large. Otherwise, noncoherent combining can be used to enhance the output SNR. Baseband schemes for matched filters can also be used that require supplying an estimate of the carrier frequency in I-Q form, that is, as $\cos \omega_0 t$, and $\sin \omega_0 t$ then baseband-match-filter each demodulated signal, and then noncoherently combine them [31].

Another method to process large time bandwidth products is to utilize serial-parallel signal processing techniques, segmenting a continuous long PN-PSK code into short blocks for recognition in a SAW programmable matched filter that is reprogrammed as each individual code segment enters it. The resulting series of correlation peaks can be coherently summed in a recirculating delay line integrator [32], [33]. More details may be found in references 31–33.

9.7 AN OPTIMUM ACQUISITION SEARCH TECHNIQUE

The purpose of this section is to provide a theory of optimization for taking into account the *a priori* signal location density function for a typical fixed dwell time or a sequential-type search to synchronize to a PN code (active correlation) [34, 35]. In other words, if the actual signal location is more likely to occur in a given set of timing locations, it seems "reasonable" that those timing locations should be searched first and/or more often.

Consider the *a priori* probability density mode that has superimposed upon it four acquisition sweeps, shown in Figure 9.33. Each sweep has a total length of L_i chips and is symmetric about the most probable timing error location. We assume that the sweep is, in fact, a discrete search in the sense that the acquisition system correlates for T_D sec, then advances, say, one-half a chip (if there is no hit) and again correlates for T_D sec. The discrete search acquisition technique is indicated in Figure 9.33 by dots.

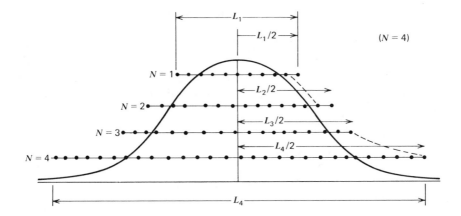

Figure 9.33 *A priori* density of true timing error with four sweeps shown.

The optimization problem is as follows: Given N sweeps, determine the optimum lengths, L_i (number of chips), to achieve a cumulative probability Q_N at the end of the Nth sweep so that the total acquisition time $L_T = \Sigma_{i=1}^{N} L_i$ is minimized.

First we formulate the cumulative probability of acquisition, assuming that the probability of signal detection is P_D, given the signal is present. At the end of the first sweep the probability of acquisition is

$$P_1 = P_D \cdot P(L_1) \tag{9.7-1}$$

where $P(L_1)$ is the probability that the signal is located in the region L_1. The probability of acquiring the signal on the second sweep is given by

$$P_2 = P(L_2 - L_1) P_D + P(L_1)(1 - P_D) P_D \tag{9.7-2}$$

where $P(L_2 - L_1)$ is the probability of being in region L_2 but not in L_1. In the same manner, denoting P_i as the probability of acquiring on the ith sweep, we obtain

$$P_3 = P(L_3 - L_2)P_D + P(L_2 - L_1)(1 - P_D)P_D + P(L_1)(1 - P_D)^2 P_D \tag{9.7-3}$$

and

$$P_i = P_D[P(L_i - L_{i-1}) + P(L_{i-1} - L_{i-2})(1 - P_D) + P(L_{i-2} - L_{i-3})(1 - P_D)^2 \\ + \cdots + P(L_2 - L_1)(1 - P_D)^{i-2} + P(L_1)(1 - P_D)^{i-1}] \tag{9.7-4}$$

where $P(L_i - L_{i-1})$ is the probability of being in region L_i but not in L_{i-1}.

Now denote Q_N as the probability of acquiring in N sweeps. Clearly

$$Q_N = \sum_{i=1}^{N} P_i \quad \text{in time} \quad T_N = \sum_{i=1}^{N} T_i \tag{9.7-5}$$

In Problem 23, it is shown that Q_N can be written as

$$Q_N = P_D \sum_{k=1}^{N} P(L_k)(1 - P_D)^{N-k} \tag{9.7-6}$$

This is obvious if we view the Nth sweep as the first one since the probability of acquisition clearly does not depend upon the order of sweeping.

PROBLEM 23

Show that Q_N can be written as

$$Q_N = \sum_{i=1}^{N} P_i = P_D \sum_{k=1}^{N} P(L_k)(1 - P_D)^{N-k}$$

With a convenient form of Q_N, it is now a simple matter to form the Lagrangian function using the Lagrange multiplier technique to find the L_i; That is, consider

$$f(\lambda) = \sum_{i=1}^{N} L_i + \lambda \, P_D(1 - P_D)^N \sum_{k=1}^{N} P(L_k)\left(\frac{1}{1 - P_D}\right)^k \tag{9.7-7}$$

where λ is the Lagrange multiplier. To proceed further, we must assume a particular distribution function of the signal location. Now it is convenient to obtain an explicit expression for the $P(L_i)$ by assuming a Gaussian *a priori* location density. By definition

$$P(L_i) = \frac{2}{\sqrt{2\pi}\,\sigma} \int_0^{L_i/2} e^{-t^2/2\sigma^2} \, dt \tag{9.7-8}$$

Denote λ' and q by

$$\lambda' = -\lambda P_D(1 - P_D)^N$$

$$q = \frac{1}{1 - P_D} \tag{9.7-9}$$

so that

$$\frac{\partial f(\lambda)}{\partial L_i} = 1 - \lambda q^i \, \frac{2}{\sqrt{2\pi}\,\sigma} \, e^{-L_i^2/8\sigma^2} \tag{9.7-10}$$

Denoting L_i/σ by l_i, we find that

$$l_i = 2\sqrt{2}\,\sqrt{C + i \ln q} \tag{9.7-11}$$

where C is a constant (absorbing λ') that is yet to be determined. Hence, the optimum value of l_i (for a given N) satisfy the above equation (for l_i), and C is obtained from the condition

$$Q_N = P_D \sum_{k=1}^{N} P(L_k)(1 - P_D)^{N-k} \tag{9.7-12}$$

by trying various values of C until the equation for Q_N is satisfied. The procedure converges quickly. The actual minimum over N must be determined by obtaining a solution for each N and picking the N that minimizes the acquisition time.

PROBLEM 24

A more conveniently implementable search procedure is asymmetric, as shown below.

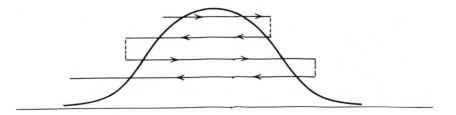

(a) Determine the Lagrange function for this problem.
(b) Determine the optimum values of l_i for a given λ.
(c) Obtain the equation that provides the determination of the actual l_i. (*Note*: The definition of the lengths is crucial in obtaining a relatively simple solution for this problem.)

When the search "range" is $\pm 3\sigma$, based on a uniform search, it has been found that a reduction to about 60% of the original time—when $P_D = 0.25$ for a Gaussian density *a priori* signal location—was obtained using the optimized sweep. And therefore optimizing can be advantageous. The actual improvement depends upon the value of P_D and the peakedness of the *a priori* signal location probability density function. It was found with the Gaussian *a priori* assumption that the improvement decreases as P_D increases. Also, as the *a priori* density function becomes less "peaked" the optimized search strategy has less benefit.

The theory developed here is quite general, the only requirement being that the *a priori* density function be unimodal and symmetric and that the $P(L_k)$ be differentiable. More details are contained in references 34 and 35.

9.8 FALSE LOCK DUE TO PARTIAL CORRELATION

In this section we briefly discuss how the predetection bandpass filter influences the correlation process during code acquisition. In Figure 9.34 the basic correlator is illustrated. We model the received rf modulated PN type code by

$$y(t) = \sqrt{2P}\, d(t)PN(t - \tau_0)\cos(\omega_0 t + \theta_0) + n(t) \qquad (9.8\text{-}1)$$

where P is the received carrier power, $d(t)$ is the data stream (± 1),

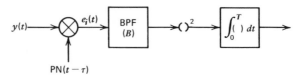

Figure 9.34   Basic PN code correlator used for acquisition.

$PN(t - \tau_0)$ is the PN type code having delay τ_0 sec, $n(t)$ is WGN, and finally ω_0 and θ_0 are the respective angular frequency and phase of the carrier. After despreading, and assuming that the code chip transition times are aligned, we obtain

$$e_i(t) = \sqrt{2P} \, d(t) \, PN(t - \tau_1) \cos(\omega_0 t + \theta_0) + n'(t) \qquad (9.8\text{-}2)$$

where $n'(t)$ is the product of $PN(t - \tau)$ and $n(t)$ and $PN(t - \tau_1)$ is some integer shift of the received PN code.

Now in order for the correlator to truly correlate the received signal with the local code, only the dc component of the signal

$$PN(t - \tau_1) = PN(t - \tau_0) \, PN(t - \tau) \qquad (9.8\text{-}3)$$

should be passed through the bandpass filter to the squarer of Figure 9.34. However, due to doppler and data considerations, it is usually necessary to "open up" the bandpass filter to accommodate the doppler shifted data spectrum. The two requirements on bandwidth are often incompatible. If the bandpass filter bandwidth is wider than twice the spectral spacing between the line components of the PN code, then without data, at least three spectral code lines will be input to the squarer. Consequently, if the postintegration time T is large, the integrator output will be proportional to the power in the sum of the three line components. Since the only desired component is the one at dc (or when multiplied by the carrier the one at ω_0), partial correlation will be present. Partial correlation and the presence of additional spectral lines are alternate ways of viewing the same problem. Hence, the wider the bandwidth of the bandpass filter, the greater the "residual" correlation due to the sum power of all the code lines within the bandpass filter bandwidth. When modulation is present each code "line" has the data modulation placed on it and must be accounted for. Also, noninteger related alignment affects the results somewhat. This residual correlation or partial correlation problem should always be considered in system design.

In effect the range of detected voltage in true lock to the detected voltage when out of code lock is diminished when the bandpass filter passes more than the spectral line at the carrier frequency. In conclusion, greater data rates or larger doppler values increase false lock susceptability for a given PN code. In order to avoid variations in the output correlation the post detection filter should be set equal to KT where K is an integer and T is the PN code length in seconds.

REFERENCES

1   Ferris, C. D., *Linear Network Theory*, Charles E. Merrill, New York, 1959.

2   Mason, Samuel J., "Feedback Theory—Some Properties of Signal Flow Graphs," *Proc. IEEE*, Vol. 41, No. 9, September 1953.

3   Mason, Samuel J., "Feedback Theory-Further Properties of Signal Flow Graphs," *Proc. IEEE*, Vol. 44, No. 7, July 1956.

4   Tustin, A., *Direct Current Machines for Control Systems*, Macmillan, New York, 1955.

5   Di Stefano, A., Stubberud, A., and Williams, S., *Feedback and Control Systems*, Schaums Outline Series, McGraw-Hill, New York, 1967.

6   Sittler, R. W., "Systems Analysis of Discrete Markov Processes," *IRE Trans. Circuit Theory*, Vol. CT-3, December 1956.

7   Huggins, W. H., "Signal Flow Graphs and Random Signals," *Proc. IRE*, January 1957.

8   Howard, R., *Dynamic Programming and Markov Processes*, Technology Press and Wiley, New York, 1960.

9   Howard, R., Class Notes for Industrial Engineering, MIT, 1963.

10 Feller, W., *An Introduction to Probability Theory and Its Applications*, 2nd ed., Vol. I, Wiley, New York, 1950.

11 Holmes, J. K., and Chen, C. C., "Acquisition Time Performance of PN Spread Spectrum Systems," *IEEE Trans. Communications*, Special Issue on Spread Spectrum Communications, Vol. COM-25, August, 1977.

12 DiCarlo, D. M., "Mean Acquisition Time and Variance for the Two-Step Acquisition System," TRW IOC 75-7131. 67-04, May 29, 1975.

13 Holmes, J. K., "Predicted Mean SSP Code Acquisition Times with the New Algorithm," TRW IOC SCTE-50-76-251/JKH, July 21, 1976.

14 Davenport, W. B., Jr., and Root, W. L., *Introduction to Random Signals and Noise*, McGraw-Hill, New York, 1958.

15 Kemeny, J., and Snell, J., *Finite Markov Chains*, Van Nostrand, New York, 1960.

16 Hopkins, P., "A Spread Spectrum Synchronization Scheme for the Space Shuttle Orbitor," Johnson Space Center, Tracking and Communications Development Division Internal Note, LEC-5982 (Rev. A), September 1975.

17 Holmes, J. K., "Code Tracking Lock Detector Mean Time to Declare Out-of-Lock," TRW IOC IUS-110, May 1979.

18 Wald, A., *Sequential Analysis*, Wiley, New York, 1947.

19 Bussgang, J. J., and Middleton, D., "Optimum Sequential Detection of Signals in Noise," *IRE Trans.*, Vol. IT-1, pp. 5–18, December 1955.

20 Wald, A., and Wolfowitz, J., "Optimum Character of the Sequential Probability Ratio Test," *Ann. Math. Statist.*, Vol. 19, p. 326, 1948.

21 Blasbalg, H., "Experimental Results in Sequential Detection," *IRE Trans.*, Vol. IT-5, No. 2, p. 41, June 1959.

22 Kendall, W. B., "Performance of the Biased Square-Law Sequential Detector in the Absence of Signal," *IEEE Trans. Information Theory*, January 1964.

23 Albert, G. E., *Ann. Math. Statist.*, Vol. 25, September 1954.

24 DiFranco, J. V., and Rubin, W. L., *Radar Detection*, Prentice-Hall, Englewood Cliffs, N.J. 1968, Chap. 16.

25 Hancock, J. C., and Wintz, P. A., *Signal Detection Theory*, McGraw-Hill, New York, 1966, Chap. 4.

26 Blasbalg, H., "Theory of Sequential Filtering and Its Application to Signal Detection and

Classification," Johns Hopkins University, Radiation Lab., Tech. Report No. AF-8, October 18, 1954.

27 Bussgang, J. J., "Sequential Detection of Signals in Noise," Ph.D. thesis, Harvard University, 1955.

28 Ricker, F., McAdam, P., and Trumpis, B., "Sequential Acquisition of PN Codes for TDRSS," paper submitted for publication.

29 Bernfeld, M., et al., "Matched Filtering, Pulse Compression and Waveform Design," Microwave J. October 1964.

30 O'Clock, G. D., Jr., "Matched Filters Boost Receiver Gain," Electronic Design, Vol. 11, May 24, 1977.

31 Cahn, C. R., "Spread Spectrum Applications and State-of-the-Art Equipments," Paper No. 5, AGARO-NATO Lecture Series No. 58 on "Spread Spectrum Communications," May 28-June 6, 1973.

32 Grant, P. M., Brown, J., and Collins, J. H., "Large Time Bandwidth Product Programmarble Saw PSK Matched Filter Module," Proceedings of the 1974 Ultrasonics Symposium.

33 Morgan, D. P., and Hannah, J. M., "Correlation of Long Spread-Spectrum Waveforms Using a Saw Convolver/Recirculation Loop Subsystem," Proceedings of the 1974 Ultrasonics Symposium.

34 Holmes, J. K., and Woo, K. T., "An Optimum PN Code Search Technique for a Given a priori Signal Location Density," NTC 78 Conference Record, Birmingham, Alabama, December 3 to 6, 1978; Section 18.6.

35 Holmes, J. K., and Woo, K. T., "An Optimum Asymmetric PN Code Search Strategy," 1979 International Telemetering Conference, pp. 319–330, San Diego, Cal. November 19 to 21, 1979.

10

PN CODE TRACKING

The purpose of this chapter is to introduce the reader to the various PN code tracking loops and to describe both their relative merits and their relative performances. There are three basic types of performance that are of concern in this chapter: (1) steady state tracking performance, (2) transient acquisition performance, and (3) loss-of-lock performance. The first performance parameter, the steady state timing error variance, only exists for the linearized model just as in the case of the ordinary phase locked loop. However, the linearized tracking error variance is an often used parameter that allows one to compare the accuracies of various loops. In the case of transient acquisition, both phase plane and transient pull-in response are of concern. Loss of lock is concerned with the time a loop can remain in lock given that initially there is no timing error. Obviously, if the time it will remain in lock is small compared to the time required to track (view time for an orbit for example), then the system will have dropouts in data and be forced to reacquire.

We will classify the code tracking loops in this chapter as delay lock loops (full-time code loops), even though they can be time shared to remove bias voltages; time-shared code tracking loops (the early and late gates are on alternately); and finally, the τ-dither* code tracking loop, which is quite similar to the time-shared loop but can be implemented with one correlator instead of two.

A typical application of a PN code tracking loop for ranging purposes is shown in Figure 10.1.

The clock provides the chip timing for the PN waveform, which is modulated onto a carrier and is transmitted via the antenna. After the PN code modulated waveform is transponded through the satellite, it is tracked in the receiver delay lock loop and compared to the transmitted code. The comparison is made with a convenient code state (epoch) such as the all-"one" state. A counter is used to count the time difference between the transmitted epoch and the received epoch, thereby providing an estimate of delay.

---

*Many times in the literature, τ-dither and time-shared code tracking loops are used interchangeably.

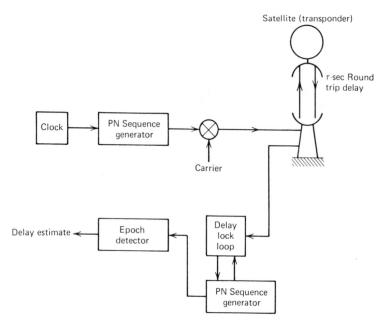

Figure 10.1 Ranging with a PN code.

10.1 THEORETICAL BASIS FOR THE EARLY-LATE CODE TRACKING LOOP

In this section we consider the motivation for the structure of a PN code tracking loop. Consider a baseband system composed of a PN waveform corrupted by white Gaussian noise (WGN). Denote this sum by $y(t)$ so that

$$y(t) = PN(t - \tau) + n(t) \tag{10.1-1}$$

where τ is the unknown delay. The maximum likelihood estimate of τ is given by the solution to the likelihood equation (see for example [4] or [5]).

$$\frac{\partial}{\partial \hat{\tau}} \{\ln \Lambda[y(t), \hat{\tau}]\} = \frac{\partial}{\partial \hat{\tau}} \left\{ \frac{1}{N_0} \int_{-T/2}^{T/2} [2y(t)PN(t - \hat{\tau}) - PN^2(t - \hat{\tau})] \right\} dt = 0 \tag{10.1-2}$$

where $\Lambda[y(t), \hat{\tau}]$ is the likelihood function and $\hat{\tau}$ is the estimate of τ. Since

$$PN^2(t - \hat{\tau}) = 1 \tag{10.1-3}$$

we have

$$\frac{2}{N_0} \int_{-T/2}^{T/2} y(t) \frac{\partial PN(t - \hat{\tau})}{\partial \hat{\tau}} dt = 0 \tag{10.1-4}$$

Equivalently, the maximum likelihood equation is the value of $\hat{\tau}$ that satisfies

$$-\frac{2}{N_0}\int_{-T/2}^{T/2} y(t)\frac{\partial PN(t-\hat{\tau})}{\partial t}\,dt = 0 \qquad (10.1\text{-}5)$$

Hence, we conclude that the optimum estimate of $\hat{\tau}$ is obtained by correlating the received waveform against the time derivative of the reference PN waveform. This leads to the closed-loop interpretation of correlating the received waveform against the derivative of the PN waveform is such a way as to drive the correlation to zero. The baseband code loop, to be discussed shortly, can be viewed as a closed-loop form of a discrete derivative approximation to the maximum likelihood estimate for the delay τ.

10.2 FULL-TIME CODE LOOP TRACKING PERFORMANCE

Our goal in this section is to obtain the timing error performance for a coherent code loop and two types of noncoherent code loops. By a *coherent code loop* we mean one in which the carrier frequency and phase are known exactly, so that the code loop itself can operate on a baseband signal. On the other hand, a *noncoherent code loop* is a code loop configuration based on the hypothesis that the carrier frequency is not known exactly (due to doppler effects, for example). This type of loop must be able to tolerate and function under specified values of carrier shift from nominal. We discuss a coherent loop and two types of noncoherent code loops. The first noncoherent loop is the standard loop, and the second one, a postdetection combining loop that can provide enhanced performance when a quadriphase signal is being tracked.

10.2.1 Baseband Code Tracking Loop

When it is possible to obtain carrier lock without the requirement for despreading the PN modulated waveform, then this carrier reference can, in turn, be used to demodulate the PN signal, which allows the use of a baseband delay lock loop. Since the baseband code tracking loop is a phase coherent device, it can also be used as a standard in which we can compare other noncoherent code loops.

Consider the waveform $y(t)$ which is a noise corrupted version of a PN waveform:

$$y(t) = \sqrt{P}\,PN(t-T) + n(t) \qquad (10.2\text{-}1)$$

with $n(t)$ a sample function of WGN, T the time delay from the transmitter, and P the signal power. A baseband, 1-Δ,* loop is shown in Figure

---

*This is the common designation in the literature. Strictly speaking, we should call this loop a one-chip or a $1T_c$ loop!

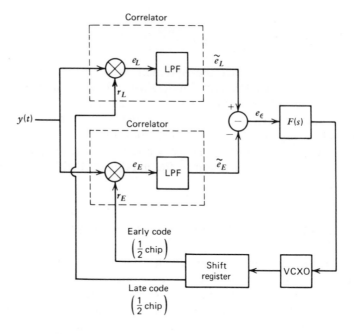

Figure 10.2 Baseband delay lock loop model.

10.2. The notation 1-Δ denotes the fact that the early and late correlators are shifted apart one chip time (T_c) are are each $T_c/2$ sec from the correct timing.

The early and late references are assumed to be one-half chip early and one-half chip late, so that

$$r_L = \text{PN}\left(t - \hat{T} - \frac{T_c}{2}\right) \tag{10.2-2}$$

$$r_E = \text{PN}\left(t - \hat{T} + \frac{T_c}{2}\right) \tag{10.2-3}$$

Where $\hat{T}$ is the code loop estimate of the delay T and T_c is the chip time.

Out of the late correlator we have, neglecting the self-noise,*

$$\tilde{e}_L(t) = \sqrt{P}\,R_{\text{PN}}\left(T - \hat{T} - \frac{T_c}{2}\right) + L(s)\left[n(t)\text{PN}\left(t - \hat{T} - \frac{T_c}{2}\right)\right] \tag{10.2-4}$$

where $L(s)[g(t)]$ denotes the Heaviside operator $L(s)$ operating on the time function $g(t)$. The notation means

$$L(s)[g(t)] = \int_{-\infty}^{\infty} l(\tau)g(t - \tau)\,d\tau = f(t) * g(t) \tag{10.2-5}$$

*See Chapter 8 for a detailed discussion of self-noise, also see Section 10.6.

with

$$L(s) = \mathcal{L}[l(t)] \tag{10.2-6}$$

and where $\mathcal{L}(l)$ denotes the Laplace transform of $l(t)$, and the $*$ denotes a convolution. Out of the early correlator we obtain (again neglecting the self-noise)

$$\tilde{e}_E = \sqrt{P}\, R_{PN}\left(T - \hat{T} + \frac{T_c}{2}\right) + L(s)\left[n(t)PN\left(t - \hat{T} + \frac{T_c}{2}\right)\right] \tag{10.2-7}$$

We assume that $L(0) = 1$, that is, the dc response of the (identical) LPFs are unity. For convenience denote

$$n_E(t) = L(s)\left[n(t)PN\left(t - \hat{T} + \frac{T_c}{2}\right)\right] \tag{10.2-8}$$

$$n_L(t) = L(s)\left[n(t)PN\left(t - \hat{T} - \frac{T_c}{2}\right)\right] \tag{10.2-9}$$

The error signal into the loop filter is given by*

$$e_\epsilon = \sqrt{P}\left[R_{PN}\left(T - \hat{T} - \frac{T_c}{2}\right) - R_{PN}\left(T - \hat{T} + \frac{T_c}{2}\right)\right] + n_E(t) - n_L(t) \tag{10.2-10}$$

The control signal or S-curve is shown in Figure 10.3. Notice that between $-T_c/2$ and $T_c/2$ the S-curve has a constant slope equal to $2/T_c$. Therefore the *linearized* loop equation is given by

$$e_\epsilon = \frac{2\sqrt{P}}{T_c}\,\epsilon + n_E - n_L \tag{10.2-11}$$

where $\epsilon = T - \hat{T}$ is the code timing error.

The code loop estimate $\hat{T}$ is given by

$$\frac{\hat{T}}{T_c} = \frac{2K\sqrt{P}\,F(s)}{s}\left[\frac{\epsilon}{T_c} + \frac{n_E - n_L}{2\sqrt{P}}\right] \tag{10.2-12}$$

where

$$K = K_M K_{VCO} \tag{10.2-13}$$

and K_M is the multiplier gain in V/chip and K_{VCO} is the VCO gain in chips/sec/V. Note that the error ϵ/T_c is in chips. Using $\epsilon = T - \hat{T}$, we obtain

$$\frac{\epsilon}{T_c} = \frac{-2\sqrt{P}\,KF(s)/s}{1 + [2\sqrt{P}\,KF(s)/s]}\left(\frac{n_E - n_L}{2\sqrt{P}}\right) + \frac{1}{1 + \{[2\sqrt{P}\,KF(s)]/s\}}\left(\frac{T}{T_c}\right) \tag{10.2-14}$$

with T the input time delay process.

*We have neglected the non dc component of correlated signal in this expression.

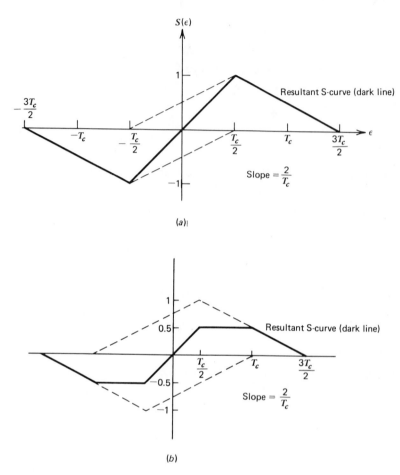

Figure 10.3 (*a*) S-Curve for baseband code loop with one-chip early-late separation. (*b*) S-Curve for baseband code loop with one-half chip separation.

Defining the closed-loop transfer function by

$$H(s) = \frac{2\sqrt{P} KF(s)}{s + 2\sqrt{P} KF(s)} \qquad (10.2\text{-}15)$$

we have

$$\frac{\epsilon}{T_c} = -H(s)\left(\frac{n_E - n_L}{2\sqrt{P}}\right) + [1 - H(s)]\frac{T}{T_c} \qquad (10.2\text{-}16)$$

Initially, we will be concerned with the noise performance only so that we will assume $T = \text{const.}$ Clearly, since $1 - H(s)$ is a highpass function, the error due to a constant input will converge to zero. Therefore, we set $T = 0$ for convenience. The linearized timing error variance from (10.2-16) is given by

$$\overline{\left(\frac{\epsilon}{T_c}\right)^2} = \frac{\sigma_\epsilon^2}{T_c^2} = \int_{-\infty}^{\infty} |H(f)|^2 \frac{\mathscr{S}_N(f)}{4P} \, df \tag{10.2-17}$$

where $\mathscr{S}_N(f)$ is the power spectral density of $N(t) = n_E(t) - n_L(t)$.

Assuming that the loop is narrow compared to the spectrum of $N(t)$, we obtain from (10.2-17)

$$\left(\frac{\sigma_\epsilon}{T_c}\right)^2 = \frac{B_L}{2P} \mathscr{S}_N(0) \tag{10.2-18}$$

Now we compute the spectral density, at $f = 0$, of $n_E(t) - n_L(t)$;

$$N(t) = n_E(t) - n_L(t) = \int_{-\infty}^{\infty} l(t-u)n(u)\text{PN}\left(u - \hat{T} + \frac{T_c}{2}\right) du$$

$$- \int_{-\infty}^{\infty} l(t-v)n(v)\text{PN}\left(v - \hat{T} - \frac{T_c}{2}\right) dv$$

$$\tag{10.2-19}$$

so that

$$R_N(\tau) = E \int\!\!\int_{-\infty}^{\infty} l(t-u)l(t+\tau-u')\text{PN}\left(u - \hat{T} + \frac{T_c}{2}\right)$$

$$\times \text{PN}\left(u' - \hat{T} + \frac{T_c}{2}\right) n(u)n(u') \, du \, du'$$

$$+ E \int\!\!\int_{-\infty}^{\infty} l(t-v)l(t+\tau-v')n(v)n(v')$$

$$\times \text{PN}\left(v - \hat{T} - \frac{T_c}{2}\right)\text{PN}\left(v' - \hat{T} - \frac{T_c}{2}\right) dv \, dv'$$

$$- E \int\!\!\int_{-\infty}^{\infty} l(t-u)l(t+\tau-v)n(u)n(v)$$

$$\times \text{PN}\left(u - \hat{T} + \frac{T_c}{2}\right)\text{PN}\left(v - \hat{T} - \frac{T_c}{2}\right) du \, dv$$

$$- E \int\!\!\int_{-\infty}^{\infty} l(t-u)l(t+\tau-v)n(u)n(v)$$

$$\times \text{PN}\left(u - \hat{T} - \frac{T_c}{2}\right)\text{PN}\left(v - \hat{T} + \frac{T_c}{2}\right) du \, dv \tag{10.2-20}$$

Assuming that the PN sequences are long so that they can be modeled as two independent random NRZ waveforms, with a transition probability of

.5, we have

$$R_N(\tau) = \frac{N_0}{2} \int_{-\infty}^{\infty} l(t-u)l(t+\tau-u)\, du + \frac{N_0}{2} \int_{-\infty}^{\infty} l(t-u)l(t+\tau-u)\, du$$

$$(10.2\text{-}21)$$

Hence

$$\mathscr{S}_N(0) = N_0 \int_{-\infty}^{\infty}\int_{-\infty}^{\infty} l(t-u)l(t+\tau-u)\, du\, d\tau \qquad (10.2\text{-}22)$$

or

$$\mathscr{S}_N(0) = N_0 \int\int_{-\infty}^{\infty} |L(\omega)|^2 e^{i\omega\tau}\frac{d\omega\, d\tau}{2\pi} \qquad (10.2\text{-}23)$$

or finally, after integrating on τ, and using

$$\frac{1}{2\pi}\int_{-\infty}^{\infty} e^{i\omega\tau}\, d\tau = \delta(\omega) \qquad (10.2\text{-}24)$$

we arrive at

$$\mathscr{S}_N(0) = |L(0)|^2 N_0 \qquad (10.2\text{-}25)$$

However, since we assumed that $L(0) = 1$, from (10.2-18) the relative chip error variance is given by

$$\left(\frac{\sigma_\epsilon}{T_c}\right)^2 = \frac{N_0 B_L}{2P} \qquad \text{chips}^2$$

or

$$\sigma_\epsilon = T_c \sqrt{\frac{N_0 B_L}{2P}} \qquad \text{sec} \qquad (10.2\text{-}26)$$

It is to be cautioned that this is the linearized standard deviation of the timing error and as such is useful for laboratory comparisons of loop performance. However, the actual system is not linear and hence the loop can slip. Therefore, the actual variance is unbounded. The actual slip phenomena will be covered in Section 10.11.

The above result for timing jitter has been obtained by previous authors: Gill [1], Spilker [2], Spilker and Magill [3], and others.

The one effect that was not considered in this analysis is the distortion of the correlation curve when PN code filtering is present. As can be seen in Figure 8.8, there is a fixed delay for $BT = 2$ of about 0.37 chips, which produces a timing error bias that must be subtracted out in the final estimate of range.* Another effect that has been neglected is the modification of the slope of the linear portion of the S-curve. However, typically this slope changes only slightly; in fact, it increases slightly, so

---

*Obviously, all filters preceeding correlation must be accounted for in ranging.

that the expression for timing error variance is quite accurate as it stands. Another problem in the design of a delay lock loop (DLL) at low SNR is the fact that the correlators are not identically equal in gain and a static timing error arises from this disparity. One solution is to "time-share" the correlators between the normal arrangement and an inverted (to correct the algebraic sign of the error signal) version, with the early and late codes interchanged. If this interchanging is done at a rate large compared to the code loop bandwidth and small compared to the BPF bandwidths, it can be shown that the timing error variance increase is negligible [6], but the offsets due to correlator imbalance are completely removed.

PROBLEM 1

Assume that the gain in one channel is $(1 + \delta)$ larger than the gain in the opposite channel for a baseband delay lock loop. Determine the timing error variance and the timing offset. Show that the variance is not affected to first order in δ but the bias is.

PROBLEM 2

Assume the reference PN generators are spaced only $\pm T_c/4$ chips apart instead of $\pm T_c/2$ apart. Show that the timing error variance is given by

$$\sigma_\epsilon = T_c \sqrt{\frac{N_0 B_L}{4P}} \quad \text{sec}$$

Sketch the S-curve and note that the region of linearity is reduced by one-half.

We see from Problem 2 that reduced reference code spacing improves small error performance, but also reduces the linear tracking region and the acquisition capability.

PROBLEM 3

Determine the tracking performance of a 1-Δ code loop when the input signal is hard limited. This can be obtained for low input SNR by determining the signal suppression after despreading and the noise spectral density without signal present.

10.2.2 Noncoherent Code Tracking Loop (Early / Late Loop)

In most instances the carrier frequency and phase are not known exactly *a priori*, so a noncoherent code loop must be utilized to track the received PN code. As might be guessed, the code tracking performance is inferior to that of the coherent baseband loop.

We consider the 1-Δ noncoherent code tracking loop shown in Figure 10.4. This loop is the most commonly used one because of its compromise between acquisition and tracking performance. We assume that the

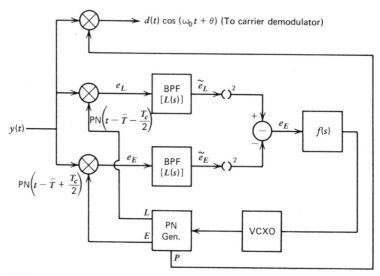

Figure 10.4 Noncoherent 1-Δ code tracking loop with total reference PN code spacing of one chip (E = early code, L = late code, and P = punctual code).

received waveform is given by

$$y(t) = \sqrt{2P}\, d(t)\text{PN}(t - T)\cos(\omega_0 t + \theta) + n(t) \qquad (10.2\text{-}27)$$

where P is the total power of the signal, $d(t)$ is the binary value (± 1) data stream having symbols of duration T_s, ω_0 is the carrier frequency, and θ is the input carrier phase. After filtering, the correlated signals are given by

$$\tilde{e}_L = \sqrt{2P}\, L(s)\left[d(t)R_{PN}\left(T - \hat{T} - \frac{T_c}{2}\right)\cos(\omega_0 t + \theta)\right]$$
$$+ L(s)\left[n(t)\text{PN}\left(t - \hat{T} - \frac{T_c}{2}\right)\right]$$
$$\tilde{e}_E = \sqrt{2P}\, L(s)\left[d(t)R_{PN}\left(T - \hat{T} + \frac{T_c}{2}\right)\cos(\omega_0 t + \theta)\right] \qquad (10.2\text{-}28)$$
$$+ L(s)\left[n(t)\text{PN}\left(t - \hat{T} + \frac{T_c}{2}\right)\right]$$

We approximate the effect of the bandpass filters on the data modulated carrier by the equivalent baseband response $L'(s)[d(t)]$. We denote this filtered data waveform by

$$\tilde{d}(t) = L'(s)[d(t)] \qquad (10.2\text{-}29)$$

Further we denote the filtered noise times PN code terms by

$$n_L(t) = L(s)\left[n(t)\text{PN}\left(t - \hat{T} - \frac{T_c}{2}\right)\right]$$
$$n_E(t) = L(s)\left[n(t)\text{PN}\left(t - \hat{T} + \frac{T_c}{2}\right)\right] \qquad (10.2\text{-}30)$$

so that we can write

$$\tilde{e}_L = \sqrt{2P}\ \tilde{d}(t) R_{PN}\left(T - \hat{T} - \frac{T_c}{2}\right)\cos(\omega_0 t + \theta) + n_L(t)$$

$$\tilde{e}_E(t) = \sqrt{2P}\ \tilde{d}(t) R_{PN}\left(T - \hat{T} + \frac{T_c}{2}\right)\cos(\omega_0 t + \theta) + n_E(t) \tag{10.2-31}$$

Squaring and forming the difference, we obtain

$$e_\epsilon = P[\tilde{d}(t)]^2\left[R_{PN}^2\left(T - \hat{T} - \frac{T_c}{2}\right) - R_{PN}^2\left(T - \hat{T} + \frac{T_c}{2}\right)\right]$$

$$+ 2\sqrt{2P}\ \tilde{d}(t)\cos(\omega_0 t + \theta)\left[n_L(t)R_{PN}\left(T - \hat{T} - \frac{T_c}{2}\right)\right.$$

$$\left. - n_E(t)R_{PN}\left(T - \hat{T} + \frac{T_c}{2}\right)\right] + n_L^2(t) - n_E^2(t) \tag{10.2-32}$$

Neglecting the filtering effect on the PN waveform, we obtain the S-curve of Figure 10.5a. For a spacing of one-half chip the S-curve is illustrated in Figure 10.5b, and for a spacing of one and one-half chips the S-curve is shown in Figure 10.5c. Hence, from (10.2-32), when $|\epsilon| \leq T_c/2$ we have the linearized equation for the input to the loop filter

$$e_\epsilon = 2P\ \tilde{d}^2(t)\frac{\epsilon}{T_c}$$

$$+ 2\sqrt{2P}\ \tilde{d}(t)\cos(\omega_0 t + \theta)\left[n_L(t)R_{PN}\left(T - \hat{T} - \frac{T_c}{2}\right)\right.$$

$$\left. - n_E(t)R_{PN}\left(T - \hat{T} + \frac{T_c}{2}\right)\right] + n_L^2(t) - n_E^2(t) \tag{10.2-33}$$

As was shown in the suppressed carrier loop chapter, the dc component of $\tilde{d}^2(t)$ is normally the only spectral component that materially affects the loop performance. Hence we denote by α the value

$$\alpha = \overline{E[\tilde{d}^2(t)]} = \int_{-\infty}^{\infty} |L'(f)|^2 \mathcal{S}_d(f)\ df \tag{10.2-34}$$

with the wavy overline denoting time averaging, $E[\ ]$ denoting an ensemble average, and $\mathcal{S}_d(f)$ denoting the spectral density of the data.

If we denote the second group of terms in the expression for e_ϵ as $n_1(t)$ and denote $n_L^2(t) - n_E^2(t)$ by $n_2(t)$, we have for the error signal

$$e_\epsilon = 2\alpha P\ \frac{\epsilon}{T_c} + n_1(t) + n_2(t) \tag{10.2-35}$$

As in the baseband loop we can close the loop to yield

$$\frac{\epsilon}{T_c} = -H(s)\left[\frac{n_1(t)}{2\alpha P} + \frac{n_2(t)}{2\alpha P}\right] + [1 - H(s)]\frac{T}{T_c} \tag{10.2-36}$$

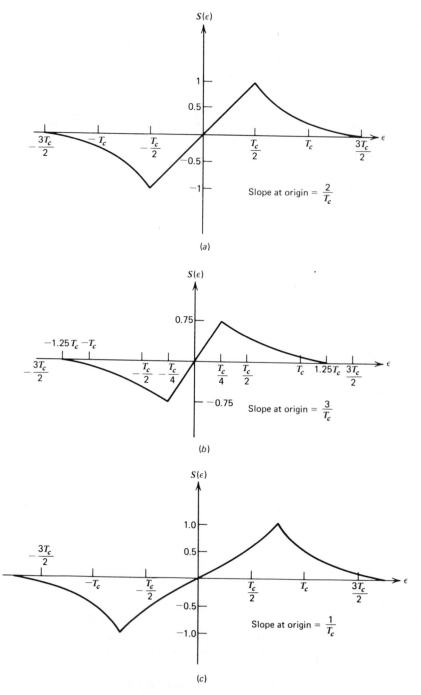

Figure 10.5 (*a*) Noncoherent delay lock loop with one chip discriminator spacing (1-Δ loop). (*b*) Noncoherent delay lock loop with one-half chip discriminator spacing (½-Δ loop). (*c*) Noncoherent lock loop with one and one-half chip discriminator spacing.

where $H(s)$ is the closed-loop transfer function;

$$H(s) = \frac{2\alpha PKF(s)/s}{1 + 2\alpha PKF(s)/s} \tag{10.2-37}$$

with $2\alpha PK$ being the total loop gain, not counting the dc gain of the loop filter $[F(0)]$. Notice that the transfer function gain now depends upon the power rather than the square root of the power, as in the coherent baseband loop. The input time process T is assumed to be a constant, as before. Therefore, the relative chip timing error variance is given by

$$\left(\frac{\sigma_\epsilon}{T_c}\right)^2 = \int_{-\infty}^{\infty} \frac{\mathcal{S}_{n_1+n_2}(f)}{4\alpha^2 P^2} |H(f)|^2 \, df \tag{10.2-38}$$

or

$$\left(\frac{\sigma_\epsilon}{T_c}\right)^2 \cong 2B_L \frac{\mathcal{S}_{n_1+n_2}(0)}{(2\alpha P)^2} \tag{10.2-39}$$

with the assumption that the noise spectral density is essentially constant near $f = 0$. First we determine the autocorrelation function of $n_1(t) + n_2(t)$. We again assume that $d(t)$ is a random NRZ signal. Since $E[\tilde{d}(t)]$ is zero for a random binary valued (± 1) waveform, we have

$$R_N(\tau) = R_{n_1}(\tau) + R_{n_2}(\tau) \tag{10.2-40}$$

First we compute the autocorrelation of $n_1(t)$. We first write the noise process in the in-phase and quadrature components:

$$n_L(t) = \sqrt{2}\, n_{L_c}(t) \cos(\omega_0 t + \theta) + \sqrt{2}\, n_{L_s}(t) \sin(\omega_0 t + \theta) \tag{10.2-41}$$

where $n_{L_c}(t)$ and $n_{L_s}(t)$ are independent baseband processes composed of baseband noise times the PN waveform:

$$n_{L_c}(t) = L(s)\left[n_c(t)PN\left(t - \hat{T} - \frac{T_c}{2}\right)\right]$$
$$n_{L_s}(t) = L(s)\left[n_s(t)PN\left(t - \hat{T} - \frac{T_c}{2}\right)\right] \tag{10.2-42}$$

$$n_{E_c}(t) = L(s)\left[n_E(t)PN\left(t - \hat{T} + \frac{T_c}{2}\right)\right]$$
$$n_{E_s}(t) = L(s)\left[n_E(t)PN\left(t - \hat{T} + \frac{T_c}{2}\right)\right] \tag{10.2-43}$$

where the input noise process can be expressed as

$$n(t) = \sqrt{2}\, n_c(t) \cos(\omega_0 t + \theta) + \sqrt{2}\, n_s(t) \sin(\omega_0 t + \theta) \tag{10.2-44}$$

Hence we can express $n_1(t)$, neglecting terms at $2\omega_0$, by

$$n_1(t) = 2\sqrt{P}\, \tilde{d}(t)\left\{n_{L_c}(t)R_{PN}\left(T - \hat{T} - \frac{T_c}{2}\right) - n_{E_c}(t)R_{PN}\left(T - \hat{T} + \frac{T_c}{2}\right)\right\} \tag{10.2-45}$$

Following the same argument of Section 10.2.1, we obtain

$$R_{n_1}(\tau) = PR_{\tilde{d}}(\tau)[R_{n_{L_c}}(\tau) + R_{n_{E_c}}(\tau)] \qquad (10.2\text{-}46)$$

or

$$R_{n_1}(\tau) = 2PR_{\tilde{d}}(\tau)R_{n_{L_c}}(\tau) \qquad (10.2\text{-}47)$$

In Section 10.2.1 we showed that

$$R_{n_{L_c}}(\tau) = \frac{N_0}{2} \int_{-\infty}^{\infty} l'(t-u)l'(t+\tau-u)\,du \qquad (10.2\text{-}48)$$

where $l'(t)$ is the impulse response of the baseband equivalent filter associated with the bandpass filters. Therefore

$$R_{n_1}(\tau) = PN_0R_{\tilde{d}}(\tau) \int_{-\infty}^{\infty} l'(t-u)l'(t+\tau-u)\,du \qquad (10.2\text{-}49)$$

Taking the Fourier transforms at $f=0$ produces

$$\mathcal{S}_{n_1}(0) = N_0PR_{\tilde{d}}(\tau) \int_{-\infty}^{\infty}\int_{-\infty}^{\infty} e^{-i\omega\tau}|L'(\omega)|^2 \frac{d\omega}{2\pi} \qquad (10.2\text{-}50)$$

or

$$\mathcal{S}_{n_1}(0) = N_0P \int_{-\infty}^{\infty} \mathcal{S}_{\tilde{d}}(f)|L'(f)|^2\,df \qquad (10.2\text{-}51)$$

where $\mathcal{S}_{\tilde{d}}(f)$ is the spectral density of the filtered data stream. But since

$$\mathcal{S}_{\tilde{d}}(f) = \mathcal{S}_d(f)|L'(f)|^2 \qquad (10.2\text{-}52)$$

we obtain finally that

$$\mathcal{S}n_1(0) = \alpha'N_0P \qquad (10.2\text{-}53)$$

where

$$\alpha' = \int_{-\infty}^{\infty} \mathcal{S}_d(f)|L'(f)|^4\,df \qquad (10.2\text{-}54)$$

Previous analyses [1, 2, 3, 7] have assumed that $n_E^2(t)$ is independent of $n_L^2(t)$; however, as is shown in the Appendix, this is not true in general. Fortunately when the bandpass filter bandwidth B is small compared to the PN chip rate, the variances are directly additive with negligible error. From the Appendix we have

$$\mathcal{S}_{n_2}(0) = N_0^2 \int_{-\infty}^{\infty} |L(f)|^4\,df = 2N_0^2B' \qquad (10.2\text{-}55)$$

where

$$2B' = \int_{-\infty}^{\infty} |L(f)|^4\,df \qquad L(f) \text{ is bandpass} \qquad (10.2\text{-}56)$$

where $B' \le B$ is a bandwidth measure somewhat smaller than the noise bandwidth B. We obtain for the timing error variance, the value

$$\left(\frac{\sigma_\epsilon}{T_c}\right)^2 = \frac{N_0B_L}{2\alpha P}\left(\frac{\alpha'}{\alpha} + \frac{2N_0B'}{\alpha P}\right) \qquad \text{chips}^2 \qquad (10.2\text{-}57)$$

For sharp cutoff filters normally used, $\alpha' \simeq \alpha$, $B' \simeq B$, so that

$$\sigma_\epsilon = T_c \sqrt{\frac{N_0 B_L}{2\alpha P}\left(1 + \frac{2N_0 B}{\alpha P}\right)} \qquad \text{sec} \qquad (10.2\text{-}58)$$

At low SNRs the optimum 3-dB bandwidth is about $1.4/T_s$ for NRZ symbols and $2.8/T_s$ for Manchester symbols, but near optimum is achieved over almost $2:1$ bandwidth changes.

We see that the noncoherent code loop suffers a degradation in tracking performance at low SNRs when compared to the baseband or coherent code tracking loop. However at high SNRs the bandwidth can be opened up somewhat so that $\alpha \to 1$ and $(1 + 2N_0 B/P) \to 1$ and there is essentially no degradation.

PROBLEM 4

Show that the timing error variance for the noncoherent code tracking loop with an early and late timing update of one-quarter chip is given by

$$\left(\frac{\sigma_\epsilon}{T_c}\right)^2 = \frac{N_0 B_L}{4\alpha P}\left(\frac{\alpha'}{\alpha} + \frac{8}{9}\frac{N_0 B'}{\alpha P}\right)$$

Notice that at low SNR, where the second term dominates, the standard deviation is reduced to approximately one-half of the value for the one-half early, one-half late noncoherent code tracking loop. Also, at high SNR, there is an advantage in performance. The pull-in range is reduced considerably, however.

It is to be pointed out that the linear S-curve region of the $\frac{1}{2}\Delta$ loop is less than the 1Δ loop (see Figure 10.5), so that in reality even though the linearized loop timing error variance is less, the actual timing error at very low SNRs (where linear theory predicts maximum advantage) would not be as much better as linear theory would suggest.

PROBLEM 5

Consider the alternate noncoherent code tracking loop configuration shown below [28]

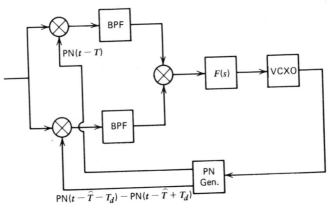

where τ_d is the code time shift. Show that

(a) The gain balance problem of the standard noncoherent loop is not present.

(b) There is a new problem with rf phase shift matching in both channels.

(c) The noise portion of the correlator outputs are statistically uncorrelated.

(d) The linearized loop mean square jitter is given by ($\tau_d = T_c/2$)

$$\left(\frac{\sigma_\epsilon}{T_c}\right)^2 = \frac{N_0 B_L}{2\alpha P}\left(\frac{\alpha'}{\alpha} + \frac{2N_0 B'}{\alpha P}\right)$$

(e) The general expression for the linearized loop mean square jitter is given by

$$\left(\frac{\sigma_\epsilon}{T_c}\right)^2 = 2\left(\frac{\tau_d}{T_c}\right)^2 \frac{N_0 B_L}{\alpha P}\left\{2\frac{\alpha'}{\alpha}\left(1 - \frac{\tau_d}{T_c}\right)\right.$$
$$\left. + \frac{2N_0 B'[1 - R_{PN}(2\tau_d)]}{\alpha P}\right\}$$

10.2.3 A Postdetection Combining Loop—Quadriphase Signaling

When quadriphase signaling is used in such a way that the code on each carrier phase is essentially independent, if the data is statistically independent, then an improvement can be obtained over using only one component of the signal.

Although the code tracking loop to be discussed here is theoretically superior in the quadriphase independent data case, there are significant imbalance problems to be compensated for in the implementation of the circuit. We will not discuss these problems in any more detail other than to suggest that a time shared circuit might be used between the early and late arms of the same code (see Figure 10.6).

We model the received quadriphase signal in the following form:

$$s(t) = \sqrt{2P_I}\, d_1(t)PN_1(t - T)\cos(\omega_0 t + \theta)$$
$$+ \sqrt{2P_Q}\, d_Q(t)PN_Q(t - T)\sin(\omega_0 t + \theta) \qquad (10.2\text{-}59)$$

This quadriphase signal $s(t)$ can be viewed as two PSK channels that are in carrier phase quadrature. The power in each channel is P_I and P_Q, respectively, with the I channel having data sequence $d_1(t)$ and PN derived sequence (for example, a PN code or a Gold code), with a delay into the code loop of T sec. The Q channel has its own data sequence $d_Q(t)$ and independent PN waveform.

It is assumed that the signal $s(t)$ is corrupted by WGN, $n(t)$. The postdetection combining code loop is shown in Figure 10.6. We shall summarize the tracking performance of the loop. The four correlator

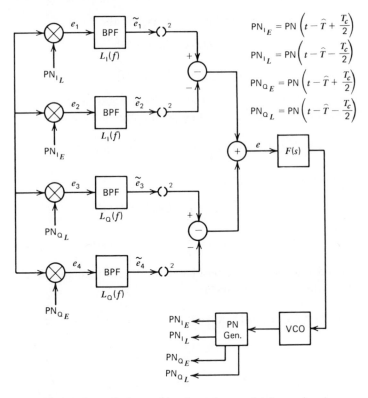

Figure 10.6 Code tracking loop for quadriphase signal.

outputs, neglecting self-noise, are given by

$$\tilde{e}_1 = \sqrt{2P_I}\, L(s) \left[d_I(t) R_{PN_I}\left(T - \hat{T} + \frac{T_c}{2}\right) \cos(\omega_0 t + \theta) \right]$$
$$+ L(s)\left[n(t) PN_I\left(t - \hat{T} + \frac{T_c}{2}\right)\right]$$
$$\tilde{e}_2 = \sqrt{2P_I}\, L(s) \left[d_I(t) R_{PN_I}\left(T - \hat{T} - \frac{T_c}{2}\right) \cos(\omega_0 t + \theta) \right]$$
$$+ L(s)\left[n(t) PN_I\left(t - \hat{T} - \frac{T_c}{2}\right)\right]$$

$$\quad (10.2\text{-}60)$$

$$\tilde{e}_3 = \sqrt{2P_Q}\, L(s) \left[d_Q(t) R_{PN_Q}\left(T - \hat{T} + \frac{T_c}{2}\right) \cos(\omega_0 t + \theta) \right]$$
$$+ L(s)\left[n(t) PN_Q\left(t - \hat{T} + \frac{T_c}{2}\right)\right]$$
$$\tilde{e}_4 = \sqrt{2P_Q}\, L(s) \left[d_Q(t) R_{PN_Q}\left(T - \hat{T} - \frac{T_c}{2}\right) \cos(\omega_0 t + \theta) \right]$$
$$+ L(s)\left[n(t) PN_Q\left(t - \hat{T} - \frac{T_c}{2}\right)\right]$$

Again we model the baseband filtered signals of the form

$$L(s)[d_1(t)\cos(\omega_0 t + \theta)] \tag{10.2-61}$$

by

$$L'(s)\{d_1(t)\}\cos(\omega_0 t + \theta) \tag{10.2-62}$$

with the assumption that $L'(s)$ is the baseband equivalent of $L(s)$.

Squaring, differencing, and adding per Figure 10.6 we obtain

$$e = 2(\alpha_I P_I + \alpha_Q P_Q)\frac{\epsilon}{T_c} + n_1(t) + n_2(t) + n_3(t) + n_4(t) \tag{10.2-63}$$

where we have used, as before,

$$\begin{aligned}
\alpha_I &= E[\tilde{d}_I^2(t)] = \int_{-\infty}^{\infty} |L_I'(f)|^2 \mathscr{S}_d(f)\, df \\
\alpha_Q &= E[\tilde{d}_Q^2(t)] = \int_{-\infty}^{\infty} |L_Q'(f)|^2 \mathscr{S}_d(f)\, df
\end{aligned} \tag{10.2-64}$$

with $L_I'(f)$ and $L_Q'(f)$ being the baseband equivalent bandpass filter transfer functions. Further, the noise terms are given by

$$\begin{aligned}
n_1(t) &= 2\sqrt{2P_I}\,\tilde{d}_I(t)\,\cos(\omega_0 t + \theta)\left[n_{L_I}R_{PN_I}\left(T - \hat{T} - \frac{T_c}{2}\right)\right. \\
&\qquad\qquad \left. - n_{E_I}R_{PN_I}\left(T - \hat{T} + \frac{T_c}{2}\right)\right] \\
n_2(t) &= 2\sqrt{2P_Q}\,\tilde{d}_Q(t)\,\cos(\omega_0 t + \theta)\left[n_{L_Q}R_{PN_Q}\left(T - \hat{T} - \frac{T_c}{2}\right)\right. \\
&\qquad\qquad \left. - n_{E_Q}R_{PN_Q}\left(T - \hat{T} + \frac{T_c}{2}\right)\right] \\
n_3(t) &= n_{L_I}^2 - n_{E_I}^2 \\
n_4(t) &= n_{L_Q}^2 - n_{E_Q}^2
\end{aligned} \tag{10.2-65}$$

where

$$\begin{aligned}
N_{L_I}(t) &= L(s)\left[n(t)\mathrm{PN}_I\left(t - \hat{T} - \frac{T_c}{2}\right)\right] \\
N_{L_Q}(t) &= L(s)\left[n(t)\mathrm{PN}_Q\left(t - \hat{T} - \frac{T_c}{2}\right)\right] \\
N_{E_I}(t) &= L(s)\left[n(t)\mathrm{PN}_I\left(t - \hat{T} + \frac{T_c}{2}\right)\right] \\
N_{E_Q}(t) &= L(s)\left[n(t)\mathrm{PN}_Q\left(t - \hat{T} + \frac{T_c}{2}\right)\right]
\end{aligned} \tag{10.2-66}$$

Denoting the VCO gain by K_{VCO}, letting K_D be the detector gain ($K = K_D K_{\mathrm{VCO}}$), we have

$$\frac{\hat{T}}{T_c} = \frac{2K(\alpha_I P_I + \alpha_Q P_Q)F(s)}{s}\left[\frac{\epsilon}{T_c} + \frac{\sum_{i=1}^{4} n_i(t)}{2(\alpha_I P_I + \alpha_Q P_Q)}\right] \tag{10.2-67}$$

Letting

$$H(s) = \frac{2(\alpha_I P_I + \alpha_Q P_Q)KF(s)/s}{1 + 2(\alpha_I P_I + \alpha_Q P_Q)KF(s)/s} \tag{10.2-68}$$

we have

$$\frac{\epsilon}{T_c} = -H(s)\frac{\sum_{i=1}^{4} n_i(t)}{2(\alpha_I P_I + \alpha_Q P_Q)} + [1 - H(s)]\frac{T}{T_c} \tag{10.2-69}$$

We again assume that the input delay process T is zero. Using the same type of arguments as before, we find that if the bandpass filter bandwidths are narrow compared to the PN code rate and are ideal in shape, the timing error variance is given by

$$\left(\frac{\sigma_\epsilon}{T_c}\right)^2 = \int_{-\infty}^{\infty} \mathcal{S}_n(f)|H(f)|^2 \, df \cong 2B_L \mathcal{S}_n(0) \tag{10.2-70}$$

where $\mathcal{S}_n(0)$ is the spectral density of the four noise terms and is given by

$$\mathcal{S}_n(0) = \frac{N_0(\alpha_I' P_I + \alpha_Q' P_Q) + 2N_0^2(B_I' + B_Q')}{4(\alpha_I P_I + \alpha_Q P_Q)^2} \tag{10.2-71}$$

Hence

$$\left(\frac{\sigma_\epsilon}{T_c}\right)^2 = \frac{2B_L[(P_I\alpha_I' + P_Q\alpha_Q')N_0 + 2N_0^2(B_I' + B_Q')]}{4(\alpha_I P_I + \alpha_Q P_Q)^2} \qquad \text{chips}^2 \tag{10.2-72}$$

where

$$2B_I' = \int_{-\infty}^{\infty} |L_I(f)|^4 \, df$$

$$2B_Q' = \int_{-\infty}^{\infty} |L_Q(f)|^4 \, df$$

$$\alpha_I' = \int_{-\infty}^{\infty} \mathcal{S}_d(f)|L_I(f)|^4 \, df$$

$$\alpha_Q' = \int_{-\infty}^{\infty} \mathcal{S}_d(f)|L_Q(f)|^4 \, df \tag{10.2-73}$$

$$\alpha_I = \int_{-\infty}^{\infty} \mathcal{S}_d(f)|L_I(f)|^2 \, df$$

$$\alpha_Q = \int_{-\infty}^{\infty} \mathcal{S}_d(f)|L_Q(f)|^2 \, df$$

For the case the two channels are at the same data rate and assuming the (ideal) bandpass filters are of the same type and bandwidth, we obtain

$$\left(\frac{\sigma_\epsilon}{T_c}\right)^2 = \frac{N_0 B_L}{4\alpha_I P_I}\left(\frac{\alpha_I'}{\alpha_I} + \frac{2N_0 B_I'}{\alpha_I P_I}\right) \qquad \text{chips}^2 \tag{10.2-74}$$

Again $B_1' \simeq B_1$ and $\alpha_1' \simeq \alpha_1$ for typical multipole bandpass filters. Observe that, compared to using only one channel of the quadriphase signal, the use of postdetection combining on both channels reduces the timing error variance to one-half! However the offset problems, including dc biases, are greatly increased because of the use of four detectors rather than two. This consideration of offsets must be considered in the design of any such loop.

Notice also that if we let

$$2P_1 = P \qquad P \text{ is total power}$$
$$\alpha_1 = \alpha \tag{10.2-75}$$
$$\alpha_1' = \alpha'$$

then the jitter can be written in terms of the total power as

$$\left(\frac{\sigma_\epsilon}{T_c}\right)^2 = \frac{N_0 B_L}{2\alpha P}\left(\frac{\alpha'}{\alpha} + \frac{4N_0 B'}{\alpha P}\right) \qquad P_1 = P_Q \tag{10.2-76}$$

By comparing (10.2-76) with (10.2-57) we see that quadriphase PN signals are not tracked as accurately as biphase PN signals. If we had a biphase signal rather than a quadriphase signal having power P, the tracking jitter would be given by

$$\left(\frac{\sigma_\epsilon}{T_c}\right)^2 = \frac{N_0 B_L}{2\alpha P}\left(\frac{\alpha'}{\alpha} + \frac{2N_0 B'}{\alpha P}\right) \tag{10.2-77}$$

Consequently, at high SNR the quadriphase loop has comparable tracking performance, but at low SNR the variance is doubled.

PROBLEM 6

Consider the quadriphase type loop shown below:

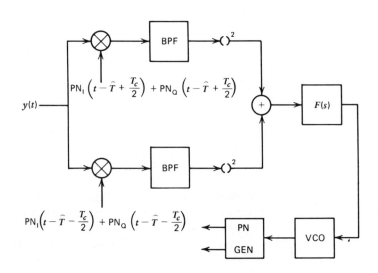

Show that this loop is uniformly inferior to the quadriphase loop of Figure 10.6 in terms of its code tracking ability.

10.3 TIME SHARED NONCOHERENT CODE LOOP TRACKING PERFORMANCE

Many times the signal to noise ratio is not especially low, so a time shared code tracking loop can be employed. This loop time-shares the use of the early-late correlators, the advantage being that only one correlator need be used in the design of the loop, and further that dc offset problems tend to be reduced.

Numerous authors have analyzed the time shared loop, including Simon [7], Stone [8], Holmes [9], Huang [6], and Hartmann [10]. Our results are in essential agreement with the above authors although we have used a slightly different approach in the analysis.

Consider the noncoherent time-shared code tracking loop illustrated in Figure 10.7. The loop alternately "time-shares" the early and late code reference signals $PN(t - \hat{T} - T_c/2)$ and $PN(t - \hat{T} + T_c/2)$ in a periodic manner so that only one of the two signals is correlating against the input $y(t)$ an any instant of time. In this case only two multipliers are needed, one for code tracking and one for despreading the signal.

We model the input process $y(t)$ by a PSK modulated waveform corrupted by WGN as before

$$y(t) = \sqrt{2P}\, d(t)PN(t - T)\cos(\omega_0 t + \theta) + n(t) \qquad (10.3\text{-}1)$$

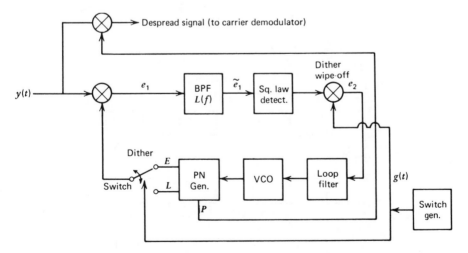

Figure 10.7 Noncoherent time-shared code tracking loop (E = early, L = late, and P = punctual).

where as before P is the total signal power, $d(t)$ is the symbol waveform having, for example, Manchester or NRZ symbols, and $n(t)$ is WGN. We model the gating function $g(t)$ as shown in Figure 10.8 along with the time-shared functions $[1 + g(t)]/2$ and $[1 - g(t)]/2$.

The waveform at e_1 is given by

$$e_1 = [\sqrt{2P}\ d(t)\text{PN}(t - T)\cos(\omega_0 t + \theta) + n(t)]$$

$$\times \left[\left(\frac{1 - g(t)}{2} \right) \text{PN}\left(t - \hat{T} + \frac{T_c}{2} \right) + \left(\frac{1 + g(t)}{2} \right) \text{PN}\left(t - \hat{T} - \frac{T_c}{2} \right) \right]$$

$$(10.3\text{-}2)$$

Out of the bandpass filter the signal is modeled as (neglecting again the self-noise term)

$$\tilde{e}_1 = \sqrt{2P}\ L(s)\left[\left(\frac{1 - g(t)}{2} \right) d(t)R_{\text{PN}}\left(T - \hat{T} + \frac{T_c}{2} \right) \cos(\omega_0 t + \theta) \right.$$

$$\left. + \left[\frac{1 + g(t)}{2} \right] d(t)R_{\text{PN}}\left(T - \hat{T} - \frac{T_c}{2} \right) \cos(\omega_0 + \theta) \right]$$

$$+ L(s)\left\{ n(t)\left[\frac{1 - g(t)}{2} \right] \text{PN}\left(t - \hat{T} + \frac{T_c}{2} \right) \right.$$

$$\left. + \left[\frac{1 + g(t)}{2} \right] n(t)\ \text{PN}\left(t - \hat{T} - \frac{T_c}{2} \right) \right\} \qquad (10.3\text{-}3)$$

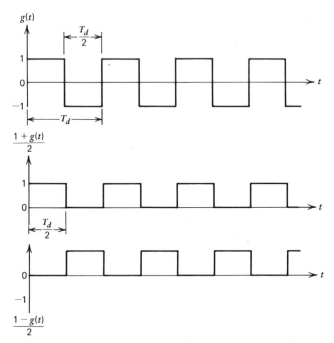

Figure 10.8 Gating functions used in analysis.

where $L(s)$ denotes the bandpass filter Heaviside operator operating on the terms in brackets. Now we assume, as in normal design practice, that $T_d \gg T_s$, with T_d being the period of $g(t)$ and T_s being the duration of one symbol of $d(t)$, the data symbol waveform. In addition, we will need to ensure that $T_d^{-1} \gg B_L$, so that discrete lines at the gating function rate do not get "into the loop," deteriorating loop performance. With these assumptions we obtain, denoting the time error by $\epsilon = T - \hat{T}$,

$$\tilde{e}_1 = \sqrt{2P}\, \tilde{d}(t) \cos(\omega_0 t + \theta) \left\{ \left(\frac{1-g(t)}{2}\right) R\left(\epsilon + \frac{T_c}{2}\right) + \left[\frac{1+g(t)}{2}\right] R\left(\epsilon - \frac{T_c}{2}\right) \right\}$$

$$+ \left[\frac{1-g(t)}{2}\right] n_E(t) + \left[\frac{1+g(t)}{2}\right] n_L(t) \tag{10.3-4}$$

where, as before,

$$n_E(t) = L(s)\left[n(t)\, \text{PN}\left(t - \hat{T} + \frac{T_c}{2}\right)\right]$$

$$n_L(t) = L(s)\left[n(t)\text{PN}\left(t - \hat{T} - \frac{T_c}{2}\right)\right] \tag{10.3-5}$$

The first term of $\tilde{e}_1$, denoted by T_1, can be rewritten as

$$T_1 = \sqrt{2P}\, \tilde{d}(t) \cos(\omega_0 t + \theta)\left[\tfrac{1}{2} R\left(\epsilon + \frac{T_c}{2}\right) + \tfrac{1}{2} R\left(\epsilon - \frac{T_c}{2}\right)\right.$$

$$\left. - \frac{g(t)}{2} R\left(\epsilon + \frac{T_c}{2}\right) + \frac{g(t)}{2} R\left(\epsilon - \frac{T_c}{2}\right)\right] \tag{10.3-6}$$

Thus, $\tilde{e}_1$ can be written, for $|\epsilon| < T_c/2$, as

$$\tilde{e}_1 = \sqrt{2P}\, \tilde{d}(t) \cos(\omega_0 t + \theta)\left[\tfrac{1}{2} + \frac{g(t)\epsilon}{T_c}\right]$$

$$+ \left[\frac{1-g(t)}{2}\right] n_E(t) + \left[\frac{1+g(t)}{2}\right] n_L(t) \qquad |\epsilon| < \frac{T_c}{2} \tag{10.3-7}$$

After squaring and multiplying by $g(t)$, we obtain after some simplications, including assuming $\epsilon \simeq 0$ for the noise terms*

$$g(t)(\tilde{e}_1)^2 = P[\tilde{d}(t)]^2 \left(\frac{\epsilon}{T_c}\right) + \left[\frac{g(t)-1}{2} n_E^2(t) + \frac{g(t)+1}{2} n_L^2(t)\right]$$

$$+ \left[\frac{g(t)-1}{2} n_E(t) + \frac{g(t)+1}{2} n_L(t)\right] \sqrt{2P}\, \tilde{d}(t) \cos(\omega_0 t + \theta) \tag{10.3-8}$$

If we denote the second and third terms as $n_1(t)$ and $n_2(t)$, then we have {again letting $\alpha = E[\tilde{d}^2]$ as is discussed in Chapter 5}

$$e_2 = \alpha P \frac{\epsilon}{T_c} + n_1(t) + n_2(t) \tag{10.3-9}$$

*We also neglect the term proportional to $g(t)$ since it is removed by the loop filter.

where

$$\alpha = \int_{-\infty}^{\infty} |L'(f)|^2 \mathscr{S}_D(f)\, df \leq 1 \qquad (10.3\text{-}10)$$

with $L'(f)$ being the baseband equivalent of the bandpass filter having transfer function $L(f)$. Again we can close the loop by denoting

$$\frac{\hat{T}}{T_c} = \frac{KF(s)}{s}\left[\alpha P\,\frac{\epsilon}{T_c} + n_1(t) + n_2(t)\right] + \frac{s}{s + \alpha KPF(s)}\left(\frac{T}{T_c}\right) \quad (10.3\text{-}11)$$

Finally we have

$$\frac{\epsilon}{T_c} = \frac{-\alpha KPF(s)}{s + \alpha KPF(s)}\left[\frac{n_1(t) + n_2(t)}{\alpha P}\right] \qquad (10.3\text{-}12)$$

We again have assumed that $T = 0$ for all time. Therefore the timing error variance is given by

$$\frac{\sigma_\epsilon^2}{T_c^2} = \int_{-\infty}^{\infty} |H(f)|^2\,\frac{\mathscr{S}_N(f)}{(\alpha P)^2}\, df \qquad (10.3\text{-}13)$$

where the closed-loop transfer function is given by

$$H(s) = \frac{\alpha KPF(s)}{s + \alpha KPF(s)} \qquad (10.3\text{-}14)$$

and

$$N(t) = n_1(t) + n_2(t) \qquad (10.3\text{-}15)$$

Since $n_1(t)$ and $n_2(t)$ are uncorrelated, we need to evaluate

$$R_N(\tau) = R_{n_1}(\tau) + R_{n_2}(\tau) \qquad (10.3\text{-}16)$$

First consider $R_{n_1}(\tau)$

$$4R_{n_1}(\tau) = \{[g(t) - 1][g(t + \tau) - 1]\}E[n_E^2(t)n_E^2(t + \tau)]$$
$$+ \{[1 + g(t)][1 + g(t + \tau)]\}E[n_L^2(t)n_L^2(t + \tau)]$$
$$+ \{[g(t) - 1][1 + g(t + \tau)]\}E[n_E^2(t)n_L^2(t + \tau)]$$
$$+ \{[1 + g(t)][g(t + \tau) - 1]\}E[n_L^2(t)n_E^2(t + \tau)] \quad (10.3\text{-}17)$$

Denote

$$\Lambda_d(\tau) = \begin{cases} \left(1 - \dfrac{2|\tau|}{T_d}\right) & |\tau| \leq \dfrac{T_d}{2} \\[2mm] 0 & \text{otherwise} \end{cases} \qquad (10.3\text{-}18)$$

The auto- and cross correlation functions of $[g(t) - 1]$ and $[g(t) + 1]$ are shown in Figure 10.9. Therefore

$$R_{1+g}(\tau) = R_{g-1}(\tau) = 2[\Lambda_d(\tau)] * \sum_{n=-\infty}^{\infty} \delta(\tau + nT_d) \qquad (10.3\text{-}19)$$

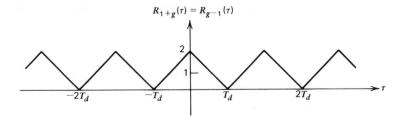

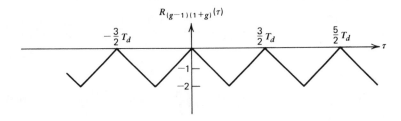

Figure 10.9   Auto- and cross correlation functions.

where $*$ denotes convolution. Finally, using symmetry, we arrive at

$$R_{n_1}(\tau) = \left\{ [\Lambda(\tau)] \underset{d}{*} \sum_{n=-\infty}^{\infty} \delta(\tau + nT_d) \right\} E[n_E^2(t)n_E^2(t+\tau)]$$

$$- \left[1 - (\Lambda(\tau)) \underset{d}{*} \sum_{n=-\infty}^{\infty} \delta(\tau + nT_d) \right] E[n_L^2(t)n_E^2(t+\tau)]$$

$$(10.3\text{-}20)$$

Now, using the assumption that $n_E(t)$ and $n_L(t)$ are essentially independent Gaussian random variables, we have that

$$E[n_E^2(t)n_E^2(t+\tau)] = E[n_E^2(t)]E[n_E^2(t)] + 2R_{n_E}^2(\tau) \qquad (10.3\text{-}21)$$

$$E[n_L^2(t)n_E^2(t+\tau)] = (N_0B)^2 \qquad (10.3\text{-}22)$$

Combining the above, we obtain

$$R_{n_1}(\tau) = 2R_{n_E}^2(\tau) \left\{ [\Lambda(\tau)] \underset{d}{*} \sum_{n=-\infty}^{\infty} \delta(\tau + nT_d) \right\} - (N_0B)^2$$

$$+ 2(N_0B)^2 \left\{ [\Lambda(\tau)] \underset{d}{*} \sum_{n=-\infty}^{\infty} \delta(\tau + nT_d) \right\} \qquad (10.3\text{-}23)$$

using

$$\mathscr{F}\left[\sum_{n=-\infty}^{\infty} \delta(\tau + nT_d) \right] = \frac{1}{T_d} \sum_{n=-\infty}^{\infty} \delta\left(f + \frac{n}{T_d} \right) \qquad (10.3\text{-}24)$$

and

$$\mathscr{F}[\Lambda(\tau)] = \frac{T_d}{2} \operatorname{sinc}^2\left(\frac{\pi f T_d}{2} \right) \qquad (10.3\text{-}25)$$

we have that the corresponding spectral density near $f = 0$ is given by

$$\mathcal{S}_{n_1}(0) = -(N_0 B)^2 \delta(f) + 2(N_0 B)^2 (\tfrac{1}{2}) \left[\delta(f) \operatorname{sinc}^2 \left(\frac{\pi f T_d}{2} \right) \right]$$

$$+ 2 \int_{-\infty}^{\infty} R_{n_E}^2(\tau) \left\{ [\Lambda(\tau)] \underset{d}{*} \sum_{n=-\infty}^{\infty} \delta(\tau + nT_d) \right\} d\tau \qquad (10.3\text{-}26)$$

where we have neglected terms at n/T_d, $n = \pm 1$, $n \pm 2$, etc., since it is assumed that they will not be passed by the closed-loop response $H(s)$. The first two terms of (10.3-26) cancel so that we can turn our attention to the last term in $\mathcal{S}_{n_1}(0)$. We assume that $B_L \ll B$, and we have

$$\mathcal{S}_{n_1}(0) = 2 \int_{-\infty}^{\infty} [\mathcal{S}_{n_E}(f) * \mathcal{S}_{n_E}(f)] \left[\frac{T_d}{2} \operatorname{sinc}^2 \left(\frac{\pi f T_d}{2} \right) \frac{1}{T_d} \sum_{n=-\infty}^{\infty} \delta \left(f + \frac{n}{T_d} \right) \right] df$$

$$(10.3\text{-}27)$$

Evaluating, we obtain

$$\mathcal{S}_{n_1}(0) = \int_{-\infty}^{\infty} \sum_{n=-\infty}^{\infty} \mathcal{S}_{n_E}(f') \mathcal{S}_{n_E} \left(\frac{n}{T_d} - f' \right) \operatorname{sinc}^2 \left(\frac{\pi n}{2} \right) df' \qquad (10.3\text{-}28)$$

Using the assumption that $T_d \gg T_s$, where T_s is the data symbol duration, we obtain

$$\mathcal{S}_{n_1}(0) \cong \left(\frac{N_0}{2} \right)^2 (2B') \left[1 + 2 \left(\frac{2}{\pi} \right)^2 \left(1 + \frac{1}{3^2} + \frac{1}{5^2} + \frac{1}{7^2} + \cdots \right) \right] \quad (10.3\text{-}29)$$

Using Jolly series no. 339 [12], we end up with

$$\mathcal{S}_{n_1}(0) \cong N_0^2 B' \qquad (10.3\text{-}30)$$

where

$$2B' = \int_{-\infty}^{\infty} |L(f)|^4 df \qquad (10.3\text{-}31)$$

Now consider $R_{n_2}(\tau)$. To this end it is convenient to express the two noise processes in the form

$$n_E(t) = \sqrt{2} \, n_{E_c}(t) \cos(\omega_0 t + \theta) + \sqrt{2} \, n_{E_s}(t) \sin(\omega_0 t + \theta)$$
$$n_L(t) = \sqrt{2} \, n_{L_c}(t) \cos(\omega_0 t + \theta) + \sqrt{2} \, n_{L_s}(t) \sin(\omega_0 t + \theta) \qquad (10.3\text{-}32)$$

with $n_{L_c}(t)$, $n_{L_s}(t)$, $n_{E_c}(t)$, and $n_{E_s}(t)$ defined in Section 10.2. Hence, we can express $n_2(t)$ by (neglecting terms at $2\omega_0$)

$$n_2(t) = \sqrt{P} \left[\frac{g(t) - 1}{2} n_{E_c}(t) + \frac{g(t) + 1}{2} n_{L_c}(t) \right] \tilde{d}(t) \qquad (10.3\text{-}33)$$

Now

$$R_{n_2}(\tau) = PR_{\tilde{d}}(\tau) E \left\{ \left[\left(\frac{g(t) - 1}{2} \right) n_{E_c}(t) + \frac{g(t) + 1}{2} n_{L_c}(t) \right] \right.$$

$$\left. \times \left[\left(\frac{g(t + \tau) - 1}{2} \right) n_{E_c}(t + \tau) + \left(\frac{g(t + \tau) + 1}{2} \right) n_{L_c}(t + \tau) \right] \right\}$$

$$(10.3\text{-}34)$$

Since the two noise terms are uncorrelated and identically distributed, we obtain

$$R_{n_2}(\tau) = \frac{PR_{\dot{a}}(\tau)}{2}\overbrace{[(g(t)-1)(g(t+\tau)-1)]}R_{n_{E_c}}(\tau) \tag{10.3-35}$$

as is clear from the upper figure (in Figure 10.9)

$$\overbrace{[g(t)-1][g(t+\tau)-1]} = 2[\Lambda(\tau)] * \sum_{n=-\infty}^{\infty} \delta(\tau+nT_d) \tag{10.3-36}$$

so that

$$\mathcal{S}_{n_2}(0) = P \int_{-\infty}^{\infty} R_{\dot{a}}(\tau)R_{n_{E_c}}(\tau)\left\{[\Lambda(\tau)] * \sum_{n=-\infty}^{\infty} \delta(\tau+nT_d)\right\} d\tau \tag{10.3-37}$$

or

$$\mathcal{S}_{n_2}(0) = \frac{P}{2} \sum_{n=-\infty}^{\infty} \int_{-\infty}^{\infty}\int_{-\infty}^{\infty} \mathcal{S}_{n_{E_c}}(f')\mathcal{S}_{\dot{a}}(f-f')\,df'\,\text{sinc}^2\left(\frac{\pi f T_d}{2}\right)\delta\left(f+\frac{n}{T_d}\right)df \tag{10.3-38}$$

or

$$\mathcal{S}_{n_2}(0) = \frac{P}{2} \sum_{n=-\infty}^{\infty} \int_{-\infty}^{\infty} \mathcal{S}_{n_{E_c}}(f')\mathcal{S}_{\dot{a}}\left(\frac{n}{T_d}-f'\right)\text{sinc}^2\left(\frac{\pi n}{2}\right)df' \tag{10.3-39}$$

Again assuming that $T_d \gg T_s$ produces

$$\mathcal{S}_{n_2}(0) = \frac{N_0 P}{4} \int_{-\infty}^{\infty} |L'(f)|^4 \mathcal{S}_{\dot{a}}(f)\,df\left[\sum_{n=-\infty}^{\infty}\text{sinc}^2\left(\frac{\pi n}{2}\right)\right] \tag{10.3-40}$$

Summing as before we obtain

$$\mathcal{S}_{n_2}(0) = \frac{\alpha' P N_0}{2} \tag{10.3-41}$$

where

$$\alpha' = \int_{-\infty}^{\infty} |L'(f)|^4 \mathcal{S}_{\dot{a}}(f)\,df \tag{10.3-42}$$

with $L'(f)$ being the baseband equivalent filter of $L(f)$, the bandpass filter.

From (10.3-12), (10.3-30), and (10.3-41), we obtain for the timing error variance the expression

$$\left(\frac{\sigma_\epsilon}{T_c}\right)^2 = \frac{N_0 B_L}{\alpha P}\left(\frac{\alpha'}{\alpha}+\frac{2N_0 B'}{\alpha P}\right) \qquad \text{chips}^2 \tag{10.3-43}$$

where

$$\alpha' = \int_{-\infty}^{\infty} |L'(f)|^4 \mathcal{S}_{\dot{a}}(f)\,df$$

$$\alpha = \int_{-\infty}^{\infty} |L'(f)|^2 \mathcal{S}_{\dot{a}}(f)\,df \tag{10.3-44}$$

$$2B' = \int_{-\infty}^{\infty} |L(f)|^4\,df$$

Again for sharp cutoff multipole bandpass filters we have to a good approximation

$$\frac{\sigma_\epsilon^2}{T_c^2} = \frac{N_0 B_L}{\alpha P}\left(1 + \frac{2N_0 B}{\alpha P}\right) \tag{10.3-45}$$

with

$$2B = \int_{-\infty}^{\infty} |L(f)|^2\, df \quad \begin{array}{l}\text{the two-sided noise bandwidth} \\ \text{of the bandpass filter}\end{array} \tag{10.3-46}$$

Again, the optimum low SNR 3-dB bandwidth is approximately given by $1.4/T_s$ for NRZ and $2.8/T_s$ for Manchester data.

Notice that the result for the timing error variance is exactly 3 dB greater for the time shared loop! Furthermore, bias problems tend to be less troublesome in this loop because of its chopping action by the gating functions. As a result this loop is desirable in many cases over the full-time code tracking loop, except at minimal SNRs.

In fact, time sharing can also be used on a full-time code tracking loop to reduce offset problems [29], [31].

PROBLEM 7

Generalize this result to the case the bandpass filter does not pass the dither signal without loss.

PROBLEM 8

Show that a time-shared coherent code tracking loop has a linearized tracking error of

$$\left(\frac{\sigma_\epsilon}{T_c}\right)^2 = \sqrt{\frac{N_0 B_L}{P}}$$

10.4 τ-DITHER CODE TRACKING LOOP

In this section we discuss an offshoot of the time-shared code tracking loop which, although similar to it, has the advantage that only one correlator is used to implement the tracking function and the despreading function.

The τ-dither code tracking loop is only useful in moderately high loop SNRs, as we will see shortly; however, its prime advantage is that only one correlator is needed to provide both the code tracking function and the despreading function.

A τ-dither loop is shown in Figure 10.10. To obtain a feel of how the error signal is developed from this loop (and as a matter of fact the time-shared code tracking loop) consider the noiseless case illustrated in Figure 10.11.

The dither signal, a squarewave signal, is used to time-share the early and late PN codes, as shown at the lower left of the figure. In Figure 10.11

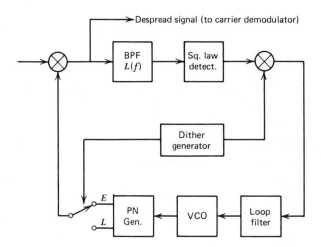

Figure 10.10   τ-Dither code tracking loop.

it was assumed that the code generator lags the input code. After correlating (multiplying and bandpass filtering) the resultant signal envelope is as shown in the upper right segment of Figure 10.11. The envelope has an amplitude modulation impressed upon it at the dither rate with an amplitude variation that depends upon the magnitude of the error $\epsilon = T - \hat{T}$. After multiplying by the dithering signal $g(t)$ and filtering, the error signal emerges to control the VCO (code generator clock) through the loop filter as seen in the lower right portion of the figure.

If, for example, the loop code generator were in time with the received code, then the dither signal would produce no envelope variation and the filtered control voltage, the error voltage, would be zero.

Alternately, if the loop code generator was ahead of the received code, it is clear that the resulting error control voltage would now be positive.

In the above discussion we have indicated how the τ-dither or time-shared code loops generate the loop control voltage. We now, with the aid of the results of a problem, describe the linearized code loop tracking error variance.

The tracking error curve generated by this loop is shown for various dither amplitudes in Figure 10.12. For the τ-dither loop typically a value of Δ equal to about $0.1T_c$ (0.1 chips) is used since the despread signal power drops off as $(1-\Delta)^2$, which is about 0.92 dB for the ideal unfiltered PN correlation curve. This relative loss becomes considerably less if the code modulated carrier is prefiltered prior to the code loop as is the normal case. However, the tracking error curve degrades also.

Another observation about the τ-dither loop that is of concern is the fact that for $\Delta = 0.1T_c$, which is commonly used, the S-curve (tracking curve), is linearly only over $\pm 0.1T_c$ chips. Hence the linearized analysis

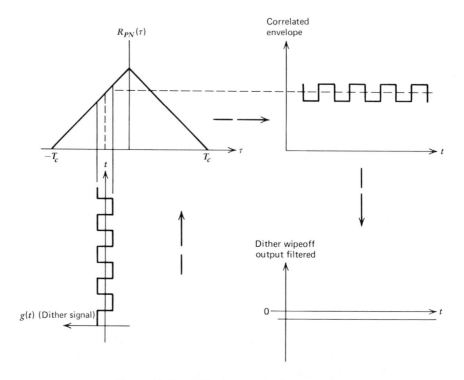

Figure 10.11 Dithering to obtain error signal.

that applies to tracking jitter performance assumes that (denoting σ as the standard deviation of timing error), $3\sigma \leq 0.1T_c$; otherwise the analysis would be a lower bound on performance and not very useful.

Subject to the assumption of a linearized loop with negligible PN code filtering, it is shown in Problem 10.9 that the timing error variance of the τ-dither loop with one-sided loop bandwidth is given by the following expression (using a square law detector)

$$\left(\frac{\sigma}{T_c}\right)^2 = \frac{N_0 B_L}{\alpha P}\left[\frac{\alpha'}{\alpha} + \frac{N_0 B'}{\alpha P}\left(\frac{1 + R^2(2\Delta)}{2[R(\Delta)]^2}\right)\right] \qquad \text{chips}^2 \qquad \Delta < T_c$$

$$(10.4\text{-}1)$$

where

$$\alpha = \int_{-\infty}^{\infty} |L'(f)|^2 \mathscr{S}_d(f)\, df \leq 1 \tag{10.4-2}$$

$$\alpha' = \int_{-\infty}^{\infty} |L'(f)|^4 \mathscr{S}_d(f)\, df \leq \alpha \leq 1 \tag{10.4-3}$$

$$2B' = \int_{-\infty}^{\infty} |L(f)|^4\, df \tag{10.4-4}$$

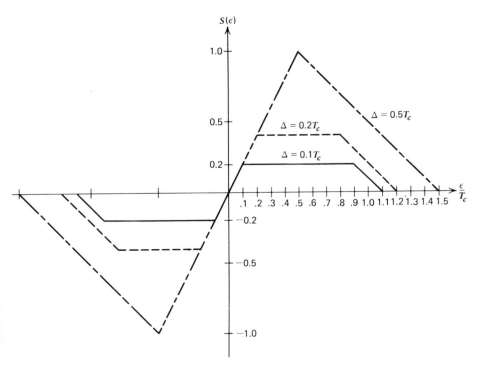

Figure 10.12 Resulting noiseless S-curves for different dither amplitudes (shown for envelope detector).

$$R(x) = \begin{cases} (1 - |x|) & |x| \le T_c \\ 0 & \text{otherwise} \end{cases} \tag{10.4-5}$$

and $L'(f)$ is the baseband equivalent translation of the bandpass filter transfer function $L(f)$.

Notice as a check when $\Delta = T_c/2$ we have the time-shared loop and our expression becomes $(\Delta = T_c/2)$

$$\left(\frac{\sigma}{T_c}\right)^2 = \frac{N_0 B_L}{\alpha P} \left(\frac{\alpha'}{\alpha} + \frac{2 N_0 B'}{\alpha P}\right) \tag{10.4-6}$$

which is precisely the value for the time-shared loop!

Finally for the case $\Delta = 0.1 T_c$ we obtain for the τ-dither loop the timing error variance result*

$$\left(\frac{\sigma}{T_c}\right)^2 = \frac{N_0 B_L}{\alpha P} \left(\frac{\alpha'}{\alpha} + \frac{N_0 B'}{\alpha P}\right) \qquad \Delta = 0.1 T_c \tag{10.4-7}$$

Note that the despread signal is always nominally 0.1 chips in error, so the signal is reduced by about 0.92 dB, which is the price one pays for the

---

*Actually, the coefficient of $N_0 B'/\alpha P$ should be 1.01.

advantage of a single correlator. This loss reduces when the filtering of the signal is considered however. Also, the tracking and pull-in range are much reduced from the time-shared code tracking loop.

PROBLEM 9

Show that the timing error variance of the time-shared code tracking loop, with Δ the early-late code time shift, and a square law detector is given by

$$\left(\frac{\sigma}{T_c}\right)^2 = \frac{N_0 B_L}{\alpha P}\left[\frac{\alpha'}{\alpha} + \frac{N_0 B'}{\alpha P}\left(\frac{1 + R^2(2\Delta)}{2[R(\Delta)]^2}\right)\right] \qquad \Delta < T_c \qquad \begin{array}{l} B_L = \text{one sided loop} \\ \text{bandwidth} \end{array}$$

where

$$\alpha = \int_{-\infty}^{\infty} |L'(f)|^2 \mathscr{S}_d(f)\, df \leq 1$$

$$\alpha' = \int_{-\infty}^{\infty} |L'(f)|^4 \mathscr{S}_d(f)\, df \leq \alpha \leq 1$$

$$2B' = \int_{-\infty}^{\infty} |L(f)|^4\, df$$

$$R(x) = \begin{cases} (1 - |x|) & |x| \leq T_c \\ 0 & \text{otherwise} \end{cases}$$

Table 10.1 Linearized timing error variance for various code loop configurations

Code Loop Type	Linearized Loop Timing Error Variance (chips²)
Baseband, $\Delta = T_c/2$	$\left(\dfrac{\sigma_\epsilon}{T_c}\right)^2 = \dfrac{N_0 B_L}{2P}$
Noncoherent, $\Delta = T_c/2$ (Standard multiplier type)	$\left(\dfrac{\sigma_\epsilon}{T_c}\right)^2 = \dfrac{N_0 B_L}{2\alpha P}\left[\dfrac{\alpha'}{\alpha} + \dfrac{2N_0 B'}{\alpha P}\right]$
Postdetection combining ($\Delta = T_c/2$) $\sigma_\epsilon\ N_0 B_L\ \ \alpha'\ \ 2N_0 B'$ (Balanced Quadriphase)	$\left(\dfrac{\sigma_\epsilon}{T_c}\right)^2 = \dfrac{N_0 B_L}{2\alpha P}\left(\dfrac{\alpha'}{\alpha} + \dfrac{4N_0 B'}{\alpha P}\right)$
Time-Shared, $\Delta = T_c/2$	$\left(\dfrac{\sigma_\epsilon}{T_c}\right)^2 = \dfrac{N_0 B_L}{\alpha P}\left(\dfrac{\alpha'}{\alpha} + \dfrac{2N_0 B'}{\alpha P}\right)$
τ-Dither, $\Delta = 0.1 T_c$	$\left(\dfrac{\sigma_\epsilon}{T_c}\right)^2 = \dfrac{N_0 B_L}{\alpha P}\left(\dfrac{\alpha'}{\alpha} + \dfrac{N_0 B'}{\alpha P}\right)$
Time gated (d_F = duty factor)	$\left(\dfrac{\sigma_\epsilon}{T_c}\right)^2 = \dfrac{N_0 B_L}{2\alpha\, d_F P}\left(\dfrac{\alpha'}{\alpha} + \dfrac{2N_0 B'}{\alpha\, d_F P}\right)$

and $L'(f)$ is the baseband equivalent filter of the bandpass filter that has transfer function $L(f)$.

The resulting code tracking (timing error variance) performance is indicated below in Table 10.1 for all the loops considered, plus the time gated loop of the next section.

10.5 TIME-GATED CODE TRACKING LOOPS

So far we have been concerned with tracking constant envelope signals. We now turn our attention to time-gated or pulsed-envelope signals such as are used in time division multiple-access (TDMA) systems. The approach presented here differs from Spilker [2], and Huff and Reinhard [12], who have also treated time-gated loops.

It is assumed that the received time-gated signal is in synchronization with the receiver time gating and that the switch rate $1/T_0$ is much greater than B_L. The gating function is shown in Figure 10.13. The gating pulse is on for $d_F T_0$ sec, where d_F, the duty factor, is off for the remaining $(1 - d_F)T_0$ sec, with T_0 the period of one cycle of the gating function. The

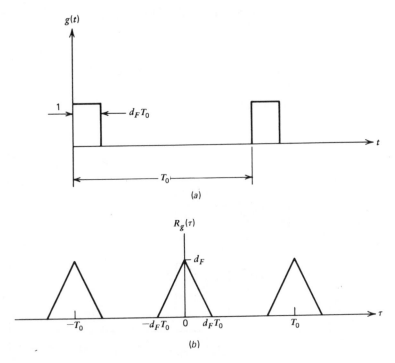

Figure 10.13 Gating function and its autocorrelation function (a) Gating function. (b) Autocorrelation function.

gated, received, signal is modeled by

$$y(t) = \sqrt{2P} \, d(t)\mathrm{PN}(t - T)g(t)\cos(\omega_0 t + \theta) + g(t)n(t) \qquad (10.5\text{-}1)$$

In Figure 10.14 the time-gated code tracking loop model is illustrated. We now briefly indicate the calculations necessary to determine the variance of the timing error variance.

The two correlator outputs are given by

$$\tilde{e}_L = \sqrt{2P} \, L(s)\left[d(t)g(t)R_{PN}\left(T - \hat{T} - \frac{T_c}{2}\right)\cos(\omega_0 t + \theta)\right] + n_L(t)$$
$$\hspace{6.5cm} (10.5\text{-}2)$$
$$\tilde{e}_E = \sqrt{2P} \, L(s)\left[d(t)g(t)R_{PN}\left(T - \hat{T} + \frac{T_c}{2}\right)\cos(\omega_0 t + \theta)\right] + n_E(t)$$

where $L(s)$ is the Heaviside operator representing the bandpass filters and

$$n_L(t) = L(s)\left[n(t)\mathrm{PN}\left(t - \hat{T} - \frac{T_c}{2}\right)\right]$$
$$\hspace{6.5cm} (10.5\text{-}3)$$
$$n_E(t) = L(s)\left[n(t)\mathrm{PN}\left(t - \hat{T} + \frac{T_c}{2}\right)\right]$$

After squaring and forming the difference, we obtain

$$e_\epsilon = PL'(s)[d^2(t)g(t)]\left[R_{PN}^2\left(T - \hat{T} - \frac{T_c}{2}\right) - R_{PN}^2\left(T - \hat{T} + \frac{T_c}{2}\right)\right]$$
$$+ 2\sqrt{2P} \, L(s)[\tilde{d}(t)g(t)\cos(\omega_0 t + \theta]$$
$$\times \left[n_L(t)g(t)R_{PN}\left(T - \hat{T} - \frac{T_c}{2}\right) - n_E(t)g(t)R_{PN}\left(T - \hat{T} + \frac{T_c}{2}\right)\right]$$
$$+ n_L^2(t)g(t) - n_E^2(t)g(t) \qquad (10.5\text{-}4)$$

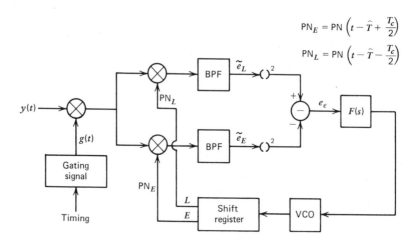

$$PN_E = PN\left(t - \hat{T} + \frac{T_c}{2}\right)$$
$$PN_L = PN\left(t - \hat{T} - \frac{T_c}{2}\right)$$

Figure 10.14 Time gated code tracking loop.

where $L'(f)$ is the baseband equivalent filter response of the bandpass filter response $L(f)$. Neglecting harmonics outside the loop bandwidth B_L allows us to write

$$e_\epsilon \cong \frac{2P\epsilon}{T_c} \alpha \, d_F + n_1(t) + n_2(t) \qquad (10.5-5)$$

where

$$n_1(t) = 2\sqrt{P} \, \tilde{d}(t)g(t)\left[n_{L_c}(t)R_{PN}\left(T - \hat{T} - \frac{T_c}{2}\right) - n_{E_c}(t)R_{PN}\left(T - \hat{T} + \frac{T_c}{2}\right)\right]$$

$$(10.5-6)$$

$$n_2(t) = n_L^2(t)g(t) - n_E^2 g(t) \qquad (10.5-7)$$

and it has been assumed that the data symbol rate $1/T_s$ is much larger than the gating rate $1/T_0$. The noise process $n_{L_c}(t)$ and $n_{E_c}(t)$ are defined in Section 10.2.2.

With the assumption that the gating process is independent of the noise processes we find that $R_{n_1}(t)$ is given by

$$R_{n_1}(\tau) = 2PR_{\tilde{d}}(\tau)R_{n_{L_c}}(\tau)R_g(\tau) \qquad (10.5-8)$$

In order to compute $\mathcal{S}_{n_1}(0)$ it is necessary to determine the autocorrelation function of $g(t)$. The autocorrelation function, $R_g(\tau)$, is shown in Figure 10.13. Therefore defining, as before,

$$\Lambda_{d_F}(\tau) = \begin{cases} \left(1 - \dfrac{|\tau|}{d_F T_0}\right) & |\tau| \leq d_F T_0 \\ 0 & \text{elsewhere} \end{cases} \qquad (10.5-9)$$

we have

$$R_g(\tau) = d_F \Lambda_{d_F}(\tau) * \sum_{n=-\infty}^{\infty} \delta(\tau + nT_0) \qquad (10.5-10)$$

Consequently

$$\mathcal{S}_{n_1}(0) = 2P \int_{-\infty}^{\infty} R_{\tilde{d}}(\tau) \left\{d_F\left[\Lambda_{d_F}(\tau)\right] * \sum_{n=-\infty}^{\infty} \delta(\tau + nT_0)\right\} d\tau \qquad (10.5-11)$$

or since

$$\mathcal{F}\{f_1(t) * f_2(t)\} = F_1(f)F_2(f) \qquad (10.5-12)$$

we have

$$\mathcal{S}_{n_1}(0) = 2P \int_{-\infty}^{\infty}\int_{-\infty}^{\infty} \mathcal{S}_{n_{L_c}}(f')\mathcal{S}_{\tilde{d}}(f - f') \, df'[d_F T_0 \, \text{sinc}^2(\pi f \, d_F T_0)]$$

$$\times \frac{1}{T_0} \sum_{n=-\infty}^{\infty} \delta\left(f + \frac{1}{T_0}\right) df \qquad (10.5-13)$$

Evaluating

$$\mathcal{S}_{n_1}(0) = 2d_F^2 P \int_{-\infty}^{\infty} \sum_{n=-\infty}^{\infty} \mathcal{S}_{n_{L_c}}(f') \mathcal{S}_d \left(\frac{n}{T_0} - f'\right) \text{sinc}^2(\pi n d_F) \, df'$$

$$(10.5\text{-}14)$$

by our assumption, that the symbol rate was much larger than the gating rate, we obtain

$$\mathcal{S}_{n_1}(0) = N_0 d_F^2 P \int_{-\infty}^{\infty} |L'(f)|^4 \mathcal{S}_d(f) \, df \left[\sum_{n=-\infty}^{\infty} \text{sinc}^2(\pi n \, d_F)\right] \quad (10.5\text{-}15)$$

Consider the summation

$$S = \sum_{n=-\infty}^{\infty} \frac{\sin^2[n(\pi d_F)]}{(n\pi d_F)^2} = \frac{1}{\pi^2 d_F^2} \sum_{n=-\infty}^{\infty} \frac{\sin^2(n\theta)}{n^2} \quad (10.5\text{-}16)$$

Now let $\theta = \pi d_F$, so that we have

$$S = \frac{1}{\pi^2 d_F^2} \left(\pi^2 d_F^2 + 2 \sum_{n=1}^{\infty} \frac{\sin^2 n\theta}{n^2}\right) \quad (10.5\text{-}17)$$

From Jolly, number 520 [11] we have

$$\sum_{1}^{\infty} \frac{\sin^2 n\theta}{n^2} = \tfrac{1}{2}\theta(\pi - \theta) \qquad 0 \leq \theta \leq \pi \quad (10.5\text{-}18)$$

So that finally we have simply

$$S = \frac{1}{d_F} \quad (10.5\text{-}19)$$

Therefore

$$S_{n_1}(0) = d_F P N_0 \alpha' \quad (10.5\text{-}20)$$

where

$$\alpha' = \int_{-\infty}^{\infty} |L'(f)|^4 \mathcal{S}_d(f) \, df \quad (10.5\text{-}21)$$

Now we consider the noise times noise term. From Section 10.2.2 and the Appendix we obtain

$$R_{n_2}(\tau) = N_0^2 \left(\int_{-\infty}^{\infty} |L(f)|^2 e^{-i2\pi fT} \, df\right)^2 R_g(\tau) \quad (10.5\text{-}22)$$

Therefore

$$\mathcal{S}_{n_2}(0) = N_0^2 \int_{-\infty}^{\infty} \int_{-\infty}^{\infty} \int_{-\infty}^{\infty} |L(f'')|^2 |L(f')|^2 e^{-i2\pi f'\tau} e^{i2\pi f''\tau} R_g(\tau) \, d\tau \quad (10.5\text{-}23)$$

Continuing

$$\mathcal{S}_{n_2}(0) = N_0^2 d_F^2 \int_{-\infty}^{\infty} \int_{-\infty}^{\infty} |L(f'')|^2 |L(f')|^2 \int_{-\infty}^{\infty} \left\{\int_{-\infty}^{\infty} e^{i(f''-f'+f)\tau} \, d\tau\right\}$$

$$\times \text{sinc}^2(\pi f d_F T_0) \sum_{n=-\infty}^{\infty} \delta\left(f + \frac{n}{T_0}\right) df \, df' \, df'' \quad (10.5\text{-}24)$$

Finally

$$\mathcal{S}_{n_2}(0) = N_0^2 d_F^2 \sum_{n=-\infty}^{\infty} \int_{-\infty}^{\infty} |L(f')|^2 \left| L\left(f' + \frac{n}{T_0}\right) \right|^2 \text{sinc}^2(\pi n d_F) \, df' \tag{10.5-25}$$

From above we then arrive at

$$\mathcal{S}_{n_2}(0) = d_F N_0^2 (2B') \tag{10.5-26}$$

with

$$2B' = \int_{-\infty}^{\infty} |L(f)|^4 \, df \tag{10.5-27}$$

Consequently the timing error variance is given by

$$\left(\frac{\sigma_\epsilon}{T_c}\right)^2 = \frac{N_0 B_L}{2\alpha(d_F P)}\left[\frac{\alpha'}{\alpha} + \frac{2N_0 B'}{\alpha(d_F P)}\right] \quad \text{chips}^2 \tag{10.5-28}$$

Notice under the assumption that $B_L \ll 1/T_0 \ll 1/T_s$, the loop performance is equivalent to a nongated loop with carrier to noise spectral density of $d_F P / N_0$! Consequently $d_F P$ is the effective continuous power in the loop. And clearly the degradation is d_F expressed in decibels.

PROBLEM 10

Show that if the noise is not gated out when the signal is not present the timing error jitter increases over that of equation 10.5-28.

10.6 SELF-NOISE EFFECTS

When assessing the ultimate performance of a PN coded spread spectrum system it is necessary to consider self-noise effects, that is, signal induced noise effects. In order to estimate the effects of the self-noise we draw upon the results of Section 8.12.2 concerning the spectra of the product of a PN waveform times a delayed version of the PN waveform.

The power spectral density of this product (baseband) despread signal is given by (unit power)

$$\mathcal{S}(f) = \left(1 - \frac{|\epsilon|}{T_c}\frac{N+1}{N}\right)^2 \delta(f) + \frac{N+1}{N}\left(\frac{\epsilon}{T_c}\right)^2 \sum_{\substack{n=-\infty \\ n \neq 0}}^{\infty} \text{sinc}^2\left(\frac{n\pi\epsilon}{T_c}\right) \delta\left(f + \frac{n}{T_c}\right)$$

$$+ \frac{N+1}{N^2}\left(\frac{\epsilon}{T_c}\right)^2 \sum_{\substack{n=-\infty \\ n \neq 0}}^{\infty} \text{sinc}^2\left(\frac{n\pi\epsilon}{NT_c}\right) \delta\left(f + \frac{n}{NT_c}\right) \tag{10.6-1}$$

where N is the code length in chips and T_c is the chip duration. Hence, if the received signal plus noise, translated to IF, is given by

$$y(t) = \sqrt{2P}\,\text{PN}(t - T)\sin(\omega_0 t + \theta) + n(t) \tag{10.6-2}$$

then the despread IF signal is given by

$$u(t) = \sqrt{P}\,\text{PN}(t - T)\text{PN}(t - \hat{T})\sin(\omega_0 t + \theta) + n(t)\text{PN}(t - \hat{T})$$

$$(10.6\text{-}3)$$

The power spectral density around f_0 is sketched in Figure 10.15, assuming that the code and carrier are not coherent. As a result we can define the signal to interference noise ratio in a bandwidth B assuming both that the line spacing $1/NT$ is much less than B Hz and that $B < 1/T_c$. Define $\epsilon = T - \hat{T}$, the timing error. Under these circumstances we obtain

$$\text{SNR} = \frac{\left(\dfrac{P}{2}\right)\left(1 - \dfrac{|\epsilon|}{T_c}\dfrac{N+1}{N}\right)^2}{\dfrac{N_0}{2}B + \left(\dfrac{P}{2}\right)\dfrac{N+1}{N^2}\left(\dfrac{\epsilon}{T_c}\right)^2\dfrac{B}{1/NT_c}} \qquad (10.6\text{-}4)$$

Hence at large $P/N_0 B$ and large N we obtain the limiting SNR due to self-noise:

$$\text{SNR}_{\text{max}} = \frac{[1 - (|\epsilon|/T_c)]^2}{(\epsilon/T_c)^2 B T_c} \qquad (10.6\text{-}5)$$

For example, if $|\epsilon|/T_c = 0.1$ and $B = 0.1/T_c$, then the thermal noise-free SNR is 27.25 dB. The above result gives the maximum signal to noise ratio for the punctual channel, which, for example, is applicable to data demodulation.

We now turn our attention to the limiting performance of a standard 1-Δ noncoherent code tracking loop. If the input is composed of WGN noise plus a PN modulated carrier, then it is easy to show that the reference multiplier output is given by

$$\text{PN}(t - \hat{T})y(t) = \sqrt{2P}\,\text{PN}(t - T)\text{PN}(t - \hat{T})\sin \omega_0 t + n'(t) \quad (10.6\text{-}6)$$

where $n'(t)$ has spectral density $N_0/2$ and the PN cross PN modulated carrier has an autocorrelation function given by

$$R_{\text{PN}\times\text{PN}}(\tau)\cos \omega_0 \tau \qquad (10.6\text{-}7)$$

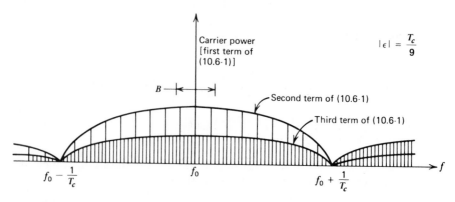

Figure 10.15 Two-sided power spectral density for $|\epsilon| = T_c/9$.

assuming that the code and carrier are not coherent. In order to obtain the limiting code tracking performance it is necessary to find the equivalent noise spectral density out of the early and late correlators. In general, for arbitrary timing error, the spectral density of the thermal noise plus self-noise is given by

$$\mathscr{S}(f_0) = \frac{N_0}{2} + P \int_{-\infty}^{\infty} R_{PN \times PN}(\tau) \left(\frac{e^{j\omega_0\tau} + e^{-j\omega_0\tau}}{2} \right) e^{j\omega_0\tau} \, d\tau$$

$$\cong \frac{N_0}{2} + \frac{P}{2} \frac{1}{N} \left(\frac{\epsilon}{T_c} \right)^2 \frac{1}{1/NT_c} \qquad |\epsilon| \le T_c \qquad (10.6\text{-}8)$$

where we have assumed that the discrete spectrum self-noise has continuous power spectral density given by the spectral line power divided by the line spacing of $\Delta f = 1/NT_c$. We assume that $1/NT_c \le B_L \ll B$ and that the data sequence is random, so it has a continuous spectral density with the data bandwidth $\gg B_L$. Since, for the 1-Δ correlator, both the early and late correlators are offset from the received code by one-half chip, the power to equivalent noise spectral density (letting $N_0 = 0$), using (10.6-8), is given by (neglecting thermal noise)

$$\mathscr{S}(f_0) = \frac{PT_c}{8} \qquad (10.6\text{-}9)$$

Using this ratio as the equivalent $P/(N_0/2)$, we have that

$$\left(\frac{P}{N_0} \right)_{eq} = \frac{4}{T_c} = 4R_c \qquad (10.6\text{-}10)$$

with R_c being the chip rate. Using the (P/N_0) equivalent value in the equation for the 1-Δ full-time code tracking loop yields for code tracking performance and infinite SNR

$$\left(\frac{\sigma_\epsilon}{T_c} \right)^2_{min} = \frac{B_L T_c}{8\alpha} \left(\frac{\alpha'}{\alpha} + \frac{BT_c}{2\alpha} \right) \qquad \text{chips}^2 \qquad (10.6\text{-}11)$$

Typically the jitter due to self-noise (equation 10.6-11) is small compared to the thermal noise induced value.

It is to be noted that oscillator phase noise of the clock has not been included in this limiting performance analysis, and neither have other sources of noise in the mixers, amplifiers, etc.

References 12 and 13 should also be consulted for additional thoughts on the subject of self-noise in code tracking loops.

10.7 SINGLE-TONE INTERFERENCE EFFECTS ON CODE TRACKING

This section is concerned with the effects of tone interference on the performance of code tracking loops. We shall assume that the loop is a noncoherent 1-Δ loop (see Figure 10.4). Boyarski [14] has considered multiple interferers in a code loop by modeling each carrier as a narrow-band Gaussian process. He neglects the influence of thermal noise. A Philco-Ford report [15] on the Global Positioning System (GPS) provides a result, without derivation, on the effect of a tone interferer on tracking jitter performance. We attempt here to estimate the effect on code tracking performance by including its effect on both the $s \times n$ and the $n \times n$ terms.

Consider again the 1-Δ noncoherent code tracking loop of Figure 10.4. We model the input waveform by $y(t)$, where

$$y(t) = \sqrt{2P}\, d(t)\text{PN}(t - T)\cos(\omega_0 t + \theta_0) + \sqrt{2P_1}\cos(\omega_1 t + \theta_1) + n(t)$$

$$(10.7\text{-}1)$$

where the unit amplitude binary valued data sequence is represented by $d(t)$, $\text{PN}(t - T)$ is the PN waveform having a delay of T sec, P and P_1 are the respective powers of the signal and the interference, and finally $n(t)$ is modeled as a WGN process.

Out of the early and late correlators we have

$$\tilde{e}_L(t) = \sqrt{2P}\, L(s)\left[d(t)R_{PN}\left(T - \hat{T} - \frac{T_c}{2}\right)\cos(\omega_0 t + \theta_0)\right] + \sqrt{2P_1}\, L(s)$$

$$\times \left[\text{PN}\left(t - \hat{T} - \frac{T_c}{2}\right)\cos(\omega_1 t + \theta_1)\right] + n_L(t)$$

$$(10.7\text{-}2)$$

$$\tilde{e}_E(t) = \sqrt{2P}\, L(s)\left[d(t)R_{PN}\left(T - \hat{T} + \frac{T_c}{2}\right)\cos(\omega_0 t + \theta_0)\right] + \sqrt{2P_1}\, L(s)$$

$$\times \left[\text{PN}\left(t - \hat{T} + \frac{T_c}{2}\right)\cos(\omega_1 t + \theta_1)\right] + n_E(t)$$

where

$$n_L(t) = L(s)\left[n(t)\text{PN}\left(t - \hat{T} - \frac{T_c}{2}\right)\right]$$

$$n_E(t) = L(s)\left[n(t)\text{PN}\left(t - \hat{T} + \frac{T_c}{2}\right)\right]$$

$$(10.7\text{-}3)$$

If we define $N_L(t)$ and $N_E(t)$ by

$$N_L(t) = \sqrt{2P_1}\, L(s)\left[\text{PN}\left(t - \hat{T} - \frac{T_c}{2}\right)\cos(\omega_1 t + \theta_1)\right] + n_L(t)$$

$$N_E(t) = \sqrt{2P_1}\, L(s)\left[\text{PN}\left(t - \hat{T} + \frac{T_c}{2}\right)\cos(\omega_1 t + \theta_1)\right] + n_E(t)$$

$$(10.7\text{-}4)$$

then letting $\epsilon = T - \hat{T}$, as before, we obtain

$$e_\epsilon = \tilde{e}_L^2 - \tilde{e}_E^2 = P[\tilde{d}(t)]^2\left[R_{PN}^2\left(\epsilon - \frac{T_c}{2}\right) - R_{PN}^2\left(\epsilon + \frac{T_c}{2}\right)\right]$$

$$+ 2\sqrt{2P}\, \tilde{d}(t)\cos(\omega_0 t + \theta_0)\left[N_L(t)R_{PN}\left(\epsilon - \frac{T_c}{2}\right)\right.$$

$$\left. - N_E(t)R_{PN}\left(\epsilon + \frac{T_c}{2}\right)\right]$$

$$+ N_L^2(t) - N_E^2(t)$$

$$(10.7\text{-}5)$$

We have used the approximate relationship

$$L(s)[d(t)\cos(\omega_0 t + \theta_0)] = \tilde{d}(t)\cos(\omega_0 t + \theta_0) \qquad (10.7\text{-}6)$$

where $\tilde{d}(t) = L'(s)[d(t)]$, with $\tilde{d}(t)$ being the baseband equivalent response due to the baseband equivalent filter transfer function $L'(s)$. Just as in the case the interference is not present, we can describe the output of the difference circuit as

$$e_\epsilon = 2\alpha P \frac{\epsilon}{T_c} + n_1(t) + n_2(t) \qquad (10.7\text{-}7)$$

with $\alpha = E[\tilde{d}(t)]^2$, where $\tilde{d}(t)$ is assumed to be an independent, equally likely, random NRZ sequence. In (10.7-7) $n_1(t)$ is the second term in (10.7-5), and $n_2(t)$ is the third term in (10.7-5).

The first noise term can be written, for ϵ small, as

$$n_1(t) = \sqrt{2P}\ \tilde{d}(t) \left[\sqrt{2}\, n_{L_c}(t)(\tfrac{1}{2}) + \sqrt{2P_I}\widetilde{PN}\left(t - \hat{T} - \frac{T_c}{2}\right)(\tfrac{1}{2}) \right.$$
$$\left. - \sqrt{2}\, n_{E_c}(t)(\tfrac{1}{2}) - \sqrt{2P_I}(\tfrac{1}{2})\widetilde{PN}\left(t - \hat{T} + \frac{T_c}{2}\right) \right] \qquad (10.7\text{-}8)$$

where we have assumed that $\omega_0 = \omega_1$ and $\theta_0 = \theta_1$ (coherent interferer). We have also used

$$\widetilde{PN}\left(t - \hat{T} - \frac{T_c}{2}\right) = L'(s)\left[PN\left(t - \hat{T} - \frac{T_c}{2}\right)\right]$$
$$\widetilde{PN}\left(t - \hat{T} + \frac{T_c}{2}\right) = L'(s)\left[PN\left(t - \hat{T} + \frac{T_c}{2}\right)\right] \qquad (10.7\text{-}9)$$

Hence the autocorrelation function of $n_1(t)$ is given by (using the respective statistical independence of the processes)

$$R_{n_1}(\tau) = PR_{\tilde{d}}(\tau)[R_{n_{L_c}}(\tau) + P_I R_{\widetilde{PN}}(\tau) + R_{n_{E_c}}(\tau) + P_I R_{\widetilde{PN}}(\tau)] \quad (10.7\text{-}10)$$

Since the PN waveforms are separated by one chip, we modeled them as independent processes. Now collecting terms yields

$$R_{n_1}(\tau) = 2PR_{\tilde{d}}(\tau)[R_{n_{L_c}}(\tau) + P_I R_{\widetilde{PN}}(\tau)] \qquad (10.7\text{-}11)$$

Therefore

$$\mathcal{S}_{n_1}(0) = 2P\left[\int_{-\infty}^{\infty} R_{\tilde{d}}(\tau)R_{n_{L_c}}(\tau)\, d\tau + P_I \int_{-\infty}^{\infty} R_{\tilde{d}}(\tau)R_{\widetilde{PN}}(\tau)\, d\tau\right]$$
$$(10.7\text{-}12)$$

If we assume the PN code is very long, then we can model it as a random, independent, NRZ process. Hence

$$\mathcal{S}_{n_1}(0) = N_0 \alpha' P + 2PP_I T_c \alpha' \qquad (10.7\text{-}13)$$

if the data sequence is assumed to have a continuous power spectrum, so that the resulting spectrum of $n_1(t)$ is continuous. We define α', as before, by

$$\alpha' = \int_{-\infty}^{\infty} \mathscr{S}_d(f)|L'f|^4 \, df \qquad (10.7\text{-}14)$$

Now we turn our attention to the second noise term, $n_2(t)$. We have

$$N_L^2(t) - N_E^2(t) = \left[\sqrt{2P_I}\,\widetilde{PN}\left(t - \hat{T} - \frac{T_c}{2}\right)\cos(\omega_0 t + \theta_0) + \sqrt{2}n_{L_c}\cos(\omega_0 t + \theta_0) \right.$$
$$\left. + \sqrt{2}n_{L_s}\sin(\omega_0 t + \theta_0) \right]^2$$
$$- \left[\sqrt{2P_I}\,\widetilde{PN}\left(t - \hat{T} + \frac{T_c}{2}\right)\cos(\omega_0 t + \theta_0) \right.$$
$$\left. + \sqrt{2}n_{E_c}\cos(\omega_0 t + \theta_0) + \sqrt{2}n_{E_c}\sin(\omega_0 t + \theta_0) \right]^2 \qquad (10.7\text{-}15)$$

where $n_{L_c}(t)$, $n_{L_s}(t)$, $n_{E_c}(t)$, and $n_{E_L}(t)$ are defined in Section 10.2.2. Hence we obtain

$$N_L^2(t) - N_E^2(t) \cong 2\sqrt{P_I}\,\widetilde{PN}\left(t - \hat{T} - \frac{T_c}{2}\right)n_{L_c}(t) + n_{L_c}^2(t) + n_{L_s}^2(t)$$
$$- 2\sqrt{P_I}\,\widetilde{PN}\left(t - \hat{T} + \frac{T_c}{2}\right)n_{E_c}(t) - n_{E_c}^2(t) - n_{E_s}^2(t) + O(2\omega_0) \qquad (10.7\text{-}16)$$

We neglect the $2\omega_0$ terms since they are filtered out in the loop filter. Approximating each noise term as an independent filtered WGN process allows us to write (using the Appendix)

$$S_{n_2}(0) = 4P_I \int_{-\infty}^{\infty} R_{\widetilde{PN}}(\tau)R_{n_{L_c}}(\tau) \, d\tau + 2N_0^2 B' \qquad (10.7\text{-}17)$$

where

$$2B' = \int_{-\infty}^{\infty} |L(f)|^4 \, df \qquad (10.7\text{-}18)$$

Again modeling the PN waveform as a random NRZ waveform with equally likely values of 1 and -1 leaves us with

$$S_{n_2}(0) = 4N_0 P_I T_c B' + 2N_0^2 B' \qquad (10.7\text{-}19)$$

with

$$B' = \int_{-\infty}^{\infty} |L'(f)|^2 \, df = \frac{1}{2}\int_{-\infty}^{\infty} |L(f)|^2 \, df \qquad (10.7\text{-}20)$$

where we have used the narrowband symmetrical BPF assumption.

We finally obtain the resulting timing error variance using (10.7-7), (10.7-3), and (10.7-19):

$$\left(\frac{\sigma_\epsilon}{T_c}\right)^2 = \frac{N_0 B_L}{2\alpha P}\left[\frac{\alpha'}{\alpha}\left(1 + \frac{2P_I T_c}{N_0}\right) + \frac{2N_0 B'}{\alpha P}\left(1 + \frac{2P_I T_c B}{N_0 B}\right)\right] \quad \text{chips}^2 \qquad (10.7\text{-}21)$$

We notice that if

$$\frac{P_I T_c}{N_0} \simeq \frac{P_I}{N_0 B_{PN}} = \text{SNR}_I$$

is small, then the degradation in terms of additional filter is negligible. Also, if $P_I = 0$, then the result reduces to the one derived in Section 10.2.2.

PROBLEM 11

Generalize this result to many tone interferers assuming that they are all of equal power and are not coherently related.

10.8 MULTIPATH EFFECTS ON CODE TRACKING

The multipath effect (see references 16 and 17) is a phenomenon associated with, for example, the transmission of a signal from a stationary satellite to an earth orbiting satellite for which the line of sight path between satellites is close to the earth, as shown in Figure 10.16. As a consequence, the reflected path is not much longer than the direct path, and in addition the

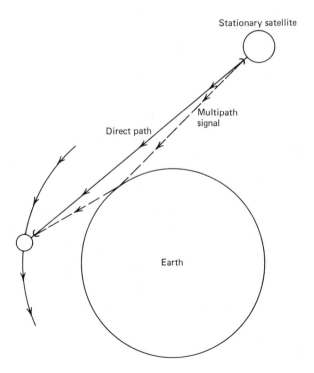

Figure 10.16 Example of multipath due to reflection.

angular difference between paths is small, which gives rise to the multipath phenomenon.

Multipath effects lead to two distinct types of interference. The first is called *specular multipath* and is modeled as a single delayed version of the PN modulated signal. The second type of multipath is called *diffuse multipath*. This type of multipath can be viewed as the sum of many delayed replicas of the received signal, with the energy spread over time considerably beyond one chip.

Specular multipath is a problem both for acquisition and tracking. In acquisition, for example, if the direct (true) path signal is missed, the specular component, being delayed in time, might be acquired, which would in turn provide a false lock point that could yield erroneous range estimates. We have assumed that the multipath signal was far enough away from the true signal in time so that either one or the other would be detected separately.

Now consider what happens when the received signal is followed by a delayed spectral multipath component. Consider the code loop detector of Figure 10.17. The correlated and detected output of both early and late channels, in the absence of noise, is given by

$$
\begin{aligned}
y_L^2(\epsilon) &= PR_{PN}^2\left(\epsilon - \frac{T_c}{2}\right) + P_I R_{PN}^2\left(\epsilon + T_d - \frac{T_c}{2}\right) + 2R_{PN}\left(\epsilon - \frac{T_c}{2}\right) \\
&\quad \times R_{PN}\left(\epsilon + T_d - \frac{T_c}{2}\right)\cos\theta_M \\
y_E^2(\epsilon) &= PR_{PN}^2\left(\epsilon + \frac{T_c}{2}\right) + P_I R_{PN}^2\left(\epsilon + T_d + \frac{T_c}{2}\right) + 2R_{PN}\left(\epsilon + \frac{T_c}{2}\right) \\
&\quad \times R_{PN}\left(\epsilon + T_d + \frac{T_c}{2}\right)\cos\theta_M
\end{aligned}
\tag{10.8-1}
$$

where the received signal is given by

$$
\begin{aligned}
y(t) &= \sqrt{2P}\,PN(t - T)\cos(\omega_0 t + \theta_0) \\
&\quad + \sqrt{2P_I}\,PN(t - T - T_d)\cos(\omega_0 t + \theta_0 + \theta_M)
\end{aligned}
\tag{10.8-2}
$$

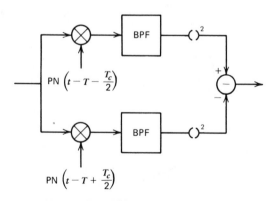

Figure 10.17   Error detector for a 1-Δ code loop.

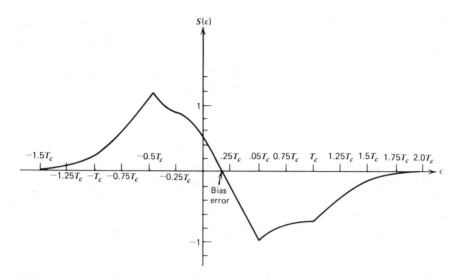

Figure 10.18 Resulting S-curve showing bias error ($P_I = 0.5P$, $\theta_M = \pi/2$, and $T_d = T_c/2$).

The received direct signal power is P, the multipath received power is P_I, and the rf phase difference between the two signals is θ_M.

The equations of (10.8-1) can be used to estimate the resulting S-curve given by

$$S(\epsilon) = y_L^2(\epsilon) - y_E^2(\epsilon) \qquad (10.8\text{-}3)$$

For example, when $\theta_M = \pi/2$, $P_I = 0.5P$, and $T_d = T_c/2$. The resulting S-curve is plotted in Figure 10.18. As can be seen for this case, an 18% static timing error results. Cahn [16] has observed that for some combinations of P_I/P, T_d, and θ_M either an apparent lead or a lag from the true time can occur. It is to be pointed out that for convenience of plotting, no precorrelation filter was assumed. With this filter, the results would be somewhat distorted. Furthermore, even without a multipath interfering signal, the precorrelation filter will produce an apparent delay of around a quarter of a chip, depending upon the filter bandwidth. This, of course, can be calibrated out.

Finally, diffuse multipath tends to appear as additional channel noise and, because of processing gain, normally would have a minimal effect on code tracking performance.

10.9 CODE LOOP TRACKING PERFORMANCE WITHOUT NOISE

So far we have discussed the performance of various tracking loops when noise is the source of error in tracking. Next we discuss the effects of dynamics, on linearized loops, based on the commonly used polynomial doppler model, which is the same as was used in the phase locked loop chapter.

It is not difficult to show that the timing error, for all code tracking loops having arbitrary order loop filters with zero input noise, is described by

$$\frac{\epsilon}{T_c} = \frac{1}{1 + GF(s)/s} \left(\frac{T_i}{T_c}\right) \tag{10.9-1}$$

where ϵ/T_c is the relative timing error, in chips; G is loop gain; $F(s)$ is the Laplace transform of the loop filter impulse response; and T_i/T_c is the relative input timing process. The gain G is the product of K_d, the S-curve gain in V/chip, and K_{VCO} in chips/sec-V.

Now consider a specific filter model for the passive second-order loop of the form

$$F(s) = \frac{1 + \tau_2 s}{1 + \tau_1 s} \tag{10.9-2}$$

Hence

$$H(s) = \frac{G[(1 + \tau_2 s)/(1 + \tau_1 s)]}{s + G[(1 + \tau_2 s)/(1 + \tau_1 s)]} \tag{10.9-3}$$

or

$$H(s) = \frac{1 + \tau_2 s}{1 + (\tau_2 + 1/G)s + (\tau_1/G)s^2} \tag{10.9-4}$$

If we define

$$\tau_2 = \frac{\sqrt{2}}{\omega_n} \quad \text{and} \quad \zeta = \tfrac{1}{2}\omega_n\left(\tau_2 + \frac{1}{G}\right) \tag{10.9-5}$$

then ζ is the loop damping factor and ω_n is the natural frequency. From (10.9-4) and (10.9-5) we have

$$H(s) = \frac{1 + 2\zeta s/\omega_n - s/G}{1 + 2\zeta(s/\omega_n) + (s/\omega_n)^2} \tag{10.9-6}$$

It follows that

$$\frac{\epsilon}{T_c} = [1 - H(s)]\frac{T_i}{T_c} = \frac{(s/\omega_n)^2 + s/G}{1 + 2\zeta(s/\omega_n) + (s/\omega_n)^2}\left(\frac{T_i(s)}{T_c}\right) \tag{10.9-7}$$

Now model the input timing signal as

$$\frac{T_i}{T_c} = T_0 + \dot{T}t + \tfrac{1}{2}\ddot{T}t^2 \quad \text{chips} \tag{10.9-8}$$

The corresponding transform is given by

$$\frac{T_i(s)}{T_c} = \frac{T_0}{s} + \frac{\dot{T}}{s^2} + \frac{\ddot{T}}{s^3} \quad \text{chips} \tag{10.9-9}$$

By use of the final value theorem we deduce that

$$\lim_{t \to \infty} \frac{\epsilon(t)}{T_c} = \lim_{s \to 0} \left\{ \frac{s[(s/\omega_n)^2 + s/G]}{1 + (2\zeta/\omega_n)s + (s/\omega_n)^2} \right\} \qquad (10.9\text{-}10)$$

Hence for large t we have asymptotically

$$\frac{\epsilon(t)}{T_c} \cong \frac{T}{G} + \frac{\ddot{T}}{\omega_n^2} + \frac{\ddot{T}t}{G} \qquad \text{chips} \qquad (10.9\text{-}11)$$

An ideal second-order loop would have $G \to \infty$, so that for large t,

$$\frac{\epsilon(t)}{T_c} = \frac{\ddot{T}}{\omega_n^2} \qquad (10.9\text{-}12)$$

The same procedure can be extended to both higher-order polynomial dynamics and higher-order loop structures.

PROBLEM 12

Determine the asymptotic expression for timing error using an input of the form

$$T_i = T_0 + \dot{T}t + \tfrac{1}{2}\ddot{T}t^2 + \tfrac{1}{6}\dddot{T}t^3$$

with the loop being third-order, so that

$$H(s) = \frac{2\omega_n s^2 + 2\omega_n^2 s + \omega_n^3}{s^3 + 2\omega_n^2 s + 2\omega_n s^2 + \omega_n^3}$$

The above calculations give an indication of what minimum loop bandwidth and loop gain are needed to adequately maintain code tracking during the period of time t in which the dynamics occur. Naturally, if the actual orbit parameters, such as range, range rate, range acceleration, jerk ($\dddot{R}$), etc., are known, they can be programmed into a digital computer that models the closed-loop response. This allows prediction of the maximum timing error.

Next the characteristics of acquisition for the code tracking loop are developed for the case in which noise is absent.

10.10 NOISELESS CODE LOOP ACQUISITION

Next we concern ourselves with the phase plane, which illustrates the relationship between the timing error and the timing error rate, thereby allowing the system designer to determine the minimum acquisition code bandwidth.

Again consider the second-order code loop with passive loop filter. In general, we can express the timing estimate $\hat{T}$ by

$$\hat{T} = \frac{GF(s)}{s}[D(\epsilon)] \qquad (10.10\text{-}1)$$

where

$$D(\epsilon) = \frac{S(\epsilon)}{S'(0)} \qquad (10.10\text{-}2)$$

with $S(\epsilon)$ being the S-curve with timing error ϵ, and $S'(0)$ the slope of $S(\epsilon)$ evaluated at $\epsilon = 0$. The term $D(\epsilon)$ is the normalized S-curve.

Letting $x = \epsilon/T_c$, $y = T/T_c$, $D(\epsilon)/T_c = D(x)$, we have from (10.10-1), after dividing both sides by T_c,

$$s(y - x) = GF(s)[D(x)] \qquad (10.10\text{-}3)$$

Define

$$g = \frac{G}{\omega_n} \qquad (10.10\text{-}4)$$

then

$$\frac{s}{\omega_n}(y - x) = gF(s)[D(x)] \qquad (10.10\text{-}5)$$

Now we evaluate $gF(s)$ for the passive filter second-order code loop. Using the equations in (10.9.5)

$$\frac{1}{G} + \tau_2 = \frac{2\zeta}{\omega_n}$$

$$\tau_1 = \frac{G}{\omega_n^2} \qquad (10.10\text{-}6)$$

with the passive filter form

$$F(s) = \frac{1 + \tau_2 s}{1 + \tau_1 s} \qquad (10.10\text{-}7)$$

Then using (10.10-4) into (10.10-7) yields (letting $G \to \infty$)

$$gF(s) = \frac{1 + 2\zeta s/\omega_n}{1/g + s/\omega_n} \qquad (10.10\text{-}8)$$

Hence

$$H(s) = \frac{1 + 2\zeta s/\omega_n}{1 + 2\zeta(s/\omega_n) + (s/\omega_n)^2} \qquad (10.10\text{-}9)$$

Defining

$$\dot{y} = \frac{1}{\omega_n}\frac{dy}{dt}$$

$$\dot{x} = \frac{1}{\omega_n}\frac{dx}{dt} \qquad (10.10\text{-}10)$$

with (10.10-5) produces

$$\frac{1}{g}\dot{y} - \frac{1}{g}\dot{x} + \ddot{y} - \ddot{x} = D(x) + 2\zeta\frac{dD(x)}{dx}(\dot{x}) \qquad (10.10\text{-}11)$$

since

$$\frac{\ddot{x}}{\dot{x}} = \frac{d\dot{x}}{dx} \qquad (10.10\text{-}12)$$

we have

$$\frac{d\dot{x}}{dx} = \frac{-D(x) - 2\zeta(dD(x)/dx)\dot{x} + \ddot{y} - (1/g)(\dot{x} - \dot{y})}{\dot{x}} \qquad (10.10\text{-}13)$$

This is the defining equation for the *phase plane*.

This equation was first derived by Spilker [19] and later used by Gill [1]. More recently, Nielsen [20] detected an error in the derivation, noting that the previous authors did not use the normalization

$$\left| \frac{dD(x)}{dx} \right|_{x=0} = 1$$

The results we derived here are based on the corrected version. Note that if $g \to \infty$, the relationship holds for the active filter second-order code loop. Notice that changing the sign of $\dot{x}$ and x simultaneously ($\ddot{y} = 0$) results in no change in $d\dot{x}/dx$. Hence, the phase plane is odd symmetric, so that determining the phase plane for $\dot{x} \geq 0$ and all x implies the phase plane for $\dot{x} \leq 0$ for all x (and $\ddot{y} = 0$).

The phase plane is generated by picking an $\omega_n \Delta t$ that implies, given $x(0)$ and $\dot{x}(0)$, that

$$\Delta x(0) = \dot{x}(0)(\omega_n \Delta t) \qquad (10.10\text{-}14)$$

Then from (10.10-13) we can compute

$$\Delta \dot{x}(0) = \frac{-D[x(0)] - 2\zeta\{dD[x(0)]/dx\}\dot{x}(0) - (1/g)[\dot{x}(0) - \dot{y}(0)]}{\dot{x}(0)} \Delta x(0)$$

$$(10.10\text{-}15)$$

So that after Δt sec

$$x(\Delta t) = x(0) + \Delta x(0)$$
$$\dot{x}(\Delta t) = \dot{x}(0) + \Delta \dot{x}(0) \qquad (10.10\text{-}16)$$

This procedure continues with an appropriately small $\omega_n \Delta t$ until the loop either moves off the restricted phase plane or it converges to the lock point ($\dot{x} = 0$). Then a new initial point $[x(0), \dot{x}(0)]$ is selected and iterated. This process is continued until a graph like Figure 10.19a is generated.

In Figure 10.19a the baseband 2-Δ infinite gain code tracking loop phase plane is illustrated. The ordinate is

$$\frac{1}{\omega_n} \frac{dx}{dt} = \frac{1}{\omega_n} \frac{d(\epsilon/T_c)}{dt}$$

so it is the code loop frequency error (chips/sec) normalized by the loop natural frequency (rad/sec). The obscissa is the code loop error $x =$

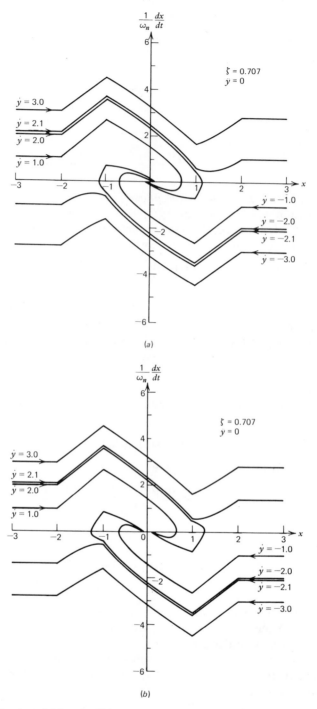

Figure 10.19 Acquisition trajectories for 2-Δ baseband loop. (*a*) Loop gain $g = \infty$. (from Huang and Holmes [6, 31]). (*b*) Loop gain $g = 10$ (from Huang and Holmes [6, 31].

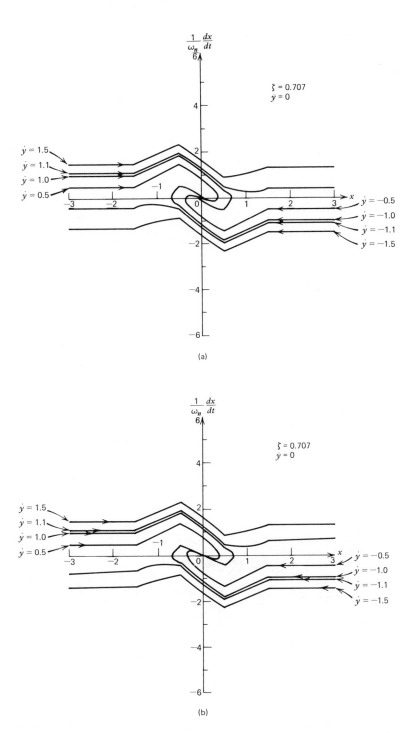

Figure 10.20 Acquisition trajectories for 1-Δ baseband loop. (*a*) Loop gain $g = \infty$ (from Huang and Holmes [6, 31]). (*b*) Loop gain $g = 10$ (from Huang and Holmes [6, 31]).

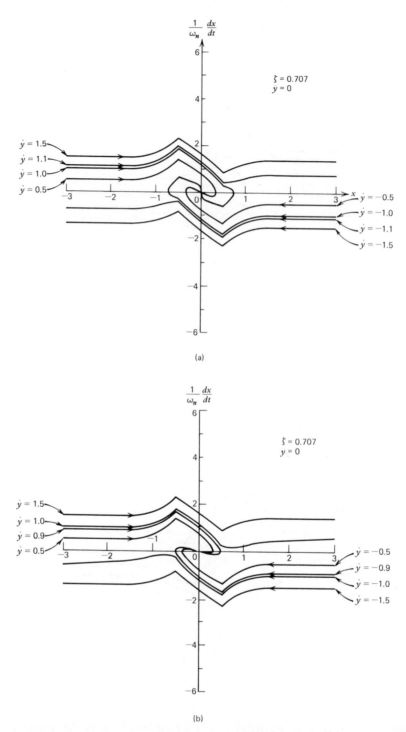

Figure 10.21 Acquisition trajectories for 1-Δ noncoherent loop. (*a*) Loop gain $g = \infty$. (*b*) Loop gain $g = 10$ (from Huang and Holmes [6, 31]).

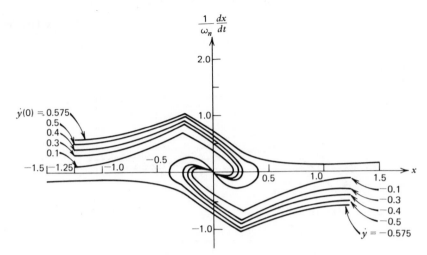

Figure 10.22 Acquisition trajectories for a 0.5-Δ noncoherent loop. Loop gain $g = \infty$ (from Huang and Holmes [6, 31]).

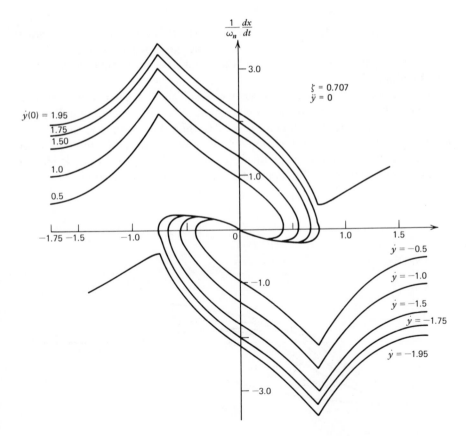

Figure 10.23 Acquisition trajectories for a 1.5-Δ noncoherent loop. Loop gain $g = \infty$ (from Huang and Holmes [6, 31]).

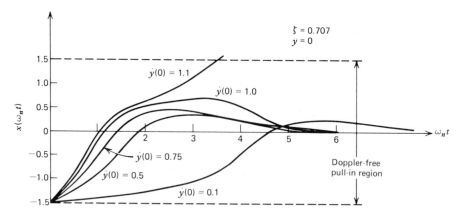

Figure 10.24 Transient response of normalized timing error $x(\omega_n t)$ for the 1-Δ noncoherent loop, with $g = \infty$ (from Huang and Holmes [6, 31]).

ϵ/T_c(chips). Note that all phase plane trajectories have *clockwise movement* since when $\dot{x}$ is greater than zero x must increase, whereas when $\dot{x}$ is less than zero x must decrease. Hence, the clockwise movement of all phase plane trajectories.

For example, if in Figure 10.19a $\dot{y} = \dot{T}/\omega_n T_c$ is 2.1 when $x = -2.0$, then following the trajectory to right we see that it just misses being pulled into lock. However, when $\dot{y} = 2.0$ and $x = -2.0$, then the trajectory (the allowed relationship between $\dot{x}$ and x), moves clockwise and settles into the point $(0, 0)$. The fact that the trajectories do not converge to $(0, 0)$ is due to accumulated errors in the computer program that generated the phase plane.

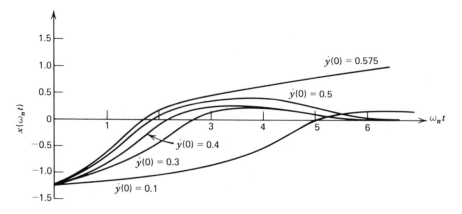

Figure 10.25 Transient response of normalized chip timing error $x(\omega_n t)$ for the 0.5-Δ noncoherent loop with $g = \infty$ (from Huang and Holmes [6, 31]).

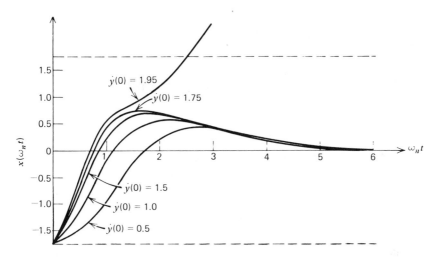

Figure 10.26 Transient response of chip timing error $x(\omega_n t)$ for the 1.5-Δ noncoherent loop, with $g = \infty$ (from Huang and Holmes [6, 31]).

In Figure 10.19 through 10-21b, phase planes are plotted for baseband and noncoherent (square law detectors) loops (see Section 10.2.2), both with values of $g = \epsilon/\omega_n$ of 10 and ∞.

In Figures 10.22 and 10.23 the phase planes for the 0.5Δ and 1.5Δ loops are illustrated from the work of Huang [6] and Huang and Holmes [31]. As is clear from the plots, the larger the early-late spacing in the loop, the better the noise-free acquisition capabilities of the code loop. However, for noise-corrupted code tracking purposes it is desirable to make the spacing as small as possible (within limits). As a consequence, code loop design is often based on a 1-Δ design utilizing a larger bandwidth for code loop acquisition and then narrowing the loop (slowly!) to the tracking bandwidth.

Although the phase plane is a useful tool in the understanding of the acquisition process of a code tracking loop, it yields no information about the time it takes to settle out the timing error to a small value. By keeping track of the updates of the phase plane trajectories in the computer, it is possible to generate the time response by plotting normalized code error versus time. Figure 10.24 to 10.26 are plots of the timing error in chips plotted against normalized time ($\omega_n t$). They are plots of infinite gain 1-Δ, 1.5-Δ and 0.5-Δ noncoherent code tracking loops. As can be seen from the plots when $t = 6/\omega_n$ or $t \simeq 4/B_L$, the loop has returned to a small timing error, given that the initial frequency error was within the allowable range. To illustrate what happens beyond the allowable range, assume for example (Figure 10.24), that $\dot{y}(0) = 1.1$ and $y(0) = -1.5$, so that after about $t = 3.5/\omega_n$ sec, the trajectory passes out beyond the region of control (± 1.5 chips), and hence acquisition is not possible.

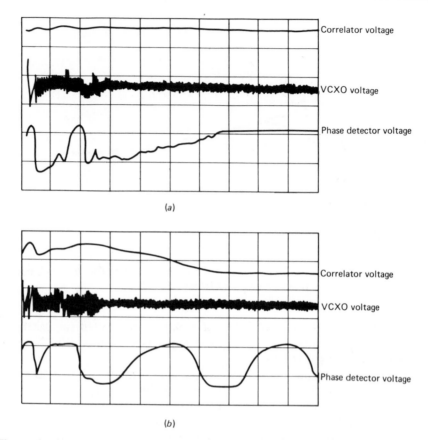

Figure 10.27 An example of noncoherent 1-Δ loop pull-in for both (*a*) success and (*b*) failure (from Hook [21]).

Actual photographs taken under the direction of Hook at TRW from the S-band shuttle code tracking loop [21] are shown in Figure 10.27, for both successful and unsuccessful acquisition. Notice that the phase detector voltage continues to oscillate when lock is not achieved but levels off when it does.

10.11 DELAY LOCK LOOP MEAN TIME TO LOSE LOCK

Threshold in a delay lock loop is difficult to define by arbitrarily specifying a specific code timing error or loop signal to noise ratio. As a consequence, we propose a more refined estimate of threshold, again arbitrary, but hopefully more satisfying. The results developed here are based on the work of Holmes and Biederman [22].

We develop the mean time to lose lock, that is, skip beyond the tracking correlation window, assuming that the code loop starts in lock with zero error. The results are derived for the first-order noncoherent code tracking loop, including both the full-time and time-shared code loop. It was assumed that the bandpass filters are ideal with rf bandwidth B Hz. The results presented here generalize a result for PLLs due to Viterbi [24].

Consider the full-time code tracking loop shown in Figure 10.28. The input signal is modeled as

$$y(t) = \sqrt{2P}\ d(t)\text{PN}(t - T)\cos(\omega_0 t + \phi) + n(t) \qquad (10.11\text{-}1)$$

where P is the average power in the signal, $d(t)$ is a binary-valued function (± 1) representing the data sequence, $\text{PN}(t - T)$ is maximal length pseudonoise code sequence (± 1) arriving at the receiver with a delay of T sec and code symbol duration of T_c sec.

The parameter ω_0 is the center frequency of the carrier (rad/sec), and $n(t)$ is the receiver noise that is modeled as WGN. The output of the upper and lower correlators, e_1 and e_2, are given by

$$e_1 = \sqrt{2P}\ d(t)\text{PN}(t - T)\text{PN}\left(t - \hat{T} + \frac{T_c}{2}\right)\cos(\omega_0 t + \phi) + n(t)\text{PN}\left(t - \hat{T} + \frac{T_c}{2}\right)$$

$$(10.11\text{-}2)$$

$$e_2 = \sqrt{2P}\ d(t)\text{PN}(t - T)\text{PN}\left(t - \hat{T} - \frac{T_c}{2}\right)\cos(\omega_0 t + \phi) + n(t)\text{PN}\left(t - \hat{T} - \frac{T_c}{2}\right)$$

where $\hat{T}$ denotes the code loop estimate of the input code all-"one" epoch time delay.

We model the bandpass filtered version of

$$d(t)\text{PN}(t - T)\text{PN}\left(t - \hat{T} + \frac{T_c}{2}\right)$$

as

$$\tilde{d}(t)R_{PN}\left(T - \hat{T} + \frac{T_c}{2}\right) \qquad (10.11\text{-}3)$$

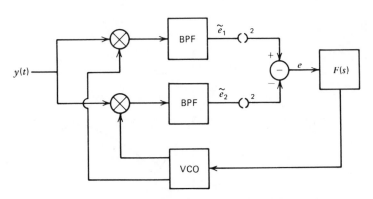

Figure 10.28 Full-time delay lock loop.

which neglects spread components, assuming the data rate is much less than $1/T_c$, and replaces the data sequence with the baseband equivalent filtered version, $\tilde{d}(t)$. With this approximation we have

$$\tilde{e}_1 = \sqrt{2P}\ \tilde{d}(t)R_{PN}\left(T - \hat{T} + \frac{T_c}{2}\right)\cos(\omega_0 t + \phi) + n(t)\overline{PN\left(t - \hat{T} + \frac{T_c}{2}\right)}$$

$$(10.11\text{-}4)$$

$$\tilde{e}_2 = \sqrt{2P}\ \tilde{d}(t)R_{PN}\left(T - \hat{T} - \frac{T_c}{2}\right)\cos(\omega_0 t + \phi) + n(t)\overline{PN\left(t - \hat{T} - \frac{T_c}{2}\right)}$$

where we have used the wavy overline to denote a filtered term.

The signal that drives the loop filter $F(s)$ is therefore given by

$$e = \alpha P\left[R_{PN}^2\left(T - \hat{T} + \frac{T_c}{2}\right) - R_{PN}^2\left(T - \hat{T} - \frac{T_c}{2}\right)\right]$$

$$+ 2\sqrt{2P}\ \tilde{d}(t)\cos(\omega_0 t + \phi)\left[n(t)\overline{PN\left(t - \hat{T} + \frac{T_c}{2}\right)}\right.$$

$$\times R_{PN}\left(T - \hat{T} + \frac{T_c}{2}\right) - \overline{PN\left(t - \hat{T} - \frac{T_c}{2}\right)}n(t)R_{PN}\left(T - \hat{T} - \frac{T_c}{2}\right)\right]$$

$$+ \left[n(t)\overline{PN\left(t - \hat{T} + \frac{T_c}{2}\right)}\right]^2 - \left(n(t)\overline{PN\left(t - \hat{T} - \frac{T_c}{2}\right)}\right)^2 \quad (10.11\text{-}5)$$

In (10.11-5) we have set $[\tilde{d}(t)]^2 = \alpha$ by neglecting all spectral components in $[\tilde{d}(t)]^2$ except the dc component as we did in the Costas loop chapter. Evaluating the bracketed portion of the first term we obtain

$$R_{PN}^2\left(T - \hat{T} + \frac{T_c}{2}\right) - R_{PN}^2\left(T - \hat{T} - \frac{T_c}{2}\right) = g(T - \hat{T}) \quad (10.11\text{-}6)$$

where $g(\epsilon)$ ($\epsilon = T - \hat{T}$ is the timing error of the loop) is the noise-free S-curve and is plotted in Figure 10.29.

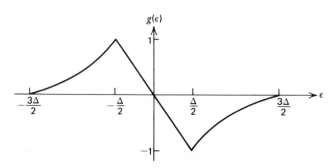

Figure 10.29 Noise-free S-curve for full-time code loop.

The error signal can be simplified by denoting

$$N_1(t) = 2\sqrt{2P}\,\tilde{d}(t)\cos(\omega_0 t + \phi)\left[n(t)\text{PN}\left(t - \hat{T} + \frac{T_c}{2}\right)R_{PN}\left(T - \hat{T} - \frac{T_c}{2}\right)\right.$$
$$\left. - \text{PN}\left(t - \hat{T} - \frac{T_c}{2}\right)n(t)R_{PN}\left(T - \hat{T} - \frac{T_c}{2}\right)\right] \qquad (10.11\text{-}7)$$

$$N_2(t) = \left[n(t)\text{PN}\left(t - \hat{T} + \frac{T_c}{2}\right)\right]^2 - \left[n(t)\text{PN}\left(t - \hat{T} - \frac{T_c}{2}\right)\right]^2 \qquad (10.11\text{-}8)$$

we have

$$e = \alpha P g(\epsilon) + N_1(t) + N_2(t) \qquad (10.11\text{-}9)$$

We will find it convenient to define a normalized S-curve $g_n(\epsilon)$ by

$$g_n(\epsilon) = \left.\frac{dg(\epsilon)}{d\epsilon}\right|_{\epsilon=0} g_n(\epsilon) \qquad (10.11\text{-}10)$$

Evaluating this expression (see Figure 10.29), we obtain

$$g(\epsilon) = 2g_n(\epsilon) \qquad (10.11\text{-}11)$$

with the slope expressed in chips (PN code symbols). We may express the error signal as

$$e = 2\alpha P g_n(\epsilon) + N_1(t) + N_2(t) \qquad (10.11\text{-}12)$$

To close the loop, denote the VCO gain by K_{VCO}, and use the Heaviside operator notation in the variable $s*$, so that the loop timing estimate is given by

$$\hat{T} = 2\alpha P K_{VCO}\frac{F(s)}{s}\left[g_n(\epsilon) + \frac{N_1(t)}{2\alpha P} + \frac{N_2(t)}{2\alpha P}\right] \qquad \text{chips} \quad (10.11\text{-}13)$$

Denoting ϵ as the tracking error, we have

$$\epsilon = T - \hat{T} \qquad (10.11\text{-}14)$$

Hence the stochastic differential equation governing the timing error is given by

$$\frac{d\epsilon}{dt} = \frac{dT}{dt} - 2\alpha P K_{VCO}F(s)\left[g_n(\epsilon) + \frac{N_1(t) + N_2(t)}{2\alpha P}\right] \qquad (10.11\text{-}15)$$

For a first-order code tracking loop, $F(s) = 1$, so that the stochastic differential equation is given by

$$\frac{d\epsilon}{dt} = -2\alpha P K_{VCO}\left[g_n(\epsilon) + \frac{N_1(t)}{2\alpha P} + \frac{N_2(t)}{2\alpha P}\right] \qquad (10.11\text{-}16)$$

---

*For example, $H(s)f(t)$ denotes $h(t)*f(t)$, with $h(t)$ the inverse Laplace transform of $H(s)$.

When the noise process driving the system has a bandwidth large compared to system bandwidth, then Stratonovich [23] shows that the probability density function is essentially a solution to the Fokker-Planck equation [24].

In order to obtain the mean time to lose lock for the first-order code tracking loop, we follow the techniques used in references 23 and 24 and the approximations developed in 23. Thus we assume that the timing error process is essentially Markovian. We define the time for the timing error ϵ to reach either $+\frac{3}{2}$ chips or $-\frac{3}{2}$ chips as the time to lose lock, since at either of these points (see Figure 10.29) the tracking error S-curve goes to (essentially) zero for long codes. By our Markovian assumption the probability density function for $|\epsilon| < \frac{3}{2}$ satisfies [23, 24]

$$\frac{\partial P(\epsilon, t)}{\partial t} = -\frac{\partial}{\partial \epsilon}[A_1(\epsilon)p(\epsilon, t)] + \frac{1}{2}\frac{\partial^2}{\partial \epsilon^2}[A_2(\epsilon)p(\epsilon, t)] \qquad (10.11\text{-}17)$$

where

$$\begin{aligned}
A_1(\epsilon) &= \lim_{\Delta t \to 0}\left[\frac{E(\Delta\epsilon \mid \epsilon)}{\Delta t}\right] \\
A_2(\epsilon) &= \lim_{\Delta t \to 0}\left\{\frac{E[(\Delta\epsilon)^2 \mid \epsilon]}{\Delta t}\right\}
\end{aligned} \qquad (10.11\text{-}18)$$

To evaluate $A_1(\epsilon)$ and $A_2(\epsilon)$ we integrate our stochastic differential equation to obtain

$$\Delta\epsilon = -2\alpha P K_{\text{VCO}}g_n(\epsilon)\Delta t - K_{\text{VCO}}\int_t^{t+\Delta t} N(u)\, du + O(\Delta t) \qquad (10.11\text{-}19)$$

where we have used

$$N(t) = N_1(t) + N_2(t) \qquad (10.11\text{-}20)$$

Therefore

$$A_1(\epsilon) = \lim_{\Delta t \to 0}\left[\frac{-2\alpha P K_{\text{VCO}}g_n(\epsilon)\Delta t + O(\Delta t)}{\Delta t}\right] = -2\alpha P K_{\text{VCO}}g_n(\epsilon) \qquad (10.11\text{-}21)$$

$$A_2(\epsilon) = \lim_{\Delta t \to 0}$$

$$\left[\frac{(2\alpha P K_{\text{VCO}}g_n(\epsilon)\Delta t)^2 + 2\alpha P K_{\text{VCO}}^2\Delta t \int_t^{t+\Delta t} N(u)\, du + K_{\text{VCO}}^2\int\int_t^{t+\Delta t} N(u)N(v)\, du\, dv}{\Delta t}\right] \qquad (10.11\text{-}22)$$

or

$$A_2(\epsilon) = \frac{N_0' K_{\text{VCO}}^2}{2}$$

We have modeled the equivalent noise process $N(t)$ as WGN, with two-sided spectral density $N_0'/2$. The solution to the first passage time problem $q(\epsilon, t)$ satisfies the following Fokker-Planck equation as long as $|\epsilon| < \frac{3}{2}$

$$\frac{\partial q}{\partial t} = 2\alpha P K_{VCO} \frac{\partial}{\partial \epsilon} [g_n(\epsilon) q(\epsilon, t)] + \frac{N_0' K_{VCO}^2}{4} \frac{\partial^2}{\partial \epsilon^2} [q(\epsilon, t)] \quad (10.11\text{-}23)$$

with the conditions

$$q(\epsilon, 0) = \delta(\epsilon)$$
$$q(\epsilon, t) = 0 \qquad \forall |\epsilon| \geq \tfrac{3}{2} \quad \text{and} \quad \forall t \qquad (10.11\text{-}24)$$
$$q(\epsilon, t) = q(-\epsilon, t) \qquad \forall t$$

The integral over ϵ of our density function $q(\epsilon, t)$ is the probability that $|\epsilon|$ has not yet exceeded $\frac{3}{2}$ chips error at time t. Hence this probability can be expressed as

$$\Phi(t) = \int_{-3/2}^{3/2} q(\epsilon, t)\, d\epsilon \leq 1 \qquad (10.11\text{-}25)$$

Since $\Phi(t)$ is the probability that the timing error has not yet exceeded $\frac{3}{2}$ of a chip, the function $\psi(t)$ defined by

$$\psi(t) = 1 - \Phi(t) \qquad (10.11\text{-}26)$$

is the probability that the boundary has been reached; that is, the error magnitude has become equal to $\frac{3}{2}$ chips. The probability density function of the time required for the error magnitude to exceed $\frac{3}{2}$ chips for the first time is given by

$$\frac{\partial \psi}{\partial t} = -\frac{\partial \Phi}{\partial t} \qquad (10.11\text{-}27)$$

Notice that $\psi(t)$ is a proper distribution function in the sense that

$$\psi(\infty) = 1$$
$$\psi(0) = 0 \qquad (10.11\text{-}28)$$

and $\psi(t)$ is a monotonically increasing function of the variable t. Consequently, the mean time to lose lock the first passage time is given by

$$\bar{T} = \int_0^\infty t \frac{\partial \psi}{\partial t}\, dt = \int_0^\infty (-t) \frac{\partial \Phi}{\partial t}\, dt \qquad (10.11\text{-}29)$$

Integrating by parts, we arrive at

$$\bar{T} = [-t\Phi(t)]\Big|_0^\infty + \int_0^\infty \Phi(t)\, dt \qquad (10.11\text{-}30)$$

Since $\Phi(t)$ must decrease asymptotically faster than $1/t$ for it to exist, the first term must be zero, and we have

$$\bar{T} = \int_0^\infty \Phi(t)\, dt \qquad (10.11\text{-}31)$$

Using the definition of $\Phi(t)$ we can express the mean time to lose lock as

$$\bar{T} = \int_0^\infty \int_{-3/2}^{3/2} q(\epsilon, t) \, d\epsilon \, dt \qquad (10.11\text{-}32)$$

In a similar manner (see [25] also) the nth moment ($n = 1, 2, 3, \ldots$) of the time to slip is given by

$$\overline{T^n} = n \int_0^\infty \int_{-3/2}^{3/2} t^{n-1} q(\epsilon, t) \, d\epsilon \, dt \qquad (10.11\text{-}33)$$

with

$$\bar{T}' = \bar{T} \qquad (10.11\text{-}34)$$

assuming that

$$t^n \Phi(t)|_0^\infty = 0 \qquad (10.11\text{-}35)$$

Since our primary goal is the mean time to lose lock, we will not use $\overline{T^n}$ in the following. Now to obtain $\bar{T}$ it is necessary to obtain $q(\epsilon, t)$. To do this we integrate the Fokker-Planck equation to obtain

$$q(\epsilon, \infty) - q(\epsilon, 0) = 2\alpha P K_{\text{VCO}} \frac{\partial}{\partial \epsilon} [g_n(\epsilon) q(\epsilon, t)] + \frac{N_0' K_{\text{VCO}}^2}{4} \frac{\partial^2}{\partial \epsilon^2} [q(\epsilon, t)] \qquad (10.11\text{-}36)$$

with

$$q(\epsilon, 0) = \delta(\epsilon)$$
$$q(\epsilon, t) = 0 \qquad \forall |\epsilon| \geq \tfrac{3}{2} \text{ chips} \qquad \text{and} \qquad \forall t \qquad (10.11\text{-}37)$$

As t becomes unbounded, all sample functions (except for a set of probability zero) reach the boundaries. As time increases without limit, it is clear that $q(t, \infty) = 0$. Therefore,

$$-\delta(\epsilon) = 2\alpha P K_{\text{VCO}} \frac{d}{d\epsilon} [g_n(\epsilon) Q(\epsilon)] + \frac{N_0' K_{\text{VCO}}^2}{4} \frac{d^2}{d\epsilon^2} [Q(\epsilon)] \qquad (10.11\text{-}38)$$

where

$$Q(\epsilon) = \int_0^\infty q(\epsilon, t) \, dt \qquad (10.11\text{-}39)$$

Notice that the boundary conditions of Q become

$$Q(\tfrac{3}{2}) = Q(-\tfrac{3}{2}) = 0 \qquad (10.11\text{-}40)$$

The solution to (10.11-38) is called the Greens function for the problem. To obtain the solution, either standard techniques involving the solution to the homogeneous differential equation can be used [26] or direct integration may be utilized. We chose to follow the latter approach. Once the Greens function is obtained, a simple quadrature produces the first slip time.

Performing an indefinite integration on (10.11-38) we obtain

$$\frac{d}{d\epsilon} Q(\epsilon) + \frac{8\alpha P}{N_0' K_{VCO}} q_n(\epsilon) Q(\epsilon) = \frac{B - u(\epsilon)}{\frac{N_0'}{4} K_{VCO}^2} \qquad (10.11\text{-}41)$$

where $u(\epsilon)$ denoted the unit step function centered at $\epsilon = 0$ and B is a constant of integration that must be evaluated from the boundary conditions. The solution can be written as

$$Q(\epsilon) = \exp\left[-\gamma \int^\epsilon g_n(\epsilon') \, d\epsilon'\right] \int_{-3/2}^\epsilon \frac{[B - u(\epsilon'')]}{\beta} \exp\left[\gamma \int^{\epsilon''} g_n(\epsilon') \, d\epsilon'\right] d\epsilon''$$

$$+ D \exp\left[-\gamma \int^\epsilon g_n(\epsilon') \, d\epsilon'\right] \qquad (10.11\text{-}42)$$

where

$$\gamma = \frac{8\alpha P}{N_0' K_{VCO}}$$

$$\beta = \frac{N_0' K_{VCO}^2}{4} \qquad (10.11\text{-}43)$$

From (10.11-42) we deduce that

$$\begin{aligned} D &= 0 & \text{satisfies} & & Q(-\tfrac{3}{2}) &= 0 \\ B &= \tfrac{1}{2} & \text{satisfies} & & Q(\tfrac{3}{2}) &= 0 \end{aligned} \qquad (10.11\text{-}44)$$

Therefore

$$Q(\epsilon) = \exp\left[-\gamma \int^\epsilon g_n(\epsilon') \, d\epsilon'\right] \int_{-3/2}^\epsilon \frac{\tfrac{1}{2} - u(\epsilon)}{\beta} \exp\left[\gamma \int^{\epsilon''} g_n(\epsilon') \, d\epsilon'\right] \qquad (10.11\text{-}45)$$

Denote $\int^\epsilon g_n(\epsilon') \, d\epsilon'$ by $G(\epsilon)$, which is an even function of ϵ, so that

$$Q(\epsilon) = \beta^{-1} e^{-\gamma G(\epsilon)} \int_{-3/2}^\epsilon [\tfrac{1}{2} - u(\epsilon')] e^{\gamma G(\epsilon')} \, d\epsilon' \qquad (10.11\text{-}46)$$

Hence, from (10.11-32) the mean time to lose lock is given by

$$\bar{T} = \frac{1}{\beta} \int_{-3/2}^{3/2} e^{-\gamma G(\epsilon)} \int_{-3/2}^\epsilon [\tfrac{1}{2} - u(\epsilon')] e^{\gamma G(\epsilon')} \, d\epsilon' \qquad (10.11\text{-}47)$$

Separating the two terms, we obtain

$$\bar{T} = \frac{1}{2\beta} \int_{-3/2}^{3/2} e^{-\gamma G(\epsilon)} \int_{-3/2}^\epsilon e^{\gamma G(\epsilon')} \, d\epsilon' \, d\epsilon$$

$$- \frac{1}{\beta} \int_{-3/2}^{3/2} e^{-\gamma G(\epsilon)} \int_0^\epsilon u(\epsilon') e^{\gamma G(\epsilon')} \, d\epsilon' \, d\epsilon \qquad (10.11\text{-}48)$$

Using the fact that $G(\epsilon)$ is an even function of ϵ, after some algebra, we obtain

$$\bar{T} = \frac{1}{2\beta} \int_0^{3/2} \int_0^{3/2} e^{-\gamma G(\epsilon)} e^{\gamma G(\epsilon')} \, d\epsilon \, d\epsilon' \qquad (10.11\text{-}49)$$

This is the final functional form for $\bar{T}$; however, it is convenient to express it in terms of the timing error variance. To this end we rewrite (10.11-16) using the Heaviside operator s to denote the derivative operation:

$$\epsilon = -\frac{2\alpha P K_{\text{VCO}}}{s}\left[g_n(\epsilon)+\frac{N_1(t)}{2\alpha P}+\frac{N_2(t)}{2\alpha P}\right] \qquad (10.11\text{-}50)$$

Now assuming that substantially all of the probability mass of the timing error probability density function is contained in the region $-\frac{1}{2}\le\epsilon\le\frac{1}{2}$, we may linearize to obtain

$$\epsilon = -H(s)\left[\frac{N_1(t)}{2\alpha P}+\frac{N_2(t)}{2\alpha P}\right] \qquad (10.11\text{-}51)$$

where

$$H(s)=\frac{2\alpha P K_{\text{VCO}}}{s+2\alpha P K_{\text{VCO}}} \qquad (10.11\text{-}52)$$

Consequently, it follows that the timing error variance is given by

$$\sigma^2 = \int_{-\infty}^{\infty}\mathscr{S}_N(f)|H(f)|^2\,df \qquad (10.11\text{-}53)$$

It is shown in the Appendix that the two noise terms of $N_2(t)$ are essentially uncorrelated when the BPF bandwidth is much less than the PN code symbol rate. Using this assumption, and assuming that the code loop bandwidth is small compared to the data rate, and in addition assuming that the bandpass filters are multipole filters (essentially ideal BPFs) we arrive at the spectral density at the origin:

$$\mathscr{S}_N(0)=\alpha P N_0+2N_0^2 B \qquad (10.11\text{-}54)$$

So that

$$\sigma^2=\frac{N_0 B_L}{2\alpha P}\left(1+\frac{2N_0 B}{\alpha P}\right) \qquad (10.11\text{-}55)$$

where B is the positive frequency noise bandwidth of the ideal bandpass filter. The result is well-known.

Now recall that we defined γ by

$$\gamma=\frac{8\alpha P}{N_0'K_{\text{VCO}}}=\frac{8\alpha P}{2(N_0\alpha P+2N_0^2 B)K_{\text{VCO}}} \qquad (10.11\text{-}56)$$

However, for a first-order loop, it is known that the one-sided closed-loop bandwidth is given by

$$B_L=\frac{A_{\text{eq}}}{4} \qquad (10.11\text{-}57)$$

With A_{eq} being the equivalent loop gain. This is simply the multiplicative factor for the error control signal when the stochastic differential equation of operation is linearized. Therefore

$$A_{eq} = 2\alpha P K_{VCO} \tag{10.11-58}$$

Hence

$$B_L = \frac{\alpha P K_{VCO}}{2} \quad \text{or} \quad K_{VCO} = \frac{2B_L}{\alpha P} \tag{10.11-59}$$

Therefore

$$\gamma = \frac{8\alpha^2 P^2}{2(N_0 \alpha P + 2N_0^2 B)2B_L} \tag{10.11-60}$$

or alternatively

$$\gamma = \frac{1}{(N_0 B_L/2\alpha P)[1 + (2N_0 B/\alpha P)]} = \frac{1}{\sigma^2} \tag{10.11-61}$$

Now notice that

$$\frac{B_L}{2\beta} = \frac{\alpha P K_{VCO}}{4\beta} = \frac{\alpha P K_{VCO}}{N_0' K_{VCO}^2} = \frac{\gamma}{8} \tag{10.11-62}$$

Noting the above and letting $W_L = 2B_L$, we obtain the final form

$$\bar{T}W_L = \frac{1}{4\sigma^2} \int_0^{3/2} \int_0^{3/2} \exp\left\{-\left[\frac{G(\epsilon)}{\sigma^2}\right]\right\} \exp\left\{\left[\frac{G(\epsilon')}{\sigma^2}\right]\right\} d\epsilon \, d\epsilon' \tag{10.11-63}$$

It can be shown that the same form for $\bar{T}W_L$ occurs for the time-shared code tracking loop except now the tracking error variance is doubled:

$$\sigma^2 = \frac{N_0 B_L}{\alpha P}\left(1 + \frac{2N_0 B}{\alpha P}\right) \tag{10.11-64}$$

If the bandpass filter is not ideal, then the 1 should be changed to α'/α. As before, α is defined by

$$\alpha = \frac{\int_{-\infty}^{\infty} \mathcal{S}_d(f)|H_{BPF}(f)|^2 \, df}{\int_{-\infty}^{\infty} \mathcal{S}_d(f) \, df} \tag{10.11-65}$$

and

$$\alpha' = \frac{\int_{-\infty}^{\infty} \mathcal{S}_d(f)|H_{BPF}(f)|^2 \, df}{\int_{-\infty}^{\infty} \mathcal{S}_d(f) \, df} \tag{10.11-66}$$

Consequently, the final expression for the mean time to lose lock applies to both time-shared and full-time code tracking loops with the appropriate expression for σ^2.

Typically, hardware considerations limit the PN spectra to essentially a bandwidth of $2/T_c$, where $1/T_c$ is the chip rate. As a consequence, the S-curve appears alightly different from the large bandwidth case of Figure 10.29 and is shown in Figure 10.30.

Using the S-curve of Figure 10.30 in the expression for $\bar{T}W_L$ yields, with the aid of a digital computer, the curve of Figure 10.31. The work developed here was applied to the S-band shuttle code tracking loop (developed at TRW Systems), which is a second-order loop, by degrading the loop SNR by 1 dB (increasing σ^2 by 1 dB). The result is plotted in

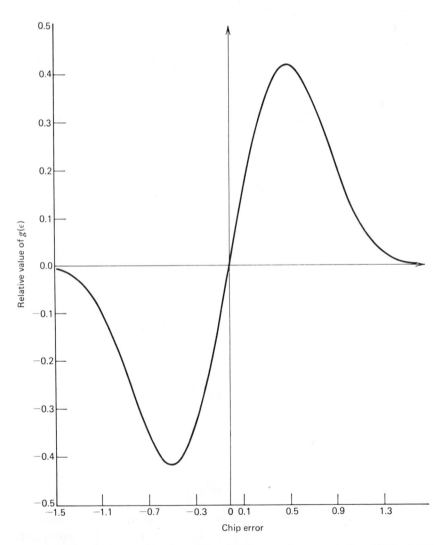

Figure 10.30   Code loop S-curve with a BPF preceding correlation ($BW = 2/T_c$).

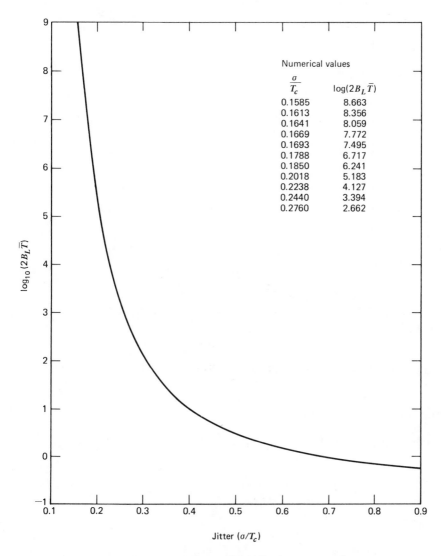

Numerical values

$\dfrac{\sigma}{T_c}$	$\log(2B_L\bar{T})$
0.1585	8.663
0.1613	8.356
0.1641	8.059
0.1669	7.772
0.1693	7.495
0.1788	6.717
0.1850	6.241
0.2018	5.183
0.2238	4.127
0.2440	3.394
0.2760	2.662

Jitter (σ/T_c)

Figure 10.31 Normalized mean slip time versus jitter.

Figure 10.32. One experimental point was measured and compared to the theoretical curve with good agreement.

For an alternate derivation of the $\bar{T}W_L$ formula, reference 27 may be consulted.

PROBLEM 13

Generalize (10.11-63) to the case when the second moment of the Fokker-Planck depends on the code error. Interpreting the differential equation in

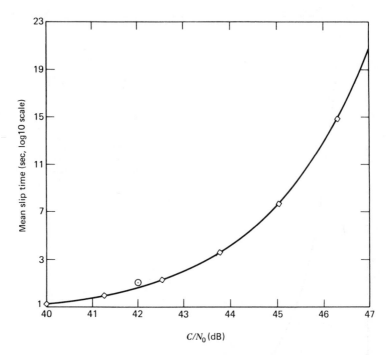

Figure 10.32   Mean slip time versus C/N_0, shuttle S-band system.

the Ito sense shows that

$$\bar{T}W_L = \frac{1}{4\sigma^2} \int_0^{3/2} \int_0^{3/2} \eta^{-1}(\epsilon) \exp\left[-\frac{\hat{G}(\epsilon)}{\sigma^2}\right] \exp\left[+\frac{\hat{G}(\epsilon')}{\sigma^2}\right] d\epsilon \, d\epsilon'$$

where

$$N_0(\epsilon) = \eta(\epsilon) N_0(0)$$

and

$$\hat{G}(\epsilon) = \int^{\epsilon} \frac{g_n(x)}{\eta(x)} dx$$

It may be shown that (10.11-63) applies to a much more general class of problems. All code loops that are first-order loops that induce approximately Markovian error processes can be characterized by (10.11-63). The only caution is that $G(\epsilon)$ is the indefinite integral of $g_n(\epsilon)$, which is normalized so that $g_n'(0) = 1$. The upper limit does not have to be $\frac{3}{2}$; in fact it depends upon the extent of the S-curve. Furthermore, (10.11-63) applies to phase lock loop structures in the form

$$\bar{T}W_L = \frac{1}{4\sigma_\phi^2} \int_0^{\phi_L} \int_0^{\phi_L} \exp\left[-\frac{G(\phi)}{\sigma_\phi^2}\right] \exp\left[\frac{G(\phi')}{\sigma_\phi^2}\right] d\phi \, d\phi' \quad (10.11\text{-}67)$$

where ϕ_L is the phase error in which a complete cycle is slipped. As an example for a cw loop, $\phi_L = 2\pi$. Then (10.11-67) equals the original result obtained by Viterbi [24].

PROBLEM 14

Using (10.11-67), show that

$$\bar{T}|_{\phi_L=2\pi} = 4\bar{T}|_{\phi_L=\pi}$$

10.12   EFFECTS OF IMAGE NOISE ON CODE TRACKING

In this section we indicate how image noise of the receiver can degrade code loop tracking performance. We make the calculation on a specific model in order to more graphically illustrate the image noise interference process. Consider the model of a full-time noncoherent code tracking loop shown in Figure 10.33.

We model the received signal after being filtered by the bandpass filter BPF$_{PN}$ by $y(t)$, where

$$y(t) = \sqrt{2P_I}\,\overbrace{PN_I(t-T)}\,d_I(t)\cos\omega_0 t + \sqrt{2P_Q}\,\overbrace{PN_Q(t-T)}\,d_Q(t)\sin\omega_0 t$$
$$+ \sqrt{2}\tilde{n}_c(t)\cos\omega_0 t + \sqrt{2}\tilde{n}_s(t)\sin\omega_0 t \qquad (10.12\text{-}1)$$

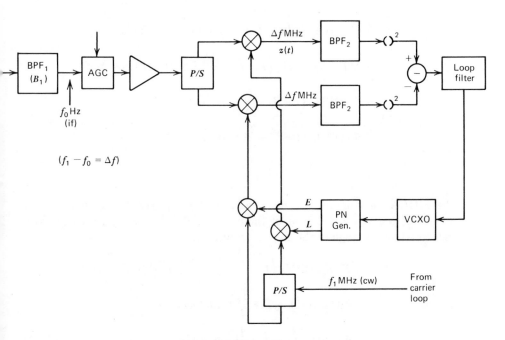

Figure 10.33   Code loop equivalent model.

where

$\overset{\displaystyle\sim}{\mathrm{PN_I}(t-T)\,d_I(t)}$: wavy line denotes filtering.

$\mathrm{PN_I}(t)$ is the I channel PN type code.

$\mathrm{PN_Q}(t)$ is the Q channel PN type code.

$\tilde{n}_c(t)$ is a filtered in-phase Gaussian noise process.

$\tilde{n}_s(t)$ is a quadrature Gaussian noise process.

$\omega_0/2\pi$ is input center frequency.

We model the first bandpass filter as an ideal filter with a B_1 Hz. Now consider the early correlator reference signal (centered at f_1 Hz). It is modeled as

$$s_E(t) = \sqrt{2}\,\mathrm{PN_Q}\left(t - \hat{T} + \frac{T_c}{2}\right)\sin\omega_1 t \qquad (10.12\text{-}2)$$

where

$\omega_1/2\pi$ is the correlator reference frequency.

$\hat{T}$ is the code timing loop estimate.

T_c is code chip duration.

Hence the resulting signal into the data bandwidth bandpass filters is given by

$$z(t) = \sqrt{P_I}\,d_I(t)\mathrm{PN_Q}\left(t - \hat{T} + \frac{T_c}{2}\right)\widetilde{\mathrm{PN}}_I(t - T)\sin(\Delta\omega t)$$

$$+ \sqrt{P_Q}\,d_Q(t)\mathrm{PN_Q}\left(t - \hat{T} + \frac{T_c}{2}\right)\widetilde{\mathrm{PN}}_Q(t - T)\cos(\Delta\omega t)$$

$$+ \tilde{n}_c(t)\mathrm{PN_Q}\left(t - \hat{T} + \frac{T_c}{2}\right)\sin(\Delta\omega t)$$

$$+ \tilde{n}_s(t)\mathrm{PN_Q}\left(t - \hat{T} + \frac{T_c}{2}\right)\cos(\Delta\omega t)$$

$$+ O(\omega_1 + \omega_0) \qquad (10.12\text{-}3)$$

$$\Delta\omega = \Delta\omega_1 - \Delta\omega_0$$

where $O(\omega_1 + \omega_0)$ denotes terms centered at $f_0 + f_1$ Hz. Since these terms are out of band, we will neglect them. Since $\mathrm{PN_I}$ and $\mathrm{PN_Q}$ are essentially uncorrelated, we neglect the first term and arrive at

$$z(t) = \sqrt{P_Q}\,d_Q(t)\mathrm{PN_Q}\left(t - \hat{T} + \frac{T_c}{2}\right)\widetilde{\mathrm{PN}}_Q(t - T)\cos(\Delta\omega t)$$

$$+ \tilde{n}_c(t)\mathrm{PN_Q}\left(t - \hat{T} + \frac{T_c}{2}\right)\sin(\Delta\omega t)$$

$$+ \tilde{n}_s(t)\mathrm{PN_Q}\left(t - \hat{T} + \frac{T_c}{2}\right)\cos(\Delta\omega t) \qquad (10.12\text{-}4)$$

Only image noise will be considered. To simplify the visualization of the image noise, model the noise as $\tilde{n}_c(t)\text{PN}_Q(t - \hat{T} + T_c/2)\cos\omega_0 t$ and the correlator as $\sin(\omega_1 t)$. The resulting product is the same, but now the correlating signal is a pure tone and the image noise is easier to visualize. The PN cross n_c and cw correlating signal are shown in Figure 10.34, along with the resulting noise for a particular case when $f_0 = 35\,\text{MHz}$, $f_1 = 37.5\,\text{MHz}$, and $\Delta\omega = 2.5\,\text{MHz}$.

Hence, noise components at both f_0 and $f_0 + 2\Delta f$ end up at Δf Hz. The component from around $f_0 + 2\Delta f$ produces the image noise and thereby degrades performance. Since there are two independent noise terms in (10.12-4), we consider one of them and double the resulting interference. Consider the noise term

$$\tilde{n}'_c(t) = \tilde{n}_c(t)\text{PN}_Q\left(t - \hat{T} + \frac{T_c}{2}\right)\sin(\Delta\omega t) \qquad (10.12\text{-}5)$$

The autocorrelation function is given by

$$2R_{\tilde{n}'_c}(\tau) = R_{\tilde{n}_c}(\tau)R_{\text{PN}_Q}(\tau)\cos(\Delta\omega t) \qquad (10.12\text{-}6)$$

Hence

$$S_{\tilde{n}'_c}(f) = [S_{n_c}(f) * S_{\text{PN}_Q}(f)] * \cos(\Delta\omega t) \qquad (10.12\text{-}7)$$

where the $*$ denotes convolution. Evaluating with the assumption that BPF_1 is an ideal filter having a bandwidth B_1, we obtain

$$S_{\tilde{n}'_c}(f) = \frac{1}{4}\int_{-B_1/2}^{B_1/2} \frac{N_0}{2}\, T_c\, \frac{\sin^2[\pi T_c(f - \Delta f)]}{[\pi T_c(f - \Delta f)]^2}\, df$$
$$+ \frac{1}{4}\int_{-B_1/2}^{B_1/2} \frac{N_0}{2}\, T_c\, \frac{\sin^2[\pi T_c(f + \Delta f)]}{[\pi T_c(f + \Delta f)]^2}\, df \qquad (10.12\text{-}8)$$

The first term is the direct noise spectral density; the second term is the

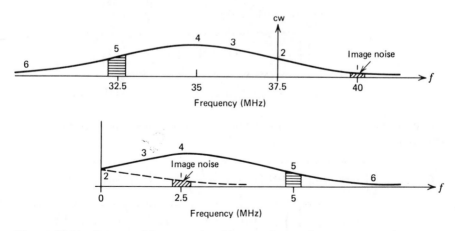

Figure 10.34 Source of image noise (the numbers relate spectral points) for a numerical example.

image noise. Evaluating the first term at Δf, we obtain

$$S_{n_{\text{direct}}}(\Delta f) = \frac{N_0}{8} \int_{-B_1 T_c/2}^{B_1 T_c/2} \frac{\sin^2 \pi x}{(\pi x)^2} \, dx \qquad (10.12\text{-}9)$$

Furthermore, the image noise term can be evaluated at $f = \Delta f$ to yield

$$S_{n_{\text{image}_1}}(\Delta f) = \frac{N_0}{8} \int_{(-B_1/2+2\Delta f)T_c}^{(B_1/2+2\Delta f)T_c} \frac{\sin^2 \pi x}{(\pi x)^2} \, dx \qquad (10.12\text{-}10)$$

It follows that the effective noise into the bandpass filters is given by the sum of (10.12-9) and (10.12-10), and not just by (10.12-9) alone. In a TDRSS MAG receiver system, the image noise amounted to about a 0.5-dB increase in thermal noise.

APPENDIX   SPECTRAL DENSITY OF $n_L^2(t) - n_E^2(t)$ AT $f = 0$ FOR THE DELAY LOCK LOOP

Recall that it is required to obtain the spectral density of $n^2(t) = n_L^2(t) - n_E^2(t)$, where

$$n_L(t) = \left[n(t)\text{PN}\left(t - \hat{T} - \frac{T_c}{2} \right) \right] * l(t) \qquad (A\text{-}1)$$

$$n_E(t) = \left[n(t)\text{PN}\left(t - \hat{T} + \frac{T_c}{2} \right) \right] * l(t) \qquad (A\text{-}2)$$

We outline the calculations to obtain the spectral density at the origin (all integrals have $-\infty$ to ∞ limits, which have been dropped for convenience)

$$
\begin{aligned}
R_{n_2}(\tau) = E \Bigg(&\left\{ \left[\int l(t-u)n(u)\text{PN}\left(t - \hat{T} + \frac{T_c}{2} \right) du \right]^2 \right. \\
&- \left. \left[\int l(t-u)n(u)\text{PN}\left(u - \hat{T} - \frac{T_c}{2} \right) du \right]^2 \right\} \\
&\times \left\{ \left[\int l(t+\tau-u)n(u)\text{PN}\left(u - \hat{T} + \frac{T_c}{2} \right) \right]^2 \right. \\
&- \left. \left[\int l(t+\tau-u)n(u)\text{PN}\left(u - \hat{T} - \frac{T_c}{2} \right) du \right]^2 \right\} \Bigg)
\end{aligned}
\qquad (A\text{-}3)
$$

This product yields four terms. From symmetry these four terms can be written as $2R_1(\tau) + 2R_2(\tau)$, where

$$
\begin{aligned}
R_1(\tau) = E \Bigg[&\int l(t-u_1)n(u_1)\text{PN}\left(u_1 - \hat{T} - \frac{T_c}{2} \right) du_1 \cdot \int l(t-u_2)\, n(u_2)\text{PN} \\
&\times \left(u_2 - \hat{T} + \frac{T_c}{2} \right) du_2 \int l(t+\tau-u_3)n(u_3)\text{PN}\left(u_3 - \hat{T} + \frac{T_c}{2} \right) du_3 \\
&\times \int l(t+\tau-u_4)\, n(u_4)\text{PN}\left(u_4 - \hat{T} + \frac{T_c}{2} \right) du_4 \Bigg]
\end{aligned}
\qquad (A\text{-}4)
$$

and

$$R_2(\tau) = E\left[\int \overbrace{l(t - u_1)n(u_1)\text{PN}\left(u_1 - \hat{T} + \frac{T_c}{2}\right)du_1} \cdot \int l(t - u_2)\right.$$

$$\times \overbrace{n(u_2)\text{PN}\left(u_2 - \hat{T} + \frac{T_c}{2}\right)du_2}$$

$$\times \int \overbrace{l(t + \tau - u_3)n(u_3)\text{PN}\left(u_3 - \hat{T} - \frac{T_c}{2}\right)du_3} \cdot \int l(t + \tau - u_4)$$

$$\left.\times \overbrace{n(u_4)\text{PN}\left(u_4 - \hat{T} - \frac{T_c}{2}\right)du_4}\right] \qquad (A\text{-}5)$$

First we consider $R_1(\tau)$ and then $R_2(\tau)$. For a WGN process, we have

$$E[n(u_1)n(u_2)n(u_3)n(u_4)] = \left(\frac{N_0}{2}\right)^2\left[\delta(u_1 - u_2)\delta(u_3 - u_4) + \delta(u_1 - u_3)\delta(u_2 - u_4)\right.$$

$$\left. + \delta(u_1 - u_4)\delta(u_2 - u_4)\right] \qquad (A\text{-}6)$$

so that

$$R_1(\tau) = \left(\frac{N_0}{2}\right)^2 \int\int l^2(t - u_1)l^2(t + \tau - u_3)\,du_1\,du_3$$

$$+ \left(\frac{N_0}{2}\right)^2 \int\int l(t - u_1)l(t + \tau - u_1)l(t - u_2)l(t + \tau - u_2)\,du_1\,du_2$$

$$+ \left(\frac{N_0}{2}\right)^2 \int\int l(t - u_1)l(t + \tau - u_1)l(t - u_2)l(t + \tau - u_2)\,du_1\,du_2$$

$$(A\text{-}7)$$

Evaluating by using the definition of the transfer function, we obtain

$$R_1(\tau) = \left(\frac{N_0}{2}\right)^2\left\{\int_{-\infty}^{\infty}|L(\omega)|^2\frac{d\omega}{2\pi}\right\}^2$$

$$+ 2\left(\frac{N_0}{2}\right)^2\left\{\int_{-\infty}^{\infty}|L(\omega)|^2 e^{-i\omega\tau}\frac{d\omega}{2\pi}\right\}^2, \qquad L(\omega) = \mathscr{F}[l(t)] \qquad (A\text{-}8)$$

since the second and third terms of $R_1(\tau)$ yield the same value. The power spectral density associated with the autocorrelation function $2R_1(\tau)$ is given by

$$\mathscr{S}_1(0) = 2\left(\frac{N_0}{2}\right)^2\left[\int|L(\omega')|^2\frac{d\omega'}{2\pi}\right]^2\delta(\omega) + 4\left(\frac{N_0}{2}\right)^2\int|L(\omega')|^4\frac{d\omega'}{2\pi} \qquad (A\text{-}9)$$

Now consider $R_2(\tau)$. Again using the property of the expectation of the product of four Gaussian random variables, we obtain

$$R_2(\tau) = \left(\frac{N_0}{2}\right)^2\left[\int\int|L(\omega')|^2\frac{d\omega'}{2\pi}\right]^2$$

$$+ 2\left(\frac{N_0}{2}\right)^2\int\int l(v)l(v + \tau)l(z)l(z + \tau)\left(1 - \frac{|v - z|}{T_c}\right)^2 dz\,dv \quad (A\text{-}10)$$

where the PN waveform has been assumed to be very long so that it could be modeled as a random NRZ (independent from symbol to symbol) sequence. The factor of 2 appears since both the second and third terms yield the same values. Taking Fourier transforms and evaluating at $f = 0$ produces

$$\mathscr{S}_2(0) = \left(\frac{N_0}{2}\right)^2 \left[\int_{-\infty}^{\infty} |L(\omega')|^2 \frac{d\omega'}{2\pi}\right]^2 \delta(\omega)$$

$$+ 2\left(\frac{N_0}{2}\right)^2 \int_{-\infty}^{\infty} \int_{-\infty}^{\infty} \int_{-T_c}^{T_c} |L(\omega_1)|^2 |H(\omega_2)|^2 e^{-ix(\omega_1+\omega_2)} \left(1 - \frac{|x|}{T_c}\right)^2 dx \frac{d\omega_1}{2\pi} \frac{d\omega_2}{2\pi}$$

$$\text{(A-11)}$$

Finally, adding all the terms up, we obtain

$$\mathscr{S}(0) = N_0^2 \int_{-\infty}^{\infty} |L(\omega)|^4 \frac{d\omega}{2\pi}$$

$$- N_0^2 \int_{-\infty}^{\infty} \int_{-\infty}^{\infty} \int_{-T_c}^{T_c} |L(\omega_1)|^2 |L(\omega_2)|^2 e^{-ix(\omega_1+\omega_2)} \left(1 - \frac{|x|}{T_c}\right)^2 dx \frac{d\omega_1}{2\pi} \frac{d\omega_2}{2\pi}$$

$$\text{(A-12)}$$

For convenience, consider the case of ideal bandpass filters having bandwidth B. Using the baseband equivalent filters, we obtain

$$\mathscr{S}(0) = N_0^2(2B) - 2N_0^2 \int_{-\pi B}^{\pi B} \int_{-\pi B}^{\pi B} \int_{-T_c}^{T_c} e^{-ix(\omega_1+\omega_2)} \left(1 - \frac{|x|}{T_c}\right)^2 \frac{d\omega_1}{2\pi} \frac{d\omega_2}{2\pi} dx$$

$$\text{(A-13)}$$

or

$$\mathscr{S}(0) = 2N_0^2 B - 2N_0^2 B^2 \int_{-T_c}^{T_c} \left[\frac{\sin(\pi x B)}{\pi x B}\right]^2 \left(1 - \frac{|x|}{T_c}\right)^2 dx \qquad \text{(A-14)}$$

Let $\pi x B = z$, so that

$$\mathscr{S}(0) = 2N_0^2 B - 2N_0^2 B^2 \int_{-\pi B T_c}^{\pi B T_c} \left(\frac{\sin z}{z}\right)^2 \left(1 - \frac{|z|}{\pi B T_c}\right)^2 \frac{dz}{\pi B} \qquad \text{(A-15)}$$

Now when $BT_c \ll 1$, which is the normal case,

$$\mathscr{S}(0) \cong 2N_0^2 B - \frac{4N_0^2}{\pi} B \int_0^{\pi B T_c} \left(1 - \frac{2z}{\pi B T_c} + \frac{z^2}{\pi^2 B^2 T_c^2}\right) dz \qquad \text{(A-16)}$$

$$\mathscr{S}(0) = 2N_0^2 B(1 - \tfrac{2}{3} B T_c) \cong 2N_0^2 B \qquad \text{(A-17)}$$

Now if $n_L(t)$ and $n_E(t)$ are assumed independent, then the value of the spectrum at $f = 0$ is just $2N_0^2 B$. Therefore, *when $BT_c \ll 1$*, it is valid to assume that $n_L(t)$ and $n_E(t)$ are statistically independent.* However, as a direct counterexample to statistical independence, when $BT_c \gg 1$, consider the following. When the filter distortion is negligible, then

$$n_L^2(t) \cong PN^2\left(t - \hat{T} - \frac{T_c}{2}\right) n^2(t) \qquad \text{(A-18)}$$

_____

*Strictly speaking, we have shown that the ac portion of the power spectral density of $n_L^2(t) - n_E^2(t)$ is the sum of the ac portions of the spectral density of each term.

and

$$n_E^2(t) \cong PN^2\left(t - \hat{T} + \frac{T_c}{2}\right) n^2(t) \qquad (A\text{-}19)$$

so that

$$n_L^2(t) - n_E^2(t) \cong 0 \qquad (A\text{-}20)$$

and $\mathscr{S}(0) = 0$. This argument is made more precise in the following: We have just shown, that for ideal bandpass filters, we have

$$\mathscr{S}(0) = 2N_0^2 B - 2N_0^2 B^2 \int_{-\pi BT_c}^{\pi BT_c} \left(\frac{\sin z}{z}\right)^2 \left(1 - \frac{|z|}{\pi BT_c}\right)^2 \frac{dz}{\pi B} \qquad (A\text{-}21)$$

If $BT_c \gg 1$, then

$$\mathscr{S}(0) \cong 2N_0^2 B - 2N_0^2 B^2 \int_{-\infty}^{\infty} \left(\frac{\sin z}{z}\right)^2 \frac{dz}{\pi B} \qquad (A\text{-}22)$$

But since

$$\int_{-\infty}^{\infty} \left(\frac{\sin z}{z}\right)^2 dz = \pi \qquad (A\text{-}23)$$

we obtain

$$\mathscr{S}(0) \cong 0 \qquad BT_c \gg 1 \qquad (A\text{-}24)$$

as was to be shown.

In conclusion, for the normal case that $BT_c \ll 1$, we have

$$\mathscr{S}(0) = N_0^2 \int_{-\infty}^{\infty} |L(\omega)|^4 \frac{d\omega}{2\pi} = 2N_0^2 B' \qquad (A\text{-}25)$$

which, for ideal bandpass filters, becomes

$$\mathscr{S}(0) = 2N_0^2 B \qquad (A\text{-}26)$$

where again B is the 3-dB (positive frequency) bandwidth.

REFERENCES

1 Gill, W. J., "A Comparison of Binary Delay-Lock Tracking-Loop Implementations," *IEEE Trans. Aerospace and Electronic Systems*, Vol. AES-2, No. 4, July 1966.

2 Spilker, J. J., *Digital Communications by Satellite*, Prentice-Hall, Englewood Cliffs, N.J., Chap. 18.

3 Spilker, J. J., and Magill, D. T., "The Delay-Lock Discriminator—An Optimum Tracking Device," *Proc. IRE*, September 1961.

4 Stiffler, J. J., *Theory of Synchronous Communications*, Prentice-Hall, Englewood Cliffs, N.J., 1971, Sec. 5.4.

5 Integrated Function Waveform Study, Magnavox Research Laboratories, RADC-TR-69-424, Vol. II, Final Technical Report, January 1970.

6 Huang, T. C., "Analysis of a Time-Shared Early/Late PN Code Tracking Loop and Signal Dropout Considerations," TRW IOC SCTE-50-75-063/TCH, September 1975.

7 Simon, M. K., "Noncoherent Pseudonoise Code Tracking Performances of Spread Spectrum Receivers," *IEEE Trans.*, Vol. COM 25, No. 3, March 1977.

8 Stone, M., "Noncoherent τ-Dither Code Tracking Loop Performance Analysis," TRW IOC 7131.50-2, June 13, 1972.

9 Holmes, J. K., Unpublished Notes on Time-Shared Code Tracking Loops, December 1973.

10 Hartmann, H. P., "Analysis of a Dithering Loop for PN Code Tracking," *IEEE Trans. Aerospace and Electronic Systems*, January 1974.

11 Jolly, L. B. W., "Summation of Series," 2nd rev. ed., Dover, New York, 1961.

12 Huff, R. J., and Reinhard, K. L., "A Delay-Lock Loop for Tracking Pulsed-Envelope Signals," *IEEE Trans. Aerospace and Electronic Systems*, May 1971.

13 Boyarsky, A., and Fukada, M., "The Effects of Self-Noise on Error Voltage of the Delay-Lock Discriminator," *IEEE Trans. Communications Technology*, August 1970.

14 Boyarsky, A., "Multiple Signal Analysis for the Delay-Lock Discriminator," *IEEE Trans. Aerospace and Electronic Systems*, March 1975.

15 "Global Positioning Systems C-Signal Multiple Access and Jamming Performance," Philco-Ford Corp., Western Development Division, Palo Alto, Cal., December 27, 1973.

16 Cahn, C. R., "Spread Spectrum Applications and State-of-the-Art Equipments," Magnavox Report No. MX-TM-3134-72, November 22, 1972.

17 Beckermann, P., and Spizzichino, A., "The Scattering of Electromagnetic Waves from Rough Surfaces," MacMillan, Pergamon, New York, 1963.

18 Meyr, H., "Delay-Lock Tracking of Stochastic Signals," *IEEE Trans. Communications*, Vol. COM-24, No. 3, March 1976.

19 Spilker, J. J., "Delay-Lock Tracking of Binary Signals," *IEEE Trans. Space Electronics and Telemetry*, Vol. SET-19, pp. 1–8, March 1963.

20 Nielsen, P. T., "On the Acquisition of Binary Delay-Lock Loops," *IEEE Trans. Aerospace and Electronic Systems*, Vol. AES-11, No. 3, May 1975.

21 Hook, W., Unpublished Notes on Code Acquisition, TRW Systems, August 1976.

22 Holmes, J. K., and Biederman, L., "Delay-Lock-Loop Mean Time to Lose Lock," *IEEE Trans. Communications*, Part I, Vol. COM-26, November 1978.

23 Stratonovich, R. L., *Topics in the Theory of Random Noise*, Vol. I, Gordon and Breach, New York, 1963, Chap. 4.

24 Viterbi, A. J., *Principles of Coherent Communication*, McGraw-Hill, New York, 1966.

25 Lindsey, W. C., *Synchronization Systems in Communication and Control*, Prentice-Hall, Englewood Cliffs, N.J., 1972.

26 Friedman, B., *Principles and Techniques of Applied Mathematics*, Wiley, New York, 1956, Chap. 3.

27 Biederman, L., and Holmes, J. K., "Analysis of the Mean First Slip Time for the S-Band Shuttle Code Loop," TRW IOC SCTE-50-76-244/LB/JKH.

28 Cahn, C., Goutmann, M., and Haefner, M., "System 621B Signal Definition Study," Magnavox Research Labs, MRL-R-4479, Final Report (code loop and analysis is unclassified but there are errors in the analysis), October 2, 1972, Vols. I and II.

29 Hopkins, P. M., "Double Dither Loop for Pseudonoise Code Tracking," *IEEE Trans. Aerospace and Electronic Systems*, Vol. 13, No. 6, November 1977.

30 Osborne, E. F., and Schonoff, T. A., "Delay-Locked Receivers with Phase Sensing of the Correlation (Error) Function," NTC 73, Atlanta, Ga., November 26–28, 1973, Vol. II.

31 Huang, T. C., and Holmes, J. K., "Performance of Noncoherent Time-Shared PN Code Tracking Loops," NTC 76, Dallas, Tex., November 29 and 30, and December 1, 1976.

11

GOLD CODES

In Chapter 7 we discussed maximal length PN codes used for single-channel communication systems. In this chapter we briefly discuss a code commonly used for code division multiple-access (CDMA) that allows for the simultaneous operation of many signals at the same carrier frequency. CDMA is a multiplexing technique in which many carriers, all at essentially the same frequency, send "different" PN type codes that have the property that the cross correlation is low between any pair of the codes. That is, the transmitted signals are of the form

$$s_i(t) = \sqrt{2P}\ d_i(t)G_i(t)\sin(\omega_0 t + \theta) \qquad i = 1, \dots, N$$

where

P is transmitted power.

$d_i(t)$ is the ith data signal (± 1).

$G_i(t)$ is the ith CDMA code.

ω_0 is the carrier radian frequency.

θ_0 is the carrier phase.

Further, it is required that

$$\left| \int_0^T G_i(t)G_j(t - KTs)\, dt \right| \ll \int_0^T G_j^2(t)\, dt \qquad i \neq j \qquad \forall k$$

so that the ith channel will not "interfere" with the jth channel for any time shift between them. This "interference" problem extends to acquisition, tracking, and data demodulation. The codes that we will be discussing will be codes named after Gold [1], [2], which have uniformly low cross correlation between different channels. CDM accessing does not require the time synchronization needed in time division multiple access (TDMA) nor the bandwidth requirement of frequency division multiple access (FDMA).

543

11.1 CORRELATION PROPERTIES OF PERIODIC BINARY SEQUENCES

Recall in Chapter 7 that we defined the periodic cross correlation function of two periodic binary sequences a and b, having N symbols, by

$$R_{ab}(k) = \sum_{0}^{N-1} a(i)b(i-k) \qquad (11.1\text{-}1)$$

We found in Chapter 7 that $R_{aa}(k)$, for a (maximal length) PN sequence, was either -1 or $+N$ and was a two-valued function. Most other codes do not possess this property. Since sequences are not transmitted, but rather, waveforms are, we now consider the cross correlation function for periodic waveforms. For a sequence a, the associated time waveform is given by

$$f_a(t) = \sum_{k=-\infty}^{\infty} a(k)s(t-kT_c) \qquad (11.1\text{-}2)$$

where

T_c is the symbol or chip duration.

$a(k)$ is the periodic sequence.

$s(t)$ is the baseband code symbol (such as NRZ) and is nonzero in $(0, T_c)$.

As an example, consider a PN code of length 7; the sequence and the waveform are shown below in Figure 11.1. Now we obtain the cross correlation function of two periodic waveforms.
    Let

$$f_a(t) = \sum_{k=-\infty}^{\infty} a(k)s_a(t-kT_c) \qquad (11.1\text{-}3)$$

$$f_b(t) = \sum_{j=-\infty}^{\infty} b(k)s_b(t-jT_c) \qquad (11.1\text{-}4)$$

We will show that the cross correlation function is given by [3]

$$R_{f_a f_b} = \frac{1}{NT_c} \sum_{l=-\infty}^{\infty} R_{ab}(l)R_{s_a s_b}(\tau - lT_c) \qquad (11.1\text{-}5)$$

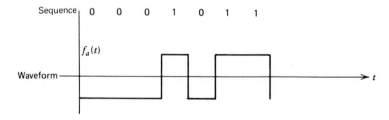

Figure 11.1 Sequence and waveform for a length 7 PN code.

where

$$R_{ab}(l) = \sum_{k=0}^{N-1} a(k)b(k-l) \qquad (11.1\text{-}6)$$

is the sequence cross correlation function and

$$R_{s_a s_b}(\tau) = \int_{-\infty}^{\infty} s_a(t)s_b(t-\tau)\,dt \qquad (11.1\text{-}7)$$

is the code symbol cross correlation function.

By definition of a periodic cross correlation function

$$R_{f_a f_b}(\tau) = \frac{1}{NT_c} \int_0^{NT_c} f_a(t)f_b(t-\tau)\,dt \qquad (11.1\text{-}8)$$

or

$$R_{f_a f_b}(\tau) = \frac{1}{NT_c} \int_0^{NT_c} \left[\sum_{k=-\infty}^{\infty} a(k)s_a(t-kT_c)\right]\left[\sum_{j=-\infty}^{\infty} b(j)s_b(t-\tau-jT_c)\right] dt$$

or equivalently

$$R_{f_a f_b}(\tau) = \frac{1}{NT_c} \sum_{j=-\infty}^{\infty} \sum_{k=0}^{N-1} a(k)b(j) \int_0^{NT_c} s_a(t-kT_c)s_b(t-\tau-jT_c)\,dt \qquad (11.1\text{-}9)$$

since the integral is zero when k is outside the interval $(0, N-1)$. Now since $s_a(t-kT_c)$ is nonzero only when $t \in (kT_c, (k+1)T_c)$, we have

$$R_{f_a f_b}(\tau) = \frac{1}{NT_c} \sum_{j=-\infty}^{\infty} \sum_{k=0}^{N-1} a(k)b(j) \int_{kTc}^{(k+1)T_c} s_a(t-kT_c)s_b(t-\tau-jT_c)\,dt \qquad (11.1\text{-}10)$$

Now let $u = t - kT_c$, then $t - T - jT_c = u - T + (k-j)T_c$, Hence

$$R_{f_a f_b}(\tau) = \frac{1}{NT_c} \sum_{j=-\infty}^{\infty} \sum_{k=0}^{N-1} a(k)b(j) \int_0^{T_c} s_a(u)s_b(u-\tau+(k-j)T_c)\,du \qquad (11.1\text{-}11)$$

Define $l = k - j$. Then

$$R_{f_a f_b}(\tau) = \frac{1}{NT_c} \left[\sum_{k=0}^{N-1} a(k)b(k-l)\right] \int_{-\infty}^{\infty} s_a(u)s_b(u-\tau+lT_c)\,du \qquad (11.1\text{-}12)$$

since $s_a(u)$ is zero outside $(0, T_c)$. Using the definition of $R_{ab}(l)$ of (11.1-6), we obtain (11.1-5).

PROBLEM 1

Determine the autocorrelation function $R_{f_a f_a}(\tau)$ for the case of the PN code of length 7 shown in Figure 11.1

11.2 MEAN AND VARIANCE OF PARTIAL CROSS CORRELATION FUNCTION

Before we discuss Gold codes, let us first discuss the partial correlation function. In a typical communication system receiver, the code acquisition subsystem (despreader) correlates the received signal against the reference code, as shown in Figure 11.2. The correlator is simply a multiplier and a bandpass filter. If the bandpass filter bandwidth (denoted by B) is very narrow, so that $(B/2)^{-1} \gg NT_c$, then the periodic cross correlation function is realized with heavy postdetection filtering.* If, however, $(B/2)^{-1} < NT_c$, then a partial correlation will be realized. In any case, the output of the correlator-squarer combination is approximately proportional to the cross correlation function defined by

$$R_{f_a f_b}(\tau) = \frac{1}{T_{\text{cor}}} \int_0^{T_{\text{cor}}} f_a(t) f_b(t + \tau) \, dt \qquad (11.2\text{-}1)$$

where T_{cor} is the correlation time (approximate integration time) of the bandpass filter. When $T_{\text{cor}} < NT_c = $ code period, it is possible that a nearly orthogonal code set (over NT_c) may be rather highly correlated over some small fraction of a code period.

Our first result concerns the average partial correlation.

Theorem 1 Let a be a maximal length sequence of length N. Then

$$\overline{R_a^W(k)} = R_a(k) \qquad W \le N \qquad (11.2\text{-}2)$$

where the partial correlation of the sequence a is given by

$$R_a^W(k) = \frac{1}{W} \sum_{i=0}^{W-1} a(i)a(i + k) \qquad (11.2\text{-}3)$$

and the averaging denoted by the overbar, is over all N starting positions.

Proof By definition:

$$\overline{R_a^W(k)} = \frac{1}{N} \sum_{j=0}^{N-1} \left[\frac{1}{W} \sum_{i=0}^{W-1} a(i + j)a(i + j + k) \right] \qquad (11.2\text{-}4)$$

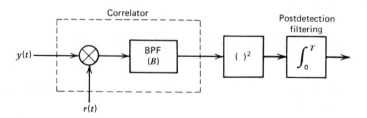

Figure 11.2 Typical acquisition circuit correlator-detector.

*That is we require the integration time to be a multiple of the code length.

Interchanging the order of summations, we obtain

$$\overline{R_a^W(k)} = \frac{1}{W} \sum_{i=0}^{W-1} \left[\frac{1}{N} \sum_{j=0}^{N-1} a(i+j)a(i+j+k) \right] \tag{11.2-5}$$

$$= \frac{1}{W} \sum_{i=0}^{W-1} R_a(k) = R_a(k) \tag{11.2-6}$$

Hence, the average partial correlation, averaged over all starting positions, is just the periodic correlation function. Next we present a result on the variance of the partial correlation.

Theorem 2 Let a be a maximal length sequence of length N. Then for $k \neq 0$ and $W < N$:

$$\text{var}[R_a^W(k)] = \frac{1}{W} \left(1 + \frac{1}{N} \right) \left(1 - \frac{W}{N} \right) \tag{11.2-7}$$

where the averaging is taken over all N starting positions.

Proof Now

$$\text{var}[R_a^W(k) = \overline{[R_a^W(k)]^2} - \{\overline{[R_a^W(k)]}\}^2 \tag{11.2-8}$$

Since we have already computed the second term, consider the first one:

$$\overline{[R_a^W(k)]^2} = \frac{1}{N} \sum_{j=0}^{N-1} \left[\frac{1}{W} \sum_{i=0}^{W-1} a(i+j)a(i+j+k) \right]^2 \tag{11.2-9}$$

so that

$$\overline{[R_a^W(k)]^2} = \frac{1}{NW^2} \sum_{j=0}^{N-1} \sum_{i=0}^{W-1} \sum_{l=0}^{W-1} a(i+j)a(i+j+k)a(l+j)a(l+j+k) \tag{11.2-10}$$

Using the shift and add property, we have

$$\overline{[R_a^W(k)]^2} = \frac{1}{NW^2} \sum_{j=0}^{N-1} \left[\sum_{i=0}^{W-1} \sum_{l=0}^{W-1} a_s(i+j+k)a_s(l+j+k) \right] \tag{11.2-11}$$

where a_s is a shifted version of the a sequence. Now break the double sum into a diagonal and an off-diagonal expansion:

$$\overline{[R_a^W(k)]^2} = \frac{1}{NW^2} \sum_{j=0}^{N-1} \left[\sum_{i=0}^{W-1} 1 + \sum_{\substack{i=0 \\ i \neq l}}^{W-1} \sum_{l=0}^{W-1} a_s(i+j+k)a_s(l+j+k) \right] \tag{11.2-12}$$

or

$$\overline{[R_a^W(k)]^2} = \frac{1}{W} - \frac{1}{NW}(W-1) \tag{11.2-13}$$

or

$$\overline{[R_a^W(k)]^2} = \frac{1}{NW^2}[NW + (-1)W(W-1)] \tag{11.2-14}$$

So that

$$\text{var}[R_a^W(k)] = \frac{1}{W} - \frac{W-1}{NW} - \left(-\frac{1}{N}\right)^2 \qquad k \neq 0 \qquad (11.2\text{-}15)$$

or

$$\text{var}[R_a^W(k)] = \frac{1}{W}\left(1 + \frac{1}{N}\right)\left(1 - \frac{W}{N}\right) \qquad (11.2\text{-}16)$$

as was to be shown.

PROBLEM 2

Show by direct calculation that $\text{var}[R_a^W(k)] = 0$ for $W = N$. Do not use (11.2-16).

Notice that for $W = N$, $\text{var}[R_a^W(k)] = 0$ and for W/N small, $\text{var}[R_a^W(k)] \cong 1/W$. Gold[3] and others have shown that the hypergeometric distribution has the same mean and variance ($k \neq 0$) of partial correlation values, and hence it can be used as a model for estimating partial correlation values.

We have seen from the above results that partial correlation values may differ drastically from full cycle (periodic) correlation values. In the next few sections, we present some periodic correlation results for preferred pairs of maximal length sequences and for Gold codes.

PROBLEM 3

Let a and b be any two binary (± 1) sequences having period N. Then show that

$$\sum_{k=0}^{N-1} R_{ab}(k) = R_a R_b$$

where

$$R_{ab}(k) = \sum_{i=0}^{N-1} a(i)b(i-k)$$

$$R_a = \sum_{i=0}^{N-1} a(i)$$

PROBLEM 4

Let a and b be any two binary (± 1) sequences having period N. Then show that

$$\sum_{k=0}^{N-1} R_{ab}^2(k) = \sum_{k=0}^{N-1} R_{aa}(k)R_{bb}(k)$$

with $R_{ab}(k)$ defined in Problem 3.

PROBLEM 5

Show that when a and b are maximal length sequences that

$$\sum_{k=0}^{N-1} R_{ab}^2(k) = N^2 + N - 1$$

We now use the results of Problem 4 to establish a lower bound on the peak value of the periodic auto- and cross correlations for two digital sequences of the same length. This result is due to Stalder and Cahn [7]. Let max R_{ab} denote the maximum value of the cross correlation between digital sequences a and b, both of length N. Further, let max R_a and max R_b be the largest autocorrelation for $k \neq 0$ of sequences a and b, respectively. From Problem 4 we have

$$\sum_{k=0}^{N-1} R_{ab}^2(k) = \sum_{k=0}^{N-1} R_{aa}(k) R_{bb}(k) \tag{11.2-17}$$

It is obvious that the left side of (11.2-17) is upper bounded by

$$N \cdot \max R_{ab}^2 \tag{11.2-18}$$

Further, the right side cannot be less than

$$N^2 - (N-1) \cdot \max R_a \cdot \max R_b \tag{11.2-19}$$

Therefore

$$N \max R_{ab}^2 \geq N^2 - (N-1) \cdot \max R_a \cdot \max R_b \tag{11.2-20}$$

It then follows that

$$\sqrt{\max R_{ab}^2 + \max R_a \cdot \max R_b} > \sqrt{N} \qquad \text{if } \max R_a \cdot \max R_b \geq 0 \tag{11.2-21}$$

When the sequences are maximal length sequences, then

$$\max R_a = \max R_b = -1 \tag{11.2-22}$$

so that

$$\max R_{ab} > \sqrt{N-1} \tag{11.2-23}$$

Therefore, the maximum cross correlation for maximal length sequences is greater than $\sqrt{N-1}$, where N is the code length.

PROBLEM 6

Show that if it is desired that the autocorrelation and cross correlation peaks be equal that

$$\max R_{ab} = \max R_{aa} = \max R_{bb} \geq \frac{\sqrt{N}}{\sqrt{2}}$$

or

$$\text{the normalized peaks} \geq \frac{1}{\sqrt{2N}}$$

11.3   PREFERRED PAIRS OF MAXIMAL LENGTH CODES

In this section, we will state a result that allows one to pick pairs of maximal length codes that have uniformly small cross correlation. Recall from Chapter 7 that there are $\phi(2^n - 1)/n$ possible distinct maximal length shift register tap connections with $\phi(\cdot)$ the Euler-ϕ function. The following result is due to Gold [3]. Preferred pairs have the property that they have the minimum cross correlation value.

Theorem 3 Let $f(x)$ be a primitive polynomial of degree n such that n is not divisible by 4. Let α be a root of $f(x)$, that is, $f(\alpha) = 0$. Let $g(x)$ be the irreducible polynomial such that

$$\alpha^{[2^{(n-1)/2}+1]} \text{ is a root of } g(x) \text{ for } n \text{ odd} \tag{11.3-1}$$

$$\alpha^{[2^{(n-2)/2}+1]} \text{ is a root of } g(x) \text{ for } n \text{ even} \tag{11.3-2}$$

Let a be a sequence generated by the shift register corresponding to the polynomial $f(x)$. Let b be any sequence generated by the shift register corresponding to the polynomial $g(x)$. Let $R_{ab}(k)$ be the periodic cross correlation function of the sequence a and b. Then $g(x)$ is a primitive polynomial of degree n and

$$\begin{array}{ll} |R_{ab}(k)| \leq 2^{(n+1)/2} + 1 & \text{for } n \text{ odd} \\ |R_{ab}(k)| \leq 2^{(n+2)/2} + 1 & \text{for } n \text{ even and not divisible by 4} \end{array} \tag{11.3-3}$$

To illustrate the use of this thoerem, we consider the case $n = 6$. From Peterson and Weldon [4, p. 476] we have as the first entry for degree 6

$$1 \qquad 103F \tag{11.3-4}$$

which denotes $\alpha^1 = \alpha$ is a root of the primitive polynomial*

$$103 \tag{11.3-5}$$

Therefore

$$f(x) = x^6 + x + 1 \tag{11.3-6}$$

We need an irreducible polynomial such that $\alpha^{2^2+1} = \alpha^5$ is a root. From reference 4, p. 476, we see that 5 147H is listed for 6th degree polynomials, which is primitive as is stated in the theorem, and further α^5 is a root of $g(x) = x^6 + x^5 + x^2 + x + 1$. It is easy to verify that $g(\alpha^5) = 0$ using $f(\alpha) = 0$. From the theorem we conclude that

$$|R_{ab}(k)| \leq 2^4 + 1 = 17 \tag{11.3-7}$$

From Gold [3, p. A-4], we find that the actual correlation values are 15 for 10 phase positions, -1 for 47 phase positions, and -17 for 6 phase positions.

   If we had not used preferred pairs and instead used the 103 and 155 $(x^6 + x^5 + x^3 + x^2 + 1)$ codes, then the cross correlation could be as high as

---

*All polynomials with E, F, G, and H on the right are primitive in Peterson and Weldon's appendix.

23. To dramatize the advantage of preferred pairs further, let $n = 13$. The cross correlation can be as high as 703 out of a possible 8191. With preferred pairs, however, the maximum cross correlation is bounded by $2^{(13+1)/2} + 1 = 129$.

Now since, for a given code length, there are only a limited number of codes that are preferred pairs, we present a code generating procedure [3] that provides $2^n + 1$ codes in the family. The previous theorem only claims low cross correlation for one pair!

11.4 GOLD CODES

In this section we present the results for Gold codes. They have been used on such projects as the global positioning satellite and the tracking data relay satellite system as multiple-access codes.

Theorem 4 Let $f(x)$ and $g(x)$ be a preferred pair of primitive polynomials of degree n whose corresponding shift registers generate maximal length sequences of period $2^n - 1$ and whose cross correlation function $R(k)$ satisfies the inequality

$$|R(k)| \leq \begin{cases} 2^{(n+1)/2} + 1 & n \text{ odd} \\ 2^{(n+2)/2} + 1 & n \text{ even} \end{cases} \quad n \neq 0 \bmod 4 \qquad (11.4\text{-}1)$$

Then the shift register corresponding to the product polynomial $f(x)g(x)$ will generate $2^n + 1$ different sequences, each of period $2^n - 1$ and such that the cross correlation of any pair satisfies the above inequality. All $2^n + 1$ distinct codes are called the family of Gold codes of period $2^n + 1$.

As an example of a product polynomial, consider our preferred pair of polynomials from the example of Section 11.3:

$$f(x) = x^6 + x + 1$$
$$g(x) = x^6 + x^5 + x^2 + x + 1 \qquad (11.4\text{-}2)$$

Then the product polynomial is given by

$$f(x)g(x) = x^{12} + x^{11} + x^8 + x^6 + x^5 + x^3 + 1 \qquad (11.4\text{-}3)$$

and the corresponding shift register will generate 65 ($2^6 + 1$) different linear sequences of period 63. The cross correlation of any pair of sequences will satisfy the inequality $|R(k)| \leq 17$. The associated normalized cross correlation is 17/63.

The $2^n + 1$ members include the $2^n - 1$ phase shifts of one member of the preferred pair relative to the other preferred pair, plus each member itself.

It is not difficult to show that the product polynomial can be realized either as a single shift register of length $2n$ or as the sum of the two

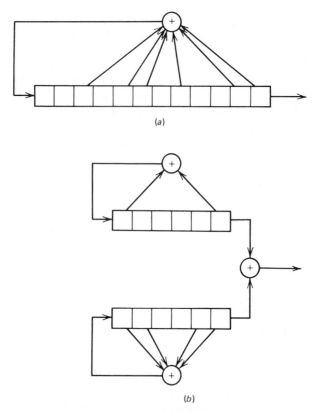

Figure 11.3 Gold code generator of length 63 symbols. (*a*) Single-shift register form (*n* = 12). (*b*) Double-shift register realization (*n* = 6).

preferred pair shift register equivalents. The two equivalent Gold code generators are shown in Figure 11.3.

Gold later established [1], after the result of Theorem 4 that Gold codes have three-level cross correlation values with cross correlation and relative frequencies shown below in Table 11.1 (see [5] also).

Thus we see that when *n* is even (and not 0 mod 4) the codes are more often of very low cross correlation ($-1/L$) for 75% of the code words correlated.

In fact Gold [6] has shown that the autocorrelation function takes on the same correlation values but with different frequencies of occurrence. It should be noted that these correlation values (both cross and auto) assume that both codes are synchronous with the same symbol timing. In a typical CDMA environment, this would not be true. However, they are useful for an initial estimate.

The cross correlation values are, in fact, the dc component of the spectra of the product of the two different Gold codes. When doppler is present (such that its magnitude is larger than the sum of the first few

Table 11.1 Three-level cross correlation properties of gold codes

n (Register Length)	Code Length	Normalized Cross Correlation	Frequency of Occurrence
n Odd	$L = 2^n - 1$	$-1/L$ $-(2^{(n+1)/2} + 1)/L$ $(2^{(n+1)/2} - 1)/L$	~ 0.50 ~ 0.25 ~ 0.25
n Even and not divisible by 4	$L = 2^n - 1$	$-1/L$ $-(2^{(n+2)/2} + 1)/L$ $(2^{(n+2)/2} - 1)/L$	~ 0.75 ~ 0.125 ~ 0.125

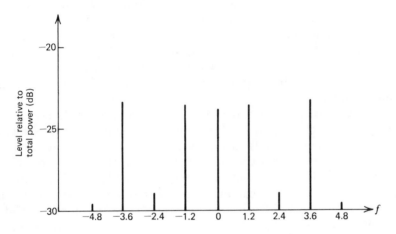

Figure 11.4 Spectral components in a 511 Gold code for a particular phase.

Fourier components of the products) they too should be considered since they will be present at the output of the bandpass filter correlation. To illustrate this point we present a portion of the product of two distinct Gold codes of length $2^9 - 1 = 511$ with a chip rate of 0.625 MBPS. The resulting spectrum is illustrated in Figure 11.4. As can be seen from the figure considerable additional power resides within $\pm 4\,\text{kHz}$ of dc.

11.5 GENERATION OF BALANCED GOLD CODES

Gold [1] has shown that Gold codes can be broken into three classes of balance. A *balanced code* is one in which the number of "ones" exceeds the number of "zeros" by one. The other two classes have an excess and deficiency of "ones" [6]. For n odd Gold [1] has shown that the number of

Table 11.2 Number of balanced and unbalanced codes
for n odd

Set	Number of "Ones" in Code Sequence	Number of Codes with This Number of "Ones"
1	2^{n-1}	$2^{n-1} + 1$
2	$2^{(n-1)} + 2^{(n-1)/2}$	$2^{(n-2)} - 2^{(n-3)/2}$
3	$2^{n-1} - 2^{(n-1)/2}$	$2^{n-2} + 2^{(n-3)/2}$

"ones" and the number of codes with that number of "ones" is as shown in
Table 11.2.

Notice that in the first set there are 2^{n-1} "ones" and therefore $2^{n-1} - 1$
"zeros" and therefore set 1 is balanced. Sets 2 and 3 are not balanced. In
set 2, for example, there are $2^{n-1} - 1$ "zeros". Because balanced codes have
more desirable spectral characteristics, we show, following Gold [6] how to
generate balanced codes; that is, we generate Gold codes of the first set.
We do this by selecting the proper relative phases of the two original
maximal sequences.

First we must determine the characteristic phase [6]. Every maximal
sequence has a characteristic phase. One important property is that if a
maximal PN sequence, in its characteristic phase, is sampled at every other
symbol, the same sequence results. Let $f(x)$ be the nth degree characteristic polynomial corresponding to the maximal PN shift register. We
have shown in Chapter 7 that any phase of the maximal sequence can be
represented by the ratio $g(x)/f(x)$, where $g(x)$ is the numerator of the
generating function and is of degree less than n. As we have seen, long
division of these polynomials results in a formal binary power series whose
binary coefficients are the symbols of the sequence generated by the shift
register. The formula for the polynomial $g(x)$ that results in the characteristic phase for the maximal length sequence has been shown by Gold [9]
to be given by

$$g(x) = \frac{d[xf(x)]}{dx} \qquad f(x) \text{ odd degree} \tag{11.5-1}$$

$$g(x) = f(x) + \frac{d[xf(x)]}{dx} \qquad f(x) \text{ even degree} \tag{11.5-2}$$

Differentiation is carried out in the usual way with coefficients interpreted,
mod 2.

As an example consider the characteristic polynomial $f(x) = 1 + x + x^3$.
We compute

$$g(x) = \frac{d(x + x^2 + x^4)}{dx} = 1 \qquad \text{mod } 2 \tag{11.5-3}$$

Therefore the characteristic phase is given by

$$G(x) = \frac{1}{1 + x + x^3} \tag{11.5-4}$$

By long division we have

$$
\begin{array}{l}
 \underline{1 + x + x^2 + x^4 + x^7 + x^8 + x^9 \cdots} \\
1 + x + x^3 \,\overline{\big|\, 1} \\
 \underline{1 + x + x^3} \\
 x + x^3 \\
 \underline{x + x^2 + x^4} \\
 x^2 + x^3 + x^4 \\
 \underline{x^2 + x^3 + x^5} \\
 x^4 + x^5 \\
 \underline{x^4 + x^5 + x^7} \\
 x^7 \\
 \underline{x^7 + x^8 + x^{10}}
\end{array}
\tag{11.5-5}
$$

We conclude that the initial conditions must have been 111 to yield the successive "ones". The sequence and the sequence represented by odd numbered symbols is given by

$$
\begin{array}{ll}
\text{sequence} & 1\ 1\ 1\ 0\ 1\ 0\ 0\ 1\ 1\ 1\ 0\ 1\ 0\ 0 \cdots \\
\text{sampled} & 1\ 1\ 1\ 0\ 1\ 0\ 0 \cdots \\
\text{sequence} &
\end{array}
\tag{11.5-6}
$$

11.5.1 Relative Phase Requirement for Balanced Codes

We shall now describe the relative phase in which the preferred pair of maximal PN sequences must be added in order to result in a balanced member of the family.

Let a and b be the preferred pair of maximal sequences in their characteristic phase. When x is of odd degree it is clear that the generator polynomial is of the form

$$G(x) = \frac{1 + c(x)}{1 + d(x)} \tag{11.5-7}$$

where the degree of $d(x)$ is n and the degree of $c(x)$ is not greater than $n - 1$. By long division it is clear that the quotient will be of the form $1 + \cdots$ so that the initial symbol of the characteristic sequence will be a "one".

Theorem 5 Any relative phase shifts of the sequence a and b (in their characteristic phase) that are obtained by shifting the sequence b until its initial "one" corresponds to a "zero" in the sequence a will result in a balanced Gold code when the two sequences are added together mod 2 [6].

Let us consider an example. Let $n = 3$ and $f(x) = x^3 + x + 1$. Now α is a root of $f(x)$. By Theorem 3 α^3 is a root of the other preferred pair. Since the polynomial for α^3 is not listed in reference 4, try the reciprocal polynomial, that is, $g(x) = x^3 + x^2 + 1$. We find that $g(\alpha^3) = 0$. Hence the two sequences are

$$a = 1\ 1\ 1\ 0\ 1\ 0\ 0 \quad \left(\frac{1}{1 + x + x^3} = 1 + x + x^2 + \cdots \right)$$

$$b = 1\ 0\ 0\ 1\ 0\ 1\ 1 \quad \left(\frac{1 + x^2}{1 + x^2 + x^3} = 1 + 0x + 0x^2 + \cdots \right) \tag{11.5-8}$$

If the sequence b is shifted cyclically three, five or six positions to the right, then the initial "one" in sequence b will be under a "zero" in sequence a. Addition of the two sequences in all cases leads to a balanced Gold code:

$$
\begin{array}{ccc}
\begin{array}{c} 1\ 1\ 1\ 0\ 1\ 0\ 0 \\ 0\ 1\ 1\ 1\ 0\ 0\ 1 \\ \hline 1\ 0\ 0\ 1\ 1\ 0\ 1 \end{array}
&
\begin{array}{c} 1\ 1\ 1\ 0\ 1\ 0\ 0 \\ 0\ 1\ 0\ 1\ 1\ 1\ 0 \\ \hline 1\ 0\ 1\ 1\ 0\ 1\ 0 \end{array}
&
\begin{array}{c} 1\ 1\ 1\ 0\ 1\ 0\ 0 \\ 0\ 0\ 1\ 0\ 1\ 1\ 1 \\ \hline 1\ 1\ 0\ 0\ 0\ 1\ 1 \end{array}
\end{array}
\tag{11.5-9}
$$

If, however the code is shifted by any other phase, say the zero shift, an unbalanced code is produced:

$$
\begin{array}{c}
1\ 1\ 1\ 0\ 1\ 0\ 0 \\
1\ 0\ 0\ 1\ 0\ 1\ 1 \\
\hline
0\ 1\ 1\ 1\ 1\ 1\ 1
\end{array}
\tag{11.5-10}
$$

11.5.2   Initial Conditions for Balanced Gold Codes

It is clear now how to proceed to produce balanced Gold codes. The Gold code generator is sketched in Figure 11.5. The initial conditions for shift register 2 are those initial conditions, of the maximal length sequence that determine the characteristic phase of the maximal PN sequence generated by shift register 2. These initial conditions are determined such that the numerator of the generating functions is determined by

$$g(x) = \frac{d[xf(x)]}{dx} \qquad n \text{ odd}$$

$$g(x) = f(x) + \frac{d[xf(x)]}{dx} \qquad n \neq 0 \bmod 4 \tag{11.5-11}$$

as before. Then the initial conditions are obtained by long division of $g(x)/f(x)$ to provide the first n coefficients, which are, in fact, the n initial conditions.

The initial conditions for the shift register 1 are only subject to the constraint that the first stage (the one on the right) contain a zero.

As an example of the above technique we shall construct a balanced

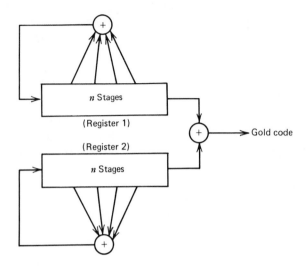

Figure 11.5   Gold code generator configuration.

Gold code of period $2^{11} - 1$. A preferred pair is found in Appendix C of reference 4:

$$1 \qquad 4005 \qquad f(x) = 1 + x^2 + x^{11} \qquad\qquad (11.5\text{-}12)$$

$$33 \qquad 7335 \qquad g(x) = 1 + x^2 + x^3 + x^4 + x^6 + x^7 + x^9 + x^{10} + x^{11}$$

since $2^{(n-1)/2} + 1 = 2^5 + 1 = 33$. The characteristic sequence generated by the shift register corresponding to polynomial 4005 is represented by the ratio

$$\frac{g(x)}{1 + x^2 + x^{11}} \qquad\qquad (11.5\text{-}13)$$

where

$$g(x) = \frac{d}{dx}(x + x^3 + x^{12}) = 1 + x^2 \qquad\qquad (11.5\text{-}14)$$

The initial conditions required for this register are found from the quotient

$$\frac{1 + x^2}{1 + x^2 + x^{11}} = 1 + x^{11} + \cdots \qquad\qquad (11.5\text{-}15)$$

The initial conditions for register 8 becomes

$$[0\ 0\ 0\ 0\ 0\ 0\ 0\ 0\ 0\ 0\ 1] \qquad\qquad (11.5\text{-}16)$$

The only constraint on the initial conditions of register A is that the entry in the first stage be zero. Hence our Gold code encoder is shown in Figure 11.6.

PROBLEM 7

Design in block diagram form a balanced Gold code of length $2^{13} - 1$.

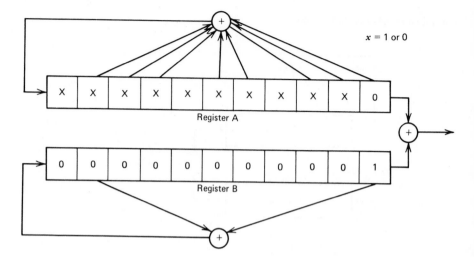

$x = 1$ or 0

Figure 11.6 Balanced Gold code of length $2^{11} - 1$.

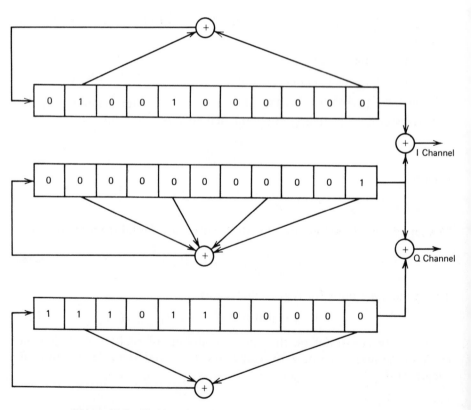

Figure 11.7 Gold code pair generator for mode 2 return link.

To generate pairs of Gold codes, all that is necessary to do is to use two distinct phases of register A and add them to register B.

An example of the TDRSS staggered quadriphase signals is shown in Figure 11.7 [10] based on the above stated extension of Figure 11.6.

11.6 FALSE CODE LOCK WITH MAXIMAL LENGTH AND GOLD CODES

The purpose of this section is to demonstrate the possibility of false code lock due to the use of staggered quadriphase PN type transmission when the transmitter filters and then limits [10, 11]. This phenomenon was apparently first reported at Magnavox [11].

Staggered quadriphase PN (SQPN) is generated in the transmitter by biphase modulating a pair of carriers in phase quadrature with two binary PN type codes. The codes are displaced one-half chip, so that the transition of one corresponds to the midpoint of the other. An SQPN signal can be represented by

$$s(t) = a(t) \cos \omega_0 t + b\left(t - \frac{T_c}{2}\right) \sin \omega_0 t \qquad (11.6\text{-}1)$$

where $a(t)$ and $b(t)$ are assumed to be NRZ waveforms taking on the values ± 1.

In our discussion here we shall assume that $a(t)$ and $b(t)$ are distinct maximal length codes or Gold codes. We shall show that there is a cross coupling between the inphase and quadrature codes that can lead to false code lock, especially under strong signal conditions when the SQPN signal is prefiltered and then limited.

We shall show that after prefiltering and limiting the resulting waveform on the $a(t)$ channel can be approximately represented by

$$\tilde{a}_L(t) \cong \tilde{a}(t)\left\{1 + \frac{p(t)}{2}\left[1 - b\left(t + \frac{T_c}{2}\right)b\left(t - \frac{T_c}{2}\right)\right]\right\} \qquad (11.6\text{-}2)$$

where the tilde over the letter denotes the filtered version and the L denotes the limiter output. Hence $\tilde{a}_L(t)$ denotes filtering and limiting. The pulse waveform $p(t)$ is dependent upon the amount of prefiltering used. In Figure 11.8 the inphase and quadrature signals are shown before and after filtering and after hard limiting.

It is clear from Figure 11.8 that a pulse amplitude modulation occurs when a transition occurs during the middle of the $a(t)$ waveform symbol. Therefore to describe the phenomenon a mathematical characterization of the occurrence of the transition must be specified. Consider Figure 11.9, which contains an unfiltered version of $b(t)$, a delayed version of $b(t)$, and the resulting product. Notice that when a transition occurs on $b(t - T_c)$ the product $b(t)b(t - T_c)$ is negative in the symbol duration prior to the

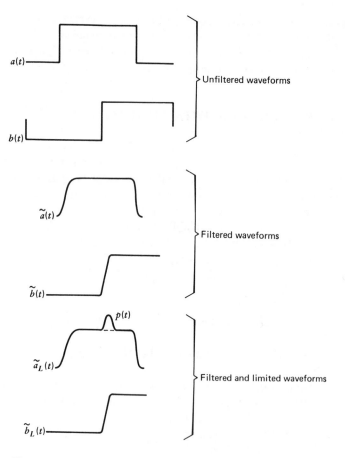

Figure 11.8 I and Q waveform after filtering and limiting.

transition as shown by the solid arrows. Alternately the product is negative in the symbol duration just after the transition in $b(t)$, as shown by the dashed arrows.

Therefore, if we denote by $p(t)$ the pulse waveform riding on $a(t)$, as shown in Figure 11.8, then we have

$$\tilde{a}_L(t) \cong \tilde{a}(t)\left\{1 + \frac{p(t)}{2}\left[1 - b\left(t + \frac{T_c}{2}\right)b\left(t - \frac{T_c}{2}\right)\right]\right\} \qquad (11.6\text{-}3)$$

Notice $\frac{1}{2}[1 - b(t)b(t - T_c)]$ is unity at a transition of $b(t)$ and zero when there is no transition. In order to simplify the following calculations we define m by

$$m = \frac{\displaystyle\int_{-T_c/2}^{T_c/2} p(t)\, dt}{T_c} \qquad (11.6\text{-}4)$$

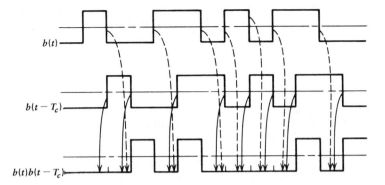

Figure 11.9 Sketch showing that $b(t)b(t - T_c)$ is a transition indicator function.

where T_c is the chip duration. Using (11.6-4) we approximate $\tilde{a}_L(t)$ by

$$\tilde{a}_L(t) \cong a(t)\left\{1 + \frac{m}{2}\left[1 + b\left(t - \frac{T_c}{2}\right)b\left(t - \frac{T_c}{2}\right)\right]\right\} \qquad (11.6\text{-}5)$$

We shall use (11.6-5) to evaluate both maximal length (PN) codes and Gold codes.

First consider the case when $a(t)$ and $b(t)$ are delayed versions of each other and are maximal length codes so that

$$b\left(t + \frac{T_c}{2}\right) = a(t - T) \qquad \begin{array}{l} T = mT_c \\ m = 1, 2, 3, \ldots \end{array} \qquad (11.6\text{-}6)$$

We obtain

$$\tilde{a}_L(t) = \left(1 + \frac{m}{2}\right)a(t) - \frac{m}{2}\,a(t)a(t - T)a(t - T - T_c) \qquad (11.6\text{-}7)$$

or

$$\tilde{a}_L(t) = \left(1 + \frac{m}{2}\right)a(t) - \frac{m}{2}a(t - kT_c) \qquad (11.6\text{-}8)$$

where we have made use of the shift-and-add property of maximal length codes. The parameter k is an integer that depends upon the particular maximal code used. Although k can be computed, *a priori*, it could lie anywhere within the code period. The implication of (11.6-8) is that false code lock could occur under higher signal conditions k chips away from true lock. The "false lock" level is about $20 \log(m/2)$ down from true lock. In reference 11 a simulation was performed with the result that for a six-pole butterworth filter having a transmitter bandpass bandwidth equal to 1.5 times the PN chip rate, the false lock was down about 20 dB. Decreasing the bandwidth makes the "false lock" more likely; conversely increasing the bandwidth decreases the chances of false lock.

Now we consider Gold codes. We model the staggered quadriphase Gold code signal with the two Gold code waveforms $a(t)$ and $b(t)$, where

$$a(t) = c(t)\,d(t)$$

$$b(t) = c\left(t - \frac{T_c}{2}\right) d\left(t - lT_c - \frac{T_c}{2}\right) \qquad (11.6\text{-}9)$$

with l an integer not equal to zero and $c(t)$ and $d(t)$ being preferred PN pairs. If the Gold codes are balanced, additional constraints are placed on admissible values of l. From (11.6-5) we have

$$\tilde{a}_L(t) = \left(1 + \frac{m}{2}\right) a(t) - \frac{m}{2} b\left(t + \frac{T_c}{2}\right) b\left(t - \frac{T_c}{2}\right) a(t)$$

or

$$\tilde{a}_L(t) = \left(1 + \frac{m}{2}\right) a(t) - \frac{m}{2}\, d(t)d(t - lT_c)c(t - T_c)d[t - (l+1)T_c]$$

or

$$\tilde{a}_L(t) = \left(1 + \frac{m}{2}\right) a(t) - \frac{m}{2}\, c(t - T_c)\, d(t - nT_c)$$

$$\tilde{a}_L(t) = \left(1 + \frac{m}{2}\right) a(t) - \frac{m}{2} g(t - qT_c)$$

where l, n, and q are known integers. Here $g(t)$ is a Gold code derived from $b(t)$. We conclude that there is a possibility of finding minor autocorrelation or cross correlation peaks when Gold codes are used.

Gold [10, section 3] has considered this problem and shown how to avoid this interference for the Gold codes used on TDRSS. Again the false lock level is about $20\log(m/2)$ down from true lock.

REFERENCES

1 Gold, R., "Maximal Recursive Sequences with 3-valued Recursive Cross-Correlation Functions," *IEEE Trans. Information Theory*, pp. 154–156, January 1968.

2 Gold, R., "Optimal Binary Sequences for Spread Spectrum Multiplexing," *IEEE Trans. Information Theory*, pp. 619–621, October 1967.

3 Gold, R., "Properties of Linear Binary Encoding Sequences," lecture notes, 1975.

4 Peterson, W. W., and Weldon, E. J., Jr., *Error Correcting Codes*, 2nd. ed., MIT Press, Cambridge, Mass., 1972.

5 Philco-Ford Corporation, "Global Positioning System C-Signal Multiple Access and Jamming Performance," Western Development Laboratories Report No. GPS-TM003C, December 27, 1973.

6 Gold, R., Associates, "TDRSS Telecommunication Systems PN Code Analysis Contract No. NAS 5-22546, Final Report," August 31, 1976.

7 Stalder, J. E., and Cahn, C. R., "Bounds for Correlation Peaks for Periodic Digital Sequences," Proc. IEEE Correspondence, October 1964.

8 LaFrieda, J. R., "Introduction to Gold Codes," Aerospace Corp. paper, undated.

9 Gold, R., "Characteristic Linear Sequences and their Cost Functions," *J. Siam. Appl. Math.*, Vol. 14, No. 5, pp. 980–985, September 1966.

10 Gold, R., "TDRSS Telecommunication System PN Code Analysis," Final Report Addendum, Contract NAS 5-22546, April 1, 1977.

11 The Magnavox Company, "TDRSS Telecommunications Study: Phase I—Final Report," Report No. R4958, September 15, 1974.

12 Lindholm, J. H., "An Analysis of the Pseudo-Randomness Properties of Subsequences of Long M Sequences," *IEEE Trans. Information Theory*, Vol. IT-14, No. 4, July 1968.

13 Dowling, T. A., and McEliece, R., "Cross-Correlations of Three Reverse Maximal-Length Shift Register Sequences," JPL SPS 37–53, Vol. III.

14 Gold, R. *et al.*, "Cross-Correlation Properties of Binary Sequences," Aeronautical System Division, Wright-Patterson AFB, Ohio, Report No. R-692, AD-431113, Vols. I–IV, January 1964.

15 Lerner, R. M., "Signals Having Good Correlation Functions," Wescon Convention Record, 1961.

16 Gold, R., "Location of Cross-Correlation Sidelobes of PN Sequences," ITC 79, San Diego, Cal. November 1979.

SYMBOL SYNCHRONIZATION FOR PSK

12

In this chapter we consider various types of symbol synchronizers. However, before we delve into the different types and the analysis of synchronization performance, let us consider the more general problems of symbol synchronization, word synchronization, and frame synchronization.

In Figure 12.1 is a typical symbol sequence of the type under discussion here. The symbols are shown grouped in units of eight (for this example), to form a *word*. A *frame* is composed of N words (N an integer) or $8N$ symbols. In order to obtain data the symbol timing must first be obtained via a symbol synchronizer. Then after symbol synchronization, word synchronization must be obtained. This is commonly accomplished by one or more words being a particular symbol sequence with desirable correlation properties such as a Barker code [1]. Barker codes have the property that the aperiodic correlation values are $\pm 1/N$, where N is the code length. However, there are only a few known Barker codes, so that for larger synchronization words, quasi-Barker codes or other types of codes must be used. The function of the frame synchronizer is to obtain, by means of correlation, the correct frame and word position. Once this has occurred, the symbol synchronizer provides the tracking function.

We will concentrate on the symbol synchronization aspects in this chapter. It should be pointed out that symbol synchronization and bit synchronization are used interchangeably in the literature. If the system is not coded, one uses a bit synchronizer; and if the system is coded, one uses a symbol synchronizer.

12.1 OPTIMUM BIT SYNCHRONIZATION

We now turn our attention to finding the optimum procedure to estimate symbol-stream-derived bit synchronization. We follow the work of McBride and Sage [2] in this section. Other related works include Stiffler [3], and Van Trees [4]. We model the observed, received, process by

$$y(t) = x(t, \tau) + n(t)$$

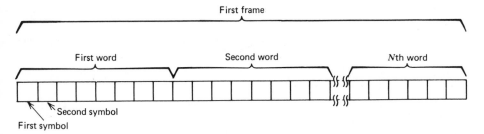

Figure 12.1 Demodulated (hard decision) symbol stream.

with τ the time uncertainty from the start of the observation to the first bit transition, with $x(t, \tau)$ the received baseband data stream, and with $n(t)$ being a sample function of WGN. In order to simplify the extension of the one-bit interval case to the more than one-bit interval case, we choose the time interval of the characterization to be

$$\tau + (j-1)T \le t \le \tau + jT \qquad \text{where } j = 0, 1, 2, \ldots, m$$

For this characterization, the sample waveforms specified for each bit interval are independent. The waveforms are shown in Figure 12.2 and it is assumed that the signals are equally likely. For further discussion of this characterization see reference 2.

We now use the Karhunen-Loeve (K.L.) expansion of a known signal in Gaussian noise to obtain the maximum *a posteriori* (MAP) estimate of τ. Each of the two signals shown in Figure 12.2 can be represented by a K.L.

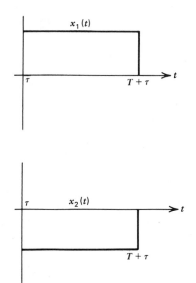

Figure 12.2 The two signals for NRZ data with starting times at the transitions.

expansion:

$$x_k(t, \tau) = \sum_{i=0}^{\infty} x_i^{(k)}(\tau)\phi_i(t) \tag{12.1-1}$$

where

$$x_i^{(k)}(\tau) = \int_{\tau+(j-1)T}^{\tau+jT} x_k(t, \tau)\phi_i(t)\, dt \tag{12.1-2}$$

where the eigenfunctions $\phi_i(t)$ satisfy

$$\lambda\phi_i(t) = \int_{\tau+(j-1)T}^{\tau+jT} R_n(t', t)\phi_i(t')\, dt' \tag{12.1-3}$$

and

$$E[n(t)n(t')] = R_n(t, t') \tag{12.1-4}$$

It can be shown that

$$R_n(t, t') = \sum_{i=1}^{\infty} \lambda_i\phi_i(t)\phi_i(t') \tag{12.1-5}$$

The eigenfunctions are orthonormal in the sense that

$$\int_{\tau+(j-1)T}^{\tau+jT} \phi_i(t)\phi_j(t)\, dt = \delta_{ij} \tag{12.1-6}$$

Further, the observed process $y(t)$ can be defined by letting

$$y_N(t) = \sum_{i=1}^{N} y_i\phi_i(t) \tag{12.1-7}$$

so that

$$y(t) = \lim_{N\to\infty} y_N(t) \tag{12.1-8}$$

The noise process can be expanded in a K.L. expansion so that

$$n(t) = \sum_{i=0}^{\infty} n_i\phi_i(t) \tag{12.1-9}$$

with the expansion constrained to produce uncorrelated coefficients n_i such that

$$E(n_i n_j) = \lambda_i\delta_{ij} \tag{12.1-10}$$

An n-dimensional representation of the observed process over the jth interval is

$$\sum_{i=0}^{N} y_i\phi_i(t) = \sum_{i=0}^{N} x_i^{(k)}(\tau)\phi_i(t) + \sum_{i=0}^{N} n_i\phi_i(t) \tag{12.1-11}$$

where clearly

$$y_i = x_i^{(k)}(\tau) + n_i \tag{12.1-12}$$

Since the random variables n_i are independent and Gaussian with density function

$$p(n_i) = \frac{1}{\sqrt{2\pi\lambda_i}} \exp\left(-\frac{n_i^2}{2\lambda_i}\right) \qquad (12.1\text{-}13)$$

the conditional density of y_i, given τ, becomes

$$p(y_i \mid \tau) = \sum_{k=1}^{2} \frac{1}{\sqrt{2\pi\lambda_i}} \exp\left[-\frac{1}{2}\frac{(y_i - x_i^{(k)}(\tau))^2}{\lambda_i}\right] p(k) \qquad (12.1\text{-}14)$$

where we have averaged over the two possible NRZ signals. We assume that $p(k) = \frac{1}{2}$; $k = 1$, 2. For an N-dimensional representation, (12.1-14) becomes

$$p(y_1, y_2, \cdots, y_N \mid \tau) = \frac{1}{2}\sum_{k=1}^{2}\left(\frac{1}{2\pi}\right)^{N/2}\prod_{i=1}^{N}\lambda_i^{-N/2}\exp\left[-\frac{1}{2\lambda_i}(y_i - x_i^{(k)}(\tau))^2\right]$$
$$(12.1\text{-}15)$$

When the noise process is white it can be shown that $\lambda_i = N_0/2$ for all i, and (12.1-15) can be written in vector notation as $y = (y_1, y_2, \ldots, y_N)$:

$$p(y \mid \tau) = \frac{1}{2}\sum_{k=1}^{2}(2\pi)^{-N/2}\left(\frac{N_0}{2}\right)^{-N/2}\exp\left\{-\frac{1}{2}(y - x^{(k)}(\tau))\left(\frac{2}{N_0}I\right)[y - x^{(k)}(\tau)]^T\right\}$$
$$(12.1\text{-}16)$$

where x^T denotes the transpose of x and where

$$x^{(k)}(\tau) = [x_1^{(k)}(\tau), x_2^{(k)}(\tau), \ldots, x_N^{(k)}(\tau)] \qquad (12.1\text{-}17)$$

and the covariance matrix is given by

$$E\left[\begin{Bmatrix} n_1 \\ n_2 \\ \vdots \\ n_N \end{Bmatrix}\{n_1, n_2, n_3, \ldots, n_N\}\right] = \frac{N_0}{2}I \qquad (12.1\text{-}18)$$

with I being an $N \times N$ unit matrix. For binary PSK we have

$$x^{(1)}(t, \tau) = -x^{(2)}(t, \tau) \qquad (12.1\text{-}19)$$

Multiplying out (12.1-16), we find that [letting $x^{(k)}(\tau) = x_k$ for convenience]

$$p(y \mid \tau) = \frac{1}{2}\sum_{k=1}^{2}(2\pi)^{-N/2}\left(\frac{2}{N_0}\right)^{N/2}\exp\left(-\frac{1}{N_0}yy^T\right)\exp\left(-\frac{1}{N_0}x_kx_k^T\right)$$
$$\times \exp\left(\frac{1}{N_0}yx_k^T\right)\exp\left(\frac{1}{N_0}x_ky^T\right) \qquad (12.1\text{-}20)$$

or

$$p(y \mid \tau) = (2\pi)^{-N/2}\left(\frac{2}{N_0}\right)^{N/2}\exp\left(-\frac{1}{N_0}yy^T\right)\exp\left(-\frac{1}{N_0}xx^T\right)$$
$$\times \left[\frac{\exp\left(\frac{2}{N_0}yx^T\right) + \exp\left(-\frac{2}{N_0}yx^T\right)}{2}\right]$$
$$(12.1\text{-}21)$$

where we have used

$$\mathbf{x} = \mathbf{x}_1 = -\mathbf{x}_2 \tag{12.1-22}$$

Recognizing the last term on the right of (12.1-21) as the hyperbolic cosine, we obtain

$$p(\mathbf{y} \mid \tau) = C_1 \cosh\left(\frac{2}{N_0} \mathbf{y}\mathbf{x}^T\right) \tag{12.1-23}$$

with C_1 a constant that does not depend upon τ. Now it can be shown that [2]

$$\frac{2}{N_0} \lim_{N \to \infty} (\mathbf{y}\mathbf{x}_1^T) = \frac{2}{N_0} \int_{\tau+(j-1)T}^{\tau+jT} y(t)x^{(1)}(t, \tau) \, dt \tag{12.1-24}$$

Furthermore, for L successive bits it can be shown that [2]

$$p(\mathbf{y} \mid \tau) = C_2 \prod_{j=1}^{L} \cosh\left[\frac{2}{N_0} \int_{\tau+(j-1)T}^{\tau+jT} y(t)x^{(1)}(t, \tau) \, dt\right] \tag{12.1-25}$$

where C_2 is a constant that does not depend upon τ. The maximum *a priori* estimate of τ, that is, $\hat{\tau}$, is the value that maximizes $p[y(t) \mid \tau]$, which, using Bayes' rule, becomes

$$p[\tau \mid y(t)] = \frac{p[y(t) \mid \tau] \, p(\tau)}{\int_\tau p[y(t) \mid \tau] \, p(\tau) \, d\tau} \tag{12.1-26}$$

and the fact that $p(\tau)$ is a density function assumed to be uniformly distributed over τ sec leads to the *a posteriori* probability density function:

$$p[\tau \mid y(t)] = C_3 \prod_{j=1}^{L} \cosh\left[\frac{2}{N_0} \int_{\tau+(j-1)T}^{\tau+jT} y(t)x^{(1)}(t, \tau) \, dt\right] \tag{12.1-27}$$

Now, since the logarithm is a monotonic function of its argument, we can find the value of τ that maximizes

$$\Lambda(\tau) = \ln[\tau/y(t)] = \sum_{j=1}^{L} \ln \cosh\left[\frac{2}{N_0} \int_{\tau+(j-1)T}^{\tau+jT} y(t)x^{(1)}(t, \tau) \, dt\right] \tag{12.1-28}$$

The open-loop synchronizer that implements (12.1-28) is shown in Figure 12.3. Note that the delay and sum allows summation over all L bits to make the decision. Since our concern is with closed-loop symbol synchronizers, we shall obtain a closed-loop (approximate) maximum likelihood structure next [5]. We now differentiate (12.2-28) with respect to τ to produce (for convenience we let the integration limits be $\pm \infty$),

$$\frac{\partial \Lambda(\tau)}{\partial \tau} = \sum_{j=1}^{L} \tanh\left[\frac{2}{N_0} \int_{\tau+(j-1)T}^{\tau+jT} y(t)x'(t, \tau) \, dt\right]\left\{\frac{2}{N_0} \int_{\tau+(j-1)T}^{\tau+jT} y(t) \frac{dx'(t, \tau)}{d\tau} \, dt\right.$$
$$\left. + \frac{2}{N_0} y(\tau + jT)\frac{dx'}{d\tau}(\tau + jT, \tau) - \frac{2}{N_0} y[\tau + (j-1)T]\frac{dx'}{d\tau}[\tau + (j-1)T, \tau]\right\} \tag{12.1-29}$$

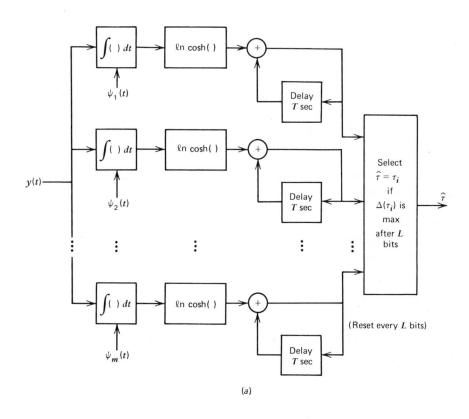

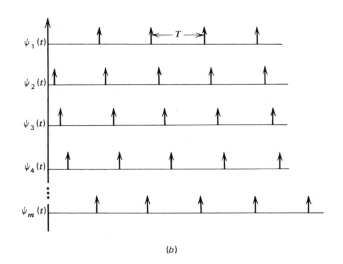

(b)

Figure 12.3 (a) Open-loop maximum *a posteriori* symbol synchronizer. (b) Associated bit timing controls.

We pick (arbitrarily)

$$\frac{dx'}{d\tau}(\tau + jT, \tau) = \frac{dx'}{d\tau}[\tau + (j-1)T, \tau] = 0$$

so that the closed-loop synchronizer suggested by (12.1-29) is shown in Figure 12.4 [5]. The effect of the differentiator is to provide an impulse, the sign of which depends upon the sign of $\tau - \hat{\tau} = \epsilon$(the timing error), so that an error signal is fed into the loop filter in such a way as to reduce the error.

One realization that is motivated by the loop of Figure 12.4 is the digital data tracking loop, which is shown in Figure 12.6. In this loop the midphase integrator provides a linear approximation to the lower channel of Figure 12.4.

Another realization that can be obtained is to approximate the derivative $\partial \Lambda(\tau)/\partial \tau$ by

$$\frac{\partial \Lambda(\tau)}{\partial \tau} \cong \frac{\Lambda(\tau + \frac{1}{2}\Delta\tau) - \Lambda(\tau - \frac{1}{2}\Delta\tau)}{\Delta\tau} \qquad (12.1\text{-}30)$$

so that from (12.1-28) we obtain

$$\frac{\partial \Lambda(\tau)}{\partial \tau} \cong \frac{1}{\Delta\tau}\sum_{j=1}^{L}\ln\cosh\left[\frac{2}{N_0}\int_{\tau+\Delta\tau/2+(j-1)T}^{\tau+\Delta\tau/2+jT} y(t)\, x^{(1)}\!\left(t, \tau + \frac{\Delta\tau}{2}\right)dt\right]$$

$$-\frac{1}{\Delta\tau}\sum_{j=1}^{L}\ln\cosh\left[\frac{2}{N_0}\int_{\tau-\Delta\tau/2+(j-1)T}^{\tau-\Delta\tau/2+jT} y(t)x^{(1)}\!\left(t, \tau - \frac{\Delta\tau}{2}\right)dt\right]$$

$$(12.1\text{-}31)$$

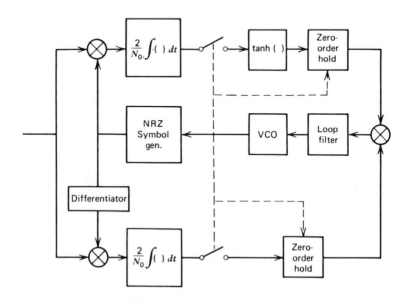

Figure 12.4 A possible closed-loop realization (approximate) of the map symbol synchronizer.

Implementations of (12.1-31) can be more easily effected if either of the following approximations are used:

$$\ln \cosh(x) = \begin{cases} \dfrac{|x|}{2} & |x| \gg 1 \\ \dfrac{x^2}{2} & |x| \ll 1 \end{cases} \qquad (12.1\text{-}32)$$

Since the argument of the cosh( ) is essentially the SNR, it suggests a low SNR and a high SNR approximation to a near optimum loop, as shown in Figures 12.5*a* and 12.5 *b*. (The zero-order hold simply holds the sample

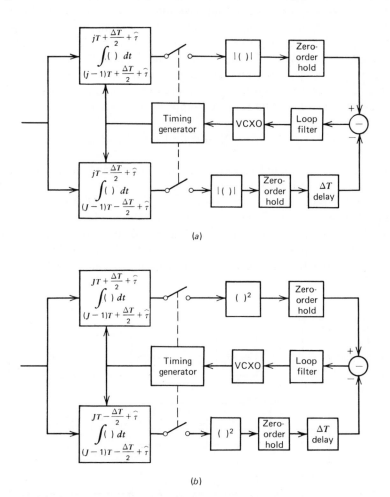

(a)

(b)

Figure 12.5 High SNR and low SNR closed-loop symbol synchronization (early-late type loop). (*a*) High SNR approximation to approximate closed-loop map estimator (early-late type loop). (*b*) Low SNR approximation to approximate closed-loop map estimator.

value for T sec.) These loops, referred to as early-late gate types of synchronizers, will be a subject of discussion in a later section.

We now analyze some of these loops in the ensuing sections.

12.2   DIGITAL DATA TRANSITION TRACKING LOOP

A digital data transition tracking loop (DTTL) was employed on the Mariner Mars 1969 mission to provide symbol synchronization. This loop has also been called the in-phase midphase symbol synchronizer since this loop utilizes both an in-phase and a midphase integrator to obtain the loop error control signal. These loops have been studied by numerous authors [6]–[9].

A DTTL for tracking NRZ symbols is shown in Figure 12.6. By appropriately modifying the integrators, the loop will track Manchester data also. The upper branch,, starting with the in-phase integrator, integrates across one symbol in order to provide an estimate of the polarity of the present symbol. The received input is modeled as

$$y(t) = \sqrt{P} \sum_{k=-\infty}^{\infty} a_k p(t - kT - \tau) + n(t) \qquad (12.2\text{-}1)$$

and where P is the signal power, τ is the transmission delay, and a_k is a random variable, independent from symbol to symbol, representing, along with the NRZ pulse waveform $p(t)$ (existing for T sec), a random symbol stream. The noise $n(t)$ is modeled as AWGN. At the end of what the synchronizer thinks is one symbol time a sample is stored. This sample is quantized to one bit and stored as $\hat{a}_k$ the estimate of a_k. (The integrators are controlled by the timing generator producing an estimate $\hat{\tau}$ of τ the input

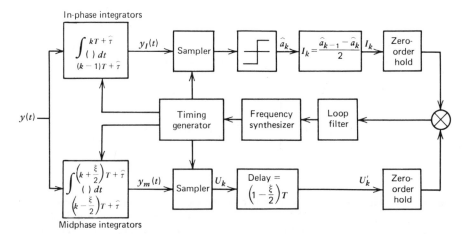

Figure 12.6 A digital data transition tracking loop for NRZ symbols.

timing relationship.) Symbol transitions are detected in the transition detector using the algorithm

$$I_k = \frac{\hat{a}_{k-1} - \hat{a}_k}{2} \tag{12.2-2}$$

Notice that when there is no transition, $I_k = 0$; and when a positive to negative symbol transition occurs, $I_k = 1$, whereas when a negative to positive symbol transition occurs, $I_k = -1$. As we shall see shortly, the lower channel will detect the timing error or the negative of it, so that when the outputs of each arm are multiplied a signal results is a proportional measure of the timing error. When the noise is large, of course, the estimate might not be very accurate; however, with the closed-loop bandwidth narrow enough the loop will tend to track the symbols very accurately.

The lower channel estimates the timing error ϵ defined by $\epsilon = \tau - \hat{\tau}$. This is accomplished by integrating across a transition with the starting and stopping of this integrator controlled by the loop itself. The width of the window is preset prior to operation. In our model we denote the width by ξT, where ξ is a fraction less than or equal to 1. Since the midphase integrator output is obtained $(1 - \xi/2)T$ sec before the in-phase integrator output, a delay of $(1 - \xi/2)T$ sec is required to "align" I_k and U_k.

In Figure 12.7 the integrator waveforms are shown illustrating how the error signal is developed for the case of when the loop error ϵ is small. In the figure, U'_k denotes U_k delayed $(1 - \xi/2)T$ sec. The resulting product extended to T sec $[(I_k U'_k)_E]$ (the input to the loop filter), is the actual error control signal when used with an analog loop filter. The loop filter can also be implemented as a digital filter. However, in the analysis to follow we shall assume an analog loop filter for convenience.

Now we determine the tracking performance of the DTTL. We assume that the input symbols have their leading edges at ... $kT + \tau$, $(k + 1)T + \tau$, $(k + 2)T + \tau, \ldots$, and that the loop generates its leading edges at ... $kT + \hat{\tau}$, $(k + 1)T + \hat{\tau}$, $(k + 2)T + \hat{\tau}, \ldots$, so that the midphase integrator output is given by (with a positive to negative transition)

$$u_k = 2\sqrt{P}\,(\tau - \hat{\tau}) + \int_{(k-\xi/2)T}^{(k+\xi/2)T} n(t)\,dt \tag{12.2-3}$$

Letting the timing error ϵ equal $\tau - \hat{\tau}$, we see that the average value of u_k, given ϵ, satisfies

$$E(u_k \mid \epsilon) = 2\sqrt{P}\epsilon \tag{12.2-4}$$

To obtain the S-curve slope at the origin in order to obtain the equivalent gain, we must compute the mean error signal conditioned on ϵ. Hence, we compute

$$E(I_k u_k \mid \epsilon) = E(\hat{\epsilon} \mid \epsilon) \tag{12.2-5}$$

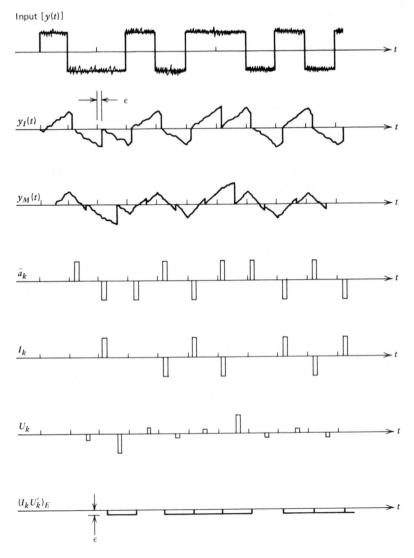

Figure 12.7 DTTL waveforms and the generation of the code loop error ϵ NRZ data.

where $\hat{\epsilon}_k = I_k u_k$. If we neglect the correlation between I_k and u_k,* we have

$$E(\hat{\epsilon} \mid \epsilon) = \tfrac{1}{2}(2\sqrt{P}\epsilon)[(1-PE)^2 - (PE)^2] \qquad (12.2\text{-}6)$$

which is due to the fact the in-phase channel produces the correct value of I_k with probability $(1-PE)^2$, zero with probability $2PE(1-PE)$, and errors (incorrect sign), with probability $(PE)^2$. We have assumed a transition

---

*Reference 6 has demonstrated by Monte Carlo simulation that this is a good approximation.

probability of $\frac{1}{2}$ here also. The symbol probability of error is given by

$$PE = \int_{-\infty}^{0} \frac{1}{\sqrt{2\pi}\,\sigma} \exp\left[\frac{(y-\mu)^2}{2\sigma^2}\right] dy \qquad (12.2\text{-}7)$$

where

$$\sigma^2 = \frac{N_0}{2}T$$

$$\mu = \sqrt{P}\,T$$

Hence

$$PE = \int_{\mu/\sigma}^{\infty} \frac{1}{\sqrt{2\pi}} e^{-z^2/2} dz = Q\left(\sqrt{\frac{2PT}{N_0}}\right) \qquad (12.2\text{-}8)$$

Simplifying (12.2-6) leads to

$$E(\hat{\epsilon}\,|\,\epsilon) = \sqrt{P}\,\epsilon(1 - 2PE) \qquad (12.2\text{-}9)$$

Now

$$1 - 2PE = 2\int_{0}^{\sqrt{R}} \frac{1}{\sqrt{\pi}} e^{-t^2} dt = \mathrm{erf}(\sqrt{R}) \qquad (12.2\text{-}10)$$

where $R = PT/N_0$. Therefore, the equivalent gain is given by

$$A = \sqrt{P}\,\mathrm{erf}(\sqrt{R}) \qquad (12.2\text{-}11)$$

Our goal is to obtain a linearized closed-loop analysis for the timing error variance.

The loop estimate of timing, $\hat{\tau}$, is given by

$$\hat{\tau} = \frac{KF(s)}{s}\,\hat{\epsilon} \qquad (12.2\text{-}12)$$

where s is the Heaviside operator and K is the loop gain relative to $\hat{\epsilon}$. Hence we write (12.2-12) in the form

$$\hat{\tau} = \frac{KF(s)}{s}[\sqrt{P}\,\mathrm{erf}(\sqrt{R})\epsilon + N(t)] \qquad (12.2\text{-}13)$$

where $N(t)$ is the T-sec extension of $N(kT)$. This process is difficult to model exactly; however, when ϵ is small it can be modeled (assuming $\epsilon = 0$) as a variable amplitude NRZ waveform that is constant over T sec and is equally likely to be nonzero or zero. The process $N(t)$ can be written as the product

$$N(t) = \left[\frac{a(t) - a(t-T)}{2}\right] u(t-T) \qquad (12.2\text{-}14)$$

where $a(t)$ and $u(t)$ are the T-sec extensions of $a(kT)$ and $u(kT)$, respectively (see Figure 12.6). The autocorrelation of the noise process, assuming that the $a(t)$ and $u(t)$ processes are uncorrelated (which they are not), is

$$R(\zeta) = \left(\frac{R_a(0) - 2R_a(T) + R_a(0)}{4}\right)\frac{N_0\xi T}{2}\left(1 - \frac{|\zeta|}{T}\right) \qquad (12.2\text{-}15)$$

Since $R_a(T) = 0$, we obtain

$$R(\zeta) = \frac{N_0\xi T}{4}\left(1 - \frac{|\zeta|}{T}\right)$$

The power spectral density at $f = 0$ is given by

$$\mathscr{S}(0) = \frac{N_0 T^2 \xi}{4} \qquad (12.2\text{-}16)$$

From (12.2-13) we obtain

$$\hat{\tau} = \frac{\operatorname{erf}(\sqrt{R})\sqrt{P}KF(s)}{s}\left[\epsilon + \frac{N(t)}{\sqrt{P}\,\operatorname{erf}(\sqrt{R})}\right] \qquad (12.2\text{-}17)$$

Rearranging, we obtain (since $\hat{\tau} = \tau - \epsilon$)

$$\epsilon = -\frac{K'F(s)}{s + K'F(s)}\left[\frac{N(t)}{\sqrt{P}\,\operatorname{erf}(\sqrt{R})}\right] + \frac{s\tau}{s + K'F(s)} \qquad (12.2\text{-}18)$$

where $K' = K\sqrt{P}\,\operatorname{erf}(\sqrt{R})$. Assuming that the input noise spectral density is constant over the loop bandwidth, we have that the (linearized loop) mean squared timing error is given by

$$\sigma_\epsilon^2 = \frac{\left(\dfrac{N_0 T^2 \xi}{4}\right)(2B_L)}{P\,\operatorname{erf}^2(\sqrt{R})} \qquad (12.2\text{-}19)$$

or

$$\sigma_\epsilon^2 = \frac{\xi T B_L}{2R\,\operatorname{erf}^2(\sqrt{R})} \qquad (12.2\text{-}20)$$

where

$$R = \frac{PT}{N_0} \qquad (12.2\text{-}21)$$

and

$$2B_L = \int_{-\infty}^{\infty}\left|\frac{K'F(j2\pi f)}{2\pi fj + K'F(j2\pi f)}\right|^2 df \qquad (12.2\text{-}22)$$

is the two-sided loop noise bandwidth. The final result, (12.2-20), is in essential agreement with Simon [8] for $B_L T \ll 1$, which is the case of interest for most systems (when $B_L T \approx 1$ the tracking performance degrades due to stability problems). Notice as ξ gets smaller, the timing variance gets smaller! This is true up to a limit where the noise excursions move the timing error outside of the midphase window a significant amount of time; then the variance starts to increase. Hurd and Anderson [6], used a window size (ξ) of $\frac{1}{2}$ and $\frac{1}{8}$ (switchable) in their hardware design. TRW's S-band shuttle bit synchronizer is capable of tracking down to a value of $E_s/N_0 = -5$ dB [12, 13].

PROBLEM 1

(a) Determine the timing error variance for the DTTL with NRZ symbols when the lower arm of Figure 12.1 is hard limited. Note that even at high SNRs there is a residual timing error due to this one-bit quantization.

(b) Now consider a DTTL with no limiting in either the upper or lower arm. Determine the timing error performance for this loop.

This loop can be used for Manchester data also with appropriate modifications, however; false lock points arise [12] that can be resolved by counting midphase transitions and comparing them to in-phase (end of bet) transitions. Since midphase transitions occur with probability 1 and in-phase bits occur at the bit transition probability rate, the true lock point can be resolved.

12.3   EARLY-LATE GATE SYMBOL SYNCHRONIZATION

In this section we consider one example of an early-late gate synchronizer utilizing square law detectors similar in structure to that shown in Figure 12.5b. This loop has been considered by Layland [10]. Simon [11] has

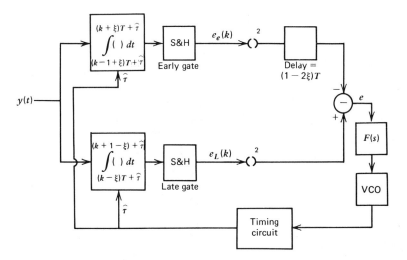

Figure 12.8   Early-late symbol synchronization for NRZ symbols with square law detectors.

considered an absolute value type of early-late gate. Lindsey and Simon have briefly considered the square law detector implementation in reference 5. Stiffler [3] has also considered these loops briefly. We take a somewhat different approach to finding the timing error variance in what follows.

The symbol synchronizer under consideration is shown in Figure 12.8. In order to better understand how the loop phase (timing error) detector works consider Figure 12.9.

In Figure 12.9a it is assumed that there exists no timing error ϵ, where $\epsilon = \tau - \hat{\tau}$, with τ being the actual delay of the signal and $\hat{\tau}$ the estimate. However, in Figure 12.9b the timing error is assumed to be nonzero so that the early and late integrations are not equal and, as we shall see, an error signal is developed.

First we determine the S-curve for small timing errors. These are two cases to be considered, no data transition and a data transition. The

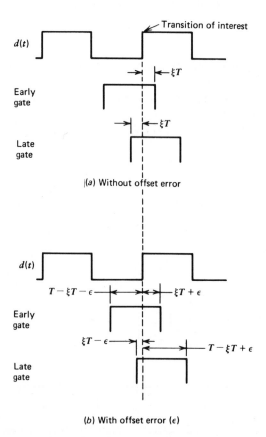

Figure 12.9 Data and gating waveforms in the early-late gate synchronizer.

algebraic sign of the data symbols is not important since the detectors are transparent to the algebraic sign.

Consider the case of no transition and assume both symbols are of level $+\sqrt{S}$ so that the early and late gate outputs are given by

$$e_E(k) = \sqrt{P}\ T + \int_{(k-1+\xi)T}^{(k+\xi)T} n(t)\ dt \qquad \text{early gate} \qquad (12.3\text{-}1)$$

$$e_L(k) = \sqrt{P}\ T + \int_{(k+\xi)T}^{(k-1+\xi)T} n(t)\ dt \quad \text{late gate} \qquad (12.3\text{-}2)$$

where the bit synchronizer input is given by

$$y(t) = \sqrt{P}\ d(t) + n(t) \qquad (12.3\text{-}3)$$

with P being the signal power, $d(t)$ the NRZ data stream, and $n(t)$ modeled as WGN. In (12.3-1) and (12.3-2) the parameter T is the symbol duration in seconds, and ξT is the early and late offset time (see Figure 12.9a).

If we denote the integrated noise term of (12.3-1) as N_E and the integrated noise term of (12.3-2) as N_L, then clearly

$$\bar{e} = -\overline{(\sqrt{P}\ T + N_E)^2} + \overline{(\sqrt{P}\ T + N_L)^2} = 0 \qquad (12.3\text{-}4)$$

where the overbar denotes the ensemble average.

Now suppose there is a minus-plus transition, as shown in Figure 12.9. The early and late gate outputs, in this case, are given by

$$e_E(k) = \sqrt{P}[-(T - \xi T - \epsilon) + (\xi T + \epsilon)] + N_E \qquad (12.3\text{-}5)$$

$$e_L(k) = \sqrt{P}[-(\xi T - \epsilon) + (T - \xi T + \epsilon)] + N_L \qquad (12.3\text{-}6)$$

after squaring, averaging, and subtracting with the appropriate delay for $e_E(k)$, we find that

$$\bar{e}_k = -P[(2\xi - 1)^2 T^2 + 4\epsilon(2\xi - 1)T + 4\epsilon^2]$$
$$+ P[(1 - 2\xi)^2 T^2 + 4\epsilon(1 - 2\xi)T + 4\epsilon^2] \qquad (12.3\text{-}7)$$

or

$$\bar{e}_k = 8\epsilon P(1 - 2\xi)T \qquad |\epsilon| \le \xi T \qquad (12.3\text{-}8)$$

Now to find the average S-curve we average over the case of no transition and a transition since both occur with probability $\frac{1}{2}$ with random data (transition probability of $\frac{1}{2}$). We have

$$E[\bar{e}_k] = 0 + \tfrac{1}{2}[8P(1 - 2\xi)T] = 4P(1 - 2\xi)T \qquad (12.3\text{-}9)$$

with $E[\ \ ]$ denoting the averaging over the data transitions. The timing estimate $\hat{\tau}$ in Heaviside operator notation is given by

$$\hat{\tau} = \frac{\bar{K}F(s)}{s}\ e \qquad (12.3\text{-}10)$$

where $\bar{K}$ is the loop gain, exclusive of the loop filter. This estimate can be written as

$$\hat{\tau} = \frac{4\bar{K}P(1-2\xi)TF(s)}{s}\left[\epsilon + \frac{N(t)}{4P(1-2\xi)T}\right] \qquad (12.3\text{-}11)$$

where the effective noise process at $\epsilon \approx 0$ is given by

$$N(t) = -2\sqrt{P}(N_e - N_L)\left[\frac{a(t) + a(t-T)}{2}\right]T - 2\sqrt{P}(N_e - N_L)T(1-2\xi)$$

$$\times \left[\frac{a(t) - a(t-T)}{2}\right] - N_E^2 + N_L^2 \qquad (12.3\text{-}12)$$

where $a(t)$ are unit amplitude NRZ symbols having duration T sec.

We will denote the first two terms as N_1 and the third and fourth terms as N_2. Closing the loop, we see that the linearized error process $\epsilon = \tau - \hat{\tau}$ satisfies

$$\epsilon = \frac{\bar{K}F(s)}{s + \bar{K}'F(s)}\left[\frac{N(t)}{4P(1-2\xi)T}\right] \qquad (12.3\text{-}13)$$

The terms $[a(t) + a(t-T)]/2$ and $[a(t) - a(t-T)]/2$ are functions that have magnitude 1 and 0 when there is no transition and 0 and 1 when there is a transition, respectively. Now let $\Delta N = N_E - N_L$, so that

$$\Delta N = \int_{(k-1+\xi)T}^{(k+\xi)T} n(t)\,dt - \int_{(k-\xi)T}^{(k+1-\xi)T} n(t)\,dt \qquad (12.3\text{-}14)$$

Letting $k = 0$ for convenience and canceling the overlapping noise process leaves us with

$$\Delta N = \int_{-(1-\xi)T}^{-\xi T} n(t)\,dt - \int_{\xi T}^{(1-\xi)T} n(t)\,dt \qquad (12.3\text{-}15)$$

Note that the two terms of (12.3-15) are independent. Since the integral is over disjoint noise processes, N_1 is independent of N_2. The autocorrelation function of N_1 evaluated at ζ is given by

$$R_{N_1}(\zeta) = 4S\left[\frac{R_a(0) + 2R_a(T) + R_a(0)}{4}\right]R_{\Delta N}(\zeta)T^2$$

$$+ 4S\left[\frac{R_a(0) - 2R_a(T) + R_a(0)}{4}\right]R_{\Delta N}(\zeta)T^2(1-2\xi)^2$$

$$(12.3\text{-}16)$$

Since $R_a(0) = 1$ and $R_a(T) = 0$, we obtain

$$R_{N_1}(\zeta) = 2ST^2 R_{\Delta N}(\zeta) + 2S(T - 2\xi T)^2 R_{\Delta N}(\zeta) \qquad (12.3\text{-}17)$$

Now let us evaluate $R_{\Delta N}(\zeta)$. For an NRZ random waveform we have

$$R_{\Delta N}(\zeta) = \sigma_{\Delta N}^2\left(1 - \frac{|\zeta|}{T}\right) \qquad (12.3\text{-}18)$$

where

$$\sigma_{\Delta N}^2 = \left(\int_{-(1-\xi)T}^{-\xi T} n(t)\,dt\right)^2 + \left(\int_{\xi T}^{(1-\xi)T} n(t)\,dt\right)^2 \qquad (12.3\text{-}19)$$

or evaluating

$$\sigma_{\Delta N}^2 = N_0 T (1 - 2\xi) \qquad (12.3\text{-}20)$$

It therefore follows that

$$R_{N_1}(\zeta) = 2PT^2[1 + (1 - 2\xi)^2]N_0 T(1 - 2\xi)\left(1 - \frac{|\zeta|}{T}\right) \qquad (12.3\text{-}21)$$

Now we consider $R_{N_2}(\zeta)$. The noise term N_2 can be written as

$$N_2 = \left(\underbrace{\int_{-(1-\xi)T}^{-\xi T} n(t)\, dt}_{n_a} + \underbrace{\int_{-\xi T}^{\xi T} n(t)\, dt}_{n_b}\right)^2$$

$$- \left(\underbrace{\int_{-\xi T}^{\xi T} n(t)\, dt}_{n_b} + \underbrace{\int_{\xi T}^{(1-\xi)T} n(t)\, dt}_{n_c}\right)^2 \qquad (12.3\text{-}22)$$

Now

$$R_{N_2}(\zeta) = \sigma_{N_2}^2\left(1 - \frac{|\zeta|}{T}\right) \qquad (12.3\text{-}23)$$

We now evaluate $\sigma_{N_2}^2$. We have

$$\sigma_{N_2}^2 = \overline{[(n_a + n_b)^2 - (n_b + n_c)^2]^2} \qquad (12.3\text{-}24)$$

Expanding

$$\sigma_{N_2}^2 = \overline{(n_a + n_b)^4} - 2\overline{(n_a + n_b)^2(n_b + n_c)^2} + \overline{(n_b + n_c)^4} \qquad (12.3\text{-}25)$$

or

$$\sigma_{N_2}^2 = 3[\overline{(n_a + n_b)^2}]^2 - 2(\overline{n_a^2\, n_b^2} + \overline{n_a^2\, n_c^2} + \overline{n_b^4} + \overline{n_b^2\, n_c^2})$$
$$+ 3[\overline{(n_b + n_c)^2}]^2 \qquad (12.3\text{-}26)$$

Using the properties of Gaussian random variables it can be shown that

$$\sigma_{N_2}^2 = N_0^2 T^2 (1 - 4\xi^2) \qquad (12.3\text{-}27)$$

Therefore

$$R_{N_2}(\zeta) = N_0^2 T^2 (1 - 4\xi^2)\left(1 - \frac{|\zeta|}{T}\right) \qquad (12.3\text{-}28)$$

Since the power spectral density at $f = 0$ is sufficient to characterize the noise process for small values of loop bandwidth, we can write

$$\sigma_\epsilon^2 = \frac{2B_L \mathcal{S}(0)}{[4P(1 - 2\xi)T]^2} \qquad (12.3\text{-}29)$$

with the denominator being the small error equivalent gain. The spectral density at the origin becomes [from (12.3-21) and (12.3-28)]

$$\mathcal{S}(0) = 4PT^3(1 - 2\xi + 2\xi^2)N_0 T(1 - 2\xi) + N_0^2 T^3(1 - 4\xi^2) \qquad (12.3\text{-}30)$$

Therefore, the linearized, small error, timing error variance is given by

$$\sigma_\epsilon^2 = \frac{(2B_L)[4PT^3(1 - 2\xi + 2\xi^2)N_0T(1 - 2\xi) + N_0^2T^3(1 - 4\xi^2)]}{16S^2(1 - 2\xi)^2T^2}$$

$$(12.3\text{-}31)$$

Simplifying yields

$$\sigma_\epsilon^2 = \frac{N_0B_LT^2}{2P}\left[\frac{1 - 2\xi + 2\xi^2}{(1 - 2\xi)} + \frac{N_0}{T}\frac{(1 - 4\xi^2)}{4P(1 - 2\xi)^2}\right] \qquad (12.3\text{-}32)$$

Typically, $\xi = 1/4$, so that

$$\sigma_\epsilon^2 = \frac{N_0B_LT^2}{2P}\left[\frac{5}{4} + \frac{N_0}{PT}\left(\frac{3}{4}\right)\right] \qquad (12.3\text{-}33)$$

By comparing the leading term of this expression, (12.3-33), with the DTTL equation, (12.2-20), we have table 12.1.

It is to be noted that the symbol ξ has a different meaning for each loop. From Table 12.1 we conclude that the early-late gate tracking loop (ELGTL) is slightly inferior to the DTTL although the difference is slight.* Both loops can be improved, in small error tracking, by decreasing ξ. However, the DTTL improves more significantly as ξ decreases. Eventually, through, the tracking performance deteriorates when an error is too large for the given value of ξ. Notice also that the leading term of (12.3-32) is minimized when $\xi = 0$, which is not practical for anything but zero errors! The decision of which loop to use depends upon the ease of circuit implementation. Based on other detector analyses, it is expected that envelope detectors, rather than square law detectors, would make little performance difference.

PROBLEM 2

Determine the linearized tracking loop error performance for the loop of Figure 12.8 with the square law detectors replaced with envelope detectors for NRZ symbols.

Table 12.1 Comparison of the two synchronizers— NRZ symbols

DTTL	ELGTL (Square Law Detectors)
$\sigma_\epsilon^2 \simeq \dfrac{N_0B_LT^2}{2P}$ $(\xi = 1)$	$\sigma_\epsilon^2 \simeq \dfrac{5}{8}\dfrac{N_0B_LT^2}{P}$ $(\xi = \frac{1}{4})$

*However, when $\xi = \frac{1}{2}$, the DTTL has only about one-half the variance of the ELGTL.

The ELGTL can also be modified to accept Manchester symbols, but again false phase lock points must be resolved.

12.4 SUBOPTIMUM BIT SYNCHRONIZERS

In previous sections we considered near optimum symbol synchronizers. In this section we discuss some suboptimum bit synchronizers. These synchronizers are not necessarily motivated by MAP estimators; however, they are, generally speaking, easier to implement than the DTTL and the ELGTL. Generally, they will not operate well at negative SNRs. Figure 12.10 illustrates a number of different synchronizers. The first system is one in which a clock component is derived from the filter and nonlinearity followed by a phase locked loop. The fundamental ($f = 1/T$) or a harmonic ($f = n/T$, $n = 2, 3, \ldots$) is tracked in a phase locked loop providing bit synchronization.

In Figure 12.10b a delay and multiply synchronizer is shown that

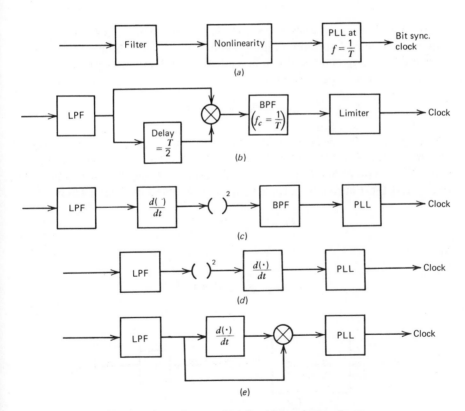

Figure 12.10   Some suboptimal bit synchronizers.

produces a spectral line at the bit rate $(f = 1/T)$ that can then be tracked by a phase locked loop,

Figures 12-10c through 12-10e are examples of differentiating type bit synchronizers that also generate spectral lines at the bit rate that in turn can be tracked by a phase-locked loop. The differentiators are normally implementated by highpass filters in practice.

As an example we shall determine the performance of the loop in Figure 12.10a with a square law nonlinearity and consider the tracker in some detail.

12.4.1 Filter-and-Square Bit Synchronizer—NRZ Data

We consider a specific type of nonlinearity in Figure 12-10a, namely a square law detector, due in part to its convenience in being analyzed [17]. See also Gardner [14] for a similar loop but without analytical results. We also assume a 50% transition density in the following $(p = 0.5)$. Our model is shown in Figure 12.11, with the lowpass filter being a one-pole RC LPF. The baseband input signal $y(t)$ is described by

$$y(t) = Ad(t) + n(t) \tag{12.4-1}$$

where A is the signal amplitude, $d(t)$ is the baseband NRZ data symbol sequence; assumed to have equally likely "ones" and "minus ones," with each symbol being statistically independent of each other. Consider now the signal $d(t)$ that can be written as

$$d(t) = \sum_{k=-\infty}^{\infty} d_k m(t - kT) \tag{12.4-2}$$

where d_k is a random variable taking on the values ± 1 with equal likelihood and is statistically independent from sample to sample, and $m(t)$ is the NRZ symbol pulse. The symbol duration is T sec, so the symbol rate is $1/T$. Let the Fourier transform of $d(t)$ be denoted by $D(\omega)$. Then

$$D(\omega) = \int_{-\infty}^{\infty} d(t)e^{-j\omega t} dt \tag{12.4-3}$$

or

$$D(\omega) = \sum_{k=-\infty}^{\infty} d_k M(\omega)e^{-jk\omega T} \tag{12.4-4}$$

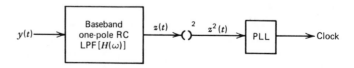

Figure 12.11   Filter-and-square bit synchronizer.

where $M(\omega)$ is the Fourier transform of $m(t)$. Therefore, the filtered baseband data sequence is given by

$$\tilde{D}(\omega) = \sum_{k=-\infty}^{\infty} d_k M(\omega) H(\omega) e^{-jk\omega T} \qquad (12.4\text{-}5)$$

with $H(\omega)$ being the lowpass filter transfer function. It follows that the filtered data sequence in the time domain is given by

$$\tilde{d}(t) = \sum_{k=-\infty}^{\infty} d_k q(t - kT) \qquad (12.4\text{-}6)$$

where

$$q(t) = \frac{1}{2\pi} \int_{-\infty}^{\infty} H(\omega) M(\omega) e^{j\omega t} \, d\omega \qquad (12.4\text{-}7)$$

and is the filtered symbol waveform. Squaring (12.4-6) produces [15]

$$[\tilde{d}(t)]^2 = \sum_{k=-\infty}^{\infty} q^2(t - kT) + \sum_{\substack{k=-\infty \\ k \neq l}}^{\infty} \sum_{l=-\infty}^{\infty} d_k d_l q(t - kT) q(t - lT) \qquad (12.4\text{-}8)$$

The first term yields a periodic spectrum and the second a continuous spectrum when $p = 0.5$. Performing the ensemble average yields

$$\overline{[\tilde{d}(t)]^2} = \sum_{k=-\infty}^{\infty} q^2(t - kT) \qquad (12.4\text{-}9)$$

Clearly $\overline{[\tilde{d}(t)]^2}$ is periodic, so it can be represented in a complex Fourier series of the form

$$\overline{[\tilde{d}(t)]^2} = \sum_{n=-\infty}^{\infty} C_n \exp\left(j\frac{2\pi n}{T} t\right) \qquad (12.4\text{-}10)$$

Evaluating, we obtain

$$C_n = \frac{1}{T} \int_0^T \sum_{k=-\infty}^{\infty} q^2(t - kT) \exp\left(-j\frac{2\pi n}{T} t\right) dt = \frac{1}{T} \int_{-\infty}^{\infty} q^2(t) \exp\left(-j\frac{2\pi n}{T} t\right) dt \qquad (12.4\text{-}11)$$

Let $Q(\omega)$ be the Fourier transform of $q(t)$. Hence the filtered baseband waveform is given from (12.4-7):

$$Q(\omega) = M(\omega) H(\omega) \qquad (12.4\text{-}12)$$

For the first-order lowpass RC filter we obtain

$$Q(\omega) = (1 - e^{-j\omega T}) \frac{1}{j\omega} \frac{1}{1 + j\left(\dfrac{\omega}{\omega_0}\right)} \qquad (12.4\text{-}13)$$

Therefore ($\omega_0 = 2\pi f_0$, with f_0 the 3-dB frequency of the filter)

$$q(t) = m(t) - \mathcal{F}^{-1}\left[(1 - e^{-j\omega T})\left(\frac{1}{\omega_0 + j\omega}\right)\right] \qquad (12.4\text{-}14)$$

Evaluating, we obtain

$$q(t) = m(t) - e^{-\omega_0 t} u(t) + e^{-\omega_0(t-T)} u(t - T) \qquad (12.4\text{-}15)$$

Finally we have that

$$\frac{1}{T} \mathcal{F}[q^2(t)] = \frac{1 - e^{-j\omega T}}{j\omega T} + \frac{1}{(\omega_0 + j\omega)T} (2e^{-\omega_0 T} e^{-j\omega T} - 2)$$

$$+ \frac{1}{(2\omega_0 + j\omega)T} (1 + e^{-j\omega T} - 2e^{-\omega_0 T} e^{-j\omega T}) \qquad (12.4\text{-}16)$$

Therefore

$$C_n = \frac{1}{T} \mathcal{F}[q^2(t)] \Big|_{\omega = 2\pi n/T} \qquad (12.4\text{-}17)$$

Evaluating (12.4-17), we obtain

$$C_n = (-1)^n \frac{\sin \pi n}{\pi n} + \frac{1}{\omega_0 T + j2\pi n} (2e^{-\omega_0 T} e^{-j2\pi n} - 2)$$

$$+ \frac{1}{2\omega_0 T + j2\pi n} (1 + e^{-j2\pi n} - 2e^{-\omega_0 T} e^{-j2\pi n}) \qquad (12.4\text{-}18)$$

Now let $\omega_0 T = 2\pi R$, with $R = f_0 T$. Then, since $\sin \pi n = 0$ (all $n \neq 0$), we have

$$C_n = (1 - e^{-2\pi R}) \left[\frac{2\pi R - j\pi n}{(2\pi R)^2 + (n\pi)^2} - \frac{4\pi R - 4jn\pi}{(2\pi R)^2 + (2\pi n)^2} \right] \qquad (12.4\text{-}19)$$

From (12.4-10) we have that the line spectral component at $f = n/T$ is given by

$$s_n(t) = C_n \exp\left(j \frac{2\pi n}{T} t \right) + C_{-n} \exp\left(-j \frac{2\pi n}{T} t \right) = 2\text{Re}\left[C_n \exp\left(j \frac{2\pi n}{T} t \right) \right] \qquad (12.4\text{-}20)$$

where $\text{Re}(z)$ denotes the real part of the variable z. Using (12.4-19) in (12.4-20), we obtain

$$s_n(t) = 2(1 - e^{-2\pi R}) \left\{ \left[\frac{2\pi R}{(2\pi R)^2 + (n\pi)^2} - \frac{4\pi R}{(2\pi R)^2 + (2\pi n)^2} \right] \cos\left(\frac{2\pi n}{T} t \right) \right.$$

$$\left. + \left[\frac{n\pi}{(2\pi R)^2 + (n\pi)^2} - \frac{4n\pi}{(2\pi R)^2 + (2\pi n)^2} \right] \sin\left(\frac{2\pi n}{T} t \right) \right\} \qquad (12.4\text{-}21)$$

The amplitude of the tone out of the squarer can be shown, after some algebra, to be given by

$$2|C_n| = \frac{2(1 - e^{-2\pi R})}{2\pi R \sqrt{\left[1 + \left(\frac{n}{2R} \right)^2 \right] \left[1 + \left(\frac{n}{R} \right)^2 \right]}} \qquad (12.4\text{-}22)$$

So that the line component at $f = n/T$ is given by

$$s_n(t) = 2|C_n| \cos\left(\frac{2\pi n}{T}t - \theta_n\right) \qquad (12.4\text{-}23)$$

where

$$\theta_n = \tan^{-1}\left\{\frac{[n\pi/((2\pi R)^2 + (n\pi)^2)] - [4n\pi/((2\pi R)^2 + (2\pi n)^2)]}{[2\pi R/((2\pi R)^2 + (n\pi)^2)] - [4\pi R/((2\pi R)^2 + (2\pi n)^2)]}\right\} \qquad (12.4\text{-}24)$$

It is to be pointed out that the C_n given in (12.4-19) is not the same C_n used in the false lock theory (reference 16, also Chapter 5).

Now notice that $z(t)$, the output of the lowpass filter, can be described by

$$z(t) = A\tilde{d}(t) + \tilde{n}(t) \qquad (12.4\text{-}25)$$

where $\tilde{n}(t)$ is the lowpass filtered noise process. Hence $z^2(t)$ is given by

$$z^2(t) = A^2[\tilde{d}(t)]^2 + 2A\tilde{d}(t)\tilde{n}(t) + [\tilde{n}(t)]^2 \qquad (12.4\text{-}26)$$

Now from (12.4-10) and (12.4-26) we see that a tone is generated at $f = n/T$, $n = 1, 2, 3, \ldots$, and therefore a phase locked loop can be used to track this tone and synchronize to the bit stream. Consequently, if we know the noise spectral density at $f = n/T$, $n = 1, 2, 3, \ldots$, and the tone power (P) out of the squarer, it follows that*

$$\sigma_\phi^2 = \frac{\left(\frac{N_0'}{2}\right)(2B_L)}{P} \qquad (P = 2A^4|C_n|^2) \qquad (12.4\text{-}27)$$

is the linearized rms phase error of the phase locked loop where $N_0'/2$ is the effective two-sided noise spectral density and we assume a narrower loop. It then follows that the rms bit synchronization timing error variance is given by

$$\sigma_\tau^2 = \frac{N_0'B_L T^2}{(2\pi)^2 P} \qquad \text{sec}^2 \qquad (12.4\text{-}28)$$

Now we determine the power spectral density of $N(t)$, where

$$N(t) = 2A\tilde{d}(t)\tilde{n}(t) + [\tilde{n}(t)]^2 \qquad (12.4\text{-}29)$$

Since the two terms are uncorrelated, we can obtain the power spectral density of each one. First consider the second term of (12.4-29). The autocorrelation function of the second term is given by

$$R_{\tilde{n}^2}(\tau) = E[\tilde{n}(t)\tilde{n}(t+\tau)\tilde{n}(t)\tilde{n}(t+\tau)] \qquad (12.4\text{-}30)$$

or, using the property of a product of four Gaussian random variables, we have

$$R_{\tilde{n}^2}(\tau) = R_{\tilde{n}}^2(\tau) + 2R_{\tilde{n}}^2(0) \qquad (12.4\text{-}31)$$

*We neglect the self noise since at low SNR's it can be shown to be negligible.

Therefore the power spectral density of $f = n/T$ is given by

$$\mathscr{S}\left(\frac{n}{T}\right) = 2\int_{-\infty}^{\infty} R_{\tilde{n}}^2(\tau)\exp\left(-j\frac{2\pi n}{T}\tau\right)d\tau + \int_{-\infty}^{\infty} R_{\tilde{n}}^2(0)\exp\left(-j\frac{2\pi n}{T}\tau\right)d\tau$$

(12.4-32)

The first term produces a spectral component at $f = n/T$, whereas the second term produces a dc component. To evaluate the component at $f = n/T$ we note that the output noise process, out of a one-pole RC lowpass filter, has an autocorrelation function given by

$$R_{\tilde{n}}(\tau) = \frac{N_0\pi f_0}{2} e^{-2\pi f_0|\tau|}$$

(12.4-33)

where the 3-dB frequency out of the lowpass filter is f_0. We have

$$\mathscr{S}_{\tilde{n}^2}\left(\frac{n}{T}\right) = 2\left(\frac{N_0\pi f_0}{2}\right)^2 \int_{-\infty}^{\infty} [\exp(-2\pi f_0\tau)]^2 \exp\left(-j\frac{2\pi n}{T}\tau\right)d\tau$$

(12.4-34)

or

$$\mathscr{S}_{\tilde{n}^2}\left(\frac{n}{T}\right) = (N_0\pi f_0)^2 \int_0^{\infty} e^{-4\pi f_0\tau} \cos\left(\frac{2\pi n}{T}\tau\right)d\tau$$

(12.4-35)

Equation 12.4-35 can be evaluated to yield

$$\mathscr{S}_{\tilde{n}^2}\left(\frac{n}{T}\right) = \frac{f_0\pi N_0^2}{4\left[1+\left(\frac{n}{2R}\right)^2\right]}$$

(12.4-36)

Now consider the first term of (12.4-29). We have*

$$R(\tau) = 4A^2 R_{\tilde{d}}(\tau)R_{\tilde{n}}(\tau)$$

(12.4-37)

Hence

$$\mathscr{S}_{\tilde{n}\times\tilde{s}}\left(\frac{n}{T}\right) = 4A^2 \int_{-\infty}^{\infty} \mathscr{S}_{\tilde{d}}(f)\mathscr{S}_{\tilde{n}}\left(\frac{n}{T}-f\right)df$$

(12.4-38)

Using the transfer function of the one-pole RC lowpass filter and the power spectral density of a random NRZ data waveform we obtain

$$\mathscr{S}_{\tilde{n}\times\tilde{s}}\left(\frac{n}{T}\right) = 4A^2T \int_{-\infty}^{\infty} \left(\frac{\sin \pi fT}{\pi fT}\right)^2 \left[\frac{1}{1+(f/f_0)^2}\right] \frac{N_0}{2}\left\{\frac{1}{1+[(f-n/T)/f_0]^2}\right\}df$$

(12.4-39)

This integral is best evaluated by contour integration. Let

$$\pi fT = x$$

(12.4-40)

$$\pi f_0 T = x_0 = \pi R$$

(12.4-41)

---

*We have approximated a cyclostationary process by a stationary process which yields negligible error at low SNR's.

which produces

$$\mathcal{S}_{\tilde{n}\times\tilde{s}}\left(\frac{n}{T}\right) = \frac{2A^2 N_0}{\pi}\int_{-\infty}^{\infty}\frac{\sin^2 x}{x^2}\left[\frac{1}{1+(x/x_0)^2}\right]\left\{\frac{1}{1+[(x-n\pi)/x_0]^2}\right\}dx$$

$$(12.4\text{-}42)$$

First observe that the poles of the integrand occur at

$$x = \pm jx_0 \qquad (12.4\text{-}43)$$

and

$$x = n\pi \pm jx_0 \qquad (12.4\text{-}44)$$

Now using the fact that

$$\frac{\sin^2 x}{x^2} = \frac{(e^{2jx}-1)+(e^{-2jx}-1)}{-4x^2} \qquad (12.4\text{-}45)$$

we have that

$$\frac{-4\pi\mathcal{S}(n/T)}{2N_0 A^2 x_0^4} = \int_{-\infty}^{\infty}\frac{(e^{2jx}-1)}{x^2}\left(\frac{1}{x^2+x_0^2}\right)\left[\frac{1}{x_0^2+(x-n\pi)^2}\right]dx$$

$$+\int_{-\infty}^{\infty}\frac{(e^{-2jx}-1)}{x^2}\left(\frac{1}{x^2+x_0^2}\right)\left[\frac{1}{x_0^2+(x-n\pi)^2}\right]dx \quad (12.4\text{-}46)$$

The required contours to evaluate each interal are shown in Figures 12.12a and 12.12b.

The first integral of (12.4-46) can be evaluated as (Figure 12-12a)

$$\int_{-\infty}^{\infty}\frac{e^{2jx}-1}{x^2}\left(\frac{1}{x_0^2+x^2}\right)\left[\frac{1}{x_0^2+(x-n\pi)^2}\right]dx + \int_{\pi}^{0}\frac{2j\epsilon e^{j\theta}\epsilon je^{j\theta}}{\epsilon^2 e^{2j\theta}x_0^2[x_0^2+(n\pi)^2]}d\theta$$

$$= 2\pi j\sum\text{ residues }(C_1) \qquad (12.4\text{-}47)$$

where ϕ is the radius of the bump at zero and θ the angle $(z = \epsilon e^{j\theta})$, where the second term is due to the bump at the origin. The residues are evualated by the formula

$$\text{residue} = \lim_{z\to z_p}(z-z_p)f(z) \qquad (12.4\text{-}48)$$

where $f(z)$ is the integrand of the integral in which the residue is to be taken. The second term of (12.4-48), J_1, can be evaluated to yield

$$J_1 = \frac{2\pi}{x_0^4+x_0^2(n\pi)^2} \qquad (12.4\text{-}49)$$

Similarly, using Figure 12.12b, the second contour integral of (12.4-46) can be evaluated to yield

$$\int_{\infty}^{-\infty}\frac{e^{-2jx}-1}{x^2}\left(\frac{1}{x_0^2+x^2}\right)\left[\frac{1}{x_0^2+(x-n\pi)^2}\right]dx + \int_{0}^{-\pi}\frac{-2j\epsilon e^{j\theta}\epsilon je^{j\theta}}{\epsilon^2 e^{2j\theta}x_0^2[x_0^2+(n\pi)^2]}d\theta$$

$$= 2\pi j\sum\text{ residues}(C_2) \qquad (12.4\text{-}50)$$

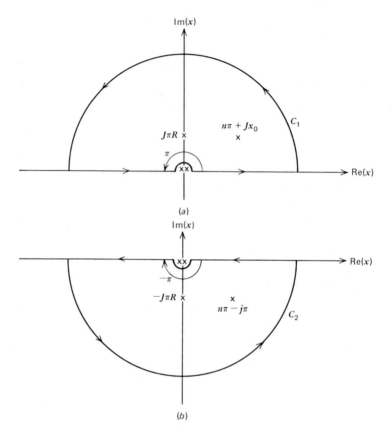

Figure 12.12   The two contours used to evaluate (12.4-46).

The second term, J_2, of (12.4-5) can be evaluated to be

$$J_2 = \frac{-2\pi}{x_0^2[x_0^2 + (n\pi)^2]} \tag{12.4-51}$$

Therefore, from (12.4-46), (12.4-47), (12.4-49), (12.4-50), (12.4-51), and evaluating the two sets of residues, we obtain

$$\frac{-2}{N_0 A^2 x_0^4} \mathscr{S}\left(\frac{n}{T}\right) = \frac{-2\pi}{x_0^4 + x_0^2(n\pi)^2} + 2\pi j\left[\frac{e^{-2x_0}-1}{x_0^2}\left(\frac{1}{n\pi}\right)\left(\frac{1}{2jx_0}\right)\left(\frac{1}{2jx_0 - n\pi}\right)\right.$$

$$\left. + \frac{e^{-2x_0}-1}{(n\pi + jx_0)^2}\left(\frac{1}{n\pi}\right)\left(\frac{1}{2jx_0}\right)\left(\frac{1}{n\pi + 2jx_0}\right)\right]$$

$$- \frac{2\pi}{x_0^4 + x_0^2(n\pi)^2} - 2\pi j\left[\frac{e^{-2x_0}-1}{x_0^2}\left(\frac{1}{2jx_0}\right)\left(\frac{1}{n\pi}\right)\left(\frac{1}{2jx_0 + n\pi}\right)\right.$$

$$\left. + \frac{e^{-2x_0}-1}{(-jx_0 + n\pi)^2}\left(\frac{1}{n\pi}\right)\left(\frac{1}{2jx_0 - n\pi}\right)\left(\frac{1}{2jx_0}\right)\right]$$

$$\tag{12.4-52}$$

After a fair amount of algebra we obtain

$$\mathcal{S}_{\tilde{n}\times\tilde{s}}\left(\frac{n}{T}\right) = \frac{2N_0A^2}{[1+(n/R)^2]} - \frac{N_0A^2[1-e^{-2\pi R}]}{8\pi R}\left\{\frac{12+2(n/R)^2+2(n/R)^4}{[1+(n/R)^2]^2[1+(n/(2R))^2]}\right\}$$

$$R = f_0T \tag{12.4-53}$$

Notice as $R \to 0$

$$\mathcal{S}_{\tilde{n}\times\tilde{s}}\left(\frac{n}{T}\right) \to \frac{2N_0A^2}{(n/R)^2} - \frac{2N_0A^2}{(n/R)^2} = 0$$

as it should! From (12.4-23) and (12.4-27) (assuming the loop is narrowband compared to the data rate), the phase locked loop phase error variance (see Figure 12.11) is given by

$$\sigma_\phi^2 = \frac{[\mathcal{S}_{\tilde{n}\times\tilde{s}}(n/T) + \mathcal{S}_{\tilde{n}\times\tilde{n}}(n/T)](2B_L)}{2A^4|C_n|^2} \tag{12.4-54}$$

From (12.4-22), (12.4-28), (12.4-36), (12.4-53) and (12.4-54) we obtain

$$\sigma_\tau^2 = \frac{N_0B_LT^2}{A^2}\left[\frac{R^2}{(1-e^{-2\pi R})^2}\right]\left(2\left[1+\left(\frac{n}{2R}\right)^2\right]-\left(\frac{1-e^{-2\pi R}}{8\pi R}\right)\right.$$

$$\left.\times\left\{\frac{12+2(n/R)^2+2(n/R)^4}{[1+(n/R)^2]}\right\}+\frac{\pi RN_0}{4A^2T}\left(1+\left(\frac{n}{R}\right)^2\right)\right)$$

$$\tag{12.4-55}$$

From (12.4-55) it is clear that $n = 1$ minimizes the absolute tracking error jitter. The coefficient of $(N_0B_L/A^2)T^2$ is tabulated versus R for four values of E_s/N_0 in Table 12.2. It should be noted that if a tone at n/T were used to obtain the clock frequency, then the expression for σ_τ^2, (12.4-55), would have to be divided by n^2 to yield the variance at the clock frequency $(1/T)$

Table 12.2 Normalized bit synchronization timing error variance for the filter-and-square synchronizer, with NRZ symbols, $p = 50\%$

$$\frac{\sigma_\tau^2}{(N_0B_LT^2/A^2)}$$

R	$\dfrac{E_B}{N_0} = 0\ \text{db}$	$\dfrac{E_B}{N_0} = 3\ \text{db}$	$\dfrac{E_B}{N_0} = 6\ \text{db}$	$\dfrac{E_B}{N_0} = 12\ \text{db}$
1/1000	60.03	50.04	45.04	41.27
1/8	0.96	0.79	0.71	0.64
→3/16	0.89	0.73	0.65	0.59
1/4	0.92	0.75	0.67	0.61
1/2	1.41	1.14	1.01	0.91
1	3.77	2.98	2.59	2.29
3/2	8.28	6.37	5.41	4.69
2	15.55	11.63	9.67	8.19

since the derived clock would have to be divided by n. Also there would be a one out of n ambiguity to resolve in this case.

From the table it is clear that the filter bandwidth must be narrower than the data rate to obtain reasonable synchronizer tracking performance of the loop. A value of R around $\frac{3}{16}$ should be selected to minimize the timing error variance. It appears from this result that this type of synchronizer compares favorably with the full aperature DTTL and the ELGTL in that the standard deviation of the timing error is only about 5–15% larger! However the DTTL with $\xi = 0.5$ is clearly better. The same methods of analysis can be applied to other pulse shapes such as raised cosine pulses, etc. as well as other transition densities. It should be noted that this type of synchronizer is simple to build and will work well at a single data rate but phase shift between the reference and the signal will occur as the data rate is varied and the bandwidth is held fixed [see (12.4-24)]. This loop can be used for Manchester data although the optimum presquaring filter will be larger.

PROBLEM 3

Determine the variance of the timing error for the filter-and-square bit synchronizer with Manchester data following the methods of Section 12.4.1.

12.4.2   Delay-and-Multiply Bit Synchronizer

This synchronizer is illustrated in Figure 12.10b. A clock component is generated at the symbol rate that in turn can be tracked by a phase-locked loop. In Section 8.12.2 the power spectral density of the product of a PN waveform times a shifted waveform is given. Results for both a PN waveform and a random NRZ waveform are developed there. The two-sided power spectral density for the product of PN waveforms is given by

$$\mathcal{S}(f, \epsilon) = \left[1 - \frac{|\epsilon|}{T_c}\right]^2 \delta(f) + \left(\frac{\epsilon}{T_c}\right)^2 \sum_{\substack{m=-\infty \\ m \neq 0}}^{\infty} \text{sinc}^2\left(\frac{m\pi\epsilon}{T_c}\right) \delta\left(f + \frac{m}{\tau_c}\right)$$

$$+ \frac{\epsilon^2}{T_c} \text{sinc}^2(\pi f \epsilon) \qquad 0 \leq |\epsilon| \leq T_c/2 \qquad (12.4\text{-}56)$$

Hence we have a discrete set of spectral lines plus a continuous spectrum. The maximum spectral line component occurs at $m = \pm 1$ (that is, at the data rate). The maximum power at this spectral component can easily be shown to occur at $|\epsilon|/T_c = \frac{1}{2}$ since

$$\left(\frac{\epsilon}{T_c}\right)^2 2 \, \text{sinc}^2\left(\frac{m\pi\epsilon}{T_c}\right) = 2 \frac{\sin^2(m\pi\epsilon/T_c)}{(m\pi)^2} \qquad (12.4\text{-}57)$$

and clearly $\sin^2(\pi\epsilon/T_c)$ is maximized when $|\epsilon|/T_c = \frac{1}{2}$. If filtering is applied to the NRZ waveforms it can be shown that a line component will still exist at the frequencies m/T_c. This line component can then be tracked by a phase locked loop. A line component also appears at the data rate when random NRZ symbols are used.

PROBLEM 4

Determine the tracking error variance of the delay-and-multiply bit synchronizer.

12.4.3 A First-Order Incremental Phase Modulator Loop

The next suboptimum bit synchronizer that we consider is a relatively simple digital first-order transition tracking loop called an incremental phase modulator (IPM) loop that can be simply implemented and provides fair tracking performance [19]. The results obtained here are applied to square wave subcarrier tracking.

The loop is designed to provide discrete phase updating. Tracking is accomplished by (1) sampling the input waveform at the points in time where the signal transitions or axis crossings occur, (2) accumulating m of these samples, and (3) incrementing the phase of the local reference (clock) in such a direction as to bring the value of the accumulation toward zero (see Figure 12.13).

Solutions for the stationary timing error variance (mod 2π) and the mean time to the first cycle slip are developed. As a consequence of the cycle slip analysis an equation is derived for the mean first slip time that generalizes an equation of Feller [18] to the case when the transition probabilities are state dependent.

The incoming waveform $y(t)$ (Figure 12.13) is assumed to be composed of a filtered square wave of amplitude A and period T_s that has passed through a WGN channel. The spectral density of the channel noise is denoted by $N_0/2$ (two-sided).

The first element in the loop is modeled as an ideal lowpass, presampling filter that passes all frequency components out to W Hz.

The sampler obtains a sufficient number of equally spaced samples per

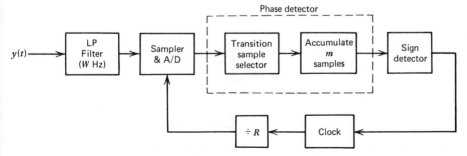

Figure 12.13 Block diagram of squarewave subcarrier tracking loop or IPM.

period to adequately represent the signal. From the sampling theorem this requires that the sampling rate be equal to $2W$. Hence there will be $2W/f_s$ samples per period where f_s is the square wave subcarrier frequency. The samples are then converted by the analog to digital converter to digital format. We assume in what follows that a sufficient number of bits are used so that digitizing introduces negligible error.

The phase detector is formed by the combination of the transition sample selector and an m sample accumulator. Figure 12.14 illustrates a square wave signal with negligible distortion for which the timing system has an error of τ sec. The timing error is the time between the negative-going transition and the corresponding sample. We see that the sampled value of the signal of the error shown in the figure with a plus sign produces a positive control signal. If the sample was taken after the transition, the signal part of the sample would be negative. Hence, the transition sampler produces an output, on the average, that is the same sign as the timing error. After the next sample is taken (the one with a minus sign), the sign of the sample must be changed to provide the correct control signal to the loop. Because noise will cause errors in updating the loop, the performance depends on the sample signal to noise ratio. The function of the accumulator in Figure 12.13 is to increase the update signal to noise ratio (SNR) by summing m transition samples before updating the timing. This update occurs after M signal periods or $MT_s = T$ sec. If no transition samples are deleted, $m = 2M$; however, with modulation it may be necessary to delete some transition samples due to transition ambiguities so that in general $m \leq 2M$.

The sum of m transition samples are then hard quantized in the sign detector. The output of the sign detector controls the addition or deletion of a clock pulse, which in turn shifts the sampling location by some fixed fraction Δ of one signal period. Since the transition sampler selects the samples the loop thinks are the transition samples, the location of the transition samples are advanced or retarded by Δ according to the sign of the update signal.

The divider following the clock divides the clock rate down to the sample rate; that is, divides by $R = f_c/2W$, where f_c is the clock frequency and $2W$ is the sampling rate.

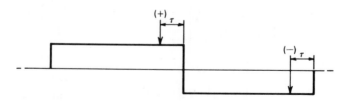

Figure 12.14 Squarewave subcarrier with negative timing error shown with 2 samples per cycle.

In the loop discussed here, timing performance is improved by decreasing W, without exceeding the allowable distortion level of the signal waveform.

White noise passed through the ideal lowpass filter of bandwidth W and then sampled at the rate of $2W$ samples per second produces independent samples. Because the updating is determined by the sum of m of these statistically independent samples and since the updating is a fixed fraction of a period Δ, the error state forms a discrete parameter, Markov chain. We define the "error state" of the loop to be the state (integer) corresponding to the existing timing error of the loop just before being updated. As a consequence, the Markov chain has a countable number of error states. Figure 12.15 illustrates the error state diagram along with the square wave signal. The error in timing is measured relative to the negative going transition of the square wave. All states are separated by Δ fractions of one period, and there are $2N = 1/\Delta$ states in one period. In the case of no noise, since the loop will never make an error, it will eventually oscillate between states -1 and 1, with timing error $\pm \Delta/2$.

In actuality the negative-going transition need not be centered between states -1 and 1 but could be closer to either state -1 or state 1. This uncentered effect can be analyzed with a slight modification of the theory [19], but its effect is not significant in the usual case when Δ is small (typically 1/128 of a period), so it will not be pursued further.

It can be shown that since there are a countable number of states in the limit as the time increases without bound, the "long run" probabilities all tend to zero. Consequently the timing error variance will be unbounded. To obtain a meaningful result we shall determine the steady state phase error distribution mod 1 period just as has been done for the continuous phase locked loop. We shall refer to this in what follows as mod 2π reduction. It is clear that the mod 2π reduced timing error variance is finite since in this case there are a finite number of states.

We can take into account the fact that errors are reduced mod 2π by

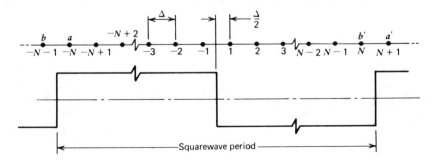

Figure 12.15 Error state diagram (above) and its relationship to the squarewave signal (below).

regarding the two outermost error nodes as reflecting states. At a reflecting state, say, state N, the probability of going to state $N - 1$ is p_N and the probability of remaining at state N is q_N. The reflecting state can be visualized as having a vertical reflecting boundary located halfway between states N and $N + 1$. At the nonreflecting states, the transition probabilities p_k and q_k are defined as follows:

$$p_k = P(\text{of reducing the error} \mid \text{system is in state } k)$$
$$q_k = P(\text{of increasing the error} \mid \text{system is in state } k)$$

and

$$p_k + q_k = 1$$

Referring to Figure 12.15, we see that the transition from state a to state $b \bmod 2\pi$ places the loop error at state b'. However, as far as the timing error magnitude is concerned we can just as well place the loop error at state a since the magnitude of timing error is the same at state b' and state a. Consequently we can model states N and $-N$ as reflecting states.

The state transition diagram can now be reduced to N total error states from $2N$, by noting that the pairs of states -1 and 1, and -2 and 2, etc., contribute the same squared error. This reduced state transition diagram is illustrated in Figure 12.16.

We have reduced the problem to a state dependent homogeneous, finite, aperiodic, Markov chain.

The general case of a periodic signal, odd symmetric about the center of every transition, with two zero crossings per period, is now considered. By allowing the lowpass filter bandwidth to be reduced from the case of negligible distortion, tracking performance can be significantly improved. However, in some applications it is required to use the samples for bit detection also so that severe distortion due to a very narrow bandwidth may not be permissible.

In the general case the signal sample amplitude, at various timing errors, is not independent of the error location as it would be in the square

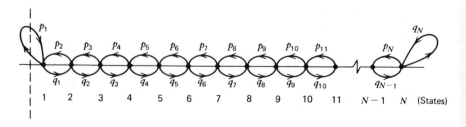

Figure 12.16 Reduced state transition diagram in the state dependent case showing the transition probabilities with reflecting states at each end (general model).

wave signal case shown in Figure 12.15. Hence, the transition probabilities are error state dependent. Let f_s denote the signal frequency. The state transition probabilities for the state dependent case, in terms of the signal and noise parameters, are given by

$$p_k = 1 - Q\left[\left(\frac{mA_k^2}{N_0W}\right)^{1/2}\right] \qquad q_k = 1 - p_k \qquad (12.4\text{-}58)$$

where

$$Q(x) = \int_{-\infty}^{x} \frac{1}{\sqrt{2\pi}} \exp(-t^2/2)dt \qquad (12.4\text{-}59)$$

and where A_k is the amplitude of the signal at state k. Now the steady state probabilities, for the positive states, must satisfy the following second-order difference equation [18]:

$$P_k = q_{k-1}P_{k-1} + p_{k+1}P_{k+1} \qquad 1 < k < N \qquad (12.4\text{-}60)$$

where p_k and q_k are the state dependent transition probabilities, see Figure 12.16. This equation is interpreted as follows: The (steady state) probability of being in state k is equal to the probability of being in state k given that the system came from state $k-1$, times the probability of being in state $k-1$, plus the probability of being in state k given that the system came from state $k+1$, times the probability of being in state $k+1$. The associated boundary conditions for (12.4-61) are

$$\begin{aligned} P_1 &= p_1P_1 + p_2P_2 \\ P_N &= q_{N-1}P_{N-1} + q_NP_N \end{aligned} \qquad (12.4\text{-}61)$$

Figure 12.16 illustrates the reduced state transition diagram for the general case.

The solution of (12.4-60) along with the boundary conditions can be shown by induction or directly [20] to be

$$P_k = P_1 \prod_{j=1}^{|k|-1} \frac{q_j}{p_{j+1}} \qquad k = -N, \ldots, -1, 1, \ldots, N \qquad (12.4\text{-}62)$$

with the definition

$$\prod_{j=1}^{0} \frac{q_j}{p_{j+1}} = 1 \qquad (12.4\text{-}63)$$

In (12.4-62) P_1 is given by

$$P_1 = \frac{1}{2}\left[1 + \sum_{k=2}^{N} \prod_{1}^{k-1} \left(\frac{q_j}{p_{j+1}}\right)\right]^{-1} \qquad (12.4\text{-}64)$$

In Appendix I it is shown that the solution is unique. Once the steady state probabilities are determined the timing error variance can be computed from

$$\sigma_{TE}^2 = 2\Delta^2 \sum_{1}^{N} (k - \tfrac{1}{2})^2 P_k \qquad (12.4\text{-}65)$$

The results for the case of a filtered squarewave signal have been plotted in Figure 12.18 and are discussed later.

We note that if we modify the timing loop to allow three possible timing corrections $+\Delta$, 0, $-\Delta$, where the zero correction occurs with probability r_k in state k, then the following difference equation for P_k is satisfied:

$$P_k = q'_{k-1}P_{k-1} + p'_{k+1}P_{k+1} \tag{12.4-66}$$

with the following boundary conditions

$$\begin{aligned} P_1 &= p'_1 P_1 + p'_2 P_2 \\ P_N &= q'_{N-1}P_{N-1} - q'_N P_N \end{aligned} \tag{12.4-67}$$

where

$$p'_k = \frac{p_k}{1 - r_{k-1}} \qquad q'_k = \frac{q_k}{1 - r_{k+1}} \qquad 1 < k < N$$

$$p'_1 = \frac{p_1}{1 - r_1} \qquad q'_N = \frac{q_N}{1 - r_N} \tag{12.4-68}$$

Therefore, this problem is formally the same as the $r_k = 0$ problem solved above except the solution replaces p_k and q_k with p'_k and q'_k as defined in (12.4-68). We have

$$P_k = P_1 \prod_{j=1}^{|k|-1} \left(\frac{q'_j}{p'_{j+1}}\right) \qquad k = -N, \ldots, -1, 1, \ldots, N \tag{12.4-69}$$

with

$$P_1 = \frac{1}{2}\left[1 + \sum_{k=2}^{N} \prod_{1}^{k-1} \left(\frac{q'_j}{p'_{j+1}}\right)\right]^{-1} \tag{12.4-70}$$

again the steady state probabilities only depend on the ratio q'_j/p'_{j+1} as mentioned before in the case when the transition probabilities were state independent.

In this section a method given in Feller [18] to determine the mean first slip time is modified to allow the transition probabilities to be state dependent. The state transition diagram of Figure 12.17 is applicable.

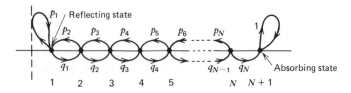

Figure 12.17 Reduced state transition diagram for the first-passage time model (general case).

Let T_k^{N+1} be the mean duration of time it takes to reach state $N+1$ starting at state k. Further denote the event incorrect update as the act of moving Δ units towards state $N+1$ and correct update as the act of moving Δ units towards state 1. Then we note that starting at k:

$$\text{the mean duration given an incorrect update} = T_{k+1}^{N+1} + 1$$
$$\text{the mean duration given a correct update} = T_{k-1}^{N+1} + 1 \tag{12.4-71}$$

and

$$P(\text{incorrect update} \mid \text{initial state is } k) = q_k$$
$$P(\text{correct update} \mid \text{initial state is } k) = p_k \tag{12.4-72}$$

Hence combining (12.4-71) and (12.4-72) we arrive at the difference equation that T_k^{N+1} satisfies:

$$T_k^{N+1} = q_k T_{k+1}^{N+1} + p_k T_{k-1}^{N+1} + 1 \tag{12.4-73}$$

With the associated boundary conditions

$$T_1^{N+1} = q_1 T_2^{N+1} + p_1 T_1^{N+1} + 1$$
$$T_{N+1}^{N+1} = 0 \tag{12.4-74}$$

State 1 is reflecting and state $N+1$ is absorbing (see Figure 12.17). The solution is derived in Appendix III for arbitrary k. The mean first slip time is given by

$$T_1^{N+1} = \sum_{l=1}^{N} \left(\sum_{j=1}^{l-1} \prod_{i=j+1}^{l} \frac{p_i}{q_i p_j} + \frac{1}{q_l} \right) \tag{12.4-75}$$

where we define, for arbitrary α_j

$$\sum_{j}^{k} \alpha_j = 0 \qquad j > k \tag{12.4-76}$$

In Appendix II it is shown that (12.4-75) is the unique solution.

If we generalize our IPM model to the case of three possible timing corrections at a given state as described previously, it can be shown that the mean first slip time is given precisely by (12.4-75) except that now for given p_k and q_k, r_k is defined by $r_k = 1 - p_k - q_k$. For example, if we let double primes on the transition probabilities denote the two-level update system and no primes denote the three-level update system, then the following holds, as is evident from (12.4-75): If

$$p_k > q_k$$
$$r_k > 0$$
$$\frac{p_k}{q_k} = \frac{p_k''}{q_k''}$$

Then

$$T_k^{N+1} < \hat{T}_k^{N+1} \tag{12.4-77}$$

Where $\hat{T}^{N+1}$ is the mean first time to slip given that $r_k > 0$. In other words, increasing r_k from the value of zero increases the mean first slip time when the ratio of p_k to q_k is kept the same.

Up until now all the results for the IPM were derived assuming no modulation was present on the subcarrier. With the introduction of modulation it is necessary to modify the loop to employ decision feedback to feedback the sign of the detected bits. The feedback removes the sign of the binary modulation.

The effect of bit errors on the loop is to degrade its timing error performance. To see why the loop is degraded we derive the transition probabilities when modulation is present. Let PE_k denote the bit error probability, given that the timing error is in a positive state k. Further denote $\hat{p}_k$ as the probability of reducing the timing error taking into account bit errors and given that the error state is k. Let $\hat{q}_k$ denote the probability of increasing the timing error (taking into account of bit errors) given that the error state is k. Then the transition probabilities with modulation are given in terms of the unmodulated ones by

$$\hat{p}_k = (1 - PE_k)p_k + PE_k q_k$$
$$\hat{q}_k = (1 - PE_k)q_k + PE_k p_k$$
(12.4-78)

Hence $\hat{p}_k < p_k$ and $q_k > \hat{q}_k$ due to bit errors.

In (12.4-78) the correlation between the bit decision statistic and timing update statistic has neen neglected. However, usually this will be a negligible effect since the correlation coefficient can be shown to be small if the number of samples per period is large enough, say, 16. The correlation coefficient is $\frac{1}{3}$ if 16 samples are used per period. So, for the results to hold with modulation, we must replace p_k and q_k with $\hat{p}_k$ and $\hat{q}_k$. Consider now what happens to PE_k as the system moves away from the transition region, that is, the timing error increases. Since the probability PE_k depends on the autocorrelation function of the modulated waveform, increasing the error decreases the correlation up to a point where it goes negative. The decrease is proportional to the reduction of E_B/N_0. At the point where the correlation goes to zero, $PE_k = \frac{1}{2}$ and by (12.4-78), $\hat{p}_k = \hat{q}_k = \frac{1}{2}$. Once the modulated waveform correlation function is computed it is not difficult to compute $\hat{p}_k$ and $\hat{q}_k$ and hence the modulated loop performance.

The average bit error rate PE is given by

$$PE = 2 \sum_{k=1}^{N} PE_k P_k$$
(12.4-79)

Because of the symmetry involved, the negative states were folded over onto the positive ones, which accounts for the 2 in the above equation.

Now we consider an example based on a low data rate command system developed at the Jet Propulsion Laboratory. The input signal is a 60-Hz squarewave subcarrier (SC) biphase modulated by a length 15 pseudonoise (PN) sequence. Each symbol of the PN sequence corresponds to one cycle of the subcarrier. The prefilter has a one-sided bandwidth of 480 Hz. Samples are taken at the Nyquist rate of 960 sample sec and then quantized. This rate provides 16 samples per subcarrier period. Four-bit quantization was used to represent the samples, which introduces negligible degradation. The accumulator adds 22 transition samples per bit. (There are 22 unambiguous transitions in a length 15 PN code.) Positive-going and negative-going transitions are resolved by the PN code generator which, by means of an internally generated PN code, multiplies the sample by a "plus one" when the transition is negative-going and by a "negative one" if the transition is positive-going. The bit detector determines whether a "one" or a "zero" was detected and changes the algebraic sign of the update signal only if a "zero" was sent.

The filtered PN code modulated subcarrier signal of this system was modeled by an unfiltered squarewave with a linear transition region. It was modeled as having a duration of 1/16 of a subcarrier period. This model agreed quite well with laboratory measurements of the filtered signal.

For modulated signals it is convenient to define the bit signal energy to noise spectral density, assuming a perfectly synchronized digital integrate-and-dump circuit, by

$$\frac{E_B}{N_0} = \frac{A^2 T_B}{N_0} \tag{12.4-80}$$

where T_B is the bit time. Equation 12.4-80 neglects the small loss of signal due to the effect of the lowpass filter. Using (12.4-62), (12.4-64), and (12.4-65) in conjunction with the modified transition probabilities of (12.4-78), we computed the timing error variance for the case $\Delta = 1/6$, 1/32, 1/64, 1/128, and 1/256 as a function of E_b/N_0. The results are plotted in Figure 12.18. Some experimental points obtained from [21] are plotted for $\Delta = 1/256$ and $\Delta = 1/128$. At high SNRs the error is due to the quantization size Δ.

It is of interest to compare the performance of this IPM with that of Cessna's [22] digital loop, which employs a midphase-inphase type of phase detector. For comparable cases, that is, $\Delta = \frac{1}{16}$, it is seen that for $E_B/N_0 \geq$ 10 dB the phase error performance is essentially identical. Below 10 dB Cessna's loop appears to produce a lower phase error. However, the phase detector and filter of the loop described here is considerably simpler and is all digital, making it very attractive for uncoded communication systems. In addition, the phase error can always be reduced by reducing Δ.

One parameter not derived in the general case but of interest is the bandwidth of our digital loop. Although the derivation [24] is too long to be included here, the result for one update per bit timing using the filtered squarewave signal model is

$$W_L = \sqrt{\frac{2}{\pi}} \frac{\rho^{1/2} \Delta}{\zeta T_B} \qquad \text{Hz} \tag{12.4-81}$$

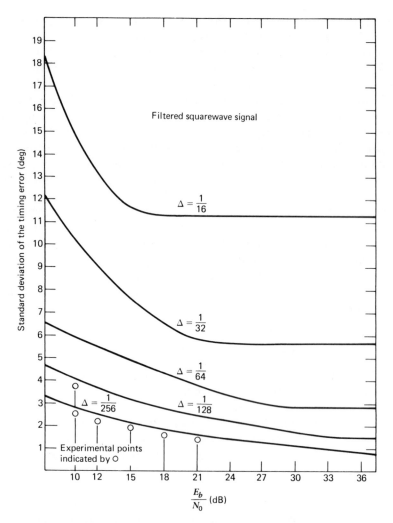

Figure 12.18 IPM timing loop error performance—squarewave subcarrier.

where ρ is defined as

$$\rho = \frac{mA^2}{N_0 W} \tag{12.4-82}$$

Also T_B is the bit time, and ζ is the ratio of the transition region duration to the total period of the SC.

For the system under consideration, $T_B = \frac{1}{4}$ sec, $\Delta = 1/256$, $\rho = 1.65$, and $\zeta = 1/16$. Using these values, we see that the digital bandwidth, from (12.4-81), is 0.256 Hz.

We now show that the general results for arbitrary signals is applicable to the case when more than one sample is obtained per transition region

[23]. Whether the timing error can be reduced by multisamples depends on the ratio A^2/N_0W as well as the signal shape around the transition. We make a reasonable practical assumption that the loop is limited to utilizing the samples obtained by the sampler so that the samples are $1/2W$ sec apart.

The effect of the multiple samples is to modify the update signal to noise ratio and consequently the transition probabilities. Since the system is digital and there are $2W/f_s$ samples per period available, the most convenient multisampler arrangement utilizes an equal number of samples located symetrically about the point the loop thinks is the zero crossing. This requires $2S+1$ samples per half period, where S is an integer so that there are $4S+2$ samples that may be used per period. These samples may be weighted according to their distance from the center or zero crossing sample to provide more nearly optimum performance. The basic equations starting at (12.4-58) are modified because of the change in update SNR. Let y_j^k, j positive, denote the sample of the signal plus noise taken j samples to the right of the sample taken at state k. When j is negative y_j^k denotes the sample taken j samples to the left (or later) of the one taken at state k. Since the signal part of y_j^k is A_j^k (using the same notation), the transition probability p_k becomes

$$p_k = 1 - Q\left\{\left[\frac{m\left(\sum_{-S}^{S} \gamma_j A_j^k\right)^2}{N_0 W \sum_{-S}^{S} \gamma_j^2}\right]^{1/2}\right\} \qquad q_k = 1 - p_k \qquad (12.4\text{-}83)$$

where γ_j is the relative weighting of each sample.

This scheme, for $S = 1$, has been investigated for the example considered with $\gamma_1 = \gamma_{-1}$ and $\gamma_0 = 1$. The value of γ_1 was optimized by the use of a computer to obtain the minimum timing variance (phase error) as a function of the signal to noise ratio. The timing error is shown in Figure 12.19 for both the single sample per transition and the triple sample per transition. As one might expect, the timing error is reduced with three samples per transition only if the timing error is sufficiently large. For $E_B/N_0 > 1$, γ_1 was found to be much less than 1 in the optimum case, so that the additional processing of adjacent samples was not worthwhile since it reduced the timing variance insignificantly.

A second-order IPM loop, a generalization of the one described here, is discussed in reference 24. The method of analysis is totally different however. Lesh [25] and Chadwick [26], have determined the acquisition behavior of both the first- and second-order loops.

PROBLEM 5

(a) In the case transition probabilities are state independent the steady state transition probabilities satisfy

$$P_k = qP_{k-1} + pP_{k+1} \qquad 2 \leq k \leq N-1$$

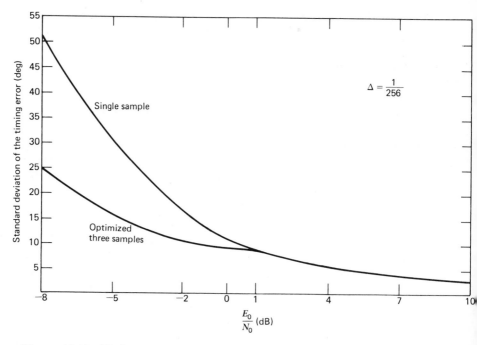

Figure 12.19 Timing error variance for the single sample and the optimized three-sample IPM system for updating.

and the boundary conditions, taking into account the reflecting barriers at states 1 and N, are

$$P_1 = pP_2 + pP_1$$
$$P_N = qP_{N-1} + qP_N$$

Show that the solution for the P_k is given by

$$P_k = \left(\frac{1}{2}\right) \frac{1 - q/p}{1 - (q/p)^N} \left(\frac{q}{p}\right)^{|k|-1} \quad k = -N, \ldots, -1, 1, \ldots, N$$

(b) Show that the timing error variance defined by

$$\sigma_{TE}^2 = 2 \sum_{1}^{N} (k - \tfrac{1}{2})^2 \Delta^2 P_k$$

is given by (letting $\alpha = q/p$)

$$\sigma_{TE}^2 = \frac{\Delta^2}{4} + \frac{\Delta^2}{1 - \alpha^N} \left\{ -N(N+1)\alpha^N + 2\left[\frac{\alpha - (N+1)\alpha^{N+1}}{1 - \alpha}\right] + 2\left[\frac{\alpha^2 - \alpha^{N+2}}{(1 - \alpha)^2}\right] \right\}$$

(c) Show, for this homogeneous case, that the mean slip time starting in state k satisfies

$$T_k^{N+1} = qT_{k+1}^{N+1} + pT_{k-1}^{N+1} + 1 \qquad k = 2, \ldots, N$$

with the two associated boundary conditions

$$T_{N+1}^{N+1} = 0$$

$$T_1^{N+1} = qT_2^{N+1} + pT_1^{N+1} + 1$$

(d) Show that the solution to (c) above is given by

$$T_1^{N+1} = T^{N+1} = -\frac{N}{p-q} + \frac{1/q + 1/(p-q)}{(p/q)-1}\left[\left(\frac{p}{q}\right)^N - 1\right]$$

which is the mean slip time.

12.5 THE EFFECT OF IMPERFECT SYMBOL TIMING ON THE BIT ERROR RATE

In this section we consider the effect of symbol timing errors on the bit error rate (BER). We first consider NRZ symbols. Figure 12.20 illustrates the two basic cases of two identical symbols and two dissimilar symbols. The other two possible cases, two consecutive negatives and a negative symbol followed by a positive symbol, are equivalent as far as degradation is concerned, so that we do not have to consider them.

Now if two successive symbols are similar, then the synchronization error causes no degradation. However, when two successive symbols are

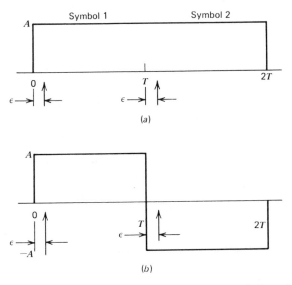

Figure 12.20 Two basic NRZ possibilities—(a) similar symbols and (b) dissimilar symbols—(two symbols).

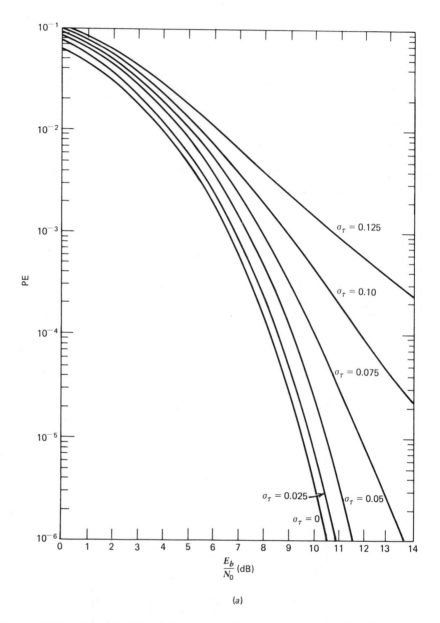

Figure 12.21 PE versus rms timing error—Gaussian error, (*a*) with NRZ data and *p* = .5

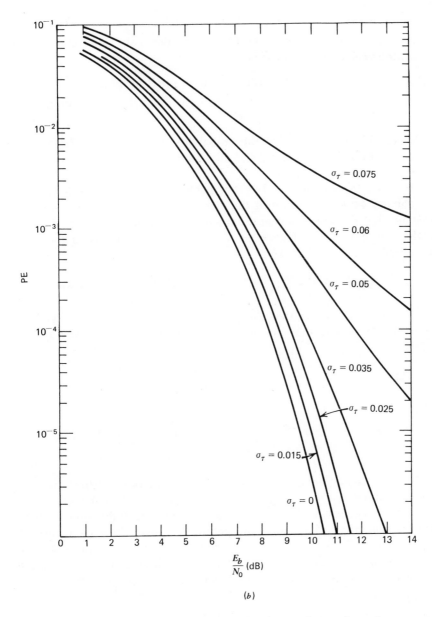

$\dfrac{E_b}{N_0}$ (dB)

(b)

Figure 12.21 (continued) (b) with Manchester data and p = .5.

dissimilar, the matched filter output voltage is reduced by the factor

$$\left[1 - \frac{2|\epsilon|}{T}\right] \tag{12.5-1}$$

where ϵ is the timing error and T is the bit duration. Conditioned on this error, the probability of error is

$$\text{PE}(\epsilon) = P(\text{error} \mid \text{transition}, \epsilon)P(\text{transition})$$
$$+ P(\text{error} \mid \text{no transition}, \epsilon)P(\text{no transition}) \tag{12.5-2}$$

From (12.5-2) it is clear that the probability of error is dependent on the transition probability. Clearly the minimum degradation occurs with zero transition probability.

In order to evaluate (12.5-2) it is necessary to have knowledge of the probability density function of the error ϵ. Since it is not unreasonable to assume that the errors are approximately Gaussianly distributed, we will make that assumption. Hence we have, for a transition probability p, that $R = E_B/N_0$

$$\text{PE}(\epsilon) = pQ\left[\sqrt{2R}\left(1 - \frac{2|\epsilon|}{T}\right)\right] + (1-p)Q(\sqrt{2R}) \qquad |\epsilon| \leq \tfrac{1}{2} \tag{12.5-3}$$

It therefore follows that the probability of error for NRZ symbols is given by (assuming that the error is essentially constant over a symbol time)

$$\text{PE} = \int_{-\infty}^{\infty} \text{PE}(\epsilon)p(\epsilon)d\epsilon$$

or

$$\text{PE} = \frac{p}{2\pi\sigma_\tau} \int_{-\infty}^{\infty} e^{-\epsilon^2/2\sigma_\epsilon^2} \int_{\sqrt{2R}(1-2|\epsilon|/T)}^{\infty} e^{-t^2/2} \, dt \, d\epsilon + \frac{1-p}{\sqrt{2\pi}} \int_{\sqrt{2R}}^{\infty} e^{-t^2/2} \, dt \tag{12.5-4}$$

This result is plotted in Figure 12.21a as a function of the timing error standard deviation σ_ϵ with $p = 0.5$. Notice that an irreducible error occurs for the bit synchronization loop as well as carrier demodulation. Generally, the bit synchronization loop can operate at a lower loop bandwidth than the carrier loop, since the dynamics are reduced by the ratio of data rate to carrier frequency. The lower loop bandwidth, in turn, usually results in a smaller contribution to BER degradation than the carrier loop.

PROBLEM 6

Using the same method as used above, show that, for Manchester data, the conditional probability of error is given by

$$\text{PE}(\epsilon) = pQ\left[\sqrt{2R}\left(1 - \frac{2|\epsilon|}{T}\right)\right] + (1-p)Q\left[\sqrt{2R}\left(1 - \frac{4|\epsilon|}{T}\right)\right] \qquad |\epsilon| \leq \tfrac{1}{4}$$

Manchester symbols produce a greater degradation than NRZ symbols (see Figure 12.21b). Also the assumption of a Tikhonov timing error density function produces a slightly greater degradation than the Gaussian assumption [5].

12.6 BIT SYNCHRONIZER LOCK DETECTOR

In this section we will consider one type of bit synchronizer lock detector that indicates whether or not the bit synchronization loop is synchronized to the data stream. Our discussion will be limited to NRZ data symbols, but the extension to Manchester and other data formats will be obvious. Figure 12.22 illustrates two ways of implementing a bit synchronization lock detector, the only difference being the type of detectors that are used. Virtually all coherent systems employ the use of a lock detector to provide lock status.

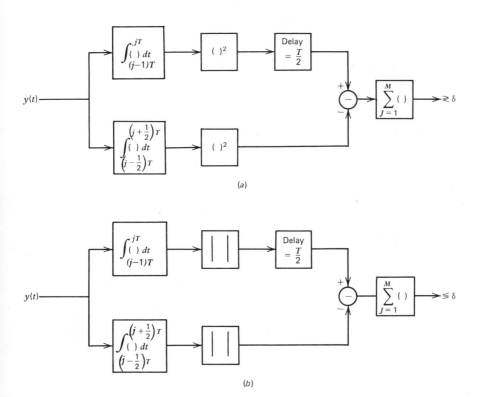

Figure 12.22 Two implementations—(a) square law type and (b) magnitude type—of one type of bit sync lock detector.

First we will discuss the way in which the lock detector works and then develop an approximate theory of performance. Consider Figure 12.23, which shows an NRZ bit sequence and the respective integration regions of the lock detector. We model the baseband data sequence as being corrupted by WGN.

The in-phase integrator, in perfect sync,* produces an output at time T, given by

$$I(T) = \pm AT + N_1 + N_2 \qquad (12.6\text{-}1)$$

where the $\pm$ sign denotes the fact that the signal can be plus or minus in the interval $(0, T)$. The midphase integrator, when in perfect sync, produces an output at time $t = 3T/2$, given by

$$Q\left(\frac{3T}{2}\right) = N_2 + N_3 \qquad \text{with probability } p$$

$$Q\left(\frac{3T}{2}\right) = \pm AT + N_2 + N_3 \qquad \text{with probability } 1 - p \qquad (12.6\text{-}2)$$

where N_i are independent, Gaussian, random variables having zero mean and a variance of $N_0T/4$, and p is the probability of a transition. Hence, if there are sufficient transitions, the I channel output power exceeds the Q channel power so that the postdetection filter (the summer) tends to increase with M. On the other hand, if the bit synchronizer is not locked, the average power out of each channel is the same, so that the postdetection filter output tends to meander around zero.

Since the detected power out of each channel is independent of the algebraic sign of the signal, we shall assume that the sign is positive. First we assume that the bit synch is synchronized to the data stream. Now we

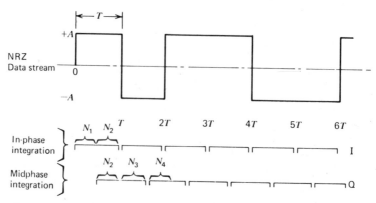

Figure 12.23 Data stream and the in-phase and midphase integrators.

*We assume perfect synchronization (sync) for our calculations since it simplifies the calculations considerably while introducing negligible error.

compute the mean per sample output of the quantity $I^2 - Q^2$, with the appropriate delay so that the I and Q samples occur at the same time. We have* denoting $E[\cdot]$ as the ensemble average the result

$$E[I^2 - Q^2] = E(AT + N_1 + N_2)^2 - pE(N_2 + N_3)^2 - (1 - p)E(AT + N_2 + N_3)^2$$
$$(12.6\text{-}3)$$

After a little algebra we obtain

$$E[I^2 - Q^2] = p(AT)^2 \qquad (12.6\text{-}4)$$

Clearly the mean is maximum with a transition every time and zero with no transitions. To obtain the variance of $I^2 - Q^2$ we use a result referenced by Woo [27] (after which this section is loosely patterned), which cites Frazer [28], concerning a central limit theorem for dependent random variables. Adjacent samples of $I^2 - Q^2$ are not statistically independent. Figure 12.24 illustrates the random variables involved before and after the half-symbol delay.

Since N_3 is present in both bit 1 and bit 2 of Figure 12.24b, we see that samples of $I_k^2 - Q_k^2$ are correlated with the next lock detector output, $I_{k+1}^2 - Q_{k+1}^2$. From References 27 and 28 we have that the lock detector sum has mean μ and variance σ^2, where

$$\frac{1}{M}\sigma^2 = \text{var}(I_k^2 - Q_k^2) + 2\,\text{cov}(I_k^2 - Q_k^2, I_{k+1}^2 - Q_{k+1}^2) \qquad (12.6\text{-}5)$$

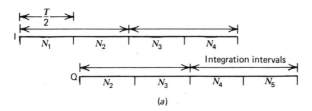

(a)

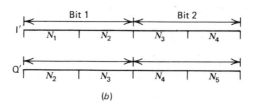

(b)

Figure 12.24 Random variables involved in the computation of the mean and variance (a) before and (b) after the balancing delay.

*For notational simplicity we assume that I and Q without subscripts denote the outputs at time $3T/2$. Subscripts will be used only when we consider sequential samples.

and

$$\frac{1}{m}\mu = E[I_k^2 - Q_k^2] \tag{12.6-6}$$

and further that the lock detector output statistic

$$\sum_{k=1}^{M} (I_k^2 - Q_k^2) \tag{12.6-7}$$

tends toward a Gaussian random variable.

To simplify the calculation somewhat, we determine the worst case variance which corresponds to the case when the transition probability p equals 1.

$$\text{var}(I_k^2 - Q_k^2) = E[(I_k^2 - Q_k^2)^2] - (AT)^4 \tag{12.6-8}$$

$$\text{var}(I_k^2 - Q_k^2) = E[(AT + N_1 + N_2)^2 - (N_2 + N_3)^2]^2 - (AT)^4 \tag{12.6-9}$$

The random variables N_i are Gaussian, independent, and have zero mean and variance equal to $N_0T/4$. Combining all the terms, we obtain

$$\text{var}(I_k^2 - Q_k^2) = 2A^2T^3N_0 + \tfrac{3}{4}N_0^2T^2 \tag{12.6-10}$$

Making the same assumption ($p = 1$) for the covariance, we obtain

$$\text{cov}(\,\cdot\,) = E\{[(AT + N_1 + N_2)^2 - (N_2 + N_3)^2 - (AT)^2][(AT + N_3 + N_4)^2$$
$$- (N_4 + N_5)^2 - (AT)^2]\} \tag{12.6-11}$$

After a fair amount of algebra we obtain

$$\text{cov}(I_k^2 - Q_k^2, I_{k+1}^2 - Q_{k+1}^2) = -\frac{N_0^2T^2}{8} \tag{12.6-12}$$

From (12.6-5), (12.6-6), (12.6-10), and (12.6-12), we obtain

$$\sigma^2 \cong M(2A^2T^3N_0 + \tfrac{1}{2}N_0^2T^2) \tag{12.6-13}$$

$$\mu = M[p(AT)^2] \tag{12.6-14}$$

Using the fact that $x_k = I_k^2 - Q_k^2$ tends to a Gaussian random variable, we find that the probability of declaring in-lock when the bit synch is synchronized to the data stream is approximately given by

$$P_L \cong \int_\delta^\infty \frac{1}{\sqrt{2\pi M\sigma^2}} \exp\left[-\frac{(x - \mu)^2}{2\sigma^2}\right] dx \tag{12.6-15}$$

where δ is threshold of the lock detector filter output. Evaluating (12.6-15) by using (12.6-13) and (12.6-14), we obtain

$$P_L = Q\left[\frac{\delta - Mp(AT)^2}{\sqrt{M}\sqrt{2A^2T^3N_0 + \tfrac{1}{2}N_0^2T^2}}\right] \tag{12.6-16}$$

where

$$Q(x) = \int_x^\infty \frac{1}{\sqrt{2\pi}} e^{-x^2/2} dx \tag{12.6-17}$$

Clearly the probability of declaring out-of-lock when the bit synch is in lock is given by

$$P_{\bar{L}} = 1 - P_L \qquad (12.6\text{-}18)$$

Notice that since, in general $\delta < Mp(AT)^2$, that as $M \to \infty$, $P_L \to 1$.

Now we shall obtain an approximation in the case the synchronizer is not synchronized to the bit stream. Clearly when the signal component of the I channel is maximum the Q channel is minimum and *vice versa*. Therefore the signal components are statistically dependent upon each other. However, for convenience in our approximate analysis, we shall model the signal components as independent and as distributed uniformly from $-AT$ to AT. Hence we approximate each channel by (over one bit time)

$$I = X + N_1 + N_2 \qquad (12.6\text{-}19)$$

$$Q = Y + N_2 + N_3 \qquad (12.6\text{-}20)$$

where X and Y are statistically independent. Furthermore, I and Q have zero mean value. As was mentioned above, the probability density functions of X and Y are, respectively,

$$p(X) = \frac{1}{2AT} \qquad -AT \le X \le AT \qquad (12.6\text{-}21)$$

$$p(Y) = \frac{1}{2AT} \qquad -AT \le X \le AT \qquad (12.6\text{-}22)$$

and the statistics of N_i are as before. We have

$$E[I^2 - Q^2] = E[(X + N_1 + N_2)^2 - (Y + N_2 + N_3)^2] = 0 \qquad (12.6\text{-}23)$$

Again we note that

$$\sigma^2 = \operatorname{var}(I_k^2 - Q_k^2) + 2 \operatorname{cov}(I_k^2 - Q_k^2, I_{k+1}^2 - Q_{k+1}^2) \qquad (12.6\text{-}24)$$

so that we first compute $\operatorname{var}(I^2 - Q^2)$. We have

$$\operatorname{var}(I^2 - Q^2) = E\{[(X + N_1 + N_2)^2 - (Y + N_2 + N_3)^2]^2\} \qquad (12.6\text{-}25)$$

After some algebra we obtain

$$\operatorname{var}(I^2 - Q^2) = \tfrac{8}{45}(AT)^4 + \tfrac{4}{3}A^2 T^3 N_0 + \tfrac{3}{4}(N_0 T)^2 \qquad (12.6\text{-}26)$$

Since we employed an approximation in computing (12.6-26), we elect to use (12.6-12) to approximate the covariance term of (12.6-21), so that we obtain

$$\sigma^2 \cong \tfrac{8}{45}(AT)^4 + \tfrac{4}{3}A^2 T^3 N_0 + \tfrac{1}{2}(N_0 T)^2 \qquad (12.6\text{-}27)$$

and from (12.6-23) we have

$$\mu = 0 \qquad (12.6\text{-}28)$$

Therefore, the probability of falsely indicating the in-lock state, with the assumption that the lock detector output after the sum of M samples has a Gaussian distribution, is given by

$$P_{FL} = \int_{\delta}^{\infty} \frac{1}{\sqrt{2\pi\sigma^2}} e^{-x^2/2\sigma^2} dx \qquad (12.6\text{-}29)$$

let $x/\sigma = t$, so that

$$P_{FL} = \int_{v}^{\infty} \frac{1}{\sqrt{2\pi}} e^{-t^2/2} dt$$

where

$$v = \left[\frac{\delta}{\sqrt{M} \sqrt{\frac{8}{45}(AT)^4 + \frac{4}{3}A^2 T^3 N_0 + \frac{1}{2}(N_0 T)^2}} \right] \qquad (12.6\text{-}30)$$

where again δ is the threshold which is proportional to M. Let

$$\delta = M\delta_M \qquad (12.6\text{-}31)$$

so that δ_M is the per sample threshold. Then (12.6-31) becomes

$$P_{FL} = Q\left(\frac{\sqrt{M}\, \delta_M}{\sqrt{\frac{8}{45}(AT)^4 + \frac{4}{3}A^2 T^3 N_0 + \frac{1}{2}(N_0 T)^2}} \right) \qquad (12.6\text{-}32)$$

Equation 12.6-32 can be rearranged to

$$P_{FL} = Q\left(\frac{\sqrt{M}\, \delta_M/(AT)^2}{\sqrt{\frac{8}{45} + \frac{4}{3}(N_0/A^2 T) + \frac{1}{2}(N_0/A^2 T)^2}} \right) \qquad (12.6\text{-}33)$$

which illustrates the fact that δ_M is on the order of $(AT)^2$. Note that as M becomes large $P_{FL} \to 0$, as it should, since δ_M is proportional to $(AT)^2$.

A third case occurs when no signal is present. In this case we are interested in the probability of falsely indicating lock P_{FL}. This is easily found from (12.6-16), when $A = 0$, so that we obtain, using (12.6-31)

$$P_{FL} = Q\left(\frac{\sqrt{M}\, \delta_M}{\sqrt{\frac{1}{2}N_0^2 T^2}} \right) \qquad (12.6\text{-}34)$$

Other types of lock detector are possible; however, this one appears to be quite efficient.

PROBLEM 7

Analyze the lock detector of Figure 12.22b by using the methods of this section and the assumption that for large M the detector statistic is Gaussian.

12.7 AN SNR ESTIMATOR

As a final topic we consider the estimation of the link SNR based on having bit synchronization. This topic is related to the bit synchronization problem and the bit synch lock detector problem. An estimate of a high SNR indicates a reliable link that includes proper carrier demodulation and bit detection as well as PN code despreading or frequency dehopping if they are present in the system.

A block diagram of the estimation procedure is shown in Figure 12.25. The estimator has been called "SNORE" at the Jet Propulsion Laboratory. (It is short for signal-to-noise ratio estimator [29, 30, 31].) The estimator considered here is of the form

$$\frac{\hat{E}_B}{N_0} = S\hat{N}R_1 = \frac{(\hat{\mu})^2}{2\hat{\sigma}^2} \tag{12.7-1}$$

where

$$\hat{\mu} = \frac{1}{n} \sum_{i=1}^{n} |x_i| \tag{12.7-2}$$

$$\hat{\sigma}^2 = \frac{1}{n-1} \sum_{i=1}^{n} (|x_i| - \hat{\mu})^2 \tag{12.7-3}$$

Therefore, $\hat{\mu}$ is the sample mean and $\hat{\sigma}^2$ is the sample variance, if the SNR is not too small so that decisions on the magnitude of x_i are equivalent to processing x_i when the mean is of the same algebraic sign for all time.

In other words, it is necessary to compute statistics on $|x_i|$, rather than x_i in general, but for high SNR this is not necessary, as we shall show. Errors using this assumption only occur when the noise causes a bit error.

It can be shown that the mean value of $\hat{\mu}$ is given by [30]

$$E[\hat{\mu}] = \sqrt{\frac{2}{\pi}} \sigma \exp\left(\frac{-\mu^2}{\sigma^2}\right) + \mu \operatorname{erf}\left(\frac{\mu}{\sqrt{2}\sigma}\right) \tag{12.7-4}$$

so that when SNR $= \mu^2/2\sigma^2$ is large enough

$$E[\hat{\mu}] = \mu \qquad \text{unbiased estimator} \tag{12.7-5}$$

where μ and σ^2 are the mean and variance of the matched filter output. For example, if $\mu^2/2\sigma^2 = 4$ dB, then from (12.7-4) it can be shown that $E[\hat{\mu}] = 0.989\mu$, which is about a 1% error. We conclude that at SNRs above 4 dB the error in taking the magnitude of the matched filter statistic is negligible.

Both Gilchrist [29] and Boyd [31] have estimated the statistics of (12.7-1), the SNR estimator. Boyd actually found a way of making the

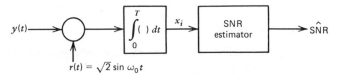

Figure 12.25   An SNR estimator model for NRZ data.

approximate analysis of Gilchrist exact whenever $E|x_i| = |E(x_i)|$, that is, at high SNR. Without going through the details, Boyd showed that the "improved" SNR estimator

$$\text{S\^NR}_2 = \frac{\frac{1}{n}\sum_{i=1}^{n}|x_i|}{\frac{1}{n-3}\sum_{i=1}^{n}(|x_i| - \hat{\mu})^2} \qquad n > 3 \qquad (12.7\text{-}6)$$

with $\hat{\mu}$ defined by (12.7-2), has a mean and variance given by

$$E(\text{S\^NR}_2) = \left(\frac{1}{2n} + \frac{\mu^2}{2\sigma^2}\right) \qquad (12.7\text{-}7)$$

$$\text{var}(\text{S\^NR}_2) = \frac{1}{2(n-5)}\left[\left(\frac{\mu}{\sigma}\right)^4 + 2\left(\frac{\mu}{\sigma}\right)^2\left(1 - \frac{1}{n}\right) + \frac{2}{n}\left(1 - \frac{1}{n}\right)\right] \qquad n > 5$$
$$(12.7\text{-}8)$$

Hence, with (12.7-7) and (12.7-8) it is possible to estimate the value of n to achieve the required SNR estimate accuracy in terms of the standard deviation of the estimate and the bias. The improved estimator has a very small bias that disappears with increasing n faster than the original estimate [31] of (12.7-1), which has a mean value given by

$$E(\text{S\^NR}_1) = \frac{n-1}{(n-3)}\left(\frac{1}{2n} + \frac{\mu^2}{2\sigma^2}\right) \qquad (12.7\text{-}9)$$

Layland [32] has investigated the estimator $\text{S\^NR}_1$ at low SNRs. Lorden [33] showed that the estimator is essentially maximum likelihood at moderate $(E_B/N_0 \geq 0 \text{ dB})$ SNRs.

APPENDIX I: UNIQUENESS OF THE STEADY STATE PROBABILITIES

We prove (see Section 12.4.3): if $q_k > 0$ for at least one k, and

$$P_k = q_{k-1}P_{k-1} + p_{k+1}P_{k+1} \qquad (I\text{-}1)$$

with

$$P_1 = p_1P_1 + p_2P_2$$
$$P_N = q_{N-1}P_{N-1} + q_NP_N \qquad (I\text{-}2)$$

$$\sum_{1}^{N} P_k = \frac{1}{2} \qquad (I\text{-}3)$$

then the solution P_k is unique.

Proof   Let two different solutions $\tilde{P}_k$ and $\hat{P}_k$ be solutions of (I-1) satisfying (I-1), (I-2), and (I-3). Define Δ_k by

$$\Delta_k = \tilde{P}_k - \hat{P}_k \qquad (I\text{-}4)$$

Then from (I-2)

$$\Delta_2 = \frac{q_1}{p_2} \Delta_1 \qquad (I-5)$$

By iterating (I-1)

$$\Delta_k = \prod_{j=2}^{k} \frac{q_{j-1}}{p_j} \Delta_1 \qquad (I-6)$$

Assume $\Delta_1 \neq 0$. Then from (I-6)

$$\sum_{k=1}^{N} \Delta_k = \sum_{k=1}^{N} \prod_{j=2}^{k} \frac{q_{j-1}}{p_j} \Delta_1 \neq 0 \qquad (I-7)$$

But from (I-3) and (I-4) we see that we have a contradiction implying $\Delta_1 = 0$, which implies, by (I-6), that $\Delta_k = 0$, for all k. This implies the solution is unique. If $q_k = 0$ for all k, then we get, for this case, the unique solution, $P_1 = \frac{1}{2}$, $P_{-1} = \frac{1}{2}$.

Q.E.D.

APPENDIX II: UNIQUENESS OF THE MEAN FIRST SLIP TIME

We prove (see Section 12.4.3): If

$$T_k^{N+1} = q_k T_k^{N+1} + p_k T_{k-1}^{N+1} + 1 \qquad 1 < k < N \qquad (II-1)$$

$$T_1^{N+1} = q_1 T_2^{N+1} + p_1 T_1^{N+1} + 1 \qquad (II-2)$$

$$T_{N+1}^{N+1} = 0 \qquad (II-3)$$

then the solution T_k^{N+1} is unique.

Proof Let $\tilde{T}_k^{N+1}$ and T_k^{N+1} be different solutions of (II-1) subject to the boundary conditions in (II-2) and (II-3). Let

$$\Delta_k = \hat{T}_k^{N+1} - \tilde{T}_k^{N+1} \qquad (II-4)$$

Then from (II-2)

$$\Delta_1 = \Delta_2 \qquad (II-5)$$

And in general

$$\Delta_k = \Delta_1 \qquad \text{for all } k \qquad (II-6)$$

From (II-3) we have

$$\Delta_{N+1} = 0 \qquad (II-7)$$

which implies $\Delta_k = 0$ for all k.

APPENDIX III: THE SOLUTION TO THE MEAN FIRST SLIP TIME

We shall obtain (see Section 12.4.3) the solution to (II-1) subject to the boundary conditions of (II-2) and (II-3). Starting with (II-1) and (II-2) we may iterate to obtain

$$T_{k+1}^{N+1} = T_k^{N+1} - \sum_{j=1}^{k-1}\left(\prod_{i=j+1}^{k}\frac{p_i}{q_iq_j}\right) - \frac{1}{q_k} \tag{III-1}$$

Let

$$\alpha_k = \sum_{j=1}^{k-1}\left(\prod_{i=j+1}^{k}\frac{p_i}{q_iq_j}\right) + \frac{1}{q_k} \tag{III-2}$$

Then using (II-3), it follows that

$$\sum_{k=1}^{N}(T_k^{N+1} - T_{k+1}^{N+1}) = T_1^{N+1} = \sum_{k=1}^{N}\alpha^k \tag{III-3}$$

In a similar manner it may be shown that

$$T_k^{N+1} = \sum_{l=k}^{N}\alpha_l$$

So that from (III-2) we have

$$T_k^{N+1} = \sum_{l=k}^{N}\left(\sum_{j=1}^{l-1}\prod_{i=j+1}^{l}\frac{p_i}{q_ip_j} + \frac{1}{q_l}\right) \tag{III-4}$$

In (III-4) the following convention was used:

$$\sum_{i=j}^{k}a_i = \prod_{i=j}^{k}b_i = 0 \qquad \text{for } j > k \tag{III-5}$$

It can be shown by some rather tedious algebra that (III-4) is the solution to (II-1) and satisfies the boundary conditions of (II-2) and (II-3). For example, we now show that the solution satisfies (II-2). From our solution we can show that

$$T_2^{N+1} - T_1^{N+1} = -\sum_{j=1}^{0}\prod_{i=j+1}^{1}\left(\frac{p_i}{q_iq_j}\right) - \frac{1}{q_1} \tag{III-6}$$

Using the conventions of (III-5) we have

$$T_2^{N+1} = T_1^{N+1} - \frac{1}{q_1} \tag{III-7}$$

which is a rearranged form of (II-2).

REFERENCES

1   Barker, R. H., "Group Synchronization of Binary Digital Systems," in Jackson, W., Ed., *Communication Theory*, Academic, New York, 1953.

2 McBride, A. L., and Sage, A. P., "Optimum Estimation of Bit Synchronization," *IEEE Trans. Aerospace and Electronic Systems*, May 1969.

3 Stiffler, J., *Theory of Synchronous Communications*, Prentice-Hall, Englewood Cliffs, N.J., 1971.

4 Van Trees, H. L., *Detection, Estimation, and Modulation Theory*, Part I, Wiley, New York, 1968.

5 Lindsey, W. C., and Simon, M. K., *Telecommunication Systems Engineering*, Prentice-Hall, Englewood Cliffs, N.J., 1973.

6 Hurd, W. J., and Anderson, T. O., "Digital Transition Tracking Symbol Synchronizer for Low SNR Coded Systems," *IEEE Trans. Communications Technology* Vol. COM-18, April 1970.

7 Lindsey, W. C., and Tausworthe, R. C., "Digital Data-Transition Tracking Loops," *JPL SPS* 37-50, Vol. III, April 1968.

8 Simon, M. K., "An Analysis of the Steady-State Phase Noise Performance of a Digital-Data-Transition Tracking Loop," *JPL SPS* 37-55, Vol. III, February 1969.

9 Simon, M. K., "Optimization of the Performance of a Digital Data Transition Tracking Loop," *IEEE Trans. Communications Technology* Vol. COM-18, October 1970.

10 Layland, J. W., "Telemetry Bit Synchronization Loop," *JPL SPS* 37-46, Vol. 3, pp. 204-215.

11 Simon, M. K., "Nonlinear Analysis of an Absolute Value Type of an Early-Late Gate Bit Synchronizer,"

12 Huey, D. C., and Fultz. G. L., "Shuttle Bit Rate Synchronizer," Final Report, TRW Systems Group, TRW No. 7333.3-360, December 1974.

13 Batson, B. H., Cellier, A., Lindsey, W. C., and Vang, H., "An All-Digital Manchester Symbol Synchronizer for Space Shuttle," NTC 74, San Diego, Cal., December 2–4, 1974.

14 Gardner, F. M., "Clock Recovery from a Nonlinear Channel," *European Satellite Agency Journal*, Vol. 2, No. 2, 1978.

15 Holmes, J. K., "Preliminary Estimates of Carrier Tracking Threshold," TRW IOC No. SCTE-50-75-027/JKH, June 1975.

16 Hedin, G., Holmes, J. K., Lindsey, W. C., and Woo, K. T., "Theory of False Lock in Costas Loops," *IEEE Trans. Communications* Vol. COM-26, January 1978.

17 Holmes, J. K., "Tracking Performance of the Filter and Square Bit Synchronizer," *IEEE Trans. Communications* Vol. COM-28, Part I, August 1980.

18 Feller, W., *An Introduction to Probability Theory and Its Applications*, Vol. I, Wiley, New York, 1957, Chap. 14.

19 Holmes, J. K., "Performance of a First-Order Transition Sampling Digital Phase-Locked Loop Using Random-Walk Models," *IEEE Trans. Communications* Vol. COM-20, April 1972.

20 Parzen, E., *Stochastic Processes*, Holden Day, San Francisco, Cal., pp. 278–281, 1962.

21 Couvillion, A., Private Communication.

22 Cessna, J. R., "Steady-State and Transient Analysis of a Digital Bit Synchronization Phase-Locked Loop," Vol. II, pp. 34–15 to 25 of Proc. of ICC 1970 Int. Conf. on Communications, San Francisco, June 1970.

23 Holmes, J. K., "A Note on the Optimality of the All-Digital Command System Timing Loop," *JPL SPS*, Vol. III, 37–65, October 1970.

24 Holmes, J. K., and Tegnelia, C., "A Second-Order All-Digital Phase-Locked Loop," *IEEE Trans. Communications*, Vol. COM-22, January 1974.

25 Lesh, J. R., "Calculating Acquisition Behavior for Completely Digital Phase-Locked Loops," *JPL DSN* Progress Report 42–29, pp. 33–45.

26 Chadwick, H. D., "A Markov Chain Technique for Determining the Acquisition Behavior of a Digital Tracking Loop," *JPL Quarterly Technical Review*, Vol. 1, No. 4, pp. 49–57, January 1972.

27 Woo, K. T., "Shuttle Bit Synch Lock Detector Performance," TRW IOC No. SCTE-50-76-184/KTW, April 5, 1976.

28 Frazer, D. A. S., *Non-Parametric Methods in Statistics*, Wiley, New York, 1957, Chap. 6.

29 Gilchrist, C. E., "Signal-to-Noise Monitoring," *JPL SPS* 37–27, Vol. IV.

30 Shottler, P. H., "Large Sample SNR Estimator for the Maximum Likelihood Estimator," *JPL SPS* 37–47, Vol. III.

31 Boyd, D. W., "Signal-to-Noise Ratio Monitoring: Error Analysis of the Signal-to-Noise Ratio Estimator," *JPL SPS* 37–39, Vol. IV.

32 Layland, J., "On S/N Estimation," *JPL SPS* 37–48, Vol. III.

33 Lorden, G., "Efficiency of the S/N Ratio Estimator," *JPL SPS*, 37–56, Vol. II.

INDEX